AF687675

Environmental Turbulence

Computation and Analysis of Turbulent Flows

Environmental Turbulence

From Fundamental Physics to Critical Applications

Edited by

Elie Bou-Zeid
Department of Civil and Environmental Engineering
Princeton University
Princeton, NJ, United States

Sutanu Sarkar
Department of Mechanical and Aerospace Engineering
University of California San Diego
San Diego, CA, United States

ELSEVIER

ACADEMIC PRESS
An imprint of Elsevier

Academic Press is an imprint of Elsevier
125 London Wall, London EC2Y 5AS, United Kingdom
50 Hampshire Street, 5th Floor, Cambridge, MA 02139, United States

Copyright © 2026 Elsevier Inc. All rights are reserved, including those for text and data mining, AI training, and similar technologies.

For accessibility purposes, images in electronic versions of this book are accompanied by alt text descriptions provided by Elsevier. For more information, see https://www.elsevier.com/about/accessibility.

Books and Journals published by Elsevier comply with applicable product safety requirements. For any product safety concerns or queries, please contact our authorised representative, Elsevier B.V., at productsafety@elsevier.com.

Publisher's note: Elsevier takes a neutral position with respect to territorial disputes or jurisdictional claims in its published content, including in maps and institutional affiliations.

No part of this publication may be reproduced or transmitted in any form or by any means, electronic or mechanical, including photocopying, recording, or any information storage and retrieval system, without permission in writing from the publisher. Details on how to seek permission, further information about the Publisher's permissions policies and our arrangements with organizations such as the Copyright Clearance Center and the Copyright Licensing Agency, can be found at our website: www.elsevier.com/permissions.

This book and the individual contributions contained in it are protected under copyright by the Publisher (other than as may be noted herein).

Notices
Knowledge and best practice in this field are constantly changing. As new research and experience broaden our understanding, changes in research methods, professional practices, or medical treatment may become necessary.

Practitioners and researchers must always rely on their own experience and knowledge in evaluating and using any information, methods, compounds, or experiments described herein. In using such information or methods they should be mindful of their own safety and the safety of others, including parties for whom they have a professional responsibility.

To the fullest extent of the law, neither the Publisher nor the authors, contributors, or editors, assume any liability for any injury and/or damage to persons or property as a matter of products liability, negligence or otherwise, or from any use or operation of any methods, products, instructions, or ideas contained in the material herein.

ISBN: 978-0-323-95873-8

For information on all Academic Press publications
visit our website at https://www.elsevier.com/books-and-journals

Publisher: Matthew Deans
Acquisitions Editor: Sophie Harrison
Editorial Project Manager: Kripa Joseph
Production Project Manager: Paul Prasad Chandramohan
Cover Designer: S Bakyalakshmi

Typeset by VTeX

ELSEVIER Book Aid International Working together to grow libraries in developing countries
www.elsevier.com • www.bookaid.org

Contents

Contributors

Guido Boffetta

Department of Physics and INFN, University of Torino, Torino, Italy

Elie Bou-Zeid

Department of Civil and Environmental Engineering, Princeton University, Princeton, NJ, United States

Marc Calaf

Department of Mechanical Engineering, University of Utah, Salt Lake City, UT, United States

Kamal Kant Chandrakar

National Center for Atmospheric Research, Boulder, CO, United States

Bicheng Chen

College of Ocean and Earth Sciences, Xiamen University, Xiamen, Fujian, China

Benoit Cushman-Roisin

Thayer School of Engineering, Dartmouth College, Hanover, NH, United States

Peter A. Davidson

Engineering Department, University of Cambridge, Cambridge, United Kingdom

Luc Deike

Princeton University, Princeton, NJ, United States

Sylvain Dupont

INRAE, Bordeaux Sciences Agro, ISPA, Villenave d'Ornon, France

Marco Giometto

Department of Civil Engineering and Engineering Mechanics, Columbia University, New York, NY, United States

Giacomo Valerio Iungo

Department of Mechanical Engineering, University of Texas at Dallas, Richardson, TX, United States

Gabriel G. Katul

Department of Civil, Construction and Environmental Engineering, University of Alabama, Tuscaloosa, AL, United States

Qi Li

Department of Atmospheric and Oceanic Sciences, Peking University, Beijing, China

Qing Li

Earth, Ocean and Atmospheric Sciences Thrust, The Hong Kong University of Science and Technology (Guangzhou), Guangzhou, Guangdong, China

Stefano Musacchio

Department of Physics and INFN, University of Torino, Torino, Italy

Prasanth Prabhakaran

Cooperative Institute for Research In Environmental Sciences (CIRES), University of Colorado, Boulder, CO, United States

Chemical Sciences Laboratory, National Oceanic and Atmospheric Administration, Boulder, CO, United States

James J. Riley

Department of Mechanical Engineering, University of Washington, Seattle, WA, United States

Sutanu Sarkar

Department of Mechanical and Aerospace Engineering, University of California San Diego, San Diego, CA, United States

University of California San Diego, La Jolla, CA, United States

Raymond A. Shaw

Department of Physics, Michigan Technological University, Houghton, MI, United States

Holger Siebert

Leibniz Institute for Tropospheric Research, Leipzig, Germany

Fan Yang

Environmental Science and Technologies Department, Brookhaven National Laboratory, Upton, NY, United States

Einara Zahn

Department of Earth and Environmental Sciences, University of Pennsylvania, Philadelphia, PA, United States

Biography

Elie Bou-Zeid

Elie Bou-Zeid's research in environmental fluid mechanics started about 25 years ago with low Reynolds number flows in the Earth's subsurface, but his focus quickly shifted to the highly turbulent flows in the lowest layer of the atmosphere, the atmospheric boundary layer. He blends numerical and experimental data to develop a deeper understanding of the physics of atmospheric flows and to derive reduced theoretical models for a range of applications. His fundamental work gravitates towards the lesser-explored dynamics that emerge over heterogeneous and complex surfaces, under unsteady, nonequilibrium conditions, and with baroclinic or other large-scale thermal forcings of the flow. His applications have mostly centered around flows of relevance to the urban environment, the hydrological cycle, and the renewable energy transition. Bou-Zeid is a Professor of Civil and Environmental Engineering at Princeton University and the Chair of the Committee on Boundary Layers and Turbulence of the American Meteorological Society. He previously served as editor of the Journal of the Atmospheric Sciences. He holds a bachelor's degree in Mechanical Engineering and a master's degree in Environmental and Water Resource Engineering from the American University of Beirut (Lebanon), and a master's degree in Mechanical Engineering and a PhD in Environmental Engineering from Johns Hopkins University (USA). He is the recipient of the 2025 Atmospheric Sciences Ascent Award from the American Geophysical Union and the 2026 Helmut E. Landsberg Award from the American Meteorological Society.

Sutanu Sarkar

Turbulence simulation, theory, and modeling have remained a major topic of inquiry for Sutanu Sarkar since his graduate days at Cornell. His research at NASA Langley and in the early years at UCSD spanned the areas of high-speed aeronautics, propulsion, combustion, and aeroacoustics. He also developed turbulence models that continue to be used in commercial software for thermofluid applications. After a few years at UCSD, his interests shifted to environmental and geophysical flows, where he develops and uses high-fidelity simulation techniques to quantify and understand small-scale processes, including turbulence, flow instabilities, nonlinear internal waves, flow-topography interactions, submesoscale eddies, and air-sea coupling. Another application concerns engineered structures in the natural environment, for example, marine submersibles and renewable energy. His work on the fundamentals has been driven by two questions: what are the routes to turbulence and mixing in a stratified, rotating fluid, and what are the properties of the ensuing turbulence? Sutanu Sarkar received his B. Tech from IIT Bombay, M.S. from Ohio State University, and

Ph. D. from Cornell University, all in Mechanical Engineering. After 4 years as a staff scientist at ICASE, NASA Langley Research Center, he joined UCSD, where he is a Distinguished Professor in the department of Mechanical & Aerospace Engineering (MAE), holds the Blasker Chair in Environmental Engineering, and is an affiliate Professor at the Scripps Institution of Oceanography (SIO). He served as Chair of MAE from 2009 to 2014. He is an associate editor of the Journal of Fluid Mechanics and has received a NASA group achievement award (1994), the Bessel Award from the Humboldt Foundation (2001), and was elected Fellow, American Physical Society (2006), Associate Fellow, AIAA (2009), and Fellow, ASME (2010).

Preface

Turbulence is often described as the last open problem in classical physics, or as the only topic that even a divine creature may not be able to explain. Nevertheless, we design and build airplanes, wind turbines, bridges, and other structures for which turbulence is a key operational element, and these human-made devices operate safely and efficiently. We also predict the weather relatively accurately days into the future, despite the fact that turbulence modulates weather and climate physics and dynamics from the cloud condensation nuclei to the global carbon and water cycles scale. So why do we suffer from a collective "imposter syndrome" when dealing with the chaotic fluid mechanics of turbulent flows?

One reason may be that we often equate understanding with the ability to explain a phenomenon in simple language. Richard Feynman argued that "If you cannot explain something in simple terms, you do not truly understand it." However, how can a system with as many degrees of freedom as turbulence be reduced to the few degrees of freedom allowed by language? We know the fundamental equations that describe turbulent flows, and we know how to solve them numerically at all scales to visualize the wonderful array of flow structures and dynamics. We even know how to decompose them into mean and turbulence or large-scale and small-scale equations, and then how to obtain reasonably accurate solutions for the mean or large scales alone. So, we can explain turbulence exactly mathematically and visually, we just cannot explain it in plain language. We may indeed never be able to map the full dynamics of turbulence to a simple system or to a string of words, and that may be why the study of turbulence will remain so fascinating, in addition to being of high practical value.

The book series this publication is part of is indeed a compilation of all we know about turbulence, and we now know much more. Humans usually develop and use technology long before understanding the underlying science (at least in classical physics). Ancient civilizations harnessed the energy of winds and water currents and developed pumps long before Euler, Navier, and Stokes developed the momentum budget equation. However, once the underlying science is understood, technologies can advance much more rapidly. Because fluid mechanics and turbulence are integral to a myriad of human activities, scientists have sought to understand and predict how fluids move for centuries to develop better water provisioning systems, extract more energy, and grow and irrigate crops more efficiently. Therefore, the earliest fluid dynamicists mostly investigated how water and air flows influence our environment and livelihoods. The Industrial Revolution and the arms races of the last 150 years shifted the focus (and funding) predominantly to the fluid dynamics of engineering applications. However, starting with the risks posed by air and water pollution and continuing with the threat of climate change and other anthropogenic perturbations of Earth's systems, environmental and geophysical applications are returning to the forefront of the discipline. These applications also introduce a suite of new physical forces and phenomena that make environmental turbulence even more challenging than its counterpart in engineering applications.

One of the most characteristic features of environmental flows is the variation in density that gives rise to buoyant forces. Buoyancy, through its stabilizing/destabilizing influence on velocity fluctuations, can qualitatively change the flow. Another complication is that flows are occurring at large enough scales such that the reference frames we use, which are almost invariably at rest with respect to the rotating Earth, can no longer be treated as inertial frames, and their rotation needs to be accounted for. The simplest way to do that, it turns out, is to introduce fictitious forces that would produce the perceived fictitious acceleration associated with the rotating frame (otherwise, one would need to fix the frame and let the boundaries move with the rotating Earth, a highly impractical approach). Environmental flows are also often heterogeneous in all directions due to the complexity of the flow domains (cities, wind farms, cropland, marine kelp farms, etc.), and they are invariably unsteady. To add further complexity, air and water are fluid mixtures, and some components of those mixtures are active, reactive, and/or inertial. Active components influence the momentum budget (e.g., salt in the water, water vapor in the air, as well as temperature, although the latter is a property, not a component of the mixture). Reactive components can be created or destroyed by chemical or physical reactions, and need to be tracked and modeled, especially if they have health or other consequential impacts. Inertial components are typically heavy particles that do not simply follow the flow, and tracking them thus requires solving their own equations of motion. Last but not least, the range of turbulent scales spans orders of magnitude (typically from $\sim$1 mm to $\sim$1 km in the atmosphere and only slightly less in the ocean). These scales can be viewed as degrees of freedom, and capturing their full range through direct numerical simulations of the governing equations remains far beyond the reach of even the most powerful computing clusters.

However, the applications are of such importance to our survival and well-being as a species that we do not have the luxury of looking away or throwing our hands up in the air and giving up. Furthermore, the more complex a scientific problem is, the more ingenuity and creativity it requires to overcome, and the more stimulating and attractive it becomes to curious minds. The increased relevance and the exciting suite of associated questions have certainly heightened the visibility of environmental fluid mechanics and turbulence in the broader field of fluid mechanics, as can be observed in the plot below that shows the percentage of papers published in the Journal of Fluid Mechanics that have the terms environ* or geophys* in the title, abstract, or keywords (the plot shows the 3-year running average for smoothing). This increasing trend will certainly continue, and the contributors to this book hope that its publication will be a key catalyst in amplifying the excitement, appreciation, and understanding of turbulence in the air and water media in which we live. We are hindered by the problem this preface opened with: there is simply no way to describe the full content of turbulence physics in simple words. But we do strive to

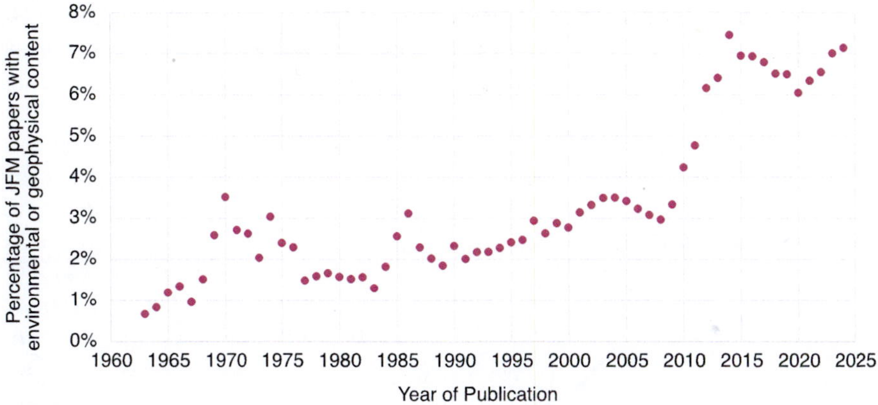

Figure 1 Three-year running average of the percentage of papers published in the Journal of Fluid Mechanics that have the terms environ* or geophys* in the title, abstract, or keywords.

make the explanations as simple as possible, and we make ample use of the language of mathematics and visualizations to better convey the splendor of environmental turbulence.

Elie Bou-Zeid, Sutanu Sarkar

Turbulence in the environment

1

An introduction

Elie Bou-Zeid[a] *and Sutanu Sarkar*[b]
[a]Department of Civil and Environmental Engineering, Princeton University, Princeton, NJ, United States, [b]Department of Mechanical and Aerospace Engineering, University of California San Diego, San Diego, CA, United States

1.1 Introduction

Whether hiking in the woods, strolling through a city, or swimming in the sea, you are immersed in an ocean of turbulence; it is your most immediate surroundings. Air and water flows shape the environmental quality we experience; dictate the weather and climate extremes we are vulnerable to; convey the pollen and pollinators we depend on for food production; impact the performance of air, land, and marine vehicles that transport us; and modulate the variability of the solar, wind, and marine energy that powers our economy. These environmental flows are almost invariably turbulent. It should thus come as no surprise that the earliest turbulence researchers were motivated by environmental and geophysical applications of these chaotic fluid dynamics. Da Vinci's first descriptions of turbulence (Marusic and Broomhall, 2021), as well as Richardson's poem on small and big whirls (Richardson, 1922), were both inspired by turbulence observations in natural systems.

Interest in environmental turbulence resurged in the last few decades, at first motivated by the interacting environmental and climatic hazards caused by anthropogenic activities. The increasing urgency to make our food and energy production more sustainable, our coasts and shores more resilient, and our buildings and cities more livable is compounding this resurgence. Turbulence is central to the physics of these human challenges, and advancing its understanding and modelling is thus central to the solutions. Environmental applications are also increasingly requiring levels of accuracy that have not been needed thus far. An error of one hour in predicting a sudden increase or decrease in wind speed (a wind ramp) might be moderately problematic in weather information provided to citizens, but it is a very consequential and costly error when this forecast is used to predict wind power generation and balance the electrical grid. Another example is the large uncertainty related to cloud processes and dynamics, and the need to reduce it given its outsized impact on the climate sensitivity of geophysical models. A deeper understanding of environmental flows is also needed to evaluate the uptake of heat and carbon dioxide by the so-called oceanic flywheel, a process mediated by oceanic turbulence. The breadth, complexity, and impact of such environmental turbulence problems are thus attracting more researchers to the field, and this book is designed to help experienced and uninitiated fluid dynamicists appreciate and understand the physical processes that make environmental turbulence so challenging, and at the same time so fascinating.

Environmental Turbulence. https://doi.org/10.1016/B978-0-32-395873-8.00008-5
Copyright © 2026 Elsevier Inc. All rights are reserved, including those for text and data mining, AI training, and similar technologies.

Density variation and Earth's rotation are two distinctive features of environmental flows. The Sun heats the Earth's surface and establishes a temperature gradient in both water and air, fluids whose densities depend on temperature. Warm light water above cold dense water is a statically stable configuration because the restoring gravitational force of buoyancy would return a vertically displaced particle to its original position. On the air side, warm below cold is a statically unstable configuration (motion will be initiated even in a fluid initially at rest), and buoyancy leads to daytime convection. The environment has temperature variations in the horizontal directions as well (for instance, the equator-to-pole temperature decrease) that generate large-scale chaotic circulations and horizontal pressure gradients in the atmosphere and oceans, which are in a leading-order balance with the rotation-induced Coriolis force, a balance that acts as a crucial component of atmospheric and oceanic flows. Although subdominant in the direct force balance of small-scale turbulent motions, the Coriolis force is nevertheless important because, first, it modulates the wind and density structure of the background mean flow and, second, its weakening at intermediate length scales results in flow instabilities that often instigate turbulence.

Turbulent perturbations in environmental flows are produced either by shear (from the mean kinetic energy) or by buoyancy (from the mean potential energy (MPE) of fluid parcels displaced from their equilibrium elevations) when the local flow is statically unstable (Stull, 1991). Turbulent kinetic energy (TKE) can then be converted back to Turbulent Potential Energy (TPE), possibly at larger length scales, or it can cascade to the smallest scales, where it is dissipated by viscosity into thermal energy (although this does not lead to significant heating of the fluid and is typically neglected in the thermal energy balance). Shear is often caused by momentum drag at the domain boundaries, while buoyancy is most often injected by fluxes of heat, salt (in water), or water vapor (in air) at the same boundaries. Therefore the most turbulent strata in environmental flows are at: (i) the interface of the atmosphere and the land, water or ice surface underneath (the atmospheric boundary layer (ABL), Wyngaard (1992)); (ii) the interface of the ocean with the air above (the ocean surface boundary layer (OSBL), D'Asaro (2014)); and (iii) the bottom of the ocean where it meets the solid bed underneath (the bottom boundary layer (BBL), Trowbridge and Lentz (2018)). These are also the layers of greatest significance to the energy, water, and nutrient cycles because they modulate exchanges among the atmosphere, cryosphere, hydrosphere, and lithosphere, and because most of the mass of the biosphere lives therein. The importance of the ABL and OSBL is further accentuated by the concentration in these layers of the overwhelming majority of humans, human activities, and the ecosystem services we rely on.

The geophysical presence and influence of turbulence, however, is not restricted to these three layers. While the ABL depth that extends only to its capping inversion at about 1 km is $\sim 1\%$ of the Kármán line height (defined as the boundary between the Earth and outer space), that layer dissipates $\sim 50\%$ of the kinetic energy of the atmosphere and acts as its main source of water vapor, cloud condensation nuclei, and heat. Clear air turbulence way above the ABL can also be triggered by strong shear between air masses moving in different directions, and is a nuisance to air transport. However, deep convection that can reach the tropopause at $\sim$ 10 km and turbulence in mid- and high-level clouds are the more consequential (meteorologically and climatologically) manifestations of atmospheric turbulence (and waves) above the ABL. The ocean accounts for about 70% of Earth's surface, and its boundary layers are also typically thin. The surface mixed layer, where vigorous turbulence approximately homogenizes density in the vertical, is of order 50–100 m (except during strong winter convection events) and is often thinner in summer, yet it is key to oceanic currents. It is bounded at the bottom by the upper ocean pycnocline with a sharp density gradient in the vertical. The bottom boundary layer is typically tens of meters thick. The turbulence in the interior is weak when averaged over ocean basins, but is nevertheless important. It has widespread and poorly understood hotspots that impact human activity and marine life by changing nutrient and oxygen levels. Interior turbulence is important to the broader ocean dynamics as well, since it influences the oceanic physical state (temperature, salinity) and, as a result, the interior stratification it generates crucially influences the turbulent ocean boundary layers (e.g., keeping them thin).

Given the large length scales ($\sim$100 – 1000 m), the elevated velocities ($\sim$ 10 ms^{-1} in the atmosphere and a factor of 10 lower in the ocean), and the low viscosities ($\sim 10^{-5}$ m^{2}s^{-1} for air and $\sim 10^{-6}$ m^{2}s^{-1} for water), the Reynolds number (Re) of turbulent environmental flows is extremely elevated ($\sim 10^7 - 10^9$). Re measures the destabilizing effects of fluid inertia relative to the stabilizing effects of fluid viscosity; therefore, turbulence dominates these environmental flows except in the thin viscous sub-layers near interfaces. The thickness of these viscous sublayers, and the associated Kolmogorov scales (the scales of the smallest turbulent eddies), are $\sim$ 1 mm in all three of these layers.

1.2 Unique features of environmental turbulence

Most readers of this textbook would be familiar with the classic literature on turbulent flows; for those who are not, this chapter offers a summary. The main goal of this chapter and the book more broadly, however, is to introduce the reader to the physics and dynamics that distinguish environmental turbulence from canonical free and wall-bounded shear flows. The first part of the book focuses on the fundamentals. The real forces resulting from density variations (buoyancy, Chapters 2 and 3) and the fictitious ones arising from the use of a frame of reference that rotates with the Earth (Coriolis, Chapters 4 and 5) are central to turbulence dynamics and mixing (Chapter 6) in environmental media (they are seldom touched upon in turbulence textbooks). However, this is not where the distinction with flows in engineered systems ends. Geometric complexity (from mountain to shrub scales), surface heterogeneity (at sub-meter to tens-of-kilometers scales), and unsteadiness (at minutes to diurnal or tidal to rotational scales) are essential features and challenges that researchers in the field have to contend with. They make the emerging flow dynamics richer in features but sometimes invalidate core assumptions that many turbulence models and analyses rely on, such as (i) equilibrium of the turbulence with the mean flow or (ii) the ability to determine the ensemble (Reynolds) averaged statistics using time averaging over steady periods or spatial averaging over homogeneous directions or areas. Finally, the set of flow variables is expanded to include the scalars that density depends on, as well as those required in various applications, such as pollutants in air and water, or nutrients in applications related to marine farming.

While there are myriad applications of environmental turbulence to overview, in the second part of this book, we focus on a selection of consequential applied problems that have garnered significant research interest over the past couple of decades. Transport of scalars (pollutants, nutrients, etc.) in environmental media is the topic of Chapter 7. Flow and turbulence inside and over urban terrain (Chapter 8), vegetation canopies (Chapter 9), and wind farms (Chapter 10) are presented next. Turbulence at the air-sea interface is then introduced at the wave scale (Chapter 11) and drop or bubble scale (Chapter 12). Finally, the complexities and implications of turbulence in clouds conclude the book (Chapter 13).

1.3 First principles

Mass, momentum, and energy conservation laws are the foundations of the dynamics of environmental turbulence. Since the Mach number typically remains low (<0.3), and vertical motions have a limited extent compared to vertical scales over which the mean pressure and density change appreciably (the latter is known as the scale height, see detailed discussion in Businger (1984)), it is often possible to write these governing equations in incompressible form, invoking the Boussinesq approximation to capture the effect of the small buoyancy-generated perturbations in density. The mass, momentum, and energy balance equations are then respectively given by:

$$\frac{\partial u_i}{\partial x_i} = 0 \tag{1.1}$$

$$\frac{\partial u_i}{\partial t} + u_j \frac{\partial u_i}{\partial x_j} = \nu \frac{\partial^2 u_i}{\partial x_j^2} - \frac{1}{\rho_r}\frac{\partial p}{\partial x_i} - g\delta_{i3} - g\delta_{i3}\left(\frac{\hat{\rho}}{\rho_r}\right) - 2\epsilon_{ijk}\Omega_j u_k \tag{1.2}$$

$$\frac{\partial \chi}{\partial t} + u_j \frac{\partial \chi}{\partial x_j} = \kappa \frac{\partial^2 \chi}{\partial x_j^2} \tag{1.3}$$

Eq. (1.1) implies that the velocity vector u_i is divergence-free, and the velocity field is thus incompressible. The position vector is x_i, and in this chapter, we will use the geophysical fluid dynamics notation where 1 or x denotes the streamwise direction; 2 or y denotes the lateral direction; and 3 or z denotes the vertical direction. The left-hand side of Eq. (1.2) is the material derivative that quantifies the acceleration of a fluid parcel as it moves in the flow; it is balanced on the right-hand side, respectively, by the forces (per unit mass) generated by molecular viscosity (denoted by ν), gradient of pressure (denoted by p), gravity (where g is the gravitational acceleration and δ_{i3} is the Kronecker delta), buoyancy (where ρ denotes the fluid density and the hat denotes its perturbation from the Boussinesq reference density ρ_r), and the Coriolis force (where ϵ_{ijk} is the alternating unit tensor and Ω_j is the Earth's angular velocity vector). Eq. (1.2), describing how net forces change velocity is hence a result of the application of Newton's second law of motion to a fluid parcel in the Earth's rotating reference frame.

In Eq. (1.3), κ is the molecular diffusivity of the transported quantity, and χ can be replaced by the density ρ for water flows in oceans or rivers (or potential density for large-scale flows where water compressibility matters), while for atmospheric air it should be replaced by the potential temperature θ (Lesieur, 2008) (a modified temperature that normalizes for the background effect of elevation and pressure differences (Brutsaert, 2005)). While its form does not clearly suggest it, this equation (for θ in air or the temperature-dependent part of ρ in water) is a simplified statement of energy conservation that derives from the application of the first law of thermodynamics to a fluid parcel. It reveals that, in environmental flows, the thermal component of energy conservation modifies temperature and, thereby, density, generating buoyant forces. In water, the temperature and salinity balance equations are both required to predict density using an equation of state. Similarly, in the atmosphere, we additionally track water (vapor, liquid, or ice) concentrations and how they, jointly with the potential temperature, modify density (Stull, 1988). The reader is referred to oceanic and atmospheric dynamics textbooks for further details about large-scale geophysical flows (Gill, 1982; Vallis, 2017) and small-scale flows in the atmosphere (Wyngaard, 2010) and the oceans (Thorpe, 2007).

Eqs. (1.1)–(1.3) describe the evolving dynamics of all turbulent scales at each point in space and time. While analytically unsolvable, their numerical integration (known as direct numerical simulation (DNS)) has emerged as a powerful tool for understanding all varieties of turbulent flows. DNS is best used for idealized problems and process studies to tease out specific turbulent flow physics. A computationally less demanding approach is to simulate the largest eddies of turbulence that tend to be the most consequential, while parameterizing the small unresolved scales; this results in the large eddy simulation (LES) technique. LES has become a popular tool in many studies of environmental flow and turbulence since it allows the modeler to match the Re of the real environment (Stoll et al., 2020). Even LES, however, remains too demanding for many geophysical applications such as weather prediction and climate projection. In such applications, one can solve for the mean flow state alone and then fully parameterize the turbulence effects on that state. The mean equations are obtained by Reynolds-averaging of Eqs. (1.1)–(1.3), where the Reynolds decomposition for any variable follows $\xi = \bar{\xi} + \xi'$; an overbar denotes the mean (ensemble average) component and the prime the turbulent component of that variable.

1.3.1 The physical basis of the Boussinesq approximation

It is useful here to revisit the Boussinesq approximation that yields the form of the buoyancy force used in the equations. The reader would first observe that we denote the density perturbation from its base state with a hat rather than a Reynolds decomposition prime. A first distinction between the two decompositions is that the Boussinesq approximation requires that perturbations from the reference state be very small ($|\hat{\rho}| << \rho_r$, implying that $\rho \approx \rho_r$), while Reynolds perturbations are not thus constrained and can be even larger than the mean. The form of the buoyancy term as written in Eq. (1.2) can then be derived by decomposing the full vertical pressure gradient force following:

$$-\frac{1}{\rho}\frac{\partial p}{\partial z} = -\frac{1}{\rho_r + \hat{\rho}}\left(\frac{\partial p_r}{\partial z} + \frac{\partial \hat{p}}{\partial z}\right) \approx -\frac{1}{\rho_r}\frac{\partial p_r}{\partial z} - \frac{1}{\rho_r}\frac{\partial \hat{p}}{\partial z} + \frac{\hat{\rho}}{\rho_r^2}\frac{\partial p_r}{\partial z} \tag{1.4}$$

where one second-order term that involves a product of the small Boussinesq perturbations was ignored. If we now require that the reference state be in a hydrostatic equilibrium such that

$$-\frac{1}{\rho_r}\frac{\partial p_r}{\partial z} = g \tag{1.5}$$

Eq. (1.4) can be further simplified into two forms

$$-\frac{1}{\rho}\frac{\partial p}{\partial z} \approx g - \frac{1}{\rho_r}\frac{\partial \hat{p}}{\partial z} - \frac{\hat{\rho}}{\rho_r}g = -\frac{1}{\rho_r}\frac{\partial p}{\partial z} - \frac{\hat{\rho}}{\rho_r}g \tag{1.6}$$

The last form is the one we invoked to obtain Eq. (1.2), while the one in the middle informs us that, in the vertical direction, gravity and the reference state vertical pressure gradient cancel out. If we had used that form, the $+g$ would cancel the $-g\delta_{i3}$ term in the vertical component of Eq. (1.2); therefore, only the perturbation $\hat{p}$ of that pressure gradient from the hydrostatic reference state is relevant. The resulting vertical momentum balance equation, accounting for this gravity-pressure balance, becomes

$$\frac{\partial w}{\partial t} + u_j\frac{\partial w}{\partial x_j} = \nu\frac{\partial^2 w}{\partial x_j^2} - \frac{1}{\rho_r}\frac{\partial \hat{p}}{\partial z} - g\left(\frac{\hat{\rho}}{\rho_r}\right) - 2\epsilon_{3jk}\Omega_j u_k \tag{1.7}$$

A second distinction between the Boussinesq and Reynolds decompositions is that the Boussinesq reference state cannot vary significantly with height (the anelastic approximation is an alternative to Boussinesq's that allows it to do so) and thus need not be identical to the mean state of a turbulent flow (which can vary significantly height). Consequently, the spatial or temporal averaging estimates of the Reynolds-averaged mean state (fundamentally an ensemble-averaged state) may not always provide a correct estimate of the Boussinesq reference state. Therefore the Boussinesq perturbations themselves should be Reynolds-decomposed for studying turbulence (e.g., $\hat{\rho} = \overline{\hat{\rho}} + \hat{\rho}'$ (Wyngaard, 2010)).

1.3.2 Nondimensional parameters

In turbulent environmental flows, all the forces on the right-hand side of Eq. (1.2) are active, but we are often interested in how their magnitudes relate to each other and whether some forces dominate under a given set of conditions. We introduce characteristic scales for the velocity (u), the length (l), and, if present, the mean vertical shear (S) in an environment that is rotating (f is a frequency that characterizes the resulting Coriolis force) and possibly has stable stratification with a buoyancy frequency (N) to be defined shortly (the quantities S, f and N have the dimensions of frequency). The Reynolds number ($Re = ul/\nu$) is the best-known dimensionless force ratio, expressing the inertial to viscous forces; inertial forces are on the left-hand side of Eq. (1.2) and reflect the acceleration forces that would result from (or are needed for) a change in the flow inertia. The gradient Richardson number ($Ri_g = N^2/S^2$, or its bulk formulation $Ri_b = N^2/(u/l)^2$) expresses the ratio of buoyant to inertial forces, and will be used in this book extensively when discussing statically stable and unstable turbulent flows. The buoyant-to-inertial force balance can alternatively be expressed by the turbulent Froude number ($Fr_t = u_{rms}/Nl$, where the root-mean-square velocity $u_{rms} = \left(\overline{u'^2}\right)^{1/2}$ characterizes the turbulent fluctuations), often used in oceanic flows with stable stratification. The ratio of inertial to Coriolis forces is given by the Rossby number ($Ro = u/fl$), which is large at the typically small scales of 3D environmental turbulence, although the effect of rotation can still be significant for the largest eddies of the ABL, and for the mean shear and structure of that layer. On the other hand, Ro is low for large values of l ($\sim 100 km$) and will be critical when discussing 2D and rotating turbulence (Chapters 4 and 5). Ro is also important in the submesoscale oceanic range (roughly between 100 m and 10 km) where it is of $\mathcal{O}(1)$: at these values, it triggers flow instabilities that link the geostrophic mesoscales to small-scale turbulence. Finally, the Froude number $Fr = u/\sqrt{gl}$ (distinct from its turbulent counterpart Fr_t defined above) estimates the ratio of (the square root of) inertial to gravity forces. Fr is important for analyzing waves and turbulence at the interface of two fluids with strong density differences (e.g., at the air-sea interface).

1.3.3 The large-scale geostrophic state

Eqs. (1.1)–(1.3) are applicable from scales smaller than a droplet to the scale of the whole planet, but we often find that some simplifications can be accurate and useful at particular scales. At scales *much larger* than the three-dimensional turbulence in the ABL, OSBL, and BBL, and particularly at mid and high latitudes, the mean horizontal flow is often dominated by the balance between the pressure and Coriolis forces, known as the geostrophic balance. The mean horizontal velocity is then given by

$$[\overline{u}, \overline{v}] \equiv [U_g, V_g] = \left[-\frac{1}{f\rho_r}\frac{\partial \bar{p}}{\partial y}, \frac{1}{f\rho_r}\frac{\partial \bar{p}}{\partial x}\right] \tag{1.8}$$

where the Coriolis parameter is a frequency defined as $f = 2|\Omega|\sin(\varphi)$, with φ the latitude and $|\Omega|$ the magnitude of the Earth rotation vector (equivalent to the planet's angular velocity of 7.2921×10^{-5} rad s^{-1}, computed using the sidereal day duration of 23 hours, 56 minutes, and 4.1 seconds). The geostrophic state challenges our intuition that the fluid should move from high- to low-pressure regions; in this state, the fluid moves along isobars.

As one approaches boundaries or interfaces, such as the land surface in the atmosphere or the upper surface of the ocean, friction forces (molecular in the viscous sublayers and turbulent farther away) clearly become important, break the geostrophic balance, and generate turbulence. The ABL, OSBL, and BBL are thus never in a geostrophic state, with the introduction of friction at boundaries leading to the Ekman force balance and the associated Ekman spiral (Kundu and Cohen, 2008; Vallis, 2017). However, it is often useful to express the mean flow therein as a departure from the geostrophic velocity. A simple example is the boundary layer beneath a horizontally homogeneous flow in geostrophic equilibrium, where the momentum Eq. (1.2) is often written with the horizontal mean pressure gradients expressed as an equivalent geostrophic wind

$$-\frac{1}{\rho_r}\frac{\partial \bar{p}}{\partial y} = fU_g \quad \text{and} \quad -\frac{1}{\rho_r}\frac{\partial \bar{p}}{\partial x} = -fV_g \tag{1.9}$$

and the deviation from the geostrophic wind corresponds to the boundary-layer flow. Even when Eq. (1.9) is used to re-express the pressure gradient in Eq. (1.2), a geostrophic flow is only expected far from the boundary and, owing to friction and boundary-layer turbulence, the flow develops an ageostrophic component near the surface, in the so-called Ekman boundary layer (see Subsection 1.4.4).

1.3.4 The thermal wind balance

Atmospheric and oceanic flows on large horizontal scales often have small vertical velocity and acceleration, leading to the hydrostatic balance described in Eq. (1.5). On the other hand, their horizontal geostrophic wind, described in the previous subsection, is a consequence of rotational control of the horizontal momentum balance, and often displays vertical variability. This variability can simply be derived by taking the vertical gradient of the geostrophic wind Eqs. (1.9), and replacing the vertical pressure gradient by its expression from the hydrostatic balance (Kundu and Cohen, 2008), resulting in

$$\frac{\partial U_g}{\partial z} \approx \frac{g}{f\rho_r}\frac{\partial \overline{\rho}}{\partial y} \qquad \text{and} \qquad \frac{\partial V_g}{\partial z} \approx -\frac{g}{f\rho_r}\frac{\partial \overline{\rho}}{\partial x} \tag{1.10}$$

The approximations in these equations are linked to the Boussinesq approximation and are very accurate when the vertical scale of motions is $<<$ the flow's scale height. More importantly, the equations above reveal that a vertical change in the geostrophic balance can only result from horizontal gradients of density, which are common in the atmosphere and oceans.

The density is a function of temperature for both air and water; i.e., these fluids are baroclinic and do not satisfy the barotropic fluid criterion of $\rho(p)$ (density only depending on pressure). This baroclinity (or baroclinicity) will manifest whenever horizontal temperature gradients are significant, leading to horizontal density gradients that, as per Eq. (1.10), then give rise to vertical variability in the geostrophic velocity. Since density gradients in air are primarily due to temperature rather than humidity, meteorologists have termed Eq. (1.10) the "thermal wind" equation, although it is, in fact, an equation for the vertical variability of the wind vector. This variability can include both shear (change in geostrophic wind magnitude) and veer (change in geostrophic wind direction).

The thermal wind equation has proved very useful for understanding large-scale atmospheric and oceanic circulations that are driven by the horizontal variability of temperature (e.g., the equator-to-pole cooling or the land-sea surface temperature contrast). It is also of relevance to environmental turbulence since the shear of the geostrophic wind is a reservoir of kinetic energy, and the associated horizontal density gradient is a reservoir of potential energy. Mesoscale strain and surface forcing by winds (drag or cooling) can drive large-scale oceanic fronts—elongated regions with a cross-front density contrast—from their low Ro thermal wind balance towards submesoscale flows with Rossby number of $\mathcal{O}(1)$ (McWilliams, 2016). The submesoscale range hosts a veritable zoo of flow instabilities (see, for example, Thomas et al. (2008)) that harness these kinetic and potential energy reservoirs to establish a down-scale energy cascade to 3D turbulence in the ocean (Taylor and Ferrari, 2010; D'Asaro et al., 2011; Skyllingstad and Samelson, 2012; Verma et al., 2019). In the atmosphere, the potential energy of horizontal density gradients is often readily converted to large thermal circulations (e.g., land-sea (Parker, 2015; Allouche et al., 2023) or urban-rural (Omidvar et al., 2020) breezes). In addition, the thermal wind can cause much higher or lower vertical shear across the whole ABL, modifying the eddy structure and the TKE balance and the power production of wind farms (Kasper et al., 2024), among others. The change in wind direction with height can be much faster or slower than what is predicted by Ekman turning (Lenschow et al., 1980; Momen et al., 2018; Ghannam and Bou-Zeid, 2021).

1.4 Foundational theories

The fundamental dynamics of turbulence that can be derived from solutions to Eqs. (1.1)–(1.3) are unaware of the application, and whether it is in an environmental, engineering, astrophysical, or biomedical setting. These dynamics only respond to the boundary and initial conditions, and to the scaling of the forces represented by the dimensionless numbers presented in the previous section. Therefore, many of the foundational theoretical developments in the field apply across all flavors of fluid turbulence, including in environmental flows. In this section, we will briefly review a handful of these core theoretical approaches that will be used in this book.

1.4.1 Kolmogorov's theory of homogeneous isotropic turbulence

In 1941, Andrey Kolmogorov published what remains one of the most consequential papers in turbulence research (English translation in Kolmogorov (1991)). His central argument is that the smaller scales in any turbulent flow form a universal regime with dynamics that are not problem- or setup-dependent. The specificities of a turbulent flow are tied to the boundary conditions of the problem, and thus the largest turbulent eddies (with a characteristic scale l_0) that are in contact with these boundaries have unique dynamics that reflect the unique problem settings.[1] These large eddies are also the ones that interact with the mean shear and buoyant plumes to convert mean and potential energy into TKE, defining the production range. However, these large eddies 'break up' and cascade their kinetic energy to smaller and smaller scales (as depicted in Richardson's poem), and this cascade process randomizes the eddy dynamics. The idea of a universal range thus postulates that the cascade causes the smaller eddies to 'forget' the boundary conditions and depend only on the fluid properties and the energy cascade rate.

Therefore at scales $l \lesssim l_0/6$ (Pope, 2000), Kolmogorov's first hypothesis postulates that the turbulence dynamics assume universal characteristics that are only dictated by: (i) the rate at which TKE is cascading through the scales, ε; (ii) the eddy scale of interest, l; and (iii) the viscosity that ultimately converts TKE to thermal energy, ν. Kolmogorov then proposed a second hypothesis on the basis that viscosity cannot act effectively to dissipate TKE at eddy scales that are much larger than a viscous length scale $\eta = (\nu^3/\varepsilon)^{1/4}$, now known as the Kolmogorov length scale. Therefore the dynamics at scales $l \gtrsim 60\eta$ (Pope, 2000), and $l \lesssim l_0/6$, known as the inertial subrange, must be solely determined by l and ε. Viscosity is then only important at scales $l \lesssim 60\eta$, known as the dissipation range. Fig. 1.1 is a schematic of the three ranges.

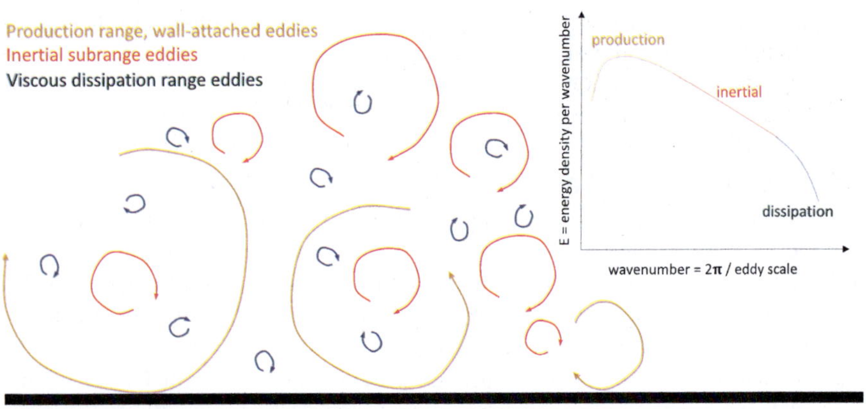

Figure 1.1 Schematic of the three ranges of eddy sizes in physical and spectral spaces for a wall-bounded flow.

It must be noted that Kolmogorov's theory assumes turbulence in both the inertial and dissipation ranges is homogeneous and isotropic. Therefore, buoyancy, rotation, and strong shear, among other processes, can further restrict the ranges of applicability since they can act directly on scales $l \lesssim l_0/6$. Furthermore, the initial theory (known as K41) assumes that only the mean TKE dissipation rate, ε, is important. The potential role of variability in the dissipation was later included in a 'refined' theory known

[1] In environmental flows, the boundary conditions can involve drag and buoyancy flux and may be set by solid boundaries (e.g., the ground or the ocean seafloor), a deformable boundary (as for the air-sea interface), or the ambient conditions (as for an isolated shear layer in the environment).

as K62 (Kolmogorov, 1962). However, the deviations of the predictions from the original K41 are small for lower-order statistical moments (like the TKE) so that the predictions of the original K41 remain widely used and useful in environmental turbulence research (for example, for simulation validation), where the accuracy of the measurements and simulations is often challenged by the complex terrain, unsteady forcing, and other features, preventing us from detecting K41–K62 differences.

Based on the two K41 hypotheses, we can determine a large number of quantitative turbulence attributes that are highly useful for testing models and analyzing experimental data across a wide range of turbulence applications, including environmental ones. The most important result may be the scaling of the density of TKE in spectral space, E. If one considers the wavenumber associated with a scale l, $k = 2\pi/l$, then the variation of E with k defines the turbulence spectrum, which, according to K41, must scale (by simple dimensional arguments) in the inertial subrange according to

$$E \sim \varepsilon^{2/3} k^{-5/3} \tag{1.11}$$

This prediction remains a standard test for turbulence simulations and measurement techniques, while also providing a means to estimate ε from observations (e.g., Chamecki and Dias (2004)).

A related prediction concerns the scale variability of the structure function (in fact, this is the original prediction since K41 was derived in terms of structure functions). The structure function of any order n for any of the velocity components is computed using data at two points with arbitrary separation r_i (to illustrate, u is used here, with a translation between the two points r_i restricted to the x direction such that $r_1 = |r| = r$; $r_2 = r_3 = 0$) following

$$S_n(r) = \overline{(u(x_i + r) - u(x_i))^n} \tag{1.12}$$

According to K41, if the separation distance between the two points r (which represents the eddy scale) falls in the inertial subrange where turbulence is homogeneous and isotropic, these structure functions must scale with the r following

$$S_n(r) \sim (\varepsilon r)^{n/3} \tag{1.13}$$

A particularly useful structure function is the third-order one, which, as noted in K41, must exactly follow

$$S_3(r) = -\frac{4}{5}\varepsilon r \tag{1.14}$$

where the $-4/5$ is a theoretically determined coefficient. This equation, and the similar one for the second-order structure function, albeit with an empirical coefficient, are extensively used in environmental turbulence observations to estimate the dissipation rate (see a comparison to the spectrum-based estimate in Chamecki and Dias (2004)). Measuring down to the Kolmogorov scale $\eta \sim 1$ mm to directly determine dissipation ε is often impossible in geophysical experiments. However, measuring velocity increments in the inertial subrange can be done at scales ~ 1 m in the ABL, for example, and allows the estimation of the dissipation via Eq. (1.14). It should be noted here that, fundamentally, this equation states that the cascade rate of TKE in the inertial subrange at a scale r must balance the ultimate dissipation at viscous scales. It is thus only accurate if no energy injection or dissipation is occurring between scale r and 60η; r should therefore be selected with care. For example, in statically stable wall-bounded flows with buoyant destruction of TKE, it should be smaller than the smallest scale affected by buoyancy, as well as significantly smaller than the distance to the wall, to avoid any production range contributions (Bou-Zeid et al., 2010).

1.4.2 Townsend's attached eddies and the law of the wall

Another foundational law of turbulent flows concerns eddies in the production range, which carry most of the fluxes and energy. For the common case of a flow over a surface (a wall-bounded flow), as discussed in the previous subsection, these scales should be able to interact with the surface. At a given elevation z, production range eddies must thus have scales $l_p \gtrsim z$. Townsend (1980) referred to these scales as 'attached eddies', meaning that they touch the wall. Therefore, at an elevation z that is (i) significantly beyond the thin surface sublayer affected by viscosity, and (ii) significantly below ($\lesssim 15\%$) the upper bound of the full turbulent flow depth δ, these attached eddies dominate the dynamics (see also review by Marusic and Monty (2019)). This is called the inertial sublayer or region.

In the absence of buoyancy, these eddies extract their energy from the mean shear created by surface friction. Their TKE content per unit volume, ρu_p^2, where u_p is the characteristic eddy fluctuating velocity (e.g., u_{rms}), must then be proportional to the surface stress magnitude τ. One can then define the friction velocity $u_* = \sqrt{\tau/\rho} \sim u_p$. With this velocity scale and the length scale z, it should be possible to normalize the mean flow and turbulence statistics across a wide range of wall-bounded flows and find universal laws. For example, the shear in the inertial sublayer should follow

$$\frac{d\overline{u}}{dz} = \frac{u_*}{\kappa z} \tag{1.15}$$

where $\kappa \approx 0.4$ is the proportionality coefficient known as the von Kármán constant. Integrating this equation between any two elevations 1 and 2 in the inertial sublayer then yields

$$\overline{u}_2 - \overline{u}_1 = \frac{u_*}{\kappa} \ln\left(\frac{z_2}{z_1}\right) \tag{1.16}$$

implying that the velocity increases with the natural logarithm of the distance from the wall z. The existence of a log-law was first formally proposed by von Kármán (1930), and has since been widely validated (at least to a very good approximation) in experiments and simulations.

A particular form of that law is obtained when one tries to integrate starting from the wall itself (assuming a no-slip wall where the velocity should be zero). The challenge in this attempt is that, as the wall is approached, (i) the assumption of negligible viscous effects (i.e., of an inviscid flow) breaks down if the wall is smooth, or (ii) large bluff obstructions such as trees or buildings are encountered if the wall is rough. Extending the logarithmic profile to the surface thus does not result in $\overline{u}(z = 0) = 0$, but it rather yields $\overline{u}(z_0) = 0$, where $z_0 > 0$ is the so-called 'roughness length' that encodes the viscous and/or roughness effects near the wall. A common form of the logarithmic velocity profile, known as the log-law of the wall, is thus written as

$$\frac{\overline{u}(z)}{u_*} = \frac{1}{\kappa} \ln\left(\frac{z}{z_0}\right) \tag{1.17}$$

We reiterate that this analysis only applies to flows without buoyancy or rotation. Under such conditions, many other flow statistics, such as vertical velocity variance and TKE dissipation, are also found to follow the (z, u_*) scaling. In particular, passive scalars (fluid-mixture properties that do not cause buoyancy or interfere otherwise with the momentum balance) are found to follow equations very similar to Eqs. (1.15)–(1.17). A notable exception is the streamwise velocity variance, and thus the full TKE, which retains a dependence on the depth of the boundary layer δ (Hultmark et al., 2013).

1.4.3 Accounting for buoyancy in the law of the wall: the Monin-Obukhov similarity theory

The log-law discussed in the previous subsection does not account for buoyant generation or destruction of turbulence, but in environmental flows, buoyancy is omnipresent. The assumptions of the $(z, u*)$ scaling in the log-law include that the flow dynamics are quasi-steady in time, homogeneous in the horizontal spatial directions, and 'local' in the vertical direction. Local here implies that the mean velocity and stress at a height z do not depend on the dynamics far from that elevation. Therefore, the attached eddies that dictate these dynamics are produced locally, and their TKE then cascades and is dissipated locally. TKE shear generation and dissipation are thus in near-balance at any point, and vertical transport, advection, or transient effects are minimal. The Monin–Obukhov similarity theory (Monin and Obukhov, 1954) was introduced to add only one factor of complexity, a quasi-steady homogeneous wall flux of buoyancy, to this canonical idealized flow. It thus also relies on all these previously listed assumptions (which are not always valid (Zahn et al., 2023)).

The theory is anchored in the budget of the turbulent kinetic energy $\overline{e} = \overline{u_i' u_i'}/2$ (that will also be referred to as TKE, K, or k in subsequent chapters). Under quasi-steady and homogeneous conditions, with minimal vertical transport and a mean flow aligned in the x direction, this equation can be simplified to

$$\frac{d\,\overline{e}}{dt} = 0 = -\overline{u'w'}\frac{d\overline{u}}{dz} - \frac{g}{\rho_r}\overline{\hat{\rho}'w'} - \varepsilon \tag{1.18}$$

The first term on the right-hand side is shear production P (almost always positive); the second is the buoyancy flux B (acts as production or destruction), and the last is viscous dissipation (always negative). The ratio $-B/P = Ri_f$ defines the flux Richardson number, a commonly used and informative parameter that measures the relative importance of buoyancy to shear on turbulence production and structure (see, e.g., Bou-Zeid et al. (2018)). The environment is neutral when $B = 0$, which is the condition where the log-law applies. Buoyancy generates turbulent perturbations ($B > 0$ and $Ri_f < 0$) when fluxes of heat, salt, or water vapor at the bottom (top) wall reduce (increase) density. For an initially neutral stratification, this results in a statically unstable environment in which a flow will self-initiate from rest (as in common environmental flows where molecular effects are weak). A simple example is a turbulent plume that emerges from a bottom horizontal wall that is heated. On the other hand, buoyancy destroys the turbulent perturbations generated by shear (in reality, it converts them to density fluctuations that store potential energy) when $B < 0$ and $Ri_f > 0$, and exerts a stabilizing influence on the flow.

Obukhov (see English translation of his original 1948 paper in Obukhov (1971)) proposed a surrogate for Ri_f with P estimated from the neutral log-law, where Eq. (1.15) and a near-constant stress assumption ($\tau/\rho = u_*^2 = -\overline{u'w'}$) are combined to yield $P \approx u_*^3/(\kappa z)$. The Monin–Obukhov stability parameter can then be computed as

$$\zeta = \frac{z}{L} = \frac{-B}{u_*^3/(\kappa z)} \tag{1.19}$$

where $L = -u_*^3/(\kappa B)$ is the Obukhov length scale. When the distance from the wall increases to become comparable to L, the effects of buoyancy become significant. The core of the associated similarity theory is that this stability parameter is sufficient to correct the log-law, and, more broadly, the inertial-layer (z, u_*) scaling, for the effect of buoyancy via L. For example, under the Monin–Obukhov similarity theory, the velocity gradient expressed in Eq. (1.15) would be adjusted to

$$\frac{d\,\overline{u}}{dz} = \frac{u_*}{\kappa z}\Psi(\zeta) \tag{1.20}$$

where Ψ is a function that has been traditionally determined empirically (note that, in boundary-layer meteorology literature, the Greek letter ϕ is usually used to denote it). More recently, theoretical models explaining its empirical form have been proposed (Li et al., 2012). Integration of this equation results in a velocity profile that is no longer exactly logarithmic due to the stability correction functions that result from Ψ. It should be noted here that the stability corrections take different forms under stable and unstable conditions (see a summary of the common ones in Li et al. (2012)).

A depiction of the effect of the stability correction relative to the neutral log-law is provided in Fig. 1.2. The left panel illustrates that the normalized profiles feature higher speeds and steeper gradients under stable conditions due to the weakening of the turbulence and the resulting momentum transfer and wall drag, with the exact opposite trends and causes under unstable conditions. The right panel (dimensional) shows the practical effect of the reduction in friction velocity under stable conditions (due to weaker downward momentum transport) and its increase under unstable conditions (stronger momentum transport). Compared to the normalized profiles, the stable profiles are shifted to the left (weaker near-wall speeds but higher speeds aloft relative to neutral), while the unstable profiles are shifted to the right (stronger near-wall speeds but lower speeds aloft relative to neutral).

As with neutral conditions, multiple other turbulence statistics have been found to follow the new (z, u_*, L) scaling, including TKE dissipation, scalar variances, vertical velocity variances, and structure functions, among others. The mean gradients and mean concentrations of scalar quantities can similarly be expressed as their neutral logarithmic profiles corrected by a function of ζ.

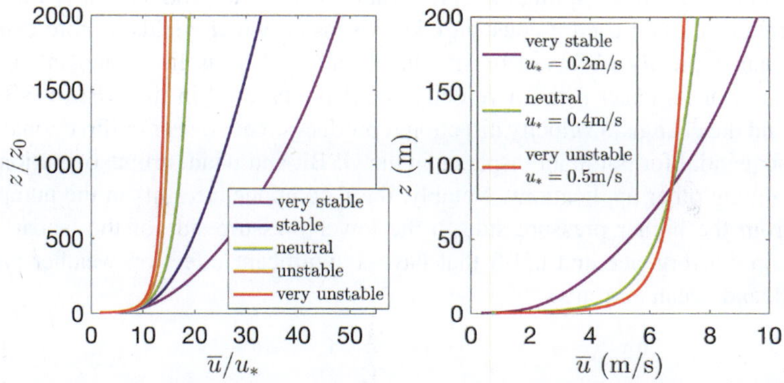

Figure 1.2 Left: the normalized velocity profile over a wall as predicted by the Monin-Obukhov similarity theory for very stable ($L = 50$ m), stable ($L = 150$ m), neutral ($L \to \infty$), unstable ($L = -100$ m), and very unstable ($L = -50$ m) conditions. The height is normalized by the roughness length (set to $z_0 = 0.1$ m) and the velocity by the friction velocity u_*. The stability functions used are the ones from Businger and Dyer under unstable conditions (as presented in Stull (1988)) and from Beljaars and Holtslag (1991) under stable conditions (note that the real profiles under very stable conditions are unlikely to follow Monin–Obukhov predictions (Mahrt, 2014)). Right: the dimensional versions of the velocity profiles accounting for the variability of u_* with stability.

1.4.4 Accounting for rotation: Ekman pumping and Rossby number similarity

The log-law and its buoyancy correction via Monin-Obukhov do not account for the effect of rotation. Monin and Obukhov's original work estimated that this restricts the elevation over which their theory is applicable in the ABL to ~ 50 m (only the bottom

5% or so). One of the first efforts to describe the velocity profiles near an interface that imposes drag on the flow, in fact, did include rotation, and it predated the work of von Kármán and the theories of Townsend and Monin and Obukhov. In 1905 Ekman (1905) examined the deviation from a geostrophic velocity field in the ocean as the surface is approached and its drag is felt by the flow. Using a constant eddy-viscosity (ν_t) model for turbulent stress that relates it to the mean velocity gradient (to compute the friction force), Ekman showed that the horizontal velocity components deviate from the geostrophic field (U_g, $V_g = 0$) following

$$\overline{u} = U_g\left(1 - e^{-z/\lambda_e}\cos(z/\lambda_e)\right) \tag{1.21}$$

$$\overline{v} = U_g\left(e^{-z/\lambda_e}\sin(z/\lambda_e)\right) \tag{1.22}$$

where $\lambda_e = \sqrt{2\nu_t/f}$ is the characteristic depth of the Ekman layer and f is the Coriolis parameter defined in Section 1.3.3. The flow speed along the isobars is $\overline{u}$, while the one normal to the isobars is $\overline{v}$. The resulting flow field recovers the geostrophic velocity when $z \gtrsim \pi\lambda_e$, while as the surface is approached, the velocity vector rotates to define a spiral, the exact details of which depend on how ν_t varies with z (originally Ekman used a constant eddy viscosity, which yields a more exuberant spiral than typically observed or simulated using LES, as illustrated in Fig. 1.3). Regardless of its exact form, a velocity spiral is observed in the ABL, OSBL, and BBL, and the change in velocity direction it produces (called veer or directional shear) is consequential for pollutant dispersal in the OSBL and wind turbine operation in the ABL, among other applications. Notably, the $\overline{v}$ component results in the pumping of fluid from the higher pressure side to the lower pressure side of the isobar, which results in convergence and uplift that have a significant effect on weather patterns, rainfall, and ocean currents.

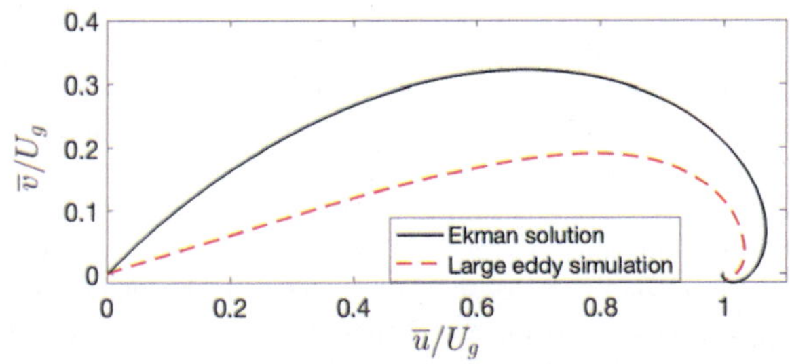

Figure 1.3 Hodographs (across ($\overline{v}$) versus along ($\overline{u}$) isobar mean flow speed) obtained from the Ekman solution in Eqs. (1.21) and (1.22) and modeled using large eddy simulation of a neutral barotropic ABL (from Ghannam and Bou-Zeid, 2021). The figure illustrates that the typical cross-stream velocity and veer are weaker in a real geophysical boundary layer than the Ekman model predicts.

The square of the dimensionless parameter z/λ_e compares the Coriolis force to the turbulent friction force. This ratio thus explains the increased importance of rotation as the spatial scales increase with distance from the wall. When the eddy viscosity is postulated to result from the product of u_* and a mixing length scale $\propto z$, one obtains

$(z/\lambda_e)^2 = fz/u_*$, which can be viewed as the inverse of a turbulent Rossby number. This parameter is the basis for the so-called Rossby number similarity, which aims to blend near-surface profiles (log-law and Monin–Obukhov similarity theory) with Ekman's spiral to predict the flow field farther from the wall. Examples of blended velocity profiles in an unstratified, rotating boundary layer are given by Blackadar and Tennekes (1968) and Tennekes (1973) for barotropic ABLs; and by Ghannam and Bou-Zeid (2021) for barotropic and baroclinic conditions.

1.4.5 Waves

Turbulence is not the only perturbation a flow can generate in a stratified environment. Gravity waves are often excited and can transfer momentum (in all directions), interact with turbulence, as well as nonlinearly evolve to generate more turbulence (whitecaps from breaking surface waves provide a familiar example). Internal gravity waves occur in statically stable regions of the flow, while surface gravity waves are present at the interface of two fluids of strikingly different densities (e.g., ocean waves on the air-sea interface). Analogous to surface waves, internal solitary waves propagate horizontally at the relatively sharp density change of the upper ocean pycnocline. In all of these cases, gravity (or, more specifically, buoyancy) is the restoring force that tries to return a displaced parcel of fluid to its neutral elevation in hydrostatic equilibrium, resulting in oscillatory motion.

For internal waves in a medium at rest and with a constant mean density stratification $\Gamma = d\overline{\hat{\rho}}/dz < 0$, a vertical displacement η (small relative to the curvature of the density gradient) of a parcel from its equilibrium generates a density perturbation $\hat{\rho} = -\Gamma\eta$ and a vertical buoyancy force $-(g/\rho_r)\hat{\rho} = (g/\rho_r)\Gamma\eta$. The balance of vertical momentum for that parcel can then be expressed in terms of the total (material) derivative of the vertical velocity changing as a result of the buoyancy force (by simplifying Eq. (1.7), assuming that the perturbation density gradient, viscous forces, and Coriolis are of second-order importance here). Since the vertical velocity is also the time derivative of the position $w = d\eta/dt$, the resulting momentum balance is then

$$\frac{dw}{dt} = \frac{d^2\eta}{dt^2} = \frac{g\Gamma}{\rho_r}\eta \tag{1.23}$$

For $\Gamma < 0$ postulated here, the solution of this ODE is an oscillatory motion with a frequency

$$N = \sqrt{-\frac{g\Gamma}{\rho_r}} \tag{1.24}$$

This is known as the buoyancy frequency or the Brunt–Väisälä frequency (derived almost concurrently and independently by David Brunt and Vilho Väisälä around 1925 (Cushman-Roisin and Beckers, 2011)).

Generalization to particle motion with a velocity at an angle ϕ with the vertical is accomplished by solving the linearized momentum equations (which assume the wave perturbation is small) to yield a dispersion relation for internal waves, see, e.g., Kundu and Cohen (2008). The resulting internal wave frequency, ω, is found to depend on the angle ϕ, but does not depend on the wavelength. In a non-rotating fluid, this frequency is $\omega = N \cos\phi$, while the inclusion of rotation leads to the following relation: $\omega^2 = N^2 \cos^2\phi + f^2 \sin^2\phi$. Internal waves span a wide range of frequencies, from a low of f to a high of N. Their propagation from the generation region occurs along the group velocity vector, which is also at the same angle ϕ with the vertical. Thus, near-inertial waves with $\omega \sim f$ propagate energy at a shallow angle to the horizontal, while near-N waves have near-vertical energy propagation.

Internal waves are triggered when topographic features (mountains, ridges, valleys) encounter atmospheric jets and oceanic currents or tides. These waves increase topographic drag on geophysical flows and, relevant to this book, also create turbulence that is often remote from the generation site. An often overlooked effect of winds on the oceans, besides creating surface waves, is driving near-inertial waves that propagate into the ocean's interior. Turbulence from these internal waves is of global importance for oceanic mixing since the ocean's interior is mostly stable and lacks the daytime convective turbulence of the atmospheric boundary layer. Such turbulence results from nonlinear processes as discussed by Sarkar and Scotti (2017) for topographic waves and by Lamb (2014) for internal solitary waves. In the atmosphere, internal waves have been linked to intermittent turbulence and mixing in the night-time stable boundary layer (Sun et al., 2015), to overturning turbulent motions in the lee of steep mountain ranges (Wurtele et al., 1996), and to clear-air turbulence (Ellrod et al., 2003).

1.5 Challenges and opportunities

This chapter briefly introduces the context of the current renaissance of interest in environmental turbulence, driven primarily by mounting environmental and climatic challenges, the urgent need to shift our resource- and energy-intensive modes of production, and the concentration of anthropogenic activities in the turbulent layers of the atmosphere and ocean. Much progress has been made over the past few decades in advancing our understanding and ability to predict the flow and transport in the atmospheric, ocean surface, and bottom boundary layers. Our understanding of interior ocean turbulence and its crucial influence on the physical and biological state of the ocean has also progressed. The rapid advances in numerical simulation techniques, computing power, and physical sensing systems have been instrumental to this progress, but environmental turbulence remains rife with stimulating challenges. Several of the following chapters offer perspectives on open questions and research opportunities in their individual subject areas. Here, we offer a broader perspective.

The need for higher fidelity in simulating and predicting turbulent environmental flows has increased with the shift in applications from describing geophysical phenomena and providing weather and climate "guidance", to ones that require high accuracy and precision for designing and operating modern engineering systems (giving rise to new fields such as meteorological engineering, and modernizing existing fields such as ocean engineering). Precision agriculture, renewable energy production both on land and offshore, urban drone flight, unmanned underwater exploration, marine farming, wildfire prediction, and weather and climate adaptation are some examples of new applications where environmental turbulence is a central challenge. The impossibility of obtaining an exact description of the real-world initial and boundary conditions, coupled with the chaotic nature of turbulence, poses a formidable hurdle that requires new ideas and approaches for examining bifurcations, designing ensembles of simulations, uncertainty quantification, and data assimilation, to name a few.

Another grand challenge is that a typical environmental flow may feature air over a terrain that is covered by a mixture of buildings and trees, with maybe some wind turbines and a wavy lake or ocean surface, and will proceed through stable and unstable periods where rotation and unsteadiness are omnipresent. That is, the various physical processes this book covers in different chapters rarely occur in isolation. Much remains to be understood about how these environmental turbulence drivers interact and what unique turbulence regimes may arise from this interaction. The chapters of this book will delve into some of these interactions involving two drivers (e.g., flow in an urban canopy with buoyancy), but this remains a wide-open research arena.

Environmental turbulence often mixes air and water, producing multiphase flows (e.g., near the ABL-OSBL interface during storms) that introduce new, fascinating physics and dynamics into the problem. Two such examples are deformable interfaces that respond to fluid flow on either side, and influence the turbulence on both sides, and breaking waves, drops, and bubbles that drastically enhance the exchange of mass and heat. A related challenge concerns the transport of particles that have negligible inertia but positive or negative buoyancy (microplastics) and inertial particles (snow crystals, pollen, raindrops, sand particles, phytoplankton, etc.) in turbulent environmental flows; some of these particles have unique anisotropic shapes that induce a response that can be very different from that of a smooth sphere. Other particles change in phase as they are transported in turbulent flows, with implications for mass, momentum, and energy budgets. Cloud turbulence is particularly influenced by phase change (and the chemical properties of aerosols), and by how it modulates radiative heat transfer, which is central to cloud dynamics.

Another research frontier is tying the statistical and structural descriptions of turbulence in environmental flows. For wall-bounded neutral flows, the existence and topology of a wide range of turbulent structures is now well established (from streamwise worm-like filament vortices in the near-wall region, to the very large scale structures that expand to $\sim$ ten times the depth of the turbulent layer (Smits et al., 2011)). The links of these structures to the various vertical sublayers and to the statistical moments of turbulence are also reasonably well understood (Gioia et al., 2010; Katul et al., 2013) in canonical wall-bounded flow settings. How rotation, buoyancy, and/or very rough or deformable boundaries, common to environmental turbulence, alter the topological properties of the coherent structures and their links to the statistical moments remains an open field of questions. The physics therein will depend on the new length or time scales introduced by the physical forces in environmental flows (buoyancy, rotation, topographic roughness, tides, large-scale eddies, among others), in relation to the integral and other scales of the simpler canonical flow problems. Static stability, for example, introduces the Ozmidov length scale (Dougherty, 1961; Ozmidov, 1965) that characterizes the lower end of the dynamic range over which buoyancy has a significant influence. The Ozmidov scale decreases with increasing stratification. If that scale becomes as small as $\sim 60\eta$, roughly the end of the inertial subrange, then the inertial subrange might be squeezed out of existence by stability. Furthermore, canonical turbulent flows may respond to stable stratification by forming flow features that are qualitatively different, such as internal gravity waves and coherent vortices. Unravelling the links between 3D turbulence, waves, and vortices in a stratified fluid is an active area of research.

Fortunately, the advances listed earlier in this section (in numerical simulation techniques, computing power, and observational systems) will accelerate and be coupled with novel analysis techniques (using new statistics and machine learning, visualizations, big data mining, and, vitally, novel theoretical ideas) to usher in a new golden age of research in environmental turbulence. The heightened importance of the applications is likely to attract more support and talent to the field. This book aspires to be a catalyst for this exciting near future.

References

Allouche, M., Bou-Zeid, E., Iipponen, J., 2023. The influence of synoptic wind on land–sea breezes. Quarterly Journal of the Royal Meteorological Society 149 (757), 3198–3219.

Beljaars, A.C.M., Holtslag, A.A.M., 1991. Flux parameterization over land surfaces for atmospheric models. Journal of Applied Meteorology and Climatology 30 (3), 327–341.

Blackadar, A.K., Tennekes, H., 1968. Asymptotic similarity in neutral barotropic planetary boundary layers. Journal of the Atmospheric Sciences 25 (6), 1015–1020.

Bou-Zeid, E., Gao, X., Ansorge, C., Katul, G.G., 2018. On the role of return to isotropy in wall-bounded turbulent flows with buoyancy. Journal of Fluid Mechanics 856, 61–78.

Bou-Zeid, E., Higgins, C., Huwald, H., Meneveau, C., Parlange, M.B., 2010. Field study of the dynamics and modelling of subgrid-scale turbulence in a stable atmospheric surface layer over a glacier. Journal of Fluid Mechanics 665, 480–515.

Brutsaert, W., 2005. Hydrology. Cambridge University Press.

Businger, J.A., 1984. Equations and concepts. In: Nieuwstadt, F.T.M., van Dop, H. (Eds.), Atmospheric Turbulence and Air Pollution Modelling. Springer, Netherlands, pp. 1–36. chapter 1.

Chamecki, M., Dias, N.L., 2004. The local isotropy hypothesis and the turbulent kinetic energy dissipation rate in the atmospheric surface layer. Quarterly Journal of the Royal Meteorological Society 130 (603), 2733–2752.

Cushman-Roisin, B., Beckers, J.-M., 2011. Introduction to Geophysical Fluid Dynamics. Physical and Numerical Aspects, second edition. Academic Press, Waltham, MA.

D'Asaro, E.A., 2014. Turbulence in the upper-ocean mixed layer. Annual Review of Marine Science 6 (1), 101–115.

D'Asaro, E.A., Lee, C.M., Rainville, L., Harcourt, R.R., Thomas, L.N., 2011. Enhanced turbulence and energy dissipation at ocean fronts. Science 332, 318–322.

Dougherty, J., 1961. The anisotropy of turbulence at the meteor level. Journal of Atmospheric and Terrestrial Physics 21 (2–3), 210–213.

Ekman, V.W., 1905. On the influence of the Earth's rotation on ocean currents. Arkiv för Matematik, Astronomi och Fysik 2, 1–53.
Ellrod, G.P., Lester, P.F., Ehernberger, L.J., 2003. Clear air turbulence. In: Holton, J.R. (Ed.), Encyclopedia of Atmospheric Sciences. Academic Press, Oxford, pp. 393–403.
Ghannam, K., Bou-Zeid, E., 2021. Baroclinicity and directional shear explain departures from the logarithmic wind profile. Quarterly Journal of the Royal Meteorological Society 147 (734), 443–464.
Gill, A.E., 1982. Atmosphere-Ocean Dynamics, vol. 30. Academic Press.
Gioia, G., Guttenberg, N., Goldenfeld, N., Chakraborty, P., 2010. Spectral theory of the turbulent mean-velocity profile. Physical Review Letters 105 (18), 184501.
Hultmark, M., Vallikivi, M., Bailey, S.C.C., Smits, A.J., 2013. Logarithmic scaling of turbulence in smooth- and rough-wall pipe flow. Journal of Fluid Mechanics 728, 376–395.
Kasper, J.H., Stieren, A., Stevens, R.J.A.M., 2024. Simulation and modeling of wind farms in baroclinic atmospheric boundary layers. Journal of Renewable and Sustainable Energy 16 (6), 063302.
Katul, G.G., Porporato, A., Manes, C., Meneveau, C., 2013. Co-spectrum and mean velocity in turbulent boundary layers. Phys. Fluids 25 (9), 91702.
Kolmogorov, A.N., 1962. A refinement of previous hypotheses concerning the local structure of turbulence in a viscous incompressible fluid at high Reynolds number. Journal of Fluid Mechanics 13 (1), 82–85.
Kolmogorov, A.N., 1991. The local structure of turbulence in incompressible viscous fluid for very large Reynolds numbers. Proceedings of the Royal Society of London Series A: Mathematical Physical and Engineering Sciences 434 (1890), 9–13.
Kundu, P.K., Cohen, I.M., 2008. Fluid Mechanics, fourth edition. Academic Press, Amsterdam, Boston.
Lamb, K.G., 2014. Internal wave breaking and dissipation mechanisms on the continental slope/shelf. Annual Review of Fluid Mechanics 46, 231–254.
Lenschow, D.H., Wyngaard, J.C., Pennell, W.T., 1980. Mean-field and second-moment budgets in a baroclinic, convective boundary layer. Journal of the Atmospheric Sciences 37 (6), 1313–1326.
Lesieur, M., 2008. Turbulence in Fluids. Fluid Mechanics and Its Applications, vol. 84. Springer, Netherlands, Dordrecht.
Li, D., Katul, G.G., Bou-Zeid, E., 2012. Mean velocity and temperature profiles in a sheared diabatic turbulent boundary layer. Physics of Fluids 24 (10), 105105–105116.
Mahrt, L., 2014. Stably stratified atmospheric boundary layers. Annual Review of Fluid Mechanics 46 (1), 23–45.
Marusic, I., Broomhall, S., 2021. Leonardo da Vinci and fluid mechanics. Annual Review of Fluid Mechanics 53 (1), 1–25.
Marusic, I., Monty, J.P., 2019. Attached eddy model of wall turbulence. Annual Review of Fluid Mechanics 51 (1), 49–74.
McWilliams, J.C., 2016. Submesoscale currents in the ocean. Proceedings of the Royal Society A. Mathematical, Physical and Engineering Sciences 472, 20160117.
Momen, M., Bou-Zeid, E., Parlange, M.B., Giometto, M., 2018. Modulation of mean wind and turbulence in the atmospheric boundary layer by baroclinicity. Journal of the Atmospheric Sciences 75 (11), 3797–3821.
Monin, A.S., Obukhov, A.M., 1954. Basic laws of turbulent mixing in the ground layer of the atmosphere. Trudy Geofizicheskogo Instituta, Akademiya Nauk SSSR 24 (151), 163–187 (in Russian).
Obukhov, A.M., 1971. Turbulence in an atmosphere with a non-uniform temperature. Boundary-Layer Meteorology 2 (1), 7–29.

Omidvar, H., Bou-Zeid, E., Li, Q., Mellado, J.-P., Klein, P., 2020. Plume or bubble? Mixed-convection flow regimes and city-scale circulations. Journal of Fluid Mechanics 897, A5.
Ozmidov, R.V., 1965. Turbulent exchange in the ocean with stable stratification. Izvestija Akademii Nauk SSSR. Fizika Atmosfery i Okeana 1 (8), 853–860.
Parker, D., 2015. Mesoscale meteorology | overview. In: North, G.R., Pyle, J., Zhang, F. (Eds.), Encyclopedia of Atmospheric Sciences, second edition. Academic Press, Oxford, pp. 316–322.
Pope, S.B., 2000. Turbulent Flows. Cambridge University Press.
Richardson, L.F., 1922. Weather Prediction by Numerical Process. Cambridge University Press.
Sarkar, S., Scotti, A., 2017. From topographic internal gravity waves to turbulence. Annual Review of Fluid Mechanics 49, 195–220.
Skyllingstad, E.D., Samelson, R.M., 2012. Baroclinic frontal instabilities and turbulent mixing in the surface boundary layer. Part I: unforced simulations. Journal of Physical Oceanography 42, 1701–1716.
Smits, A.J., McKeon, B.J., Marusic, I., 2011. High–Reynolds number wall turbulence. Annual Review of Fluid Mechanics 43 (1), 353–375.
Stoll, R., Gibbs, J.A., Salesky, S.T., Anderson, W., Calaf, M., 2020. Large-eddy simulation of the atmospheric boundary layer. Boundary-Layer Meteorology 177 (2–3), 541–581.
Stull, R.B., 1988. An Introduction to Boundary Layer Meteorology. Springer, Netherlands.
Stull, R.B., 1991. Static stability—an update. Bulletin of the American Meteorological Society 72 (10), 1521–1529.
Sun, J., Nappo, C.J., Mahrt, L., Belušić, D., Grisogono, B., Stauffer, D.R., Pulido, M., Staquet, C., Jiang, Q., Pouquet, A., Yagüe, C., Galperin, B., Smith, R.B., Finnigan, J.J., Mayor, S.D., Svensson, G., Grachev, A.A., Neff, W.D., 2015. Review of wave-turbulence interactions in the stable atmospheric boundary layer. Reviews of Geophysics 53, 956–993.
Taylor, J.R., Ferrari, R., 2010. Buoyancy and wind-driven convection at mixed layer density fronts. Journal of Physical Oceanography 40, 1222–1242.
Tennekes, H., 1973. The logarithmic wind profile. Journal of the Atmospheric Sciences 30 (2), 234–238.
Thomas, L.N., Tandon, A., Mahadevan, A., 2008. Submesoscale processes and dynamics. In: Ocean Modeling in an Eddying Regime. In: Geophysical Monograph, vol. 177, pp. 17–38.
Thorpe, S.A., 2007. An Introduction to Ocean Turbulence. Cambridge University Press.
Townsend, A.A., 1980. The Structure of Turbulent Shear Flow. Cambridge University Press.
Trowbridge, J.H., Lentz, S.J., 2018. The bottom boundary layer. Annual Review of Marine Science 10 (1), 397–420.
Vallis, G.K., 2017. Atmospheric and Oceanic Fluid Dynamics. Cambridge University Press.
Verma, V., Pham, H.T., Sarkar, S., 2019. The submesoscale, the finescale and their interaction at a mixed layer front. Ocean Modelling 140, 101400.
von Kármán, T., 1930. Mechanical Similitude and Turbulence. Nachrichten von der Gesellschaft der Wissenschaften zu Gottingen, vol. 5. Von der Gesellschaft der Wissenschaften zu Gottingen, pp. 58–76. English translation at https://ntrs.nasa.gov/citations/19930094805.
Wurtele, M.G., Sharman, R.D., Datta, A., 1996. Atmospheric Lee waves. Annual Review of Fluid Mechanics 28, 429–476.
Wyngaard, J., 1992. Atmospheric turbulence. Annual Review of Fluid Mechanics 24 (1), 205–233.
Wyngaard, J.C., 2010. Turbulence in the Atmosphere. Cambridge University Press.
Zahn, E., Bou-Zeid, E., Dias, N.L., 2023. Relaxed eddy accumulation outperforms Monin-Obukhov flux models under non-ideal conditions. Geophysical Research Letters 50 (7), e2023GL103099.

Unstably stratified shear flows

2

Gabriel G. Katul[a], *Einara Zahn*[b], *and Elie Bou-Zeid*[c]
[a]Department of Civil, Construction and Environmental Engineering, University of Alabama, Tuscaloosa, AL, United States, [b]Department of Earth and Environmental Sciences, University of Pennsylvania, Philadelphia, PA, United States, [c]Department of Civil and Environmental Engineering, Princeton University, Princeton, NJ, United States

2.1 Introduction

The exchange of momentum, heat, water vapor, carbon dioxide, and many gases between the land- or ocean- surface and the lower atmosphere is complicated by the coexistence of shear- and buoyancy-generated turbulence. Buoyancy forces often destabilize the flow, as may occur when the surface is heated from below. Such flows differ from pure convection and shear-only in many ways, as the turbulence generation mechanisms act in both the horizontal (shearing of adjacent fluid layers due to the presence of a boundary) and vertical (fluid density gradients) directions (Stull, 1988; Kaimal and Finnigan, 1994; Wyngaard, 2010; Bou-Zeid et al., 2018). Turbulence that is excited from multiple directions and at different wavelengths is more diffusive compared to its shear-only counterpart (Mellor and Yamada, 1982; Li, 2021). The scale separation between the macro- and micro- (or Kolmogorov) eddy scales also widens, providing more degrees of freedom within the turbulent kinetic energy cascade. The macro-eddy sizes become difficult to characterize with a single integral scale, and their degree of organization becomes more diverse in size and topology (Salesky and Anderson, 2018). This chapter deals with these issues, with a focus on the unstable atmospheric surface layer (ASL).

The ASL is a region within the atmospheric boundary layer (ABL) that is above the roughness sublayer (RSL) but below the so-called mixed layer (Brutsaert, 1982), as represented in Fig. 2.1. Its proximity to the land (or ocean) surface allows neglecting the Coriolis forces (except for their role in rotating the velocity profiles). It is also the layer where the much-studied law of the wall, and buoyancy corrections to it, form some of its key foundational theories (Brutsaert, 1982). This layer is significant because human life and the ecosystem services that support it primarily reside within it. Wind turbines aim to harvest kinetic energy from the flow within the ASL. Dispersion from smoke stacks within urban environments discharges waste into the ASL. The 'handshake' between the climate system and the land or ocean surface is mathematically described using ASL formulations for the mean velocity, mean air temperature, and scalar concentrations.

2.2 Unique features of convective environmental turbulence

The general study of turbulent convection commences with a perturbation from a quiescent state of a fluid bounded by a cold top and/or a heated bottom. This perturbation allows hot fluid to rise and cold fluid to settle, thereby releasing some

Environmental Turbulence. https://doi.org/10.1016/B978-0-32-395873-8.00009-7
Copyright © 2026 Elsevier Inc. All rights are reserved, including those for text and data mining, AI training, and similar technologies.

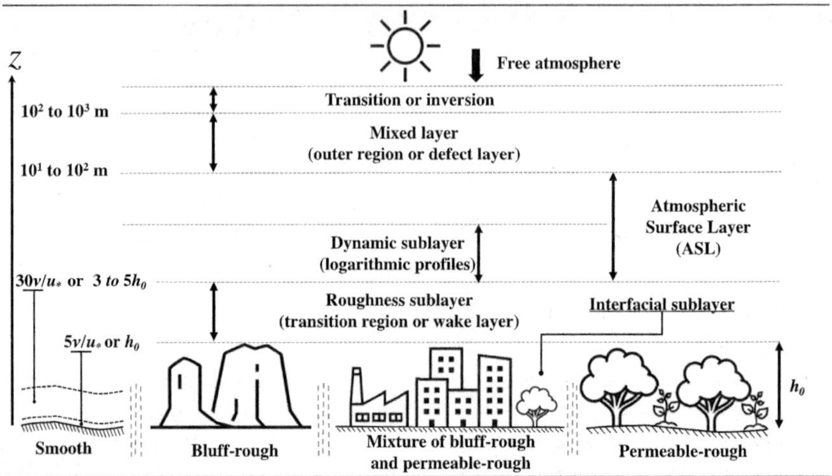

Figure 2.1 Division of the atmospheric boundary layer (ABL) into sublayers, where h_0 is the mean height of the roughness elements, ν is the kinematic air viscosity, and u_* is the friction velocity. The diagram represents a typical sunny day when solar radiation heats the ground surface (and roughness elements). This heating enhances turbulence generation by buoyancy. The atmospheric surface layer (ASL) is the region where Monin–Obukhov similarity theory (MOST) is valid, and is the focus of this chapter. Note that the vertical axis is not to scale.

potential energy. When viscous forces are not large enough to dissipate the released potential energy, the fluid vertical acceleration destabilizes the quiescent state and sets up Bénard convection cells. As the temperature difference between the bottom and top is further increased, these Bénard cells become unstable — and a state known as turbulent convection arises. The presence of a solid boundary, with no-slip at its surface (smooth or rough), also leads to shearing between fluid layers. Shear between fluid layers can be studied in isolation using laboratory experiments with concentric rotating cylinders. These studies suggest that increasing rotational speed leads to the emergence of Taylor vortices that are, like Bénard cells, organized modes of instabilities. These vortices, initially deterministic, are further destabilized with increasing rotational speed in a manner analogous to the destabilization experienced by Bénard cells with increasing temperature difference between top and bottom layers. A fully developed turbulent state also results in further increases in rotational speed, with determinism entirely lost. While the ASL does not experience such rotational speed and Taylor instabilities, shearing does occur because time-averaged velocity in the free atmosphere is finite and substantial (e.g., geostrophic winds), whereas near the ground it is small. What makes atmospheric turbulent flows that are sheared and heated from the bottom (e.g., the land surface during the daytime) a rich topic in environmental fluid mechanics is the coexistence and interaction among these instability modes. Both shearing and buoyancy instabilities occur and interact to shape the mean velocity and mean air temperature profiles, as well as the velocity and air temperature variances, among other flow statistics.

Theories that describe the shape of the mean velocity, mean air temperature, and mean water vapor concentration profiles based on these interactions have been proposed to predict regional and watershed-scale heat and evaporative fluxes (Brutsaert and Mawdsley, 1976; Brutsaert, 1986; Parlange et al., 1995; Parlange and Katul, 1995; Brutsaert, 1999). These theories also enabled measurements of chemical compounds (e.g., biogenic hydrocarbons) and the estimation of their mass exchange with the underlying surface using what is now labeled as the aerodynamic method (Fuentes et al., 2000). The aforementioned method employs the so-called flux-gradient approximation to represent turbulent fluxes as a function of the mean concentration gradient using an eddy diffusivity related to the eddy viscosity. Interpreting meteorological measurements and contributing source areas (colloquially labeled as the footprint problem) is one line of inquiry where, once again, these theories feature (Hsieh et al., 2000; Schmid, 2002; Vesala et al., 2008; Hsieh and Katul, 2009). The interplay between temporal measurements and the size-frequency distribution of unstable vortices opened up new explanations and inquiries into the much-discussed energy balance closure at the land surface (Wilson et al., 2002; Kanda et al., 2004; Oncley et al., 2007; Foken, 2008; Li et al., 2018a; Mauder et al., 2020; Liu et al., 2021). This problem emerged after demonstrating again and again that the measured available energy at the land surface (net radiative surface flux minus conductive flux into the ground) consistently exceeds the sum of the measured turbulent sensible and latent (evaporative) heat fluxes by some 20%–30% during daytime.

Convection even arises in cold environments where it may not be expected as long as a negative vertical temperature gradient exists. Airflow over ice sheets during summertime, for example, was shown to be unstable some 20%–40% of the time over a 2-month period (Cullen and Steffen, 2001). These unstable flow states appear to play a first-order role in controlling katabatic forcing over ice sheets. The role of turbulent eddies in air pollution, especially in describing gaseous dry deposition during the daytime, has a long tradition in micrometeorology (Wesely and Hicks, 2000) and continues to be a subject of research. Today, models for deposition velocities still use K-theory in unstable atmospheric conditions and require models for eddy diffusivity. More recent reviews that cover settling of dust, snow, aerosols, micro-plastics, pollen grains, among others, also highlight the role of shear-driven convective eddies (Farmer et al., 2021). It will be remiss if the role of ASL turbulence in linking the land (or ocean) surface to climate models (and the concomitant ABL growth and cloud-formation representation) is not flagged here (Bogenschutz et al., 2013; Waterman et al., 2022). Listing all these topics is offered to provide a preview and motivation for the significance of unstably stratified and sheared turbulence, and its representation in the unstable atmospheric boundary layer.

2.3 First principles

The starting point is a continuum description of turbulence, using the Reynolds-averaged longitudinal momentum balance and temperature continuity equations for an incompressible flow, subject to the Boussinesq approximation (see Chapter 1). These mean equations are derived from the instantaneous momentum and transport equations presented in Chapter 1 after ignoring the Coriolis force and are repeated here for convenience as

$$\frac{\partial U}{\partial t} + U_j \frac{\partial U}{\partial x_j} = \nu \frac{\partial^2 U}{\partial x_j \partial x_j} - \frac{\partial \overline{u' u'_j}}{\partial x_j} - \frac{1}{\rho}\frac{\partial P}{\partial x} \tag{2.1}$$

$$\frac{\partial \overline{T}}{\partial t} + U_j \frac{\partial \overline{T}}{\partial x_j} = D_m \frac{\partial^2 \overline{T}}{\partial x_j \partial x_j} - \frac{\partial \overline{u'_j T'}}{\partial x_j} \tag{2.2}$$

where t is time, $x_j = (x, y, z)$ represents the longitudinal ($= x$), lateral ($= y$), and vertical ($= z$) directions, respectively. The longitudinal direction is aligned along the mean wind direction so that the mean lateral velocity is zero. The U, $\overline{T}$, and P represent the Reynolds-averaged longitudinal velocity, temperature, and pressure, respectively; ρ is the mean air density (overline dropped for notational simplicity); ν is the air kinematic viscosity; D_m is the molecular diffusivity of temperature T in air; $u'_i = (u', v', w')$ are the component-wise turbulent velocity excursions in direction x_i; T' is the turbulent temperature fluctuation; and, unless otherwise stated, primed quantities represent turbulent excursions from the Reynolds-averaged mean state represented by overbar or capital letter symbol. Hence, the instantaneous velocity and temperature can be expressed as $u_i = U_i + u'_i$ and $T = \overline{T} + T'$. Because t indicates time, a capital symbol for instantaneous T and $\overline{T}$ for Reynolds-averaged T are adopted. The use of these equations to predict basic quantities such as U, $\overline{T}$, or heat fluxes frames the scope here.

The chapter is organized as follows: a reference state labeled as the idealized ASL is first introduced to highlight conditions where the turbulent stresses and fluxes are constant in z. Within this reference state, the effects of buoyancy on dU/dz and $d\overline{T}/dz$, and their deviations from the so-called law-of-the wall (see Chapter 1), are considered using dimensional analysis, directional dimensional analysis, and phenomenological links based on foundational ideas of locally homogeneous and isotropic turbulence covered in Chapter 1 (but adjusted by results from experiments on how integral scales vary with thermal stratification). A discussion on the success of these theories beyond the mean state is also presented, leading to new theories for measuring sensible heat flux. The effects of buoyancy on dU/dz and $d\overline{T}/dz$ are then considered using momentum and heat flux budgets. The goal of this analysis is to delineate regimes where gradient-diffusion approximations are appropriate and to predict turbulent stresses and fluxes from dU/dz and $d\overline{T}/dz$ when only a local balance between stress generation and its destruction due to pressure-velocity interactions. Thus the roles of flux transport terms and buoyancy terms arising from the complex interactions between u' and T' are presented. The same approach is then revisited from a spectral perspective to highlight the roles of multiple eddy sizes and their associated energetics in eddy viscosity and eddy diffusivity. The chapter concludes with a number of open questions and challenges.

2.3.1 A reference state: the idealized ASL

An idealized ASL assumes that the flow is characterized by (i) high Reynolds and Peclet numbers (molecular viscosity and diffusivity are neglected relative to their turbulent counterparts); (ii) stationarity (i.e., $\partial(.)/\partial t = 0$) and planar-homogeneity (i.e., $\partial(.)/\partial x = \partial(.)/\partial y = 0$) with no subsidence (i.e., $\overline{U_3} = \overline{W} = 0$); and (iii) near-zero mean horizontal pressure gradient (i.e., $\partial P/\partial x = 0$). For these idealized conditions, the aforementioned mean longitudinal momentum balance and the mean temperature budget reduce to

$$\frac{\partial \overline{w'u'}}{\partial z} = 0, \quad \frac{\partial \overline{w'T'}}{\partial z} = 0 \tag{2.3}$$

Hence, the turbulent stresses (i.e. $\overline{w'u'}$) and heat fluxes (i.e. $\overline{w'T'}$) do not vary with z. It is for this reason that the idealized ASL is labeled as the constant-stress or constant-flux layer or region (Stull, 1988). Integrating Eq. (2.3) with respect to z from $z = 0$ (or the ground surface) to $z = z_r$, a reference height in the ASL, yields $\overline{w'u'} = -u_*^2$ and $\overline{w'T'} = -T_* u_*$ being constant and set by the flux boundary condition at $z = 0$, where u_* is referred to as the friction (or shear) velocity, and T_* is a characteristic temperature scale formed from the sensible heat flux and u_*.

Because the mean velocity and mean temperature gradients arise frequently, they are labeled as $S = \partial U/\partial z$ and $\Gamma = \partial \overline{T}/\partial z$. The associated turbulent momentum flux ($\overline{w'u'}$) and heat flux ($\overline{w'T'}$) budgets in the idealized ASL are (Stull, 1988)

$$\frac{\partial \overline{u'w'}}{\partial t} = 0 = -\overline{w'w'}S - \frac{\partial \overline{w'w'u'}}{\partial z} - \frac{1}{\rho}\overline{u'\frac{\partial p'}{\partial z}} + \beta\overline{u'T'} - \epsilon_{uw} \tag{2.4}$$

$$\frac{\partial \overline{w'T'}}{\partial t} = 0 = -\overline{w'w'}\Gamma - \frac{\partial \overline{w'w'T'}}{\partial z} - \frac{1}{\rho}\overline{T'\frac{\partial p'}{\partial z}} + \beta\overline{T'T'} - \epsilon_{wT} \tag{2.5}$$

where p' is the turbulent pressure fluctuation, $\beta = g/\overline{T}$ is known as the buoyancy parameter, g is the gravitational acceleration, and ϵ_{uw} and ϵ_{wT} are the decorrelation sink terms due to the action of molecular viscosity and thermal diffusivity. The molecular Prandtl number of air is defined as $Pr_m = \nu/D_m \approx 0.72$ (order unity). The terms on the left-hand side of the equations represent time changes in covariances (i.e., $\overline{u'w'}$ and $\overline{w'T'}$). The terms on the right-hand side of the equations represent (in order): production terms due to the presence of mean velocity and air temperature gradients ($-\overline{w'w'}S$ and $-\overline{w'w'}\Gamma$), flux-transport terms that represent turbulent transport of momentum and heat fluxes ($-\partial\overline{w'w'u'}/\partial z$ and $-\partial\overline{w'w'T'}/\partial z$), pressure-decorrelation terms due to interactions between pressure and velocity and pressure and temperature, buoyancy terms arising from thermal stratification in the ASL ($\beta\overline{u'T'}$ and $\beta\overline{T'T'}$), and molecular destruction terms involving ν and D_m. At high Reynolds numbers, the decorrelation terms due to molecular processes are small and will not be further considered (unlike their counterparts in the turbulent kinetic energy or temperature variance equations). A plausibility argument, which will be elaborated upon later in spectral space, is that those molecular destruction terms are primarily active in the dissipation range. At such small scales, turbulent motion is already in a near-isotropic state, and differing flow variables tend to be already (in a turbulence cascade sense) statistically decorrelated from each other.

The quantities $\overline{u'w'}$ and $\overline{w'T'}$ are statistical quantities (covariances) originating from averaging the nonlinear advective acceleration term in the Navier-Stokes equations. The terminology used for describing the physical processes associated with their budgets on the right-hand side of Eqs. (2.4) and (2.5) reflects whether these process, on average, enhance covariances (i.e., production), weaken covariances (i.e., destruction or decorrelation), or simply act to transport covariances across different layers without adding or removing heat and momentum when spatially averaged (or depth-averaged here) across the entire flow domain. When S or Γ are externally imposed, parcels of air displaced towards some reference height z_r in the ASL with vertical kinetic energy σ_w^2 create excursions from the mean state at z_r. Parcels moving upwards ($w' > 0$) carry with them, on average, the mean state in U and T below z_r. For an $S > 0$ (typical for wall-bounded flows) and a $\Gamma < 0$ (warmer surface), the upward vertically displaced parcel experiences a background at z_r that is elevated in mean velocity and colder in mean temperature, as schematically displayed in Fig. 2.2. Hence, this upward motion most likely leads to $u' < 0$ and $T' > 0$, and with $w' > 0$,

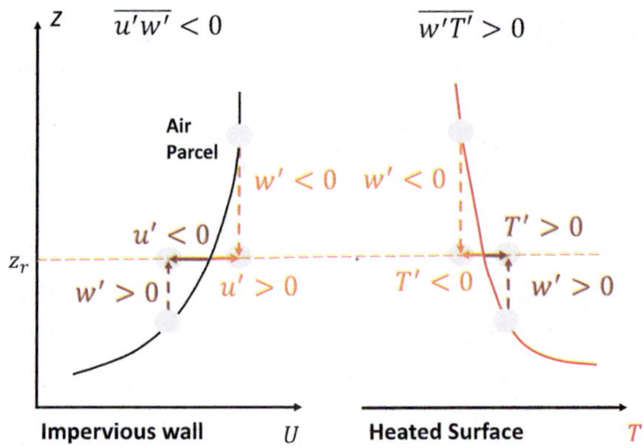

Figure 2.2 Air parcels exchanging momentum and heat at a reference height z_r. Upward moving parcels ($w' > 0$) towards z_r carry with them depleted velocity ($u' < 0$) and elevated temperature ($T' > 0$) thereby contributing to $u'w' < 0$ and $w'T' > 0$. Conversely, downward moving parcels ($w' < 0$) carry with them higher velocity ($u' > 0$) and reduced temperature ($T' < 0$) to z_r thereby contributing, once again, to $u'w' < 0$ and $w'T' > 0$.

results in a negative $u'w'$ and a positive $w'T'$ at z_r, as shown in Fig. 2.2. Likewise, parcels of air experiencing a downward displacement from z_r due to a $w' < 0$ also contribute to a $u' > 0$ and $T' < 0$ when compared to the new background state they experience at z_r as in Fig. 2.2. Hence, this downward motion towards z_r, once again, results in a negative $u'w'$, and a positive $w'T'$ (on average). Therefore this mechanical displacement of air parcels with energy σ_w^2 acting along S and Γ consistently contributes to downgradient transport (downward momentum fluxes and upward sensible heat fluxes) and is labelled as a mechanical production (since it will always be positive) term in the turbulent stress and heat flux budgets. For well-mixed conditions with $S = 0$ and $\Gamma = 0$, mechanical production ceases to exist in the aforementioned turbulent stress and heat flux budgets.

Unsurprisingly, $\overline{u'w'}$ and $\overline{w'T'}$ are likely to depend on $\overline{w'^2}$, S, and Γ and their signs are dictated by the signs of $-S$ and $-\Gamma$. The triple correlation terms, which are the transport of stresses and fluxes by w', are presumed to be zero at the boundaries of the entire flow domain (ground and top of ABL) because turbulence is damped at those boundaries. These triple moments do not produce or remove momentum and heat from the entire flow domain and are thus labeled as transport terms. Because fluid pressure is isotropic, interactions between flow variables and pressure-related terms tend to act in a manner to enhance isotropization. In the context of turbulent stresses and fluxes, isotropization leads to decorrelation among flow variables.

Likewise, the turbulent kinetic energy (TKE) budget is given by

$$\frac{\partial \overline{e}}{\partial t} = 0 = -\overline{u'w'}S + \beta\overline{w'T'} - \epsilon - \left[\frac{\partial \overline{w'e}}{\partial z} + \frac{1}{\rho}\frac{\partial \overline{w'p'}}{\partial z}\right] \tag{2.6}$$

where $e = (u'_i u'_i)/2$ is the instantaneous TKE; $\overline{e} = \overline{u'_i u'_i}/2$ is the mean TKE; ϵ is the mean TKE dissipation rate; and the quantities in [.] are turbulent and pressure transport of TKE, respectively. These transport terms, similar to the ones in the covariance budgets, do not produce or dissipate TKE, on average, across the boundary layer depth and abide by

$$\int_0^h \left(-\frac{\partial \overline{w'e}}{\partial z} - \frac{1}{\rho}\frac{\partial \overline{w'p'}}{\partial z}\right) dz = 0 \tag{2.7}$$

where h is the ABL height. When these transport terms are locally small relative to other terms at a given z, $\epsilon = P_m + B_p$, where $P_m = -\overline{u'w'}S$ is labeled as mechanical production and $B_p = \beta\overline{w'T'}$ is labeled as buoyant production of TKE. The flux Richardson number $Ri_f = -B_p/P_m$ is routinely used to assess the relative importance of these two TKE-generating mechanisms. Because $B_p > 0$ in the unstable ASL and near the boundary, $P_m > 0$ due to a large S, the overall TKE production and dissipation rates tend to be large. The injection of TKE occurs within different velocity components and can be understood from the component-wise half-variance budgets. In the absence of flux transport terms (i.e., vertical gradients of triple moments are momentarily ignored for simplicity), these budgets are given by (Stull, 1988; Bou-Zeid et al., 2018)

$$\frac{\partial \overline{u'u'}}{\partial t} = P_m + R_u - \epsilon_u = 0 \tag{2.8}$$

$$\frac{\partial \overline{v'v'}}{\partial t} = R_v - \epsilon_v = 0 \tag{2.9}$$

$$\frac{\partial \overline{w'w'}}{\partial t} = B_p + R_w - \epsilon_w = 0 \tag{2.10}$$

where $R_u = \rho^{-1}\overline{p'\partial u'/\partial x}$, $R_v = \rho^{-1}\overline{p'\partial v'/\partial y}$, and $R_w = \rho^{-1}\overline{p'\partial w'/\partial z}$ are pressure redistribution terms in the longitudinal, lateral, and vertical directions, and ϵ_u, ϵ_v, and ϵ_w are the viscous dissipation rates of their respective half-variances $\sigma_u^2/2$, $\sigma_v^2/2$, and $\sigma_w^2/2$, where $\sigma_s^2 = \overline{s's'}$ is the variance of an arbitrary flow variable s.

The component-wise half variance budget equations illustrate how mechanical production produces $\overline{u'u'}$ through P_m, and buoyancy production produces $\overline{w'w'}$ through B_p. Because the component-wise kinetic energy dissipation rates occur at scales commensurate with the Kolmogorov microscale and are thus nearly isotropic, it will be assumed hereafter that $\epsilon_u = \epsilon_v = \epsilon_w = (1/3)\epsilon$ for simplicity. For an incompressible flow, the instantaneous continuity equation $(\partial u_i'/\partial x_i = 0)$ ensures that

$$R_u + R_v + R_w = \frac{1}{\rho}\overline{p'\left[\frac{\partial u'}{\partial x} + \frac{\partial v'}{\partial y} + \frac{\partial w'}{\partial z}\right]} = 0 \quad (2.11)$$

When summed, the redistribution terms do not produce or dissipate TKE. They only act to isotropize turbulent energy among velocity components. From definitions, the strength of these redistribution terms depends on Ri_f by Bou-Zeid et al. (2018)

$$\frac{R_u}{P_m} = \frac{1}{3}\left(-2 - Ri_f\right), \quad \frac{R_v}{P_m} = \frac{1}{3}\left(1 - Ri_f\right), \quad \frac{R_w}{P_m} = \frac{1}{3}\left(1 + 2Ri_f\right) \quad (2.12)$$

In the near-neutral limit when $Ri_f = 0$, mechanical energy is extracted from the σ_u^2 budget and injected into σ_w^2 and σ_v^2 budgets equally (in practice, wall blockage will reduce redistribution into σ_w^2). For this reason, these redistribution terms are responsible for the so-called 'return-to-isotropy', which is a 'macro-energetic' state defined by the equi-partition of kinetic energy among turbulent velocity components trending towards $\sigma_u^2 = \sigma_v^2 = \sigma_w^2 = (2/3)\overline{e}$. When $Ri_f < 0$, energy is injected in the σ_u^2 and σ_w^2 budgets, and at $Ri_f << 0$, the redistribution terms act in a manner to extract energy from the σ_w^2 budget and provide energy to both the σ_u^2 and σ_v^2 budgets.

2.3.2 Scalewise momentum and heat transport

The prior budgets, while foundational, do not explicitly reveal the role of turbulent eddies in the transport of momentum, heat, and TKE. To bring eddy sizes explicitly into the idealized ASL formulation, co-spectral and spectral definitions are introduced and given by

$$\overline{u'w'} = \int_0^\infty F_{uw}(k)\,dk\,, \quad \overline{w'T'} = \int_0^\infty F_{wT}(k)\,dk\,, \quad \overline{u'T'} = \int_0^\infty F_{uT}(k)\,dk, \quad (2.13)$$

and

$$\overline{w'w'} = \int_0^\infty F_{ww}(k)\,dk\,, \quad \overline{u'u'} = \int_0^\infty F_{uu}(k)\,dk\,, \quad \overline{T'T'} = \int_0^\infty F_{TT}(k)\,dk, \quad (2.14)$$

where $F_{uw}(k)$ and $F_{wT}(k)$ are the co-spectra of w' and u', and w' and T', respectively; $F_{ww}(k)$, $F_{uu}(k)$, and $F_{TT}(k)$ are the spectra of vertical- and longitudinal-velocities and temperature, respectively; and k is a one-dimensional wavenumber in the streamwise direction. This definition of k is selected here to be consistent with the myriad of ASL experiments (Kaimal et al., 1972; Kaimal, 1973; Kaimal and Finnigan, 1994; Wyngaard and Coté, 1972) that report spectra and co-spectra from single point time series measurements converted to one-dimensional streamwise wavenumber using Taylor's frozen turbulence hypothesis (Taylor, 1938).

The associated co-spectral budgets in the idealized ASL can be simplified to (Panchev, 1971; Bos et al., 2004; Bos and Bertoglio, 2007; Canuto et al., 2008; Katul et al., 2013b):

$$\frac{\partial F_{uw}(k)}{\partial t} = P_{uw}(k) + T_{uw}(k) + \pi_u(k) + \beta F_{uT}(k) - 2\nu k^2 F_{uw}(k) \tag{2.15}$$

$$\frac{\partial F_{wT}(k)}{\partial t} = P_{wT}(k) + T_{wT}(k) + \pi_T(k) + \beta F_{TT}(k) - (\nu + D_m)k^2 F_{wT}(k) \tag{2.16}$$

The terms on the left-hand side of the equations represent temporal changes in the co-spectra at each wavenumber k. The terms on the right-hand side of the equations represent (in order): production (P), flux-transfer across scales (T_{ab}), pressure-decorrelation (π), buoyancy (βF_{uT} and βF_{TT}), and molecular destruction. The advantage of establishing links between spectral and co-spectral terms is that Kolmogorov's theory for locally homogeneous and isotropic turbulence within the inertial subrange, which describes a broad range of eddy sizes in the unstable ASL, can then be employed to parts of the $F_{ww}(k)$ and $F_{TT}(k)$ spectra (see discussion on Kolmogorov's theory in Chapter 1).

2.4 Foundational theories

2.4.1 Dimensional and directional dimensional analysis: flux-profile relations

One of the earliest attempts to include the effects of thermal stratification on the mean velocity gradient S was conducted by Laikhtman and later by Budyko (Monin and Obukhov, 1954). This effort laid the foundations for what is now labeled as Monin–Obukhov similarity theory (MOST) for the surface layer. It is instructive to ask why earlier theories sought to describe dU/dz instead of U directly (including the law of the wall in Chapter 1). The answer to this question is not straightforward or obvious from standard textbooks. A plausibility argument is that dU/dz is Galilean invariant, whereas $U(z)$ is not. Galilean invariance here occurs when the laws describing S are identical in all inertial frames of reference. Thus the laws describing dU/dz should not depend on whether an observer is moving at a constant velocity and observing dU/dz, or is doing so while stationary. Clearly, these two observers will disagree about the laws describing U, but not dU/dz. Another plausible reason for starting from dU/dz

is that the asymptotic approaches that result in the law of the wall, in fact, must match not only the velocity, but also the gradients in the overlap region. Returning to theories for dU/dz, the starting point is to revise the dimensional analysis leading to the law of the wall covered in Chapter 1 by including some measure of thermal stratification (e.g., Ri_f). This revision may be achieved by introducing another dimensionless group formed by z and a new length scale reflecting the significance of thermal stratification on turbulence generation. These revisions, which are the backbone of MOST, are presented as

$$\frac{dU}{dz}\frac{\kappa z}{u_*} = \phi_m(\zeta) \quad , \quad \frac{d\overline{T}}{dz}\frac{\kappa z}{T_*} = \phi_T(\zeta) \tag{2.17}$$

where $\phi_m(\zeta)$ and $\phi_T(\zeta)$ are referred to as the momentum and heat stability correction functions, respectively; $\zeta = z/L$ is the stability parameter; and L is the Obukhov length scale.

The L directly emerges from considerations of the TKE budget when expressed as

$$\left(\frac{\kappa z}{u_*^3}\right)\epsilon = [\phi_m(\zeta) - \zeta], \quad L = -\frac{u_*^3}{\kappa\beta\overline{w'T'}} \tag{2.18}$$

The relation between the earlier derived Ri_f and ζ is nonlinear and given by $Ri_f = \zeta/\phi_m(\zeta)$. Physically, ζ is in fact an estimate of Ri_f where the shear production is assumed to be the same as under neutral conditions and approximated using the law of the wall. This removes the need for any information about S when computing L; therefore ζ (instead of Ri_f) is more useful in theories and models of mean velocity and mean air temperature profiles. Eq. (2.18) ignores the transport of TKE by turbulence and pressure and assumes a local balance between ϵ and the two TKE production (i.e., mechanical and buoyancy) terms at a given z in the ASL. Experimental support for this approximation for the unstable ASL has been provided by a number of field campaigns (Charuchittipan and Wilson, 2009; Salesky et al., 2013), where ϵ was independently estimated from $F_{uu}(k)$ after identifying the $-5/3$ scaling (see Chapter 1, Kolmogorov's theory) and the production terms were directly measured. More recent work directly linked the failure of MOST during some periods to the departure from that equilibrium and the presence of significant transport or unsteadiness (Zahn et al., 2023). Because mechanical production is reduced with increasing z but buoyant production is insensitive to z variations in the idealized ASL, L can be interpreted as roughly the height at which buoyant production first becomes commensurate with mechanical production.

In the limit of free convection, it is often assumed that $u_* \to 0$, $|L| \to 0$, and $|\zeta| \to \infty$ (Kaimal and Finnigan, 1994). Conversely, in the limit of near-neutral conditions when $\overline{w'T'} \to 0$, $|L| \to \infty$, and $|\zeta| \to 0$. To recover the law-of-the wall for mean velocity and temperature, $\phi_m(0) = 1$ and $\phi_T(0) = 1$ (when assuming a unity turbulent Prandtl number). Dimensional analysis alone cannot predict the entire shape of $\phi_m(\zeta)$ and $\phi_T(\zeta)$ but may still offer constraints. Results from the Kansas experiment, now labeled as the Businger–Dyer (BD) stability correction functions

(Businger et al., 1971; Businger, 1988), suggest that

$$\phi_m(\zeta) = (1-16\zeta)^{-1/4} \ , \ \phi_T(\zeta) = [\phi_m(\zeta)]^2 \ , \ Pr_t = \frac{\phi_T(\zeta)}{\phi_m(\zeta)}. \tag{2.19}$$

The Pr_t is the turbulent Prandtl number and also varies with $-\zeta$. According to these BD functions, as $-\zeta \to \infty$, $Pr_t = \phi_m(\zeta) \to 0$ (since $\phi_T \to 0$ faster than ϕ_m). The BD functions are routinely featured in textbooks and used in operational models. Some Large Eddy Simulations (LES) support their shape and universal character as discussed elsewhere (Khanna and Brasseur, 1997; Maronga and Reuder, 2017). When the Kansas $\phi_m(\zeta)$ and $\phi_T(\zeta)$ estimates were confirmed in separate experiments (e.g., in Minnesota), it led some to optimistically state that *with proper non-dimensionalization, all flow statistics in the idealized surface layer can be reduced to a set of universal curves*. However, turbulence almost never turns out to be such an open book.

One challenge to these BD functions is in the limit $|\zeta| \to \infty$ where it is routinely argued that $\phi_m(\zeta) \sim (-\zeta)^{-1/3}$ to eliminate dependence on u_* in Eq. (2.17) for dU/dz (even when large thermal cells create some surface friction and a non-zero u_*, that u_* is no longer an input parameter that should appear in the dimensionless laws). That $\phi_m \sim (-\zeta)^{-1/4}$ instead of $\phi_m \sim (-\zeta)^{-1/3}$ for large $-\zeta$ in these BD functions was explained by the limited $\zeta \in [-2, 0]$ range interrogated within the Kansas experiment (i.e. no convective limit was actually reached where $|\zeta| > 5$). Several attempts have since been proposed to reconcile the $\phi_m(\zeta)$ scaling with $(-\zeta)^{-1/3}$ for large $|\zeta|$, but two stand out. They are elaborated upon here because they reflect theoretical tactics not conventionally covered in standard textbooks (Brutsaert, 1982; Stull, 1988; Sorbjan, 1989; Garratt, 1992; Kaimal and Finnigan, 1994; Wyngaard, 2010). The first is the OKEYPS equation (Foken, 2006; Li, 2021) and the second is directional dimensional analysis (Betchov and Yaglom, 1971; Zilitinkevich, 1973; Kader and Yaglom, 1990). Other attempts remain empirical and include (Wilson, 2001)

$$\phi_m(\zeta) = \left[1 + 3.59(-\zeta)^{2/3}\right]^{-1/2} \tag{2.20}$$

Eq. (2.20) does recover $\phi_m(\zeta) \sim (-\zeta)^{-1/3}$ as $-\zeta \to \infty$ by construction. For small $-\zeta$, $\phi_m(\zeta) = 1 - 1.8(-\zeta)^{2/3}$ is nonlinear in ζ whereas the BD formulation is linear and given by $\phi_m(\zeta) = 1 - 4(-\zeta)$.

• *The OKEYPS equation* (after Obukhov, Kazansky, Ellison, Yamamoto, Panofsky, and Sellers): This equation is given by (Plate, 1971; Kaimal, 1973; Businger, 1988; Foken, 2006; Li, 2021)

$$[\phi_m(\zeta)]^4 - \gamma_o\zeta\,[\phi_m(\zeta)]^3 = 1 \tag{2.21}$$

where γ_o is an empirical constant that must be inferred from experiments, first proposed by Obukhov (Obukhov, 1946; Businger and Yaglom, 1971). The OKEYPS equation was intended as an interpolation function that recovers the BD scaling for mildly unstable conditions (for small $-\zeta$) while ensuring that $\phi_m(\zeta) \sim (-\zeta)^{-1/3}$

for very large $-\zeta$. The theoretical justification for the OKEYPS equation remained heuristic, and early attempts assumed a constant heat-to-momentum eddy diffusivity, an assumption negated by the Kansas experiment, thereby preventing its widespread acceptance (Katul et al., 2011; Li, 2021). A framework that predicts the shape of $\phi_m(\zeta)$ without invoking equality between momentum and heat eddy diffusivities was later proposed (Katul et al., 2011).

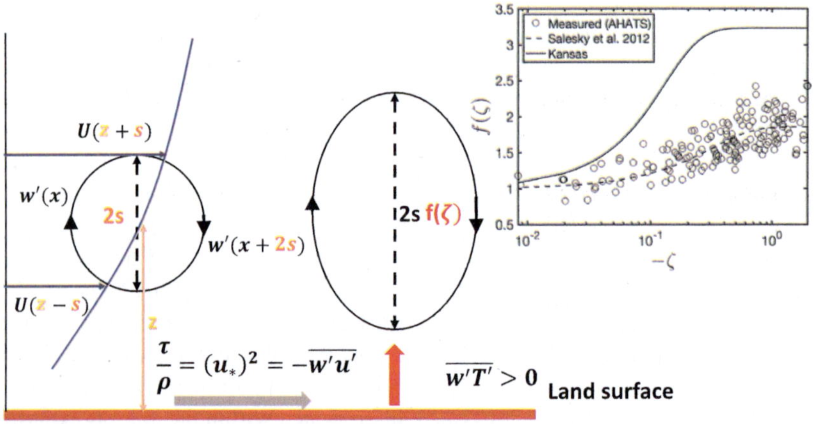

Figure 2.3 Momentum exchange $\overline{w'u'}$ for a sheared and heated ASL. The momentum exchange at any level is set by u_*^2 assuming a constant flux layer. The most efficient eddy is the one where $s = z$ (eddy touching the ground). The effect of heating is to 'elongate' eddies with respect to their near-neutral state by a stretching function $f(\zeta)$. The estimates of $f(\zeta)$ from recent experiments using vertical velocity spectra are shown here (Salesky et al., 2013) along with the estimate from the Kansas experiments in unstable conditions as in Eq. (2.27). The limit of $f(\zeta) = 3.2$ for large $-\zeta$ from the Kansas experiment agrees with other measurements (Li et al., 2012).

In this framework, the momentum flux exchanged by the most effective momentum transporting eddy at a distance from the ground z as shown in Fig. 2.3 is given by (Gioia et al., 2010)

$$u_*^2 = \kappa_\tau \upsilon(s)\left[U(s+z) - U(s-z)\right] \approx \kappa_\tau \upsilon(s)\left[\frac{dU}{dz}2s\right] \tag{2.22}$$

where $U(s+z) - U(s-z)$ is the net momentum per unit mass exchanged at height z due to eddies of size $2s$, $\upsilon(s) = |w(x+2s) - w(x)|$ is the turnover velocity characterized by the magnitude of the vertical velocity difference, and κ_τ is a proportionality constant. In near-neutral conditions, the eddy size that contributes most efficiently to u_*^2 is an eddy 'touching' the ground surface, resulting in $s = z$ (Townsend, 1976), thereby simplifying Eq. (2.22) to

$$2\frac{\kappa_\tau}{\kappa}\frac{\upsilon(z)}{u_*}\left[\frac{dU(z)}{dz}\frac{\kappa z}{u_*}\right] = 2\frac{\kappa_\tau}{\kappa}\frac{\upsilon(z)}{u_*}\left[\phi_m\right] = 1 \tag{2.23}$$

An estimate of $v(z)$, the turnover velocity, is now needed to describe $\phi_m(\zeta)$. This estimate may be provided from Kolmogorov's 4/5th law for the third-order velocity structure function (Frisch, 1995; Monin and Yaglom, 1975), according to which $v(z) = [\kappa_\epsilon \epsilon z]^{1/3}$, where $\kappa_\epsilon = 4/5$, as already discussed in Chapter 1. This estimate of $\epsilon(z)$ is only exact for locally homogeneous and isotropic turbulence (Monin and Yaglom, 1975) away from the dissipation range or macro-scales. It is used here for a first-order analysis.

Using Eq. (2.6) for ϵ and ignoring transport terms, after some algebra, it can be shown that

$$\beta_1 \, [\phi_m(\zeta)]^4 \left[1 - \frac{\zeta}{\phi_m(\zeta)}\right] = 1 \quad , \quad \beta_1 = \frac{2^3 \kappa_\tau^3 \kappa_\epsilon}{\kappa^4} \tag{2.24}$$

Imposing $\phi_m(0) = 1$ results in $\beta_1 = 1$ and $\kappa_\tau = (1/2)(\kappa^4/\kappa_\epsilon)^{1/3} \approx 0.16$. This linkage between the constant κ_τ, κ, and the Kolmogorov's 4/5th law (through κ_ϵ) is an outcome of attached eddies of size z still following inertial subrange scaling reasonably (i.e., where $F_{ww}(k) \sim k^{-5/3}$), an assumption that requires revision discussed later on. When $\overline{w'T'} > 0$, this equation recovers all the scaling exponents observed in $\phi_m(\zeta)$ for small and large $|\zeta|$. For small values of ζ, the $-1/4$ power-law for unstable conditions is recovered by noting that the term $[\zeta/\phi_m(\zeta)] \approx \zeta$ so that

$$[\phi_m(\zeta)]^4 \, [1 - \zeta] \approx 1; \Rightarrow \phi_m(\zeta) \approx (1 - \zeta)^{-1/4} \tag{2.25}$$

For large $-\zeta$, the $-1/3$ power-law for convective scaling (Priestley and Swinbank, 1947; Kaimal et al., 1976) is recovered when noting that the term $[-\zeta/\phi_m(\zeta)] \gg 1$ so that

$$[\phi_m(\zeta)]^4 \left[\frac{-\zeta}{\phi_m(\zeta)}\right] \approx 1; \Rightarrow \phi_m(\zeta) \approx (-\zeta)^{-1/3} \tag{2.26}$$

This model representation recovers the OKEYPS equation (Eq. (2.21) when $\gamma_o = 1$) without assuming equality of heat and momentum eddy diffusivity and without employing heuristic gradient-diffusion closure arguments to the vertical velocity variance (Businger, 1988; de Bruin, 1999; Fleagle and Businger, 1963).

This analysis also shows that $\gamma_o = 1$ is linked to the assumption that eddy sizes are dictated by $s = z$ without being amplified in the vertical by thermal stratification. Reported γ_o values range from 5 to 18, with $\gamma_o = 9$ employed by some authors (Plate, 1971; Businger, 1988). The BD form $\phi_m(\zeta) = (1 - 16\zeta)^{-1/4}$ also suggests that $\gamma_o = 16$. To explain why γ_o exceeds unity, it is constructive to recall that in thermally stratified flow, the characteristic eddy size is no longer z and must be stretched in the vertical to include the effects of thermal stratification (Salesky et al., 2013). To account for this effect in $v(s')$, it may be convenient to assume that $s' = zf(\zeta)$, where $f(\zeta)$ is a dimensionless anisotropy function such that $f(0) = 1$ to recover the log-law. Thus, with $v(s') = [\kappa_\epsilon \epsilon z f(\zeta)]^{1/3}$, and $f(\zeta)$ determined from measurements of the pre-multiplied vertical velocity spectral peaks (Kaimal et al., 1972), the OKEYPS equation can be recovered in unstable stratification (Katul et al., 2011; Salesky et al.,

2013). An approximate expression for $f(\zeta)$ in unstable conditions derived from the Kansas experiment is given by (Katul et al., 2011)

$$f(\zeta) = \frac{1}{1 - a_f\left[1 - \exp(b_f \zeta)\right]}, \tag{2.27}$$

where the choices $a_f = 0.69$ and $b_f = 15$ recover the Kansas experiments (Kaimal and Finnigan, 1994). The solution to Eq. (2.22) along with $v(s') = [\kappa_\epsilon \epsilon z f(\zeta)]^{1/3}$, and $f(\zeta)$ given by Eq. (2.27) is labeled as the spectral link (Gioia et al., 2010; Katul et al., 2011). The form of Eq. (2.27) agrees with independent measurements of the peaks of the vertical velocity spectra collected at two heights across a wide range of ζ discussed elsewhere (Salesky et al., 2013). However, these measurements suggest that $a_f = 0.48$ (instead of 0.69) and $b_f = 4.82$ (instead of 15) and thus fall below the Kansas experiment estimate of $f(\zeta)$, as shown in Fig. 2.3. Other experiments across multiple land cover report $f(\zeta) = 3.2$ for large $-\zeta$ in agreement with the Kansas experiments, and $a_f = 0.69$ in Eq. (2.27). While the precise values of a_f and b_f remain to be uncovered, this phenomenological model is suggestive that the empirical parameter γ_o in the OKEYPS equation or the factor 16 in $\phi_m(\zeta)$ of BD encodes information about how changes in the peaks of $kF_{ww}(k)$ with ζ occur as ζ is altered. A similar analysis, not repeated here, was conducted for $\phi_T(\zeta)$ and is reviewed elsewhere along with some discussions about stable conditions and approach towards near-neutral conditions from the stable stability conditions (Li et al., 2012; Li, 2019).

• *Directional Dimensional Analysis (DDA)*: This approach was pioneered for the ASL in the early 1970s (Betchov and Yaglom, 1971; Zilitinkevich, 1973) and formalized in 1990 after the collection of an extensive record spanning a broad range of atmospheric stability conditions (Kader and Yaglom, 1990). The DDA extends MOST by assuming that turbulence generation in the horizontal (P_m) and the vertical (B_p) can be decoupled. DDA proceeds to associate separate horizontal and vertical length scales (L_x and L_z) with each generation mechanism. That is, length scales associated with generating u' are characterized by L_x, whereas processes associated with generating w' are associated with L_z. A drawback of this approach is that it ignores interactions between these two components that may occur due to pressure redistribution and return to isotropy (Bou-Zeid et al., 2018). Nonetheless, when the generation mechanisms of u' and w' occur at scales much larger than the scales at which return to isotropy becomes effective, the assumptions behind DDA may still hold. Some support for this conjecture was recently reported using ASL experiments in the context of scale-wise return to isotropy of the stress tensor from production to inertial (Brugger et al., 2018; Stiperski et al., 2021b).

Defining a single characteristic time t_c, and accepting momentarily a 'decoupling' between L_x and L_z, implies that $u_*^2 = -\overline{u'w'}$ is associated with $L_x L_z t_c^{-2}$. Thus u_* does not 'qualify' as being strictly a horizontal velocity scale. DDA argues that a characteristic vertical velocity having dimensions L_z/t_c, but encoding buoyancy sources, can be defined by $w_* = \left(\beta \overline{w'T'} z\right)^{1/3}$ for the ASL. With these arguments, DDA proposes a horizontal velocity that must be formed from L_x/t_c, and a logical choice would be a $u_{**} = u_*^2/w_*$. Likewise, the ASL temperature scale T_* must be replaced by $T_{**} = \overline{w'T'}/w_*$.

The ensuing analysis (Kader and Yaglom, 1990) proposes that the ASL be decomposed into 3 sub-layers: dynamic (or near-neutral), dynamic-convective, and free convective. For the dynamic sublayer, the standard form of the law-of-the-wall for near-neutral conditions is assumed to hold and is given by Eqs. (2.17) with $\phi_m(0) = 1$ and $\phi_T(0) = Pr_t(0)$, where $Pr_t(0) \in [0.75 - 1]$ is the turbulent Prandtl number (Kays, 1994; Li, 2019).

For the dynamic-convective layer,

$$\frac{dU}{dz} = B_U \frac{u_{**}}{z} \quad , \quad \frac{dT}{dz} = -B_T \frac{T_{**}}{z} \tag{2.28}$$

where B_U and B_T are dimensionless constants. This analysis leads to $\phi_m(\zeta) = B_U(-\zeta)^{-1/3}$ and $\phi_T(\zeta) = B_T(-\zeta)^{-1/3}$. For the free convective layer,

$$\frac{dU}{dz} = C_U \frac{w_*}{z} \quad , \quad \frac{dT}{dz} = -C_T \frac{T_{**}}{z} \tag{2.29}$$

where C_U and C_T are dimensionless constants that can differ from B_U and B_T. Focusing momentarily on $\phi_m(\zeta)$, Eq. (2.29) suggests that

$$\frac{dU}{dz} = \frac{u_*}{\kappa z}\phi_m(\zeta) = C_U\left(\beta \overline{w'T'}\right)^{1/3} z^{-2/3} \tag{2.30}$$

That is, DDA predicts $\phi_m(\zeta) = C_U(-\zeta)^{+1/3}$ (instead of $-1/3$ as exponent). Partial LES support for the $+1/3$ scaling is presented elsewhere (Maronga and Reuder, 2017), but other LES runs did not agree well with this prediction (Li et al., 2018b). Likewise, DDA predicts a $\phi_T(\zeta) = C_T(-\zeta)^{-1/3}$ for this limit.

An approximate form that recovers all scaling laws across the three sublayers and agrees with experiments was proposed (Kader and Yaglom, 1990). After correcting for κ in the definition of L in the original source, they are given by (Hsieh and Katul, 1997)

$$\phi_m(\zeta) = \left[\frac{1 + 0.625(-\zeta)^2}{1 + 7.5(-\zeta)}\right]^{1/3} \quad , \quad \phi_T(\zeta) = 0.64\left[\frac{3 + 2.5(-\zeta)}{1 + 10(-\zeta) + 50(-\zeta)^2}\right]^{1/3} \tag{2.31}$$

This expression, labeled KY90, is verified by long-term experiments conducted at the Tsimlyansk field station operated by the Moscow Institute of Atmospheric Physics. A Comparison between $\phi_m(\zeta)$ predicted by all these methods is featured in Fig. 2.4. This figure shows that the expressions of $\phi_m(\zeta)$ all agree up to $-\zeta = 2$, but appreciably diverge for $-\zeta > 2$. Likewise, the KY90 data diverges from the weighty Kansas experiments when $-\zeta > 2$. These departures have not been discussed in the literature or textbooks. The textbook argument is that in the limit of free convection, $C_U = 0$ and $dU/dz = 0$. However, KY90's analysis considered this possibility but precluded its occurrence for a finite u_*, no matter how small.

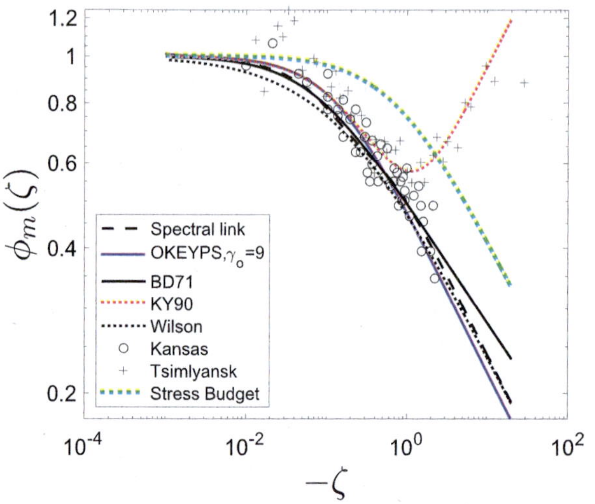

Figure 2.4 Comparison of the stability correction functions for momentum $\phi_m(\zeta)$ as a function of the stability parameter ζ for unstable stratification in the idealized ASL. The models include the spectral link with $f(\zeta)$ specified by Eq. (2.27), the OKEYPS equation with $\gamma_o = 9$, the BD $\phi_m(\zeta) = (1 - 16\zeta)^{-1/4}$, the approximation derived by KY90 from DDA in Eq. (2.31), and Eq. (2.20). Two main data sets—the Kansas experiment (determining BD) and the Tsimlyansk long-term record near Moscow (determining KY90)– are also featured. The $\phi_m(\zeta) \sim (-\zeta)^{+1/3}$ is only predicted by DDA in the limit of free convection. Partial LES support for this $+1/3$ scaling is presented elsewhere (Maronga and Reuder, 2017). The estimation of $\phi_m(\zeta)$ using a flux budget where stress production is balanced by the Rotta-modeled decorrelation term is presented, reflecting the contribution to ϕ_m from local gradient-diffusion only processes.

Another possibility is that w_* scales with the ABL height h as $w_* = \left(\beta\overline{w'T'}h\right)^{1/3}$ and the z dependency in w_* is lost as the free convection limit is approached. Thus

$$\phi_m = \kappa^{2/3} C_U \left(\kappa\beta\frac{\overline{w'T'}}{u_*^3}h\right)^{1/3} \tag{2.32}$$

resulting in $\phi_m = \kappa^{2/3} C_U (h/L)^{1/3}$ instead of $(z/L)^{1/3}$. The measurements from the Tsimlyansk field station (reproduced in Fig. 2.4) are ensemble-averaged in ζ bins and hint that $\phi_m \to 0.9$ when $-\zeta > 5$ (i.e., a constant). Yet another plausible argument is that, in the limit of free convection, the constant flux assumption no longer holds and is replaced by the constant mean potential temperature even in the lowest layers of the ASL. Thus a z dependency may emerge in the analysis of $\phi_m(\zeta)$ in the limit of free convection, but this dependency may be due to $\overline{w'T'}(z)=\overline{w'T'}_s(1 - z/h)$, where $\overline{w'T'}_s$ is the surface heat flux.

The Pr_t also offers another dynamically interesting perspective on this argument, especially in light of recent high-resolution LES experiments targeting the transition from near-neutral to near-convective limits (Maronga and Reuder, 2017). Fig. 2.5

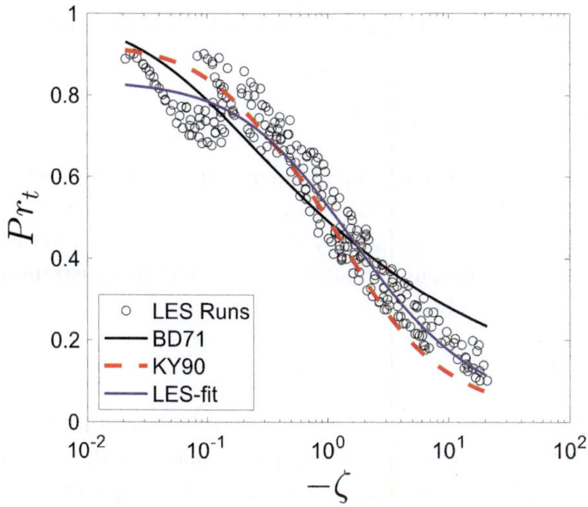

Figure 2.5 Comparison between LES reported Pr_t (Maronga and Reuder, 2017), BD, and KY90, as a function of the stability parameter ζ for unstable stratification in the idealized ASL. The best-fit $Pr_t(\zeta)$ function to all the LES runs is given by Eq. (2.33) and shown here for reference.

compares simulated Pr_t for all the LES runs against those computed from BD and KY90. It is clear that Pr_t predicted from KY90 agrees better with these recent LES runs when compared to the conventional BD formulations. The best-fit function to the LES runs is given by

$$Pr_t = 0.53\,[0.49 - 0.51(\zeta)]^{-0.64} \tag{2.33}$$

and further suggest that Pr_t decays as $(-\zeta)^{-0.64}$ with increasing $-\zeta$ instead of $(-\zeta)^{-1/4}$ as in BD. In the near-neutral limit, a $Pr_t = 0.84$ emerges from the LES runs and appears to approach values reported from a number of laboratory and field experiments (Li, 2019).

2.4.2 Dimensional and directional dimensional analysis: flux-variance relations

While the focus up to this point has been on flux-profile relations for dU/dz, extensions of MOST to second-order statistics has been met with some successes in the unstable ASL for the temperature variance σ_T^2. This success has led to a new method to operationally infer $\overline{w'T'}$ from high-frequency air temperature measurements only. Within the confines of MOST, the σ_T^2 can be related to T_* and ζ using

$$\phi_{TT}(\zeta) = \frac{\sigma_T}{|T_*|}, \tag{2.34}$$

and is now formulated as a 'flux variance' (instead of flux-profile) method given by

$$\overline{w'T'} = \left(\frac{\kappa \beta z}{|\zeta|}\right)^{1/2} \left(\frac{\sigma_T}{\phi_{TT}(\zeta)}\right)^{3/2} \tag{2.35}$$

There are several formulations for ϕ_{TT} (Tillman, 1972; Sorbjan, 1989), and as before, this function is presumed to be constrained as follows: (i) when $|\zeta| \to \infty$, $\phi_{TT} \to (-\zeta)^{-1/3}$ to eliminate the u_* dependency in the limit of free convection, and (ii) when $|\zeta| \to 0$, $\phi_{TT} \to C_3$, a constant. An interpolation function satisfying these limits has been proposed and is given by

$$\phi_{TT} = C_1 (C_2 - \zeta)^{-1/3} \tag{2.36}$$

where C_1 and C_2 are similarity constants. A $C_1 = 0.99$ (after adjustment of the originally used $\kappa = 0.35$ to the widely accepted value of 0.4), $C_3 = 1.77 - 2.5$, and thus a $C_2 = (C_1/C_3)^3$ have been reported by the pioneering work of Tillman (1972). For the limiting case of the free convection $-\zeta \gg C_2$ and thus $\phi_{TT} \approx C_1(-\zeta)^{-1/3}$ (Tillman, 1972; Albertson et al., 1995). Hence, for free convective conditions, the relation between a turbulent sensible heat flux and temperature variance can be expressed as

$$\overline{w'T'} = \left(\frac{\sigma_T}{C_1}\right)^{3/2} (\kappa \beta z)^{1/2} \tag{2.37}$$

independent of u_* with $C_1 = 0.99$. A number of studies have shown that the free convection scaling yields surprisingly accurate predictions of $\overline{w'T'}$ from measured σ_T even when the atmospheric stability conditions are mildly unstable. This result popularized the use of the flux-variance method in operational approaches (Tillman, 1972; Weaver, 1990; de Bruin et al., 1993; Padro, 1993; Albertson et al., 1995; Hsieh et al., 1996; Katul and Hsieh, 1999; Asanuma and Brutsaert, 1999; Wesson et al., 2001; Zahn et al., 2016) and in parametrization of Earth System models (Waterman et al., 2022). In fact, DDA also predicted a $\phi_{TT} \approx (-\zeta)^{-1/3}$ for $-\zeta > 0.1$, which is confirmed by experiments.

While the link between $\overline{w'T'}$ and σ_T^2 marked a success in extending MOST to higher-order statistics, the scaling of σ_u^2/u_*^2 with ζ remains a 'thorny issue'. MOST and DDA predict that for near-neutral conditions, σ_u/u_* is a constant determined from experiments to be about 2.4. This assertion has been contested. It has been known for some time now that Townsend's attached eddy hypothesis (Townsend, 1976; Marusic et al., 2010b; Smits and Marusic, 2013; Marusic and Monty, 2019) and more recent spectral theories (Banerjee and Katul, 2013; Huang and Katul, 2022) show that

$$\left(\frac{\sigma_u}{u_*}\right)^2 = B_{To} - A_{To} \log\left(\frac{z}{h}\right) \tag{2.38}$$

where $B_{To} = 1.56 - 2.3$ and $A_{To} = 1.21 - 1.3$ are the Perry–Townsend constants derived from laboratory experiments. That σ_u/u_* depends on z/h for near-neutral conditions could not have been foreshadowed from MOST or DDA. For the dynamic-convective sublayer, DDA predicts that σ_u/u_{**} should be a constant and this outcome inevitably leads to $\sigma_u \propto u_*^2(\beta\overline{w'T'}z)^{-1/3}$, whereas MOST predicts $\sigma_u/u_* = 2.7(1-3\zeta)^{+1/3}$. Again, recent spectral theories and measurements suggest a much more nuanced view (Banerjee et al., 2015) except for moderately unstable conditions. Thus the σ_u/u_* analysis points out that the key length scales involved are, at minimum, z, L, and h. Thus far, neither DDA nor MOST accommodates all three length scales.

While it may be tempting to extend Eq. (2.38) to dynamic-convective and free-convective conditions by assuming B_{To} and A_{To} vary with ζ, a number of cautionary notes are in order. To begin with, there is an association between the term $\log(z/h)$ in σ_u/u_* and the occurrence of an $F_{uu}(k) \sim k^{-1}$ regime for $k \in [a_u' 1/h, 1/z]$ in the spectra, where a_u' is a proportionality constant of order unity. The integration of the portion of $F_{uu}(k) \sim k^{-1}$ is one reason why a log-scaling emerges in $\sigma_u^2 = \int_0^\infty F_{uu}(k)dk$ (i.e., the attached eddies contribute to σ_u^2 via $\int_{1/h}^{1/z} k^{-1}dk$ to yield a $\log(z/h)$ dependency). The $F_{uu}(k) \sim k^{-1}$ only holds for near-neutral conditions and mildly unstable conditions and has been linked to Townsend's attached eddy hypothesis (Townsend, 1976). Deviations from $F_{uu}(k) \sim k^{-1}$ scaling begin to occur in the dynamic-convective with $F_{uu}(k) \sim k^{-5/3}$ in the free-convective limit. This shift in scaling laws from a k^{-1} to a $k^{-5/3}$ with increased instability precludes the $\log(z/h)$ scaling in σ_u/u_* as discussed elsewhere (Banerjee et al., 2015). A second hurdle is that large and very large-scale motion, with scales much larger than h, begins to dominate $F_{uu}(k)$. Thus neither z nor h may be appropriate descriptors of these types of eddy motion. A fresh insight into this issue was recently proposed, where σ_u/u_* can be reconciled with MOST by the addition of another local scale related to anisotropy in the stress tensor (Stiperski et al., 2021a). This finding is among the first to suggest that anisotropy in the stress tensor, as determined by its invariants, may offer corrections to MOST for σ_u/u_* (even for nonideal situations).

2.4.3 Eddy viscosity theories

Eqs. (2.17) can be expressed as formulations for the eddy viscosity and eddy diffusivity, often labeled as gradient-diffusion theory, and are given as

$$\overline{u'w'} = -\nu_t \frac{dU}{dz}, \overline{w'T'} = -\frac{\nu_t}{Pr_t}\frac{d\overline{T}}{dz}, \nu_t = u_* \left[\frac{\kappa z}{\phi_m(\zeta)}\right], Pr_t = \frac{\phi_h(\zeta)}{\phi_m(\zeta)} \quad (2.39)$$

These formulations are also the essence of the aerodynamic method routinely employed to infer turbulent fluxes of scalars in the ASL. They show that turbulent diffusion coefficients are based on a characteristic velocity u_* and a characteristic mixing length $\kappa z/\phi_m(\zeta)$. Because $\phi_m(\zeta) < 1$, they illustrate why the effective eddy sizes increase above and beyond κz in the unstable ASL. A logical question to ask then is when a turbulent flux can be represented by a gradient-diffusion formulation similar to Eq. (2.39).

To address this question, models for the pressure-velocity and pressure-temperature interactions are needed. A model of maximum simplicity is the Rotta closure scheme, featured here for the turbulent stress as illustration, and is given by (Launder et al., 1975; Pope, 2000)

$$\frac{-1}{\rho}\overline{u'\frac{\partial p'}{\partial z}} = -C_R\frac{\overline{w'u'}}{\tau_R} + C_I\overline{w'^2}S \tag{2.40}$$

where $C_R = 1.8$ is the Rotta constant; $C_I = 3/5$ is a constant related to the 'isotropization' of the stress production term (and can be predicted from rapid distortion theory); and τ_R is a decorrelation time between u' and w'. That is, τ_R is related to the decay time for $\overline{u'w'} \to 0$ when the production terms in the $\partial(\overline{u'w'})/\partial t$ budget are suppressed. The first term on the right-hand side of Eq. (2.40) is the 'slow' component, whereas the second term on the right-hand side of Eq. (2.40) is the 'rapid' component and acts rapidly on the stress production term. That is, the fast part of the Rotta scheme appears to reduce the mechanical production term by a factor C_I at time scales much faster than τ_R. In effect, when mechanical production generates $\overline{u'w'}$, the pressure decorrelation acts rapidly (relative to τ_R) and causes a portion of the mechanically produced $\overline{u'w'}$ to be decorrelated. Eq. (2.40) represents a general tendency of all turbulent flows– not only boundary layers—in which a return-to-isotropy at finer scales occurs where w' becomes decorrelated from u' because of pressure perturbations at those fine scales. Inserting Eq. (2.40) into Eq. (2.4), assuming the molecular stress destruction term is small in the stress budget compared to the pressure-decorrelation term (Katul et al., 2013a) at high Reynolds number, and rearranging, yields

$$\overline{w'u'} = \frac{\tau_R}{C_R}\left[-(1-C_I)\overline{w'w'}S - \frac{\partial\overline{w'w'u'}}{\partial z} + \beta\overline{u'T'}\right] \tag{2.41}$$

Therefore when the stress transport by turbulence (i.e., the term involving $\overline{w'w'u'}$) and the buoyancy $\beta\overline{u'T'}$ effects are ignored, a 'local' eddy viscosity can be derived and given by

$$\overline{w'u'} = -\nu_t\frac{dU}{dz}, \quad \nu_t = \left(\frac{1-C_I}{C_R}\right)\tau_R\,\overline{w'w'} \tag{2.42}$$

Failure of gradient-diffusion theory may be attributed to the significance of the sum of the transport or buoyancy processes relative to the turbulent stress production term. Ignoring the sum of these two processes enables the determination of how much momentum flux is carried by the 'local mean gradient' alone, vis-a-vis nonlocal transport associated with the triple moment term or new buoyancy forces that vary with ζ.

A number of choices for specifying τ_R are available. A plausible choice elaborated upon here is $\tau_R = 2\sigma_w^2/\epsilon$. This choice originates from the fact that (i) the wavenumbers at which the peaks in pre-multiplied vertical velocity spectra (i.e., $kF_{ww}(k)$) and co-spectra ($kF_{uw}(k)$) occur are closer than the wavenumbers at which the peaks in premultiplied TKE and co-spectra occur (Kaimal and Finnigan, 1994), and (ii) what dissipates the energy in w' at the aforementioned peak wavenumber is the energy transfer across scales (assumed to be ϵ). Employing this formulation for τ_R yields an eddy-viscosity given by (Mortarini et al., 2023)

$$\nu_t = \left(\frac{1-C_I}{C_R}\right)\frac{2\sigma_w^4}{\epsilon} \tag{2.43}$$

Defining a macro-scale dissipation length $L_d = u_*^3/\epsilon$ (Davidson and Krogstad, 2014; Pan and Chamecki, 2016; Chamecki et al., 2017; Ghannam et al., 2018; Mortarini et al., 2023) provides a link between ν_t and L_d (instead of z) given by

$$\nu_t = 2\left(\frac{1-C_I}{C_R}\right)\left(\frac{\sigma_w}{u_*}\right)^4 (u_* L_d). \tag{2.44}$$

For near-neutral conditions, $\sigma_w/u_* \in [1.2 - 1.3]$ and the simplified TKE budget is approximated by $\epsilon = u_*^3/(\kappa z)$ and $L_d = \kappa z$. It is to be noted that for $C_I = 3/5$, $C_R = 1.8$, and $\sigma_w/u_* = 1.22$, the numerical value of the combination of constants is unity, resulting in $\nu_t = u_* L_d$ (i.e., the law of the wall). The justification for introducing L_d is that ϵ reflects a transfer of energy across scales (see Chapter 1) conserved in the energy cascade. While L_d is labeled as a dissipation length scale, it is emphasized that L_d must be viewed as an integral measure much larger than the Kolmogorov micro-scale $\eta = (\nu^3/\epsilon)^{1/4}$ (see Chapter 1). In fact, $L_d/\eta = (u_*/v_{ko})^3$ where $v_{ko} = (\nu\epsilon)^{1/4}$ is the Kolmogorov velocity satisfying $Re_k = v_{ko}\eta/\nu = 1$, Re_k is the micro-scale Reynolds number formed using v_{ko}, η and ν. A macro-scale Reynolds number can then be defined as $Re_d = u_* L_d/\nu = (u_*/v_{ko})^4$. Hence, $L_d/\eta = Re_d^{3/4}$, consistent with the Reynolds number dependent separation between macro- and micro-scales in many turbulent flows (Tennekes and Lumley, 1972; Mortarini et al., 2023). It is also independent of how ϵ is determined and emerges strictly from definitions.

As a bridge to prior dimensional analysis, one can define $\phi_w(\zeta) = \sigma_w/u_*$ so that

$$\frac{\nu_t}{\kappa z u_*} = \frac{1}{\phi_m} = 2\left(\frac{1-C_I}{C_R}\right)\frac{[\phi_w(\zeta)]^4}{[\phi_m(\zeta)-\zeta]}. \tag{2.45}$$

This expression can be rearranged to yield

$$\phi_m(\zeta) = \frac{\zeta}{1-2\left(\frac{1-C_I}{C_R}\right)[\phi_w(\zeta)]^4}. \tag{2.46}$$

Eq. (2.46) must be complemented by a separate condition that $\phi_m(0) = 1$ as both numerator and denominator approach zero when $\zeta \to 0$. For large $-\zeta$, it may be expected that $\phi_w(\zeta) \to (-\zeta)^{1/3}$ and $\phi_m(\zeta) \to (-\zeta)^{-1/3}$, consistent with DDA in the dynamic-convective sublayer but with three caveats. The first is that for moderate to large $-\zeta$, the flux transport term by turbulence is not necessarily small, and the local mean velocity gradient S only carries a modest portion of the momentum flux (Ghannam et al., 2018). The second is that the buoyancy term, formed by $\overline{u'T'}$ and its ζ dependence, is much less studied. DDA predicts $|\overline{u'T'}/\overline{w'T'}| = 4$ for near-neutral conditions. However, for $-\zeta > 0.1$, the ratio of the horizontal to vertical heat flux declines rather rapidly at a rate commensurate with $(-\zeta)^{-2/3}$. The third is the linearity of the Rotta scheme, and the role of the isotropization of the production term may be questioned. Even when adopting a linear Rotta scheme as done here, the isotropization of the stress production must include both mechanical production and $\beta\overline{u'T'}$ in ways yet to be fully uncovered. Nonetheless, the analysis here sheds light on what processes the 'empirically' determined $\phi_m(\zeta)$ encodes. It must include local and nonlocal contributions to momentum transport, with nonlocal contributions dominating this transport with increased $-\zeta$. A comparison between Eq. (2.46) (labeled as stress budget) and the prior methods is also featured in Fig. 2.4 when setting $\phi_w(\zeta) = a_1(1-3\zeta)^{1/3}$, where $a_1 = [(1/2)C_R/(1-C_I)]^{1/4} = 1.22$ when $C_R = 1.8$ and $C_I = 3/5$. This numerical value of a_1 is close to those reported in field experiments $a_1 \in [1.2 - 1.3]$ as discussed elsewhere (Stull, 1988; Kaimal and Finnigan, 1994; Kader and Yaglom, 1990). The difference between these predictions and empirically determined $\phi_m(\zeta)$ may be viewed as a coarse estimate of 'nonlocal' gradient transport, which is routinely observed in the mixed layer (Ghannam et al., 2018). It is to be noted that the choice of $\phi_w(\zeta) = a_1(1-3\zeta)^{1/3}$ here is supported by DDA and numerous field experiments reviewed elsewhere (Kader and Yaglom, 1990).

2.4.4 Eddy viscosity theories from co-spectral budgets

A link between the energetics of eddies (kinetic and potential) and scale-wise eddy viscosity is established. The molecular destruction terms in the turbulent stress and flux budgets are assumed to be small since they are only significant at sufficiently large wavenumbers. These large wavenumbers are in the range of $1/\eta$, where η is the Kolmogorov microscale scale (0.1–1.0 mm in the ASL). Hence, as a starting point, the co-spectral budgets of momentum and heat fluxes here involve only the most basic terms common to all theories and are given by (Li and Katul, 2017)

$$P_{uw}(k) + T_{uw}(k) + \pi_u(k) = 0 \tag{2.47}$$

and

$$P_{wT}(k) + T_{wT}(k) + \pi_T(k) + \beta F_{TT}(k) = 0 \tag{2.48}$$

Here, $F_{uT}(k)$ is momentarily ignored in Eq. (2.47) as its decay rate in $k-$ space (i.e., $k^{-7/3}$) is much faster than $F_{TT}(k)$ (i.e., $k^{-5/3}$) as discussed elsewhere (Katul et al., 2014).

The production terms can be obtained from the direct Fourier transforming $-\overline{w'w'}S$ and $-\overline{w'w'}\Gamma$ and are expressed as (Bos et al., 2004; Bos and Bertoglio, 2007; Cava and Katul, 2012)

$$P_{uw}(k) = -F_{ww}(k)\,S \quad , \quad P_{wT}(k) = -F_{ww}(k)\,\Gamma \tag{2.49}$$

The flux-transfer terms act to move turbulent stresses and fluxes away from the peak of the co-spectra and hence are parametrized using a spectral gradient-diffusion model given by (Bos et al., 2004; Bos and Bertoglio, 2007; Cava and Katul, 2012)

$$T_{uw}(K) = -A_{UU}\frac{\partial}{\partial k}\left(\epsilon^{1/3}\, k^{5/3} F_{uw}(k)\right) \tag{2.50}$$

and

$$T_{wT}(K) = -A_{TT}\frac{\partial}{\partial k}\left(\epsilon^{1/3}\, k^{5/3} F_{wT}(k)\right) \tag{2.51}$$

where A_{UU} and A_{TT} are model constants. The term 'flux-transport' is used in the context of Reynolds-averaged equations, while 'flux-transfer' is used here to represent the same processes in spectral space. It is also conceivable that the flux-transport terms are negligible in an idealized ASL, but the flux-transfer terms can be significant at any given k. To model the pressure decorrelation terms, a Rotta-type parameterization is invoked in the spectral domain and is given by

$$\pi_u(k) = -A_U\frac{F_{uw}(k)}{\tau_{uw}(k)} - C_{IU}P_{uw}(k) \tag{2.52}$$

and

$$\pi_T(k) = -A_T\frac{F_{wT}(k)}{\tau_{wT}(k)} - C_{IT}P_{wT}(k) \tag{2.53}$$

where A_T and $A_U = C_R$ (≈ 1.8) are the Rotta constants (Pope, 2000), C_{IT} and C_{IU} ($= 3/5$) are constants associated with isotropization of production terms as before. The $\tau_{uw}(k)$ and $\tau_{wT}(k)$ are scale-dependent decorrelation time scales for momentum and heat fluxes. A widely used formulation for $\tau_{uw}(k)$ and $\tau_{wT}(k)$ is given by (Corrsin, 1964; Bos et al., 2004; Bos and Bertoglio, 2007; Cava and Katul, 2012; Katul et al., 2014; Li and Katul, 2017)

$$\tau_{uw}(k) = \tau_{wT}(k) = \epsilon^{-1/3}k^{-2/3} \tag{2.54}$$

These formulations are based on the Kolmogorov inertial subrange (ISR) scaling argument or K41 (i.e., the dynamically relevant variables in the ISR are k for length and ϵ for length and time as in Chapter 1).

With these parameterizations, Eqs. (2.47) and (2.48) become

$$\frac{\partial F_{uw}(k)}{\partial k} + \left(\frac{A_U}{A_{UU}} + \frac{5}{3}\right)k^{-1}F_{uw}(k) = -\frac{(1-C_{IU})S}{A_{UU}\epsilon^{1/3}}F_{ww}(k)\,k^{-5/3} \tag{2.55}$$

and

$$\frac{\partial F_{wT}(k)}{\partial k} + \left(\frac{A_T}{A_{TT}} + \frac{5}{3}\right)k^{-1}F_{wT}(k) = \frac{k^{-5/3}\left[-(1-C_{IT})\Gamma F_{ww}(k) + \beta F_{TT}(k)\right]}{A_{TT}\epsilon^{1/3}} \tag{2.56}$$

The above first-order differential equations for $F_{uw}(k)$ and $F_{wT}(k)$ can be solved once $F_{ww}(k)$ and $F_{TT}(k)$ are known or externally supplied. These expressions may be viewed as a sort of 'spectral flux-variance' relation because they link co-spectra to the energetics of eddies at wavenumber k. They also highlight why the shape of $F_{uw}(k)$ differs from $F_{wT}(k)$, given the active role of $F_{TT}(k)$ in the $F_{wT}(k)$ budget. To illustrate a few key properties regarding this spectral link, the solution of $F_{uw}(k)$ is considered, as this budget only requires the specification of $F_{ww}(k)$.

The general solution to Eq. (2.55) may be viewed as a superposition of a homogeneous component given by

$$F_{uw}(k) = E_1 k^{-D_1} \quad , \quad D_1 = \frac{A_U}{A_{UU}} + \frac{5}{3}, \tag{2.57}$$

where E_1 is an integration constant and a nonhomogeneous component that depends on the precise shape of $F_{ww}(k)$. For the special case when $A_U/A_{UU} \gg 5/3$, the homogeneous contribution decays rapidly with increasing k and the solution is only governed by the nonhomogeneous term given by

$$F_{uw}(k) = -\frac{(1 - C_{IU})}{A_U} F_{ww}(k) (k L_{co})^{-2/3}, \tag{2.58}$$

where $L_{co} = (\epsilon/S^3)^{1/2}$ is known as the Corrsin length (Saddoughi and Veeravalli, 1994) not to be mistaken with the Obukhov–Corrsin scalar microscale. From definitions, $L_{co} = \kappa z$ in the near-neutral limit, but is larger for unstable conditions.

The eddy viscosity results derived from DDA and other scaling analyses are considered using the scale-by-scale outcomes here. To do so, the co-spectral budgets without the flux transfer terms are now scale-wise integrated and rearranged to yield an estimate of the eddy-viscosity ($\nu_t = -\overline{u'w'}/S$) given as (Mortarini et al., 2023)

$$\nu_t = \frac{1 - C_I}{C_R} \int_0^\infty \tau_{uw}(k) F_{ww}(k) dk. \tag{2.59}$$

Eq. (2.59) relates the fluctuations in vertical velocity at all scales through $F_{ww}(k)$ to a macroscopic turbulent viscosity (Goldenfeld and Shih, 2017). While the fluctuation-dissipation here arises from a co-spectral budget model that is in local equilibrium (i.e., stress production balances covariance destruction at each k while ignoring the transfer terms), the $F_{ww}(k)$ contains all the complexities in turbulence associated with nonequilibrium phenomena (Sreenivasan, 1999). That is, the production of $F_{ww}(k)$ will occur at scales much larger than the Kolmogorov micro-scales (see Chapter 1), setting an energy cascade from large to small eddies. This transfer of energy across scales (set to ϵ here) may be accommodated in a K41 theory for the ISR or the energy transfer term in the $F_{ww}(k)$ budget. The Reynolds number effects on ν_t arise when the ISR is censored by viscous effects (e.g., Karman–Howarth equation) or truncated at $1/\eta$ as a leading approximation. Thus the scale-wise integration of $F_{ww}(k)$ will involve macro-scales (transition from production to ISR) and viscous cutoff at η. The scalewise separation between these macro-scales (e.g., z) and η is linked to the Reynolds number as discussed elsewhere (Katul et al., 2013a).

Returning to Eq. (2.59), it may be possible to approximate the scale-wise integration by defining an effective decorrelation time scale τ_{eff} so that

$$\nu_t = \frac{1 - C_I}{C_R} \sigma_w^2 \tau_{\mathrm{eff}} \tag{2.60}$$

When $\tau_{\mathrm{eff}} = 2\sigma_w^2/\epsilon$, Eq. (2.43) is recovered. The work here explores the scalewise approximation leading to Eq. (2.43) and the consequences of ignoring the spectral transfer terms (i.e., $A_U/A_{UU} \gg 5/3$ when $A_{UU} \to 0$).

2.5 Challenges and opportunities

Over the course of some 60 years now, improved experimentation for canonical turbulent boundary layers revealed regularity in large-scale structures for neutral conditions (Kovasznay et al., 1970). More recently, experiments and simulations have shown that these large-scale structures dominate the energetics of the flow and contribute to momentum, heat, and scalar exchanges (Monty et al., 2007; Marusic et al., 2010a; Shah and Bou-Zeid, 2014; Salesky et al., 2017). Connecting these large eddies to phenomenological models is spawning renewed interest in properties of large-scale motion (Hutchins and Marusic, 2007; Marusic et al., 2010b), as well as how they evolve across different Ri_f. So, how can knowledge of these large-scale eddies be used to 'upgrade' MOST, DDA, or spectral theories used to predict the mean velocity gradient, S? It is well established that σ_u^2 is the most sensitive velocity component to the presence of large-scale motion (Guala et al., 2006). The P_m is injected into σ_u^2 and interactions between turbulent pressure, and velocity gradients redistribute energy across scales and velocity components. The aforementioned attainment of equi-partitioning of turbulent kinetic energy at large scales is prohibited by the presence of the ground surface—at least in the absence of B_p. Possible geometry of eddies that maintain such anisotropy while recovering the shape of S are hairpin vortex structures presumed to be attached to the wall.

In this framework, large-scale motions (or LSM) are formed due to streamwise coalescence of hairpin vortices formed near the surface into hairpin packets, and very large scale motion (VLSM) form due to yet another level of streamwise coalescence of LSMs. These VLSM are labeled as very large because they are over an order of magnitude larger than h (Smits et al., 2011). Moreover, the collection of these structures (hairpins to VLSMs) appears to be self-similar and are characterized by a population density that remains inversely related to the distance from the boundary. These results describe simultaneously energy anisotropy, size anisotropy, inclination with the wall, and several rotational properties in the limit of $Ri_f = 0$. When B_p is present and becomes sufficiently large, LSMs and VLSMs undergo a gradual transformation. LES have recently revealed how coalescent hairpin vortices transition to convection cells with increased ζ (Salesky et al., 2017; Salesky and Anderson, 2018), as shown in Fig. 2.6. In this figure, the VLSMs transition from long, quasi-linear updrafts of size comparable to $6h$ to open cellular patterns analogous to turbulent Rayleigh-Bénard convection cells. The inclusion of these transformations

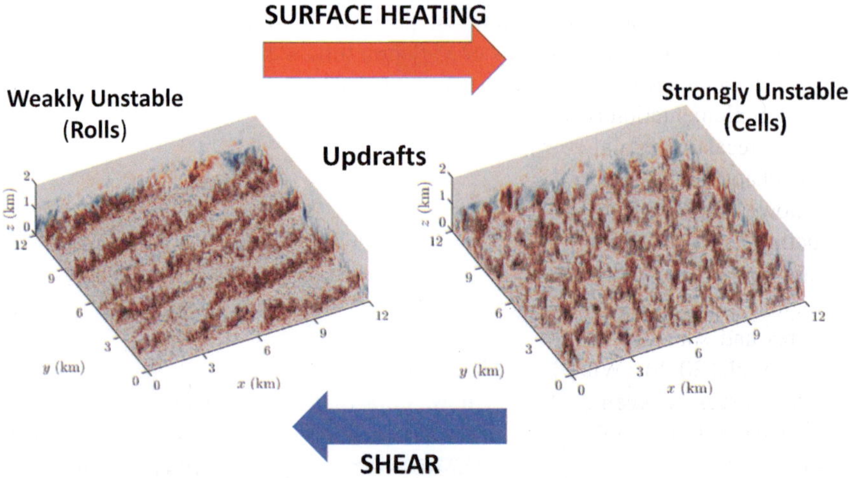

Figure 2.6 Updrafts ($w' > 0$) in a mildly heated (left) and strongly heated (right) atmospheric boundary layer. This figure is reproduced from Katul (2019) with permission from Cambridge University Press.

into MOST and DDA, or their use to propose alternative conceptual pictures of turbulence in the convective ABL (Zilitinkevich et al., 2021), is certainly a new challenge and a new line of inquiry that will advance models of ASL turbulence in the foreseeable future. Moreover, experiments and simulations that use the concept of uniform momentum zones are beginning to offer a new perspective on how to proceed (Heisel et al., 2020).

The presence of heterogeneity at the land surface invalidates the idealized state of the ASL. Depending on the particularity of the heterogeneity, advective terms will arise in the mean momentum budgets of U, W, and T. These advective terms are in direct response to a heterogeneity being 'sensed' first by the mean pressure gradients (longitudinal or vertical) in U and W, or heterogeneity in surface temperature that is then communicated via $\partial p'/\partial z$ and that may set baroclinic vertical variability in the horizontal pressure gradients driving the flow as discussed in Chapter 1 and other work (Momen et al., 2018). A number of recent reviews have emerged on this topic (Bou-Zeid et al., 2020) and some theoretical tactics have been proposed, including the use of dispersive stresses and fluxes to accommodate their effect on bulk flow properties (Calaf et al., 2022). Recent studies also showed the potential of dimensionless approaches for generalizing the findings on flow and turbulence dynamics over heterogeneous surfaces (Omidvar et al., 2020; Margairaz et al., 2020).

Reynolds decomposition in the ASL assumes a spectral gap between turbulence and meso-scale motion. The presence of this gap is needed to ensure that the mean flow evolves on time scales much slower than turbulence. A number of studies have now reported the absence of such a gap when analyzing high-frequency time series runs over months (Voronovich and Kiely, 2007). The effect on turbulence is particularly pronounced where the variability time scale of the forcing or mean flow is of the same order (not much smaller nor larger) than the integral turbulence scale (Momen and Bou-Zeid, 2017). A triple decomposition where the flow variable is decomposed into a long-term mean, meso-scale motion, and turbulent motion has been proposed, and its practical implementation is now being a subject of inquiry and discussion in numerical weather forecasting (Calaf et al., 2022).

This chapter only covered a selection of the many questions related to unstable shear flows. As dependence on the lowest layer of the atmosphere for energy, food production, and climate-proofing various socio-economic and infrastructure systems increases, more questions will arise. Among them are: how does turbulence evolve and mix air in the hurricane boundary layer (Momen et al., 2021)? How are turbulent structures and statistics modified by the presence of clouds at the top of the ABL (Heinze et al., 2015)? What are the differences in heat exchange and the resulting buoyancy forces between smooth, porous-rough, and bluff-rough surfaces (Li et al., 2020)? How to translate these theoretical advances into succinct parametrizations for unstably stratified shear flows in coarse geophysical models (Frassoni et al., 2018), and many more.

References

Albertson, J., Parlange, M., Katul, G., Chu, C.-R., Stricker, H., Tyler, S., 1995. Sensible heat flux from arid regions: a simple flux-variance method. Water Resources Research 31 (4), 969–973.

Asanuma, J., Brutsaert, W., 1999. Turbulence variance characteristics of temperature and humidity in the unstable atmospheric surface layer above a variable pine forest. Water Resources Research 35 (2), 515–521.

Banerjee, T., Katul, G., 2013. Logarithmic scaling in the longitudinal velocity variance explained by a spectral budget. Physics of Fluids 25 (12), 125106.

Banerjee, T., Katul, G., Salesky, S., Chamecki, M., 2015. Revisiting the formulations for the longitudinal velocity variance in the unstable atmospheric surface layer. Quarterly Journal of the Royal Meteorological Society 141 (690), 1699–1711.

Betchov, R., Yaglom, A., 1971. Notes on similitude theory for turbulence in an unstably stratified fluid. Izvestiâ Akademii Nauk SSSR. Fizika Atmosfery i Okeana 7, 1270–1279.

Bogenschutz, P.A., Gettelman, A., Morrison, H., Larson, V.E., Craig, C., Schanen, D.P., 2013. Higher-order turbulence closure and its impact on climate simulations in the community atmosphere model. Journal of Climate 26 (23), 9655–9676.

Bos, W., Bertoglio, J., 2007. Inertial range scaling of scalar flux spectra in uniformly sheared turbulence. Physics of Fluids 19, 025104.

Bos, W., Touil, H., Shao, L., Bertoglio, J., 2004. On the behavior of the velocity-scalar cross correlation spectrum in the inertial range. Physics of Fluids 16, 3818–3823.

Bou-Zeid, E., Anderson, W., Katul, G.G., Mahrt, L., 2020. The persistent challenge of surface heterogeneity in boundary-layer meteorology: a review. Boundary-Layer Meteorology 177 (2), 227–245.

Bou-Zeid, E., Gao, X., Ansorge, C., Katul, G.G., 2018. On the role of return to isotropy in wall-bounded turbulent flows with buoyancy. Journal of Fluid Mechanics 856, 61–78.

Brugger, P., Katul, G.G., De Roo, F., Kröniger, K., Rotenberg, E., Rohatyn, S., Mauder, M., 2018. Scalewise invariant analysis of the anisotropic Reynolds stress tensor for atmospheric surface layer and canopy sublayer turbulent flows. Physical Review Fluids 3 (5), 054608.

Brutsaert, W., 1982. Evaporation into the Atmosphere: Theory, History, and Applications. Environmental Fluid Mechanics. Kluwer Academic Publishers, Dordrecht.

Brutsaert, W., 1986. Catchment-scale evaporation and the atmospheric boundary layer. Water Resources Research 22 (9S), 39S–45S.

Brutsaert, W., 1999. Aspects of bulk atmospheric boundary layer similarity under free-convective conditions. Reviews of Geophysics 37 (4), 439–451.

Brutsaert, W., Mawdsley, J.A., 1976. The applicability of planetary boundary layer theory to calculate regional evapotranspiration. Water Resources Research 12 (5), 852–858.

Businger, J.A., 1988. A note on the Businger-Dyer profiles. Boundary-Layer Meteorology 42 (1–2), 145–151.

Businger, J.A., Wyngaard, J.C., Izumi, Y., Bradley, E.F., 1971. Flux-profile relationships in atmospheric surface layer. Journal of the Atmospheric Sciences 28 (2), 181–189.

Businger, J.A., Yaglom, A.M., 1971. Introduction to Obukhov's paper on 'Turbulence in an atmosphere with a nonuniform temperature'. Boundary-Layer Meteorology 2, 3–6.

Calaf, M., Vercauteren, N., Katul, G.G., Giometto, M.G., Morrison, T.J., Margairaz, F., Boyko, V., Pardyjak, E.R., 2022. Boundary-layer processes hindering contemporary numerical weather prediction models. Boundary-Layer Meteorology 186, 43–68.

Canuto, V.M., Cheng, Y., Howard, A., Isau, I., 2008. Stably stratified flows: a model with no Ri_{cr}. Journal of the Atmospheric Sciences 65, 2437–2447.

Cava, D., Katul, G., 2012. On the scaling laws of the velocity-scalar cospectra in the canopy sublayer above tall forests. Boundary-Layer Meteorology 145, 351–367.

Chamecki, M., Dias, N.L., Salesky, S.T., Pan, Y., 2017. Scaling laws for the longitudinal structure function in the atmospheric surface layer. Journal of the Atmospheric Sciences 74 (4), 1127–1147.

Charuchittipan, D., Wilson, J., 2009. Turbulent kinetic energy dissipation in the surface layer. Boundary-Layer Meteorology 132 (2), 193–204.

Corrsin, S., 1964. Further generalization of Onsager's cascade model for turbulent spectra. Physics of Fluids 7, 1156.

Cullen, N.J., Steffen, K., 2001. Unstable near-surface boundary conditions in summer on top of the Greenland Ice Sheet. Geophysical Research Letters 28 (23), 4491–4493.

Davidson, P., Krogstad, P.-Å., 2014. A universal scaling for low-order structure functions in the log-law region of smooth-and rough-wall boundary layers. Journal of Fluid Mechanics 752, 140–156.

de Bruin, H., 1999. A note on Businger's derivation of nondimensional wind and temperature profiles under unstable conditions. Journal of Applied Meteorology 38, 626–628.

de Bruin, H., Kohsiek, W., Van Den Hurk, B., 1993. A verification of some methods to determine the fluxes of momentum, sensible heat, and water vapour using standard deviation and structure parameter of scalar meteorological quantities. Boundary-Layer Meteorology 63 (3), 231–257.

Farmer, D.K., Boedicker, E.K., DeBolt, H.M., 2021. Dry deposition of atmospheric aerosols: approaches, observations, and mechanisms. Annual Review of Physical Chemistry 72, 375–397.

Fleagle, R., Businger, J., 1963. An Introduction to Atmospheric Physics. Academic Press, New York.

Foken, T., 2006. 50 years of the Monin-Obukhov similarity theory. Boundary-Layer Meteorology 119, 431–447.

Foken, T., 2008. The energy balance closure problem: an overview. Ecological Applications 18 (6), 1351–1367.

Frassoni, A., Castilho, D., Rixen, M., Ramirez, E., de Mattos, J.G.Z., Kubota, P., Calheiros, A.J.P., Reed, K.A., da Silva Dias, M.A.F., da Silva Dias, P.L., de Campos Velho, H.F., de Roode, S.R., Doblas-Reyes, F., Eiras, D., Ek, M., Figueroa, S.N., Forbes, R., Freitas, S.R., Grell, G.A., Herdies, D.L., Lauritzen, P.H., Machado, L.A.T., Manzi, A.O., Martins, G., Oliveira, G.S., Rosário, N.E., Sales, D.C., Wedi, N., Yamada, B., 2018. Building the next generation of climate modelers: scale-aware physics parameterization and the "grey zone" challenge. Bulletin of the American Meteorological Society 99, ES185–ES189.

Frisch, U., 1995. Turbulence. Cambridge University Press, Cambridge, England.

Fuentes, J.D., Lerdau, M., Atkinson, R., Baldocchi, D., Bottenheim, J., Ciccioli, P., Lamb, B., Geron, C., Gu, L., Guenther, A., et al., 2000. Biogenic hydrocarbons in the atmospheric boundary layer: a review. Bulletin of the American Meteorological Society 81 (7), 1537–1575.

Garratt, J.R., 1992. The Atmospheric Boundary Layer. Cambridge University Press.

Ghannam, K., Katul, G.G., Bou-Zeid, E., Gerken, T., Chamecki, M., 2018. Scaling and similarity of the anisotropic coherent eddies in near-surface atmospheric turbulence. Journal of the Atmospheric Sciences 75 (3), 943–964.

Gioia, G., Guttenberg, N., Goldenfeld, N., Chakraborty, P., 2010. Spectral theory of the turbulent mean-velocity profile. Physical Review Letters 105, 184501.

Goldenfeld, N., Shih, H.-Y., 2017. Turbulence as a problem in nonequilibrium statistical mechanics. Journal of Statistical Physics 167 (3), 575–594.

Guala, M., Hommema, S., Adrian, R., 2006. Large-scale and very-large-scale motions in turbulent pipe flow. Journal of Fluid Mechanics 554, 521–542.

Heinze, R., Mironov, D., Raasch, S., 2015. Second-moment budgets in cloud topped boundary layers: a large-eddy simulation study. Journal of Advances in Modeling Earth Systems 7 (2), 510–536.

Heisel, M., De Silva, C.M., Hutchins, N., Marusic, I., Guala, M., 2020. On the mixing length eddies and logarithmic mean velocity profile in wall turbulence. Journal of Fluid Mechanics 887, R1.

Hsieh, C., Katul, G., 1997. Dissipation methods, Taylor's hypothesis, and stability correction functions in the atmospheric surface layer. Journal of Geophysical Research. Atmospheres 03, 16391–16405.

Hsieh, C.-I., Katul, G., 2009. The Lagrangian stochastic model for estimating footprint and water vapor fluxes over inhomogeneous surfaces. International Journal of Biometeorology 53 (1), 87–100.

Hsieh, C.-I., Katul, G., Chi, T.-w., 2000. An approximate analytical model for footprint estimation of scalar fluxes in thermally stratified atmospheric flows. Advances in Water Resources 23 (7), 765–772.

Hsieh, C.-I., Katul, G.G., Schieldge, J., Sigmon, J., Knoerr, K.R., 1996. Estimation of momentum and heat fluxes using dissipation and flux-variance methods in the unstable surface layer. Water Resources Research 32 (8), 2453–2462.

Huang, K.Y., Katul, G.G., 2022. Profiles of high-order moments of longitudinal velocity explained by the random sweeping decorrelation hypothesis. Physical Review Fluids 7 (4), 044603.

Hutchins, N., Marusic, I., 2007. Large-scale influences in near-wall turbulence. Philosophical Transactions of the Royal Society of London A: Mathematical, Physical and Engineering Sciences 365 (1852), 647–664.

Kader, B.A., Yaglom, A.M., 1990. Mean fields and fluctuation moments in unstably stratified turbulent boundary-layers. Journal of Fluid Mechanics 212, 637–662.

Kaimal, J., Finnigan, J., 1994. Atmospheric Boundary Layer Flows: Their Structure and Measurement. Oxford University Press, New York.

Kaimal, J.C., 1973. Turbulence spectra, length scales and structure parameters in the stable surface layer. Boundary-Layer Meteorology 4, 289–309.

Kaimal, J.C., Izumi, Y., Wyngaard, J.C., Coté, R., 1972. Spectral characteristics of surface-layer turbulence. Quarterly Journal of the Royal Meteorological Society 98 (417), 563–589.

Kaimal, J.C., Wyngaard, J.C., Haugen, D.A., Coté, O.R., Izumi, Y., Caughey, S.J., Readings, C.J., 1976. Turbulence structure in the convective boundary layer. Journal of the Atmospheric Sciences 33, 2152–2169.

Kanda, M., Inagaki, A., Letzel, M.O., Raasch, S., Watanabe, T., 2004. LES study of the energy imbalance problem with eddy covariance fluxes. Boundary-Layer Meteorology 110 (3), 381–404.

Katul, G., 2019. The anatomy of large-scale motion in atmospheric boundary layers. Journal of Fluid Mechanics 858, 1–4.

Katul, G., Konings, A., Porporato, A., 2011. Mean velocity profile in a sheared and thermally stratified atmospheric boundary layer. Physical Review Letters 107, 268502.

Katul, G., Porporato, A., Manes, C., Meneveau, C., 2013a. Co-spectrum and mean velocity in turbulent boundary layers. Physics of Fluids 25, 091702.

Katul, G.G., Hsieh, C.-I., 1999. A note on the flux-variance similarity relationships for heat and water vapour in the unstable atmospheric surface layer. Boundary-Layer Meteorology 90 (2), 327–338.

Katul, G.G., Li, D., Chamecki, M., Bou-Zeid, E., 2013b. Mean scalar concentration profile in a sheared and thermally stratified atmospheric surface layer. Physical Review. E 87 (2), 023004.

Katul, G.G., Porporato, A., Shah, S., Bou-Zeid, E., 2014. Two phenomenological constants explain similarity laws in stably stratified turbulence. Physical Review. E 89 (2), 023007.

Kays, W., 1994. Turbulent Prandtl number: where are we? Journal of Heat Transfer 116 (2), 284–295.

Khanna, S., Brasseur, J.G., 1997. Analysis of Monin-Obukhov similarity from large-eddy simulation. Journal of Fluid Mechanics 345, 251–286.

Kovasznay, L., Kibens, V., Blackwelder, R., 1970. Large-scale motion in the intermittent region of a turbulent boundary layer. Journal of Fluid Mechanics 41 (2), 283–325.

Launder, B., Reece, G., Rodi, W., 1975. Progress in the development of a Reynolds-stress turbulence closure. Journal of Fluid Mechanics 68, 537–566.

Li, D., 2019. Turbulent Prandtl number in the atmospheric boundary layer: where are we now? Atmospheric Research 216, 86–105.

Li, D., 2021. The O'KEYPS equation and 60 years beyond. Boundary-Layer Meteorology 179 (1), 19–42.

Li, D., Katul, G., Bou-Zeid, E., 2012. Mean velocity and temperature profiles in a sheared diabatic turbulent boundary layer. Physics of Fluids 24 (10), 105105.

Li, D., Katul, G.G., 2017. On the linkage between the $k^{-5/3}$ spectral and $k^{-7/3}$ cospectral scaling in high-Reynolds number turbulent boundary layers. Physics of Fluids 29 (6), 065108.

Li, D., Katul, G.G., Liu, H., 2018a. Intrinsic constraints on asymmetric turbulent transport of scalars within the constant flux layer of the lower atmosphere. Geophysical Research Letters 45 (4), 2022–2030.

Li, Q., Bou-Zeid, E., Grimmond, S., Zilitinkevich, S., Katul, G., 2020. Revisiting the relation between momentum and scalar roughness lengths of urban surfaces. Quarterly Journal of the Royal Meteorological Society 146 (732), 3144–3164.

Li, Q., Gentine, P., Mellado, J.P., McColl, K.A., 2018b. Implications of nonlocal transport and conditionally averaged statistics on Monin–Obukhov similarity theory and Townsend's attached eddy hypothesis. Journal of the Atmospheric Sciences 75 (10), 3403–3431.

Liu, H., Gao, Z., Katul, G.G., 2021. Non-closure of surface energy balance linked to asymmetric turbulent transport of scalars by large eddies. Journal of Geophysical Research. Atmospheres 126 (7), e2020JD034474.

Margairaz, F., Pardyjak, E.R., Calaf, M., 2020. Surface thermal heterogeneities and the atmospheric boundary layer: the relevance of dispersive fluxes. Boundary-Layer Meteorology 175 (3), 369–395.

Maronga, B., Reuder, J., 2017. On the formulation and universality of Monin–Obukhov similarity functions for mean gradients and standard deviations in the unstable surface layer: results from surface-layer-resolving large-eddy simulations. Journal of the Atmospheric Sciences 74 (4), 989–1010.

Marusic, I., Mathis, R., Hutchins, N., 2010a. Predictive model for wall-bounded turbulent flow. Science 329 (5988), 193–196.

Marusic, I., McKeon, B., Monkewitz, P., Nagib, H., Smits, A., Sreenivasan, K., 2010b. Wall-bounded turbulent flows at high Reynolds numbers: recent advances and key issues. Physics of Fluids 22 (6), 065103.

Marusic, I., Monty, J., 2019. Attached eddy model of wall turbulence. Annual Review of Fluid Mechanics 51, 49–74.

Mauder, M., Foken, T., Cuxart, J., 2020. Surface-energy-balance closure over land: a review. Boundary-Layer Meteorology 177 (2), 395–426.

Mellor, G.L., Yamada, T., 1982. Development of a turbulence closure model for geophysical fluid problems. Reviews of Geophysics 20 (4), 851–875.

Momen, M., Bou-Zeid, E., 2017. Mean and turbulence dynamics in unsteady Ekman boundary layers. Journal of Fluid Mechanics 816, 209–242.

Momen, M., Bou-Zeid, E., Parlange, M., Giometto, M., 2018. Modulation of mean wind and turbulence in the atmospheric boundary layer by baroclinicity. Journal of the Atmospheric Sciences 75 (11), 3797–3821.

Momen, M., Parlange, M.B., Giometto, M.G., 2021. Scrambling and reorientation of classical atmospheric boundary layer turbulence in hurricane winds. Geophysical Research Letters 48 (7), e2020GL091695.

Monin, A., Obukhov, A., 1954. Basic laws of turbulent mixing in the surface layer of the atmosphere. Transactions of the Academy of Sciences of the USSR, Geophysical Institute 24 (151), 163–187.

Monin, A., Yaglom, A.M., 1975. Statistical Fluid Mechanics, vol. 2. MIT Press, Cambridge.

Monty, J., Stewart, J., Williams, R., Chong, M., 2007. Large-scale features in turbulent pipe and channel flows. Journal of Fluid Mechanics 589, 147–156.

Mortarini, L., Katul, G.G., Cava, D., Dias-Junior, C.Q., Dias, N.L., Manzi, A., Sorgel, M., Araújo, A., Chamecki, M., 2023. Adjustments to the law of the wall above an Amazon forest explained by a spectral link. Physics of Fluids 35 (2), 025102.

Obukhov, A., 1946. Turbulence in thermally inhomogeneous atmosphere. Transactions of the Academy of Sciences of the USSR, Geophysical Institute 1, 95–115.

Omidvar, H., Bou-Zeid, E., Li, Q., Mellado, J.-P., Klein, P., 2020. Plume or bubble? Mixed-convection flow regimes and city-scale circulations. Journal of Fluid Mechanics 897, A5.

Oncley, S.P., Foken, T., Vogt, R., Kohsiek, W., DeBruin, H.A., Bernhofer, C., Christen, A., Gorsel, E.v., Grantz, D., Feigenwinter, C., et al., 2007. The energy balance experiment EBEX-2000. Part I: overview and energy balance. Boundary-Layer Meteorology 123 (1), 1–28.

Padro, J., 1993. An investigation of flux-variance methods and universal functions applied to three land-use types in unstable conditions. Boundary-Layer Meteorology 66 (4), 413–425.

Pan, Y., Chamecki, M., 2016. A scaling law for the shear-production range of second-order structure functions. Journal of Fluid Mechanics 801, 459–474.

Panchev, S., 1971. Random Functions and Turbulence. Pergamon Press, Oxford.
Parlange, M.B., Eichinger, W.E., Albertson, J.D., 1995. Regional-scale evaporation and the atmospheric boundary-layer. Reviews of Geophysics 33 (1), 99–124.
Parlange, M.B., Katul, G.G., 1995. Watershed scale shear stress from tethersonde wind profile measurements under near neutral and unstable atmospheric stability. Water Resources Research 31 (4), 961–968.
Plate, E., 1971. Aerodynamic Characteristics of the Atmospheric Boundary Layer. AEC Critical Review Series. U.S. Atomic Energy Commission.
Pope, S.B., 2000. Turbulent Flows. Cambridge University Press, Cambridge, UK.
Priestley, C.H.B., Swinbank, W.C., 1947. Vertical transport of heat by turbulence in the atmosphere. Proceedings of the Royal Society of London: A 189 (1019), 543–561.
Saddoughi, S., Veeravalli, S., 1994. Local isotropy in turbulent boundary layers at high Reynolds number flow. Journal of Fluid Mechanics 268, 333–372.
Salesky, S., Anderson, W., 2018. Buoyancy effects on large-scale motions in convective atmospheric boundary layers: implications for modulation of near-wall processes. Journal of Fluid Mechanics 856, 135–168.
Salesky, S., Chamecki, M., Bou-Zeid, E., 2017. On the nature of the transition between roll and cellular organization in the convective boundary layer. Boundary-Layer Meteorology 163 (1), 41–68.
Salesky, S.T., Katul, G.G., Chamecki, M., 2013. Buoyancy effects on the integral lengthscales and mean velocity profile in atmospheric surface layer flows. Physics of Fluids 25 (10), 105101.
Schmid, H.P., 2002. Footprint modeling for vegetation atmosphere exchange studies: a review and perspective. Agricultural and Forest Meteorology 113 (1–4), 159–183.
Shah, S., Bou-Zeid, E., 2014. Very-large-scale motions in the atmospheric boundary layer educed by snapshot proper orthogonal decomposition. Boundary-Layer Meteorology 153 (3), 355–387.
Smits, A., Marusic, I., 2013. Wall-bounded turbulence. Physics Today 66 (9), 25–30.
Smits, A.J., McKeon, B.J., Marusic, I., 2011. High–Reynolds number wall turbulence. Annual Review of Fluid Mechanics 43, 353–375.
Sorbjan, Z., 1989. Structure of the Atmospheric Boundary Layer. Prentice Hall.
Sreenivasan, K.R., 1999. Fluid turbulence. Reviews of Modern Physics 71 (2), S383.
Stiperski, I., Chamecki, M., Calaf, M., 2021a. Anisotropy of unstably stratified near-surface turbulence. Boundary-Layer Meteorology 180 (3), 363–384.
Stiperski, I., Katul, G.G., Calaf, M., 2021b. Universal return to isotropy of inhomogeneous atmospheric boundary layer turbulence. Physical Review Letters 126 (19), 194501.
Stull, R.B., 1988. An Introduction to Boundary Layer Meteorology. Springer, Netherlands.
Taylor, G.I., 1938. The spectrum of turbulence. Proceedings of the Royal Society of London: A 164, 476–490.
Tennekes, H., Lumley, J., 1972. A First Course in Turbulence. MIT Press, Cambridge.
Tillman, J., 1972. The indirect determination of stability, heat and momentum fluxes in the atmospheric boundary layer from simple scalar variables during dry unstable conditions. Journal of Applied Meteorology and Climatology 11 (5), 783–792.
Townsend, A.A., 1976. The Structure of Turbulent Shear Flow, vol. 2. Cambridge University Press, Cambridge.
Vesala, T., Kljun, N., Rannik, Ü., Rinne, J., Sogachev, A., Markkanen, T., Sabelfeld, K., Foken, T., Leclerc, M.Y., 2008. Flux and concentration footprint modelling: state of the art. Environmental Pollution 152 (3), 653–666.
Voronovich, V., Kiely, G., 2007. On the gap in the spectra of surface-layer atmospheric turbulence. Boundary-Layer Meteorology 122 (1), 67–83.

Waterman, T., Bragg, A.D., Katul, G., Chaney, N., 2022. Examining parameterizations of potential temperature variance across varied landscapes for use in Earth system models. Journal of Geophysical Research. Atmospheres 127 (8), e2021JD036236.

Weaver, H.L., 1990. Temperature and humidity flux-variance relations determined by one-dimensional eddy correlation. Boundary-Layer Meteorology 53 (1), 77–91.

Wesely, M., Hicks, B., 2000. A review of the current status of knowledge on dry deposition. Atmospheric Environment 34 (12–14), 2261–2282.

Wesson, K.H., Katul, G., Lai, C.-T., 2001. Sensible heat flux estimation by flux variance and half-order time derivative methods. Water Resources Research 37 (9), 2333–2343.

Wilson, D.K., 2001. An alternative function for the wind and temperature gradients in unstable surface layers. Boundary-Layer Meteorology 99 (1), 151–158.

Wilson, K., Goldstein, A., Falge, E., Aubinet, M., Baldocchi, D., Berbigier, P., Bernhofer, C., Ceulemans, R., Dolman, H., Field, C., et al., 2002. Energy balance closure at FLUXNET sites. Agricultural and Forest Meteorology 113 (1–4), 223–243.

Wyngaard, J.C., 2010. Turbulence in the Atmosphere. Cambridge University Press.

Wyngaard, J.C., Coté, O.R., 1972. Cospectral similarity in the atmospheric surface layer. Quarterly Journal of the Royal Meteorological Society 98 (417), 590–603.

Zahn, E., Bou-Zeid, E., Dias, N.L., 2023. Relaxed eddy accumulation outperforms Monin-Obukhov flux models under non-ideal conditions. Geophysical Research Letters 50, e2023GL103099.

Zahn, E., Dias, N.L., Araújo, A., Sá, L.D., Sörgel, M., Trebs, I., Wolff, S., Manzi, A., 2016. Scalar turbulent behavior in the roughness sublayer of an Amazonian forest. Atmospheric Chemistry and Physics 16 (17), 11349–11366.

Zilitinkevich, S., 1973. Shear convection. Boundary-Layer Meteorology 3 (4), 416–423.

Zilitinkevich, S., Kadantsev, E., Repina, I., Mortikov, E., Glazunov, A., 2021. Order out of chaos: shifting paradigm of convective turbulence. Journal of the Atmospheric Sciences 78 (12), 3925–3932.

Stratified shear flows

3

Sutanu Sarkar
University of California San Diego, La Jolla, CA, United States

3.1 Introduction

As discussed previously, density variation is a notable environmental feature arising from temperature and water vapor gradients in the atmosphere, and temperature and salinity gradients in the ocean. Background density stratification—specifically, lighter fluid overlying heavier fluid—creates an energetically stable configuration that modifies, and may even control, turbulence properties in environmental flows. This chapter focuses on the fundamentals of stratified shear flows that underpin many of the applications discussed in subsequent chapters. We demonstrate how buoyancy in a stratified fluid modifies turbulence—its intensity and structure—by analyzing several canonical problems.

The buoyancy frequency (N with units of s^{-1}), introduced in Chapter 1 as a measure of stratification, is defined as $N^2 = -(g/\rho_r)\partial\rho_b/\partial z$, where $\rho_b(\mathbf{x}, t)$ is the background density profile and ρ_r is the reference density. In a turbulent shear flow, the dynamical importance of stratification is determined by comparing N with the characteristic vertical shear $(\partial u/\partial z)$. This comparison can be performed using the mean, Reynolds-averaged shear (S), yielding the gradient Richardson number $Ri_g = N^2/S^2$. The 'mean' in the case of environmental flows is unsteady, for example, on diurnal, inertial, or tidal time scales. Therefore S corresponds to a relevant averaging time scale that is sufficiently large relative to the turbulent scales but small compared to the longer-term unsteadiness. Buoyancy effects become important when Ri_g is not small, typically $Ri_g > O(0.01)$. The Miles–Howard criterion, based on linear, inviscid, stability analysis of a parallel stratified shear flow, states that small disturbances do not grow exponentially if $Ri_g > 0.25$ everywhere in the flow. This finding has established the concept of a critical Richardson number, $Ri_{cr} = 0.25$.

Alternatively, one may compare N with an estimate of the fluctuating shear (u_h/l_h, where u_h is the root-mean-square of the horizontal velocity fluctuation and l_h is its horizontal integral length scale). This leads to the turbulent Froude number, $Fr_h = u_h/Nl_h$. The use of Fr_h is convenient and sometimes necessary in the scaling analysis of stratified turbulence; buoyancy effects *must be considered* if $Fr_h \leq O(1)$.

Density stratification tends to stabilize turbulent flow, thereby reducing TKE. A primary challenge in studying stratified turbulence is that it differs substantially from classical turbulence. *Briefly, stratified turbulence tends to be anisotropic in velocity components and length scales, with those in the vertical reduced with respect to the horizontal; patchy and layered in space; intermittent in time; and interspersed with distinctive coherent structures and internal gravity waves.* These features may be mild under moderate stratification, but with increasing relative strength of buoyancy, some

Environmental Turbulence. https://doi.org/10.1016/B978-0-32-395873-8.00010-3
Copyright © 2026 Elsevier Inc. All rights are reserved, including those for text and data mining, AI training, and similar technologies.

or all become important. We will illustrate these features in simple shear flows such as a mixing layer, wake, channel, and an Ekman boundary layer.

There is significant interest in the stratified turbulence problems originating from internal waves (Lamb, 2014; Sun et al., 2015; Sarkar and Scotti, 2017) and submesoscale flows with $O(1)$ Rossby number (Taylor and Ferrari, 2010; Skyllingstad and Samelson, 2012; McWilliams, 2016; Sarkar et al., 2016; Sullivan and McWilliams, 2018). While insights into these problems have been made possible by extending the simple shear flows discussed here to include oscillatory motion and rotation, these extensions will not be discussed further in this chapter.

3.2 Scales of stratified turbulence

In the multiscale state that is turbulence, buoyancy effects are scale-dependent. A central question is the relation of these scale-dependent effects to the Kolmogorov description of high-Re turbulence.

Generalize the turbulent Froude number ($Fr_h = u_h/Nl_h$) to a length-scale-dependent Froude number, $Fr(l) = u(l)/Nl$ where the subscript h is implicit. In turbulent flows, large-l eddies in the inertial subrange also have smaller values of $u(l)/l \sim \varepsilon^{1/3} l^{-2/3}$, correspondingly lower $Fr(l)$, and therefore have stronger buoyancy effects. Physically, one can understand this by direct examination of the vertical momentum equation that was given in Chapter 1. Let $\hat{\rho}$ denote the density deviation from the constant reference density (ρ_r). An upward-moving eddy motion with vertical length scale l experiences a vertical buoyancy force per unit mass,

$$-\frac{g}{\rho_r}\hat{\rho} \sim -\frac{g}{\rho_r}\frac{-\partial \rho_b}{\partial z} l \sim -N^2 l \tag{3.1}$$

which is a deceleration that increases in magnitude with increasing l. The condition that the magnitude of buoyant acceleration ($N^2 l$) exceeds the inertial acceleration (u^2/l) requires $l > u/N$ or $Fr(l) < 1$. The implication for high-Re turbulence with an inertial subrange is as follows. Using the inviscid scaling of dissipation rate, $\varepsilon \sim u^3/l$, to substitute for the scale-dependent u in the definition of $Fr(l)$ leads to

$$\frac{u}{Nl} < 1 \quad \Longrightarrow \quad \frac{(\varepsilon l)^{1/3}}{Nl} < 1 \Longrightarrow l > (\frac{\varepsilon}{N^3})^{1/2} \tag{3.2}$$

Eq. (3.2) leads to the *Ozmidov length*, $l_{oz} = \sqrt{\varepsilon/N^3}$, as a distinguished length scale at which $Fr(l) \sim 1$. The Ozmidov scale estimates the largest horizontal scale that can overturn in the presence of a restoring buoyancy. Motion at length scales above the Ozmidov scale (l_{oz}) feels the effect of buoyancy that is manifested as a damping of the vertical velocity (w) and is reflected as an increased Reynolds stress anisotropy: $\overline{w'^2}(l)$ becomes smaller than its horizontal counterparts as $Fr(l)$ decreases. Moving to length scales smaller than l_{oz}, if there is sufficient scale separation between l_{oz} and the Kolmogorov scale ($\eta = (\nu^3/\varepsilon)^{1/4}$), the down-scale energy cascade can support an inertial subrange with $k^{-5/3}$ scaling of the energy spectrum.

From the definitions of l_{oz} and η, it follows that

$$\frac{l_{oz}}{\eta} = \left(\frac{\varepsilon}{\nu N^2}\right)^{3/4} \tag{3.3}$$

motivating the introduction of the buoyancy Reynolds number, defined by $Re_b = \varepsilon/\nu N^2$. This 3/4 power-law is analogous to the relationship, $l_h/\eta \sim Re_t^{3/4}$, that determines the dynamic range of length scales in unstratified turbulence between its integral length scale (l_h) and the smallest scale η that is dominated by viscous effects. Another interpretation of Re_b is

$$Re_b = \frac{\varepsilon}{\nu N^2} = \frac{u_h^3/l_h}{\nu N^2} = \frac{u_h l_h}{\nu}\frac{u_h^2}{l_h^2 N^2} = Re_h Fr_h^2 \tag{3.4}$$

which shows that Re_b represents a buoyancy modification of the customary turbulence Reynolds number. **Strongly stratified turbulence** (SST) is the regime where $Fr_h << O(1)$ and $Re_h >> O(1)$ so that Re_b is sufficiently large. In practice, fluctuations with a broadband spectrum are seen at $Re_b > 1$ (equivalently, $Re_h Fr_h^2 > 1$). Plotting the evolution of a stratified shear flow in (Re_h, $Re_h Fr_h^2$) *phase space* is a potential method to gauge the strength of buoyancy effects on a unified basis. SST, its scalar and momentum mixing properties, its manifestation in physical space, and also the route to SST from geophysical flows are subjects of active research.

3.3 Turbulence energetics

The turbulent kinetic energy (TKE or $K = \langle u_i' u_i' \rangle/2$) satisfies the following transport equation in stratified flow:

$$\frac{\partial K}{\partial t} + U_i \frac{\partial K}{\partial x_i} = P - \varepsilon + B - \frac{\partial T_i}{\partial x_i} \tag{3.5}$$

where $U_i = \langle u_i \rangle$ is the mean (Reynolds-averaged) velocity, T_i is the turbulent transport, P is the turbulent production, ε is the turbulent dissipation rate, and B is the turbulent buoyancy flux with definitions as follows:

$$P = -\langle u_i' u_j' \rangle \frac{\partial U_i}{\partial x_j}, \qquad \varepsilon = 2\nu \langle s_{ij}' s_{ij}' \rangle, \qquad B = -\frac{g}{\rho_0}\langle \rho' u_3' \rangle \tag{3.6}$$

Here, gravity acts in the negative x_3 (or z) direction and $s_{ij}' = (\partial_j u_i' + \partial_i u_j')/2$ is the fluctuating strain rate tensor.

Production (P) acts as a source in the transport equation, extracting energy from the mean kinetic energy (MKE) reservoir and generating an equivalent amount of TKE. Dissipation (ε) serves as a sink of TKE. Buoyancy flux (B) exchanges energy between the TKE and Turbulent Potential Energy (TPE) reservoirs, with a positive buoyancy flux converting TPE into TKE, and vice versa. Unlike the aforementioned source and sink terms, turbulent transport (T_i) is responsible for the spatial redistribution of TKE and is given by

$$T_i = \frac{1}{2}\langle u'_i u'_j u'_j \rangle + \langle u'_i p' \rangle - 2\nu \langle u'_j s'_{ij} \rangle \tag{3.7}$$

3.3.1 Buoyancy considerations in turbulence energetics

The only explicit buoyancy term in Eq. (3.5) is the buoyancy flux (B). It is therefore tempting to attribute the decrease of TKE in stratified flow entirely to B as was done in some early RANS turbulence models. But, such a modeling approach is incorrect. In flows with vertical mean shear (dU_1/dx_3), it is the progressive reduction of the Reynolds shear stress ($\langle u'_1 u'_3 \rangle$) magnitude with increasing value of Ri_g (Jacobitz et al., 1997) that leads to decreased TKE. The Ri_g-dependent decrease in $\langle u'_1 u'_3 \rangle$, although for a nonlinear turbulent flow, is physically related to the linear stability theory result that maximal perturbation growth rate decreases with increasing Ri_g until reaching zero at $Ri_g = 0.25$. In practice, current turbulence models in RANS or lower-resolution LES approaches employ a Ri_g-based correction of the eddy viscosity to account for stratification.

In turbulent flows, the pressure-velocity correlation ($\langle u'_i p' \rangle$) is typically lumped together with turbulent transport ($(1/2)\langle u'_i u'_j u'_j \rangle$) and has not received separate attention. In contrast, the term, $\langle u'_i p' \rangle$, is fundamental to stratified flows since it represents the internal wave energy flux that either enters or exits the control volume of interest. The nonlinear evolution or breaking of an incident linear internal wave to create a patch of turbulence and the generation of an internal wave field by a turbulent flow are two examples where $\langle u'_i p' \rangle$ is important. The character of $\langle u'_i p' \rangle$ changes from the turbulent region, where it obeys the nonlinear scaling of $\rho_0 u^3$, to the wave propagation region, where it is equal to $\mathbf{c}_g E$ as per linear wave scaling. Here $\mathbf{c}_g$ is the wave group velocity vector and E is the total (sum of kinetic and potential energies per unit volume) wave energy.

Stratified flows not only have kinetic energy but also potential energy, of which there is a fraction, the available potential energy (APE), that can be converted to flow kinetic energy and scalar mixing. The first step to APE evaluation is the definition of a reference state (equilibrium depth) $z_r(\rho)$ for each fluid parcel of density ρ. One popular definition of APE (Holliday and McIntyre, 1981; Winters et al., 1995) of a fluid parcel with $\rho(x, y, z, t)$ is as follows:

$$E_{APE} = \int_{z_r}^{z} \frac{g}{\rho_0}(\rho - \rho_r(z'))dz' \tag{3.8}$$

where $\rho_r(z)$ is the density profile of the background state, $z_r(\rho)$ being its inverse mapping. A quadratic approximation for APE in a linearly stratified constant-N fluid can be defined (Kang and Fringer, 2010) and decomposed into a mean and turbulent component, each with its transport equation (Scotti and White, 2014):

$$\langle E_{APE} \rangle = MPE + TPE; \quad MPE = \frac{g^2 \langle \rho^* \rangle^2}{2\rho_0^2 N^2}, \; TPE = \frac{g^2 \langle \rho^{*\prime 2} \rangle}{2\rho_0^2 N^2} \tag{3.9}$$

Here, ρ^* is the density deviation with respect to the vertically varying background density, $\rho_b(z)$. For brevity, we have dropped the term "available" from the mean and turbulent potential energy, MPE and TPE, respectively. TPE, given by Eq. (3.9), is a *locally defined* equivalent of TKE, and its transport equation can be written as follows, e.g., Puthan et al. (2019):

$$\frac{D(TPE)}{Dt} = P_\rho - \varepsilon_\rho + T_\rho \tag{3.10}$$

where

$$\begin{aligned}
&\text{TPE production } (P_\rho): \frac{-g^2}{\rho_0^2 N^2}\left\{\langle u'_j \rho^{*\prime}\rangle \frac{\partial \langle \rho^* \rangle}{\partial x_j} + \langle \rho^{*\prime} u'_3 \rangle \frac{d\rho_b}{dx_3}\right\} \\
&\text{Transport of TPE } (T_\rho): \frac{g^2}{2\rho_0^2 N^2}\left\{\kappa \frac{\partial^2 \langle \rho^{*\prime 2}\rangle}{\partial x_j^2} - \frac{\partial \langle u'_j \rho^{*\prime 2}\rangle}{\partial x_j}\right\} \\
&\text{TPE dissipation } (\varepsilon_\rho): \frac{g^2}{\rho_0^2 N^2}\left\{\kappa \left\langle \frac{\partial \rho^{*\prime}}{\partial x_j} \frac{\partial \rho^{*\prime}}{\partial x_j}\right\rangle\right\}
\end{aligned} \tag{3.11}$$

The TPE production term, P_ρ in Eq. (3.11), can be further simplified when the mean density varies only in the vertical and N^2 is allowed to evolve from its initial background value, resulting in

$$P_\rho = \frac{g}{\rho_0}\langle \rho' u'_3 \rangle = -B\,. \tag{3.12}$$

Comparison of Eqs. (3.6) and (3.12) shows that the buoyancy flux (B), which can take either sign, transfers energy between the TKE and TPE reservoirs. When $B < 0$, energy flows from TKE, specifically its vertical component, to TPE, i.e., turbulence stirs the buoyancy field. Conversely, when $B > 0$, velocity fluctuations are energized at the expense of density fluctuations.

B is overall negative in mechanically driven stratified turbulence and has an irreversible component that acts to molecularly mix the density profile and thus the background potential energy of the flow. Owing to density stratification and rotation, turbulence is generally weak in the ocean interior. Nevertheless, as discussed in the introduction as well as Chapter 1, the mixing accomplished by turbulence is crucial to maintain the physical state of the ocean (density, temperature, and salinity profiles). Characterization of the stable atmospheric boundary layer and its reduced mixing of particulate matter and other constituents are similarly of interest. Thus accurate parameterization or modeling of B is necessary in environmental flows because of its importance to mixing, in addition to its presence in the TKE transport equation. Chapter 6 on "Turbulent mixing" can be consulted for more information on mixing in turbulent flows, including the situation with a stable stratification.

The turbulent diffusivity of momentum (K_m or ν_t) and density (K_ρ or D_t) for vertically sheared stratified flow are given by

$$\langle u'w' \rangle = -K_m \frac{\partial \langle u \rangle}{\partial z} \quad , \quad \langle \rho' w' \rangle = -K_\rho \frac{\partial \langle \rho \rangle}{\partial z} \tag{3.13}$$

and characterize the turbulent fluxes using gradient transport. Buoyancy effects (in both stable and unstable stratification) on K_m and K_ρ, or the turbulent Prandtl number ($Pr_t = K_m/K_\rho$), are incorporated in different ways depending on the level of complexity of the turbulence model. Typically, the turbulent diffusivity is given by the product of characteristic length (l_t) and velocity (u_t) scales of the energy-containing scales of motion. Buoyancy effects are incorporated by allowing the model-specific choices of l_t and u_t to change as a function of nondimensional measures of stratification, e.g., the local value of Ri_g in stable shear flows.

B is often compared to other terms in the TKE balance to gauge the strength of buoyancy effects and also to constrain turbulence models. The flux Richardson number, $Ri_f = -B/P$, that was introduced in Chapter 1, is used in the stable ABL in connection with Monin–Obukhov theory, also introduced in Chapter 1, as a constraint on turbulence models. In high-Re statistically steady turbulence and ignoring transport for now, P is mostly balanced by ε so that Ri_f is not large, typically between 0.1 and 0.2. The flux and gradient Richardson numbers are related by the turbulent Prandtl number,

$$Ri_f = \frac{-B}{P} = \frac{(g/\rho_0)\langle \rho' w' \rangle}{-\langle u'w' \rangle \partial U/\partial z} = Ri_g \frac{K_\rho}{K_m} = \frac{Ri_g}{Pr_t} \tag{3.14}$$

If Ri_f stays between 0.1 and 0.2, Pr_t varies approximately as Ri_g. There is some evidence from laboratory and numerical experiments that Ri_f is bounded from above and does not exceed 0.2. Nevertheless, it is possible that the transport term, $-\partial T_i/\partial x_i$ in Eq. (3.5) transports energy from a "more turbulent" region to a stratified region with high Ri_g and low P. In this case, because of the additional source of turbulence through transport, Ri_f may exceed 0.2.

The *mixing efficiency* ($\hat{\eta}$), defined by

$$\hat{\eta} = \frac{\langle \varepsilon_\rho \rangle}{\langle \varepsilon \rangle + \langle \varepsilon_\rho \rangle} \tag{3.15}$$

is the fraction of total irreversible energy loss within the fluid that goes towards mixing the density field. In the ocean sciences community, $\hat{\eta}$ or the flux coefficient ($\gamma = -B/\varepsilon$) is widely used for relating B to measurements of ε (directly obtained with microstructure profilers or inferred from overturning regions in density profiles). Since,

$$K_\rho = -\frac{\langle \rho' w' \rangle}{\partial \langle \rho \rangle/\partial z} = -\frac{B}{N^2} = \gamma \frac{\varepsilon}{N^2}, \tag{3.16}$$

an assumed value of γ leads to a quantitative estimate of K_ρ. Osborn (1980) suggested $\gamma \leq 0.2$ based on the assumption of statistically steady, homogeneous turbulence where the TKE balance reduces to

$$P = \varepsilon - B \implies \gamma = -B/\varepsilon = \frac{Ri_f}{1 - Ri_f}, \qquad (3.17)$$

and additionally assuming that $Ri_{f,\max} = 0.16$. For concreteness, oceanic estimates of K_ρ and B have been made with $\gamma = 0.2$, but these estimates have uncertainty because several stratified flow examples have shown that γ can deviate significantly from 0.2 depending on other governing parameters such as Fr_t and Re_b, as well as the flow configuration.

3.4 Free shear flows

The competition between destabilizing vertical shear ($\partial u/\partial z$) and stabilizing buoyancy is a common aspect of turbulent shear flows. Planar free shear flows with mean vertical shear include mixing layers, jets, and plumes in a density-stratified environment, as well as internal waves that do not have a counterpart in a homogeneous fluid. This competition will be illustrated with the classic example of a mixing layer with inflectional du/dz in Section 3.4.1. The concept of marginal instability, which arises in the context of mixing layers that are subject to external forcing, will be introduced in Section 3.4.2. Homogeneous vertical shear—likely the simplest model of stratified shear flow—will be discussed in Section 3.4.3.

In a horizontal shear flow, the destabilizing shear is due to horizontal ($\partial u/\partial y$, $\partial v/\partial x$) gradients, e.g., in a shear layer formed by flow separation past a protruding element on the coastline, in one formed between two adjacent streams, or in a vertical flow with predominantly vertical vorticity (ω_z). Since the vertical buoyancy force is orthogonal to the horizontal plane of the shear flow, buoyancy effects in horizontal shear flow are somewhat different from those in the classical vertical shear cases of a mixing layer or homogeneous shear, as will be illustrated in Section 3.4.4.

3.4.1 Mixing layer with vertical shear

The temporally evolving mixing layer between two streams with opposite velocities and different densities is a prototypical example of a stratified shear layer. In a frame moving with the average velocity, there is a layer of inflectional shear with a jump (ΔU) in the streamwise (x) velocity over a vertical distance (δ), and a similar profile for the density *deviation*, $\rho(z)$, from the reference density (say, background density at $z = 0$):

$$u(z) = \frac{-\Delta U}{2}\tanh\left(\frac{2z}{\delta}\right) \quad ; \quad \rho(z) = \frac{-\Delta\rho}{2}\tanh\left(\frac{2z}{\delta}\right) \qquad (3.18)$$

The overall stratification measure is the bulk Richardson number,

$$Ri_b(t) = \frac{\Delta\rho g \delta(t)}{\rho_0 \Delta U^2} \qquad (3.19)$$

Vertical shear ($S(z)$) and buoyancy frequency ($N(z)$) of the initial profiles peak at the center ($z = 0$) and decrease to zero at the top and bottom boundaries. The minimum value of $Ri_g(z) = N^2(z)/S^2(z)$ occurs at $z = 0$ and is initially equal to Ri_b. For the $u(z)$ profile of Eq. (3.18), δ is also the vorticity thickness ($\delta_\omega = \Delta U/S_{max}$), often used as the thickness metric for turbulent mixing layers. The governing parameters are the initial values of N^2/S^2, the Reynolds number ($Re = \Delta U\delta/\nu$), and the molecular Prandtl number ($Pr = \nu/D$).

Linear stability analysis with the inviscid, infinite vertical domain assumption (Hazel, 1972), shows that the fastest growing wave, the Kelvin–Helmholtz (KH) mode, has a wavenumber that increases from $k\delta = 0.86$ to 1 as Ri_b increases from 0 to the stable value of 0.25. The nonlinear evolution of this flow has been the subject of several DNS studies, e.g., Caulfield and Peltier (2000); Smyth and Moum (2000); Mashayek and Peltier (2012) that were initialized with small-amplitude disturbances. The flow evolves through a sequence of instabilities. The energy at the KH wavelength grows exponentially, and the vorticity rolls up to form coherent KH billows that are connected by thin braids. The path to 3D turbulence proceeds through a sequence of secondary shear instabilities and convective instabilities associated with heavy fluid being transported locally over light fluid. Fig. 3.1 shows the progression to turbulence through qualitatively similar instabilities for a somewhat different initial configuration of constant N.

In applications, a shear layer may develop from an almost-quiescent background, as for a steepening internal wave in the ocean interior, but it may also develop in a turbulent background. Brucker and Sarkar (2007) utilized turbulent initial conditions and did not find quasi-2D KH billows in their simulations which spanned $0 \leq Ri \leq 0.1$. The local mixing efficiency, $\hat{\eta}(t)$, was consistently lower than the high values that have been found in the pre-turbulent stage realized with a laminar initial state. Kaminski and Smyth (2019) systematically increased the initial turbulence level, finding the suppression and eventual disappearance of KH billows that was noted by Brucker and Sarkar (2007). They found a progressive decrease in the cumulative mixing efficiency from about 0.3 to 0.13 with increasing level of initial turbulence.

In a homogeneous fluid, the turbulent shear layer grows linearly with time. In contrast, its stratified counterpart eventually stops growing, and turbulence collapses. This can be understood by taking the centerline value of Ri_g, equivalently Ri_b defined by Eq. (3.19), as the characteristic value of gradient Richardson number in the shear layer. Since $\delta(t)$ increases with time, so do Ri_b and the minimum Ri_g, and shear instability subsides eventually when Ri_g exceeds 0.25 everywhere in the flow. The amount of shear layer growth by entrainment is measured by the final $Ri_{b,f}$ for a fixed initial Ri_b. The governing parameters are the Reynolds number ($Re = \Delta U\delta(t = 0)/\nu$), the Richardson number ($Ri = Ri_b(t = 0)$) and the molecular Prandtl number ($Pr = \nu/\alpha$). Here, ν is the kinematic viscosity and α the thermal diffusivity. Brucker and Sarkar (2007) assessed the dependence of the final value of $Ri_{b,f}$ on its initial value as well as on Re, Pr and initial Ri. All cases exhibited an asymptotic constant value of $Ri_{b,f}$ that was case-dependent. These values spanned $Ri_{b,f} = 0.32 \pm 0.06$, the range that has been found in laboratory experiments and

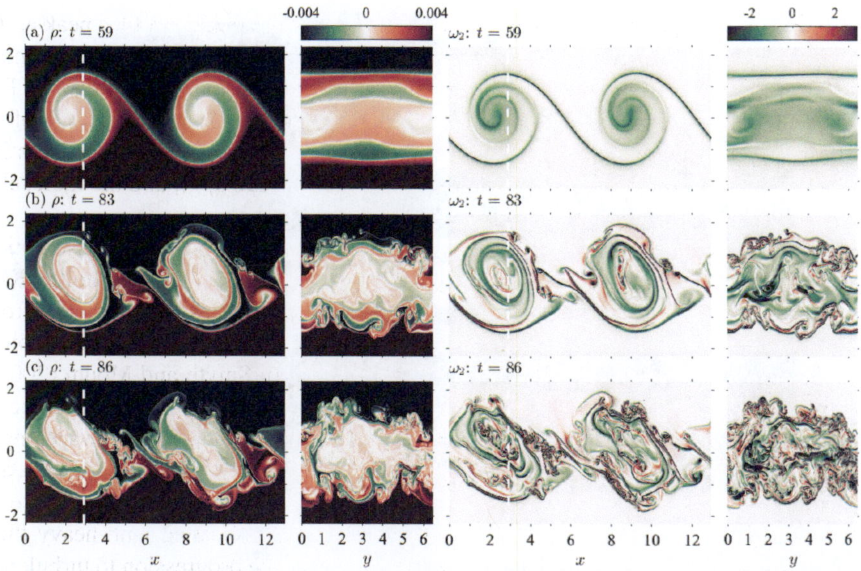

Figure 3.1 Progression of instabilities in the $Ri = 0.04$, $Re = 24,000$ vertical shear flow of VanDine et al. (2020). Cross-sectional snapshots of the density deviation (ρ) and spanwise vorticity (ω_2) fields: (a) $t = 59$, (b) $t = 83$, and (c) $t = 86$. Here t is in units of $\delta/\Delta U$. From left to right, the columns show: streamwise cross-section of ρ at $y = L_y/2$, spanwise cross-section of ρ along the dotted line shown in the first column, streamwise cross-section of ω_2 at $y = L_y/2$, and spanwise cross-section of ω_2 along the dotted line shown in the third column. Here, x, y, and z are the streamwise, spanwise, and vertical coordinates, respectively, made nondimensional using δ. From VanDine et al. (2020), reprinted with permission from Cambridge University Press.

other simulations, and showed a systematic dependence on the governing parameters. For example, $Ri_{b,f}$ was low at smaller values of Re, but Re-sensitivity became insignificant at sufficiently high Re. $Ri_{b,f}$ tended to decrease with increasing Ri. This effect, analogous to increased stability of the flow at high Ri, is related to a suppression of Reynolds shear stress and, thereby, TKE production.

When turbulence is sufficiently strong (buoyancy Reynolds number, $Re_b \geq O(10)$), the direct effect of molecular diffusion and thus molecular Pr on the energy-containing scales is weak. However, in the later stages of evolution, when turbulence has considerably weakened because $Ri_g(z,t)$ has increased sufficiently (say, up to $Ri_g = 0.2$), differential molecular diffusion between density and momentum can play a role. Net entrainment and the final value of Ri_b have been found to decrease with increasing Pr in the turbulent mixing layer (Brucker and Sarkar, 2007; Salehipour et al., 2015). The reduced mixing at high Pr is attributed to the sharper late-time $N^2(z)$ profiles associated with the weaker diffusion of density at high Pr. This Pr effect is likely relevant to oceanic mixing since the value of Pr (used interchangeably here with Schmidt number Sc) is about 700 in salt-stratified water.

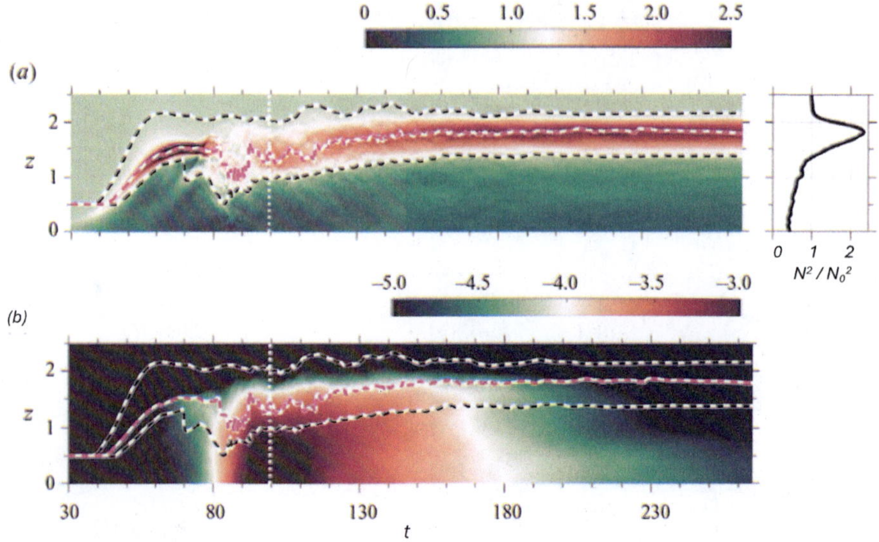

Figure 3.2 Time evolution of $N^2/N_0^2(z)$ (top) and $\varepsilon(z)$ (bottom) in the $Ri = 0.04$, $Re = 24,000$ vertical shear flow of VanDine et al. (2020). The upper half ($z \geq 0$) of the shear layer is shown. The inner (TL_i) and outer (TL_o) transition layer boundaries are each identified using a black dashed line, while the location of maximum N^2/N_0^2 inside the shear layer (TL_m) is shown with a magenta dashed line. Panel on the right side of the top row shows $N^2/N_0^2(z)$ at $t \approx 250$ when the turbulence has subsided. The dotted white lines at $t = 100$ indicate the time of maximum ε. From VanDine et al. (2020), reprinted with permission from Cambridge University Press.

Stratification and shear profiles in the environment are more complex than the two-stream configuration of Eq. (3.18). When the density layer is significantly thinner than the shear layer ($R = \delta_u/\delta_\rho > O(1)$), as may occur for a high-Pr stratified fluid, a Holmboe instability in the form of a pair of waves with equal and opposite-signed phase speed emerges. Nonlinear evolution of the Holmboe waves leads to cusp-like features with spanwise vorticity that break down by secondary instabilities. Although its initial growth rate is smaller than the KH growth rate, Holmboe instability lasts longer and is able to lead to overall mixing that is comparable to mixing by the KH instability (Smyth and Winters, 2003; Salehipour et al., 2016). Holmboe instability achieves mixing by scouring the edges of the density layer in a "slow burn" in contrast to the rapid mixing by the overturning of the entire layer thickness in a KH billow. Additionally, stratification can be asymmetric, e.g., its center can be vertically displaced with respect to the shear layer center (Carpenter et al., 2007; Olsthoorn et al., 2023), say by a height d, or its profile may be different, e.g., a thermocline between a low-stratification (buoyancy frequency N_1) and a high-stratification (N_2) region (Pham et al., 2014). Depending on the specific values of parameters, the asymmetrically stratified cases may be more like (symmetric) Holmboe or more like KH in their nonlinear evolution and mixing characteristics. Smaller d/δ_u tends to make the evolution more Holmboe-like and, for the thermocline profile, increasing the value of $N_2\delta_u/\Delta U$ makes for a more Holmboe-like instability.

Continuous stratification outside a region of localized shear is common in the ocean and introduces new effects. Energy and momentum can be transported out as propagating internal gravity waves, thus providing a nonlocal pathway to turbulent mixing in the interior. Pham et al. (2009) linked internal wave (IW) properties (phase lines, wavelength) quantitatively to the shear layer parameters in a problem with a tanh $N^2(z)$ profile that links a smaller N^2 in the sheared zone to a larger N^2 below it. The wave generation process is conceptualized as being forced by the corrugations at the edges of the shear layer that are introduced by flow instabilities and turbulence. This is roughly analogous to IW generation by a corrugated surface, e.g., wind-generated mountain waves. Internal waves have a frequency ω that satisfies $f < \omega < N$, implying stratification in the wave-propagation zone must be sufficiently large: $N > Uk_x$ where U is the velocity relative to the corrugated surface and k_x is the wavenumber of the corrugation. The radiated wave flux has been found to be a significant fraction (15%–30%) of the turbulent dissipation integrated across the shear layer.

A localized shear layer with $u(z)$ as in Eq. (3.18) but with a constant initial buoyancy frequency ($N(z) = N_0$) was considered by Watanabe et al. (2018); VanDine et al. (2020). The DNS of VanDine et al. (2020) was conducted at a high initial value of $Re = 24,000$ made possible by a $2304 \times 1152 \times 1408$ grid. The route to 3D turbulence from small-amplitude initial fluctuations was found to be similar to the well-studied two-stream problem with both velocity and density obeying Eq. (3.18). Fig. 3.1 is an illustration of the progression from a quasi-2D KH billow to 3D turbulence. Concerning IW, they are evanescent (do not reach and propagate in the far field) in this problem because N is not sufficiently large. Instead, IW is found in a limited region outside the turbulent zone, and the IW flux is a significant contributor to the local TKE balance. Transition layers (TL), defined as layers with elevated stratification (relative to the constant background N), form at the shear-layer edges (VanDine et al., 2020) as shown in Fig. 3.2 and exhibit local N values much larger than at the edges of the two-stream problem. The TL, which also has elevated shear, contains stratified turbulence with higher mixing efficiency than in the core and a substantial fraction (as large as 30%) of the time-integrated dissipation of potential energy. Thus, the TL contributes significantly to mixing.

3.4.2 Marginal instability

Suppose there is a localized region of vertical thickness δ with velocity difference ΔU and density difference $\Delta\rho$ that has sufficiently small Ri_g so that there is shear-driven turbulence. The resultant turbulent diffusion tends to increase $\delta(t)$ so that the local value of $Ri_g = N^2/S^2 \approx (\Delta\rho g/\rho_0 \Delta U^2)\delta(t)$ also increases with time until it exceeds the critical value of 0.25 so that shear production of turbulence is shut off. Once residual turbulence and its attendant mixing subsides, Ri_g does not increase anymore. If external forcing is present so that shear again increases locally, Ri_g can decrease to a subcritical value that permits shear instability and turbulence. The resulting mixing increases Ri_g to a value above 0.25, where it remains until the cycle restarts with another nudge towards instability from the next episode of external forcing. As a consequence of the feedback mechanisms between Ri_g and turbulence, the forced stratified shear layer has a propensity for Ri_g to lie in a relatively narrow band around the critical value of 0.25, i.e., in a state of *marginal instability*.

An excellent example of marginal instability is provided by the upper flank of the equatorial undercurrent (EUC) in the eastern Pacific basin, where Ri_g is found to cluster around 0.25 despite more than an order-of-magnitude variability in $N^2(z)$ and $S^2(z)$ (Smyth and Moum, 2013). The Pacific EUC is a long (several 100 km) and vertically thin (about 100 m) shear flow that is set up by the westward blowing trade winds and the eastward return flow. The trade winds provide a consistent surface forcing that is modulated by the solar heat flux throughout the diurnal cycle. Observations (Smyth et al., 2013) and simulations (Pham et al., 2013) show that marginal instability of the Pacific EUC is accompanied by turbulent mixing that occurs cyclically in nightly bursts of so-called deep cycle turbulence. Marginal instability of stratified shear flow has been extended to the notion that self-organized criticality is potentially a unifying feature of geophysical turbulence (Smyth et al., 2019).

3.4.3 Homogeneous vertical shear

Uniform shear or homogeneous shear, which is a flow where the initial velocity and density profiles are linear, has been the subject of DNS study (Gerz et al., 1989; Holt et al., 1992; Jacobitz et al., 1997; Shih et al., 2000). Since the turbulence is statistically homogeneous in all directions, the initial values of N, S, and Ri_g remain constant, and the statistics are solely a function of time. Unlike the mixing layer with inflectional shear, uniform shear flow is linearly stable (no KH instability), but it does support nonmodal transient growth from small perturbations, followed by nonlinear growth. The initial growth of turbulence during the period with small $Ri_g(t)$ in a mixing layer is inevitably followed by the rundown of turbulence when Ri_g becomes sufficiently large, and the flow becomes linearly stable. In contrast, uniform shear flow exhibits growth or decay of TKE separated by a critical value ($Ri_{g,s}$) where the TKE remains stationary in time. The TKE equation for $K(t)$ simplifies considerably for homogeneous shear flow to

$$\frac{dK}{dt} = P - \varepsilon + B \tag{3.20}$$

$$\frac{1}{\varepsilon}\frac{dK}{dt} = \frac{P}{\varepsilon} - 1 + \frac{B}{\varepsilon} \tag{3.21}$$

TKE exhibits growth/decay/stationarity depending on whether P exceeds/equals/is less than $\varepsilon - B$. Jacobitz et al. (1997) find $Ri_{g,s} = 0.16 - 0.17$ at high values of Reynolds number. For all initial values of Ri_g, nondimensional second-order statistics asymptote to time-invariant values. In particular, stationary turbulence at $Ri_{g,s}$ has the following properties: $-B/\varepsilon = \gamma \approx 0.2$, magnitude of b_{13} (Reynolds shear stress anisotropy) ≈ 0.1 and $SK/\varepsilon \approx 6$. As noted earlier, γ has been found to vary considerably across flows in experiments and simulations. It is worth noting that the value of $\gamma = 0.2$ found for $Ri_{g,s}$ in homogeneous shear flow is also the value often assumed by oceanographers to infer the buoyancy flux and eddy diffusivity from observations of ε.

In a recent DNS of homogeneous stratified shear flow (Portwood et al., 2019) with additional *forcing*, the TKE was kept stationary by a mass-spring-damper system that relaxes TKE towards an equilibrium value, and, operationally, the system is recast as a first-order control law for Ri_g. Irrespective of the value of the equilibrium TKE and kinematic viscosity, Ri_g was found to reach a value of $Ri_{g,s} \approx 0.16$ and $B/\varepsilon = 0.2$ very close to the corresponding values found without additional forcing.

It is worth noting that although homogeneous shear flow exhibits an equilibrium at $Ri_{g,s} \approx 0.16$ with stationary TKE, the specific equilibrium may depend on the nature of the flow and its forcing. For instance, the EUC example of Section 3.4.2 is a stratified shear flow where night-time turbulence is sustained by boundary forcing (wind and surface buoyancy flux). Here the values of the gradient Richardson number cluster around $Ri_g \approx 0.25$.

3.4.4 Horizontal shear

As the name implies, the mean shear in this type of flow arises from the horizontal gradient ($\partial U/\partial y = S_h$) of the mean velocity ($U$) instead of its vertical gradient ($\partial U/\partial z = S_v$). Horizontal shear is widely prevalent in coastal flows, coherent geophysical vortices, and at fronts between different water masses. Since gravity and stratification are orthogonal to the plane of the flow, three-dimensional simulations are necessary to capture buoyancy effects—there is no simpler 2D counterpart as in the vertical shear layer.

Jacobitz and Sarkar (1998) conducted DNS of uniform horizontal shear (HS). Two notable findings were as follows: (i) vertical transport and mixing efficiency were larger than for mean vertical shear of the same magnitude, and (ii) for sufficiently large stratification so that $N^2/S_h^2 \gtrapprox 2$, turbulence eventually decayed. In HS flow, energy is extracted by horizontal eddies that are not directly affected by buoyancy, and, because of the vertical variability of these eddies, a pressure gradient $\partial p/\partial z$ develops and leads to secondary motions with vertical velocity (w). Furthermore, the decoupling between layers with horizontal shear leads to instantaneous vertical shear, $\partial u/\partial z$, that promotes shear instabilities if the local value of $Ri_g = N^2/(\partial u/\partial z)^2$ is small. Suppose the energetic 3D motions of the unstratified homogeneous shear flow have a vertical length scale l_v. As N is increased, the local Ri_g becomes sufficiently large, and buoyancy is able to stabilize the flow. For constant S_h, it appears that such stabilization occurs at $N^2/S_h^2 \approx 2$, an order of magnitude larger than $N^2/S_v^2 \approx 0.16$ for its counterpart with constant S_v.

A mixing layer between two adjacent streams with velocity difference ΔU over a horizontal distance of δ in the lateral y-direction in a constant-N vertically stratified background is the horizontal-shear analog of the localized layer with vertical shear (Watanabe et al., 2018; VanDine et al., 2020) discussed earlier. A measure

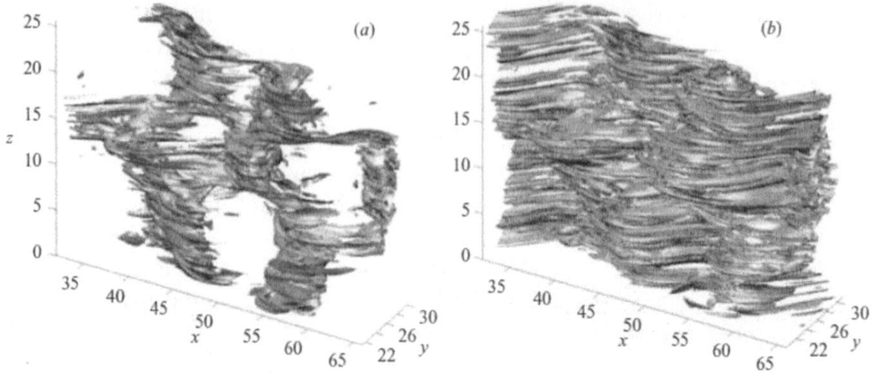

Figure 3.3 Horizontal mixing layer. Case A2 of Basak and Sarkar (2006) at a late time when $N^2/S_h^2(t) = 3.46$ and $Fr_h(t) = 0.16$: (a) magnitude of vertical vorticity $|\omega_z|$, (b) magnitude of horizontal vorticity $|\omega_h|$. From Basak and Sarkar (2006), reprinted with permission from Cambridge University Press.

of buoyancy effects is $N^2/S_h^2(t)$ defined with $S_h = \Delta U/\delta(t)$ that is the squared ratio of flow and buoyancy time scales. The flow, unconstrained by buoyancy in the horizontal, continues to spread horizontally over a long time. As the mixing layer spreads laterally, local u'_h progressively decreases, and the 3D overturning turbulent motions are damped by buoyancy. In the first DNS study of this configuration, Basak and Sarkar (2006) changed the initial value $N^2/S_h^2(t = 0)$ among cases and observed 3D turbulence with high mixing efficiency (calculated using y-integrated $\langle\varepsilon\rangle$ and $\langle\varepsilon_\rho\rangle$ in Eq. (3.15)) values of $\hat{\eta} = 0.3 - 0.4$ during the time interval over which $Ri(t)$ did not become too large—not greater than approximately 1.5. They found rich dynamics, including coherent vortices, vortical instabilities, internal wave emission, and intermittent lateral intrusions that exported mixed fluid into the surrounding ambient. A staggered system of columnar vortex cores emerged with segments of a characteristic vertical height l_v interacting between adjacent vortices. As buoyancy effects progressively strengthened, the segment height of $l_v \approx 6\Delta U/N$ became substantially smaller than the horizontal scale that kept increasing due to vortex pairing. In addition, there were diffusive-scale dislocations between the vortices that had even smaller vertical thickness $\approx 15\sqrt{D/N}$, with D the kinematic mass diffusivity. Fig. 3.3 (a) illustrates the segmented columnar vortex cores, and Fig. 3.3 (b) shows strips of horizontal vorticity that are mostly wound around the framework of vortex cores. These strips of horizontal vorticity are associated with layers of dissipation (Fig. 3.4 (a)) and the thin dislocations with viscous scale (Fig. 3.4 (b)) are evident at the low $Fr_h = 0.05$.

Recently, Lewin and Caulfield (2024) explored the evolution of a horizontally sheared mixing layer that is perturbed with small-amplitude $u(z)$, $v(z)$ having a prescribed vertical wavenumber k_z instead of small broadband perturbations used by Basak and Sarkar (2006). The initial Reynolds number was also an order of magnitude larger. The perturbations grow rapidly in the study of Lewin and Caulfield (2024)

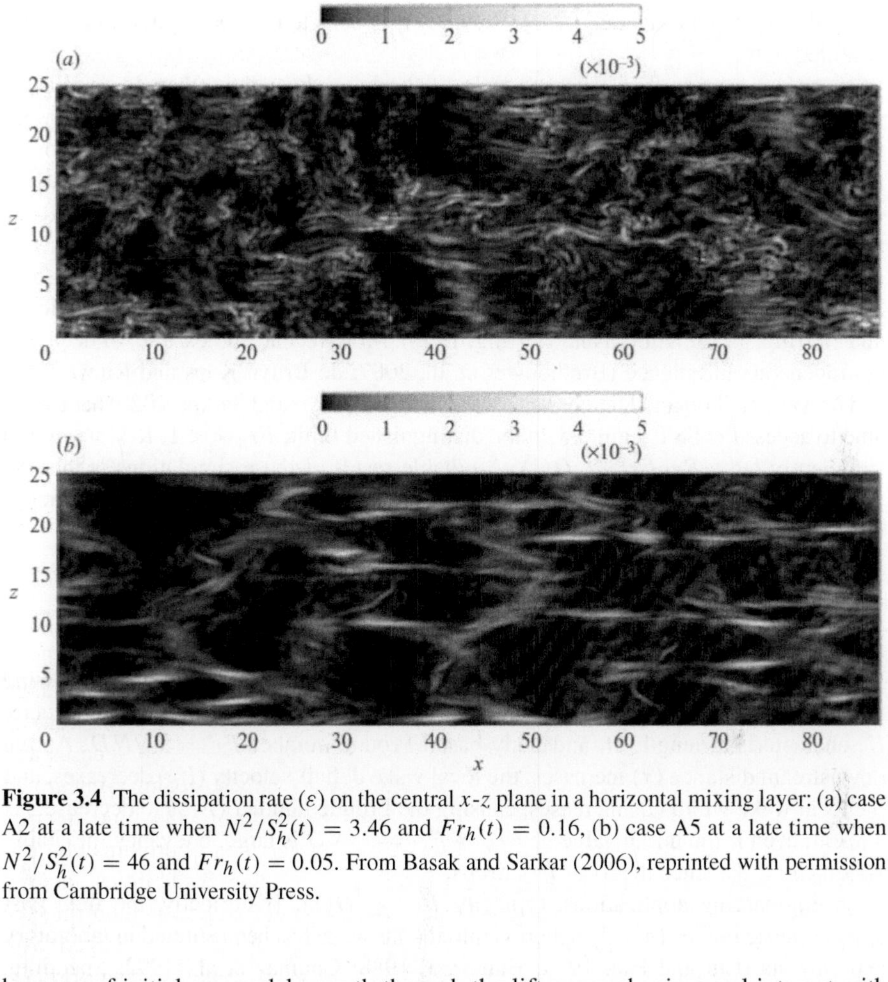

Figure 3.4 The dissipation rate (ε) on the central x-z plane in a horizontal mixing layer: (a) case A2 at a late time when $N^2/S_h^2(t) = 3.46$ and $Fr_h(t) = 0.16$, (b) case A5 at a late time when $N^2/S_h^2(t) = 46$ and $Fr_h(t) = 0.05$. From Basak and Sarkar (2006), reprinted with permission from Cambridge University Press.

because of initial nonmodal growth through the lift-up mechanism and interact with the developing shear instability of the mixing layer to facilitate turbulence. As a result, the flow exhibited an SST regime with $Re_b \geq O(1)$ and $Fr_h < 0.02$ for cases with relatively large N, e.g, $N^2/S_h^2(t = 0) = 1$ and 2.

3.4.5 Layering

Vertically thin layers of quasi-horizontal eddying motions which have $l_v << l_h$ constitute a generic feature of strongly stratified flows (Riley and Lelong, 2000; Caulfield, 2021). This feature is not only observed in experiments and simulations but also has theoretical support through scaling analysis of the inviscid Boussinesq form of the NS equations by Billant and Chomaz (2001), who show that, in the limit of $Fr_h = u_h/Nl_h << 1$, the equations are invariant with respect to zN/u_h. Here u_h and l_h characterize the horizontal velocity and its horizontal length scale. Thus a vertical length scale, namely $l_v = O(u_h/N)$, is selected in the regime of strongly stratified flow.

If ΔU is a prescribed velocity scale of the base shear flow, then $l_v = \Delta U/N$ is the natural choice for the vertical length scale for these eddying motions in flows such as the horizontal shear layer (Basak and Sarkar, 2006) discussed above, a vertically oriented counter-rotating vortex pair (Billant and Chomaz, 2000a,b), a co-rotating vortex pair (Otheguy et al., 2006), a Taylor-Green vortex array (Riley and deBruynKops, 2003) and a wake (Spedding, 2002) among others. These layers also arise spontaneously in decaying turbulence subject to stratification as seen in laboratory experiments (Fincham et al., 1996; Praud et al., 2005) and DNS (Métais and Herring, 1989; Kimura and Herring, 1996). More recent studies, e.g., of decaying homogeneous turbulence (Brethouwer et al., 2007; de Bruyn Kops and Riley, 2019) and of wakes (Zhou and Diamessis, 2019; Chongsiripinyo and Sarkar, 2020) have been able to access the SST regime with the distinguished limit: $Fr_h << 1$, $Re_h >> 1$ and sufficiently large $Re_h Fr_h^2 \geq O(1)$. A notable result of these simulations is that the theoretical scaling of $l_v = O(u/N)$ in SST is realized, and $Fr_v = u/Nl_v$ approaches an $O(1)$ constant.

3.5 Wakes

Engineered systems in a fluid—aerial or underwater vehicles, buildings, and wind turbines—generate wakes. Consider the wake behind a body with relative velocity U, characteristic length D, and body-based Froude number $Fr = U/ND$. As the downstream distance (x) increases, the local wake deficit velocity (U_d) decreases, and the local wake width (L) increases, causing the Froude number (Fr_h) to decrease. As a result, even if the initial value of $Fr_h \approx Fr = U/ND$ is large, buoyancy inevitably affects the wake since the local Fr_h drops.

In engineering applications, typically $Fr \geq O(1)$. Historically, the flow past axisymmetric bodies (mostly spheres) into the far wake has been studied in laboratory experiments (Lin and Pao, 1979; Hanazaki, 1988; Chomaz et al., 1992; Spedding, 1997), in body-exclusive simulations using a temporal wake approximation (Gourlay et al., 2001; Brucker and Sarkar, 2010; Diamessis et al., 2011; De Stadler and Sarkar, 2012; Abdilghanie and Diamessis, 2013) and, more recently, in body-inclusive simulations (Orr et al., 2015; Pal et al., 2017; Chongsiripinyo and Sarkar, 2020). Stratified wakes persist longer than their unstratified counterpart (Spedding, 2014) in terms of the defect velocity and also their coherent structures, such as the vortex shedding mode of a disk (Nidhan et al., 2022). The mean stratified wake (U_d and L) and turbulence statistics are typically expressed as a function of $Nt = Nx/D$. Spedding (1997) identified three regimes in the evolution of a stratified wake:

1. The three-dimensional (3D) regime: Until $Nt \approx 1$, buoyancy has a negligible influence on the wake.
2. The nonequilibrium (NEQ) regime: In this subsequent phase ($1 < Nt < 25-50$), the decay of U_d slows relative to the 3D regime and the anisotropic stabilization of turbulence begins, characterized by the preferential damping of vertical velocity

and vertical length scale. The power law governing this decay depends on the wake generator; for example, $U_d \sim x^{-0.25\pm0.04}$ for a sphere (Spedding, 1997) and a 6:1 prolate spheroid (Ortiz-Tarin et al., 2023), while $U_d \sim x^{-0.18}$ for a disk (Chongsiripinyo and Sarkar, 2020). For bodies with $Fr \sim O(1)$, the NEQ decay rate is preceded by a stage of oscillatory modulation of U_d linked to lee waves (Pal et al., 2017).

3. The quasi two-dimensional (Q2D) regime: In this final regime, the wake decay accelerates (e.g., $U_d \sim x^{-3/4}$). The Q2D wake progressively organizes into horizontally meandering vortices (Gourlay et al., 2001; Dommermuth et al., 2002; Brucker and Sarkar, 2010), taking a layered appearance (Spedding, 2002).

Recently, wake evolution has been studied from the perspective of the state of fluctuations, such as in $(Fr_h, ReFr_h^2)$ phase space, similar to mixing layers and decaying turbulence. As noted in section 3.4.5, recent wake simulations (Zhou and Diamessis, 2019; Chongsiripinyo and Sarkar, 2020) have accessed the SST regime, finding $l_v = O(u/N)$.

Geophysical wakes are familiar through the von Kármán vortices revealed by cloud cover over mountain ranges. The vortices of these high-Re wakes are accompanied by turbulence. It has been hypothesized that steep underwater seamounts and hills act as "stirring rods," constituting an important source of turbulent mixing in the ocean. For geophysical wakes, buoyancy effects become consequential close to or at the wake generator since $Fr < O(1)$ or $\ll O(1)$.

Most numerical studies of wakes have used ocean models, such as the hydrostatic ROMS model (Dong et al., 2007; Perfect et al., 2018; Srinivasan et al., 2021), in which the entire turbulence is modeled using an eddy diffusivity scheme. Recent high-resolution LES (Puthan et al., 2021, 2022) of an isolated conical model of a seamount has shown that stratified turbulence in the wake has large overall turbulent dissipation with vertically thin layers (height $\propto U/N$) of ε.

Rotational effects come into play for large obstacles away from the equator. Liu et al. (2025) systematically varied the Froude ($Fr = U/Nh$) and Rossby ($Ro = U/fD$) numbers of a conical seamount with base diameter D and height h. They found that turbulent dissipation and scalar mixing in the interior of the stratified wake are dominated by: (i) KH instability, especially within the oblique dislocations of Kármán vortices at low Fr, and (ii) Centrifugal/inertial instability (CI), which peaks at intermediate rotation rates (Rossby number order unity, submesoscale regime) and relatively weaker stratification. The KH instability at dislocations is reminiscent of turbulence in the dislocated vortex array of a horizontal shear layer (Basak and Sarkar, 2006) while the CI instability resembles that of rotating shear flows (Yanase et al., 1993; Kloosterziel et al., 2007; Arobone and Sarkar, 2012; Carnevale et al., 2013).

3.6 Wall-bounded shear flows

Wall-bounded flow has a near-wall layer where viscosity is important to the mean velocity variation and an outer layer where it is not. The near-wall layer has a large mean velocity shear, a small vertical scale (distance to the wall), and higher turbulence levels, in contrast to the weak shear and larger vertical length scales of the outer layer. Therefore for a given buoyancy difference (potential temperature difference, $\Delta\theta$) across the boundary layer, buoyancy effects (e.g., as measured by local Ri_g) are weaker in the near-wall layer even though the outer layer may be modified or controlled by buoyancy. Nevertheless, the inner and outer layers are coupled and, for sufficiently large $\Delta\theta$, the entire boundary layer can be strongly modified in the stable case. Friction velocity will be denoted by u_τ (common in the turbulence literature) or u_* (common in atmospheric/ocean sciences) interchangeably.

Night-time radiative cooling of the ground can create a near-ground stratified layer in the atmosphere. Winter-time conditions in polar regions and the advection of denser air from land over a cool ocean are other examples leading to a stable ABL. The stratified ABL differs from its neutral and convective counterparts, as reviewed by Mahrt (2014), who classified the stable ABL into two states: the *weakly stable boundary layer* and the *very stable boundary layer*. The weakly stable boundary layer, which exhibits continuous but weakened turbulence, obeys Monin–Obukhov similarity theory (MOST) introduced in Chapter 1, with its stability function differing from that in the neutral and convective cases. The very stable boundary layer occurs during weak winds and strong radiative cooling under clear skies. It exhibits *temporal intermittency* wherein ABL turbulence is significantly reduced for long time intervals of about 30 minutes to a few hours, and also *spatial intermittency* with large spatially localized patches of turbulence or laminar flow. Fig. 3.5, which will be discussed in Section 3.6.2, illustrates the spatial intermittency and temporal intermittency that occur under strong stability.

MOST is based on the Obukhov length,

$$L = -\frac{u_*^3}{\kappa B_0} = -\frac{u_*^3 \rho_0 c_p}{\kappa \beta g q_0} \tag{3.22}$$

where κ is the von-Kármán constant, u_* is the friction velocity, B_0 is the surface buoyancy flux corresponding to the applied surface cooling flux (q_0 that has a negative value), c_p is the specific heat at constant pressure, and β is the thermal expansion coefficient for air. Under MOST, the mean velocity and temperature gradients are given by

$$\frac{d\langle u\rangle}{dz} = \frac{u_*}{\kappa z}\phi_m(\zeta) \quad , \quad \frac{d\langle \theta\rangle}{dz} = -\frac{\theta_*}{\kappa z}\phi_h(\zeta) \tag{3.23}$$

where $\zeta = z/L$, u_* is the friction velocity and $\theta_* = q_0/u_*$ is its temperature analog, while ϕ_m and ϕ_h are the stability functions for momentum and heat transport, respectively. The values of κ range within $0.36 - 0.41$ in most studies. Stability is accounted for by a z-less scaling of mean shear since L is the relevant length scale, i.e., the expression for mean shear in outer-layer variables does not depend on z, which implies that ϕ_m must be a linear function of z. Empirical stability functions (Högström, 1996), formulated using $\kappa = 0.4$, for the stable dry atmosphere are as follows:

$$\phi_m(\zeta) = 1 + 5.3\zeta, \ \ \phi_h(\zeta) = 1 + 8\zeta \quad \text{for } \zeta < 0.5 \tag{3.24}$$

Högström (1996) notes that there is variability in the pre-multipliers of the ζ dependence in various observational datasets and their values in Eq. (3.24) is an

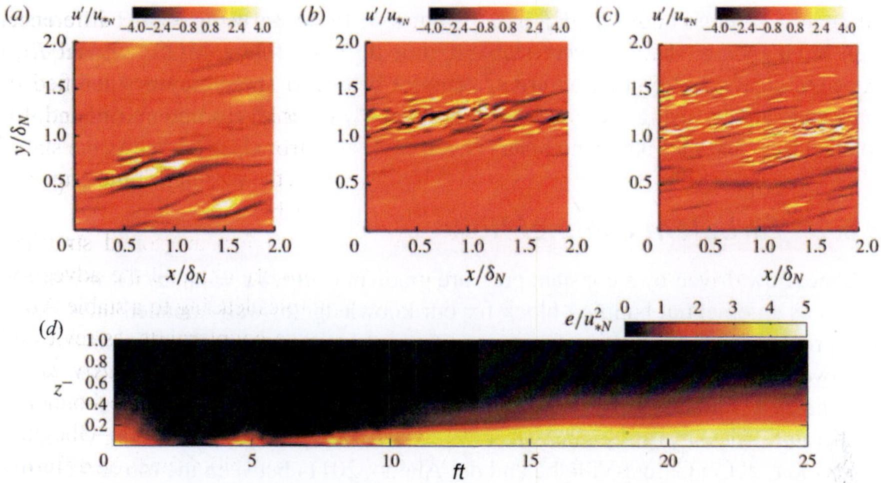

Figure 3.5 A visual representation of the key concepts of spatial and temporal intermittency in the very stable boundary layer, as discussed in the text. Spatial intermittency (top row of panels) and temporal intermittency (bottom panel) in a stratified Ekman boundary layer. Instantaneous streamwise velocity fluctuations at a near-surface location ($z^+ = 15$) in three cases with $L^+ = 700$ and different Reynolds number: (a) case with $Re_* = 700$ at $ft = 2.2$, (b) case with $Re_* = 1100$ at $ft = 1.87$ and (c) Case with $Re_* = 1700$ at $ft = 1.21$. Panel (d) shows the temporal evolution of TKE profiles in the case with $Re_* = 1100$. From Gohari and Sarkar (2018), reprinted with permission from Cambridge University Press.

average among the datasets. Integration of Eq. (3.23) with these stability functions leads to a *log-linear* profile for the mean velocity and temperature.

The very stable ABL has distinctive features. One is the super-geostrophic velocity at shallow height as in a low-level jet (Smedman et al., 1993; Cuxart et al., 2000; Banta et al., 2002) with intensified shear. Another is the collapse of surface-layer turbulence for long periods when the near-surface layer appears to decouple from the upper regions of the ABL (Poulos et al., 2002; Banta et al., 2007). Large values of the bulk Richardson number (Ri_b) are indicative of strong stability, e.g., periods with $Ri_b > 2$ in the Cooperative Atmosphere-Surface Exchange Study in 1999 (CASES-99) observational campaign corresponded to apparent decoupling of the surface layer (Poulos et al., 2002). TKE changes qualitatively to a twin-peaked profile with two layers – one above and the other below a local minimum of TKE (Banta et al., 2007; Cuxart and Jiménez, 2007).

Environmental boundary layers are often bounded on one side by a stable stratification. For example, the ABL is generally capped by the stratification of a thermal inversion. The bottom boundary layer (BBL) on the ocean floor is subject to the overlying stratification of the ocean interior. The ocean surface boundary layer is capped by the underlying stratified water column. In each of these three examples, entrainment from the stratified region into the boundary layer is accomplished by stratified, sheared turbulence.

There is a large body of LES-based studies of the ABL using MOST as a near-wall turbulence model. The very stable regime does not follow MOST and requires fine numerical resolution to capture its complex behavior. Therefore we will illustrate buoyancy effects using two canonical problems with high-resolution simulations: channel flow and the Ekman boundary layer.

3.6.1 Turbulent channel flow

Channel flow driven by a constant pressure gradient ($-\partial p/\partial x = \rho_0 u_\tau^2/h$), which has served as an essential building block for our knowledge of wall-bounded turbulence, also provides a simple setting for its stratified counterpart since the friction velocity and overall boundary-layer height are fixed, irrespective of the stratification level. Stratification has been considered in closed or open channel flow with a fixed temperature (density) difference (Komori et al., 1983; Garg et al., 2000; Armenio and Sarkar, 2002; García-Villalba and del Álamo, 2011) between the top and bottom boundaries, open channel flow with a cooling flux at the bottom (Flores and Riley, 2011) or a warming heat flux at the top (Taylor et al., 2005; Issaev et al., 2022).

Consider neutral turbulent channel flow at $Re_\tau = u_\tau h/\nu$ with a ΔT (equivalently, $\Delta\rho$ or Δb) difference across its height of $2h$. The friction Richardson number (Ri_τ) and the bulk Richardson number (Ri_b) are defined by

$$Ri_\tau = \frac{\Delta\rho g h}{\rho_0 u_\tau^2} = \frac{\Delta b h}{u_\tau^2} \quad , \quad Ri_b = \frac{\Delta b h}{2U_b^2} \tag{3.25}$$

where U_b is the bulk (average) velocity. Ri_τ and Ri_b are overall measures of buoyancy; typically, Ri_τ and Re_τ are varied in parametric studies. Insofar as $Ri_g(z)$, it takes a small value at the bottom wall. A log-law region would exhibit $Ri_g \propto z/h$ because of the $1/z$ dependence of $S(z)$ and $N^2(z)$ in that region.

The surface value of the buoyancy gradient, $(\partial\langle b\rangle/\partial z)_0$, is related to Δb through the Nusselt number,

$$Nu = \frac{(\partial\langle b\rangle/\partial z)_0}{\Delta b/2h}$$

and the surface buoyancy flux is

$$B_0 = D(\partial\langle b\rangle/\partial z)_0 = \frac{\nu}{Pr} Nu \frac{\Delta b}{2h}$$

where D is the scalar diffusivity (say, temperature for thermal stratification), leading to the following useful relationships for the Obukhov length in outer and inner scales as well as the surface Richardson number

$$L^- = \frac{L}{h} = \frac{2Pr\, Re_\tau}{\kappa Nu\, Ri_\tau} \quad , \quad L^+ = \frac{L}{\nu/u_\tau} = L^- Re_\tau \quad , \quad Ri_{g,0} = \frac{N_0^2}{S_0^2} = Pr(\kappa L^+)^{-1} \tag{3.26}$$

It can be inferred from Eq. (3.26) that an increase of Re_τ would need an increase of Ri_τ to keep either estimate of nondimensional L constant and thus have some similarity in buoyancy effects on the turbulence. Ri_b can also be related to friction-based quantities by

$$Ri_b = \frac{Ri_\tau C_f}{4} \quad , \quad C_f = 2\left(\frac{u_\tau}{U_b}\right)^2 \tag{3.27}$$

Since, C_f is small (e.g., found to be 5.9×10^{-3} in DNS of a $Re_\tau = 550$ channel) in neutral flow and decreases significantly with increasing Ri_τ, it follows that Ri_τ has to be significantly large for the $Ri_b = O(1)$ values that translate to significant stabilization of free shear flows.

Based on the studies reviewed below, the effect of buoyancy on stratified channel flow depends significantly on the Richardson number (Ri_τ) and Reynolds number (Re_τ). At moderate stratification (lower Ri_τ), there is an unaffected/weakly affected near-wall region while the central region experiences strongly reduced turbulent fluxes and internal wave activity. There is scale-selective suppression of turbulence. At strong stratification (higher Ri_τ), there is direct near-wall suppression, spatial intermittency, and temporal collapse/rebirth of turbulence. We now expand on this summary of stratified channel turbulence.

Armenio and Sarkar (2002) found an initial transient where TKE decreases to a small value (almost zero for large Ri_τ), which in all cases is followed by recovery to a new flow state. At $Re_\tau = 180$ and moderate $Ri_\tau \leq 60$, near-wall streaks are unaffected, and there is a large region of essentially unstratified or weakly-stratified turbulence, but the buoyancy-controlled central core has strongly reduced turbulent fluxes and also internal waves. For larger $Ri_\tau \geq 120$, the central region reaches towards the wall and reduces near-wall turbulence production. For all cases, the correlation coefficient between u' and w' exhibits a remarkable decrease in the region with $Ri_g(z) = 0.15 - 0.2$, which is also the region where the slope of the $Ri_g(z)$ curve increases sharply from its near-wall constant value. At a higher $Re_\tau = 550$ investigated by García-Villalba and del Álamo (2011), all cases with $Ri_\tau \leq 960$ have the near-wall streaks of u that are typical of wall-bounded turbulence, a near-wall region ($z/h < 0.1$) with unstratified turbulence, and a region ($0.1 < z/h < 0.75$) with weakly-stratified turbulence that has $Re_b >> 1$ and $Fr_h = 0.2 - 1$. In agreement with Armenio and Sarkar (2002), the following features were found: (1) the central core had sharply reduced turbulent fluxes, internal wave activity and viscous stratified turbulence with $Re_b < O(1)$ and $Fr_h < 0.1$, (2) the $Ri_g(z)$ profile departed sharply from linear behavior at $Ri_g = 0.15 - 0.2$, (3) $Ri_f \approx Ri_g$ until Ri_g exceeded about 0.2, and (4) C_f and Nu decreased substantially with increasing Ri_τ. The $Re_\tau = 550$ case needs larger values of Ri_τ (about 1000 instead of 100 for $Re_\tau = 180$) to have comparably strong buoyancy effects on turbulence. This is consistent with the inference made from and stated after Eq. (3.26).

The $Ri_\tau = 1980$ case of García-Villalba and del Álamo (2011) has $Ri_b = 0.46$ and shows disruption of near-wall turbulence production, leading to spatial intermittency – large laminar/turbulent patches. They also examined 2D spectra, $E_{uu}(\lambda_x, \lambda_y)$ and their v and w counterparts. The global modes or very large structures with $\lambda_x > 10h$ and $\lambda_y > 2h$ progressively disappear with increasing stratification because of the suppression of vertical transport. However, the spectra at local scales in the near-wall region (the overlap region with the incipient log law) are unaffected (weakly affected) by buoyancy.

Flores and Riley (2011) applied a constant surface buoyancy flux (B_0) to the bottom wall of a neutral, turbulent open channel at $Re_\tau = u_\tau h/\nu = 360 - 560$, and parametrically varied $L^+ = Lu_\tau/\nu$ (equivalently $L^- = L/h$). The surface layer responded at a time scale of L/u_τ, exhibiting reduced levels of w', TKE production, and TKE. For sufficiently large B_0, there was turbulence collapse with TKE reaching almost zero. Importantly, Flores and Riley (2011) found that the criterion for the initial collapse is a sufficiently small L^+, when there is insufficient scale separation between the Obukhov and viscous scales, and not L^-. Specifically, collapse occurred when the local value $L^+(t) \lesssim 100$ for a case with initial $L^+ = 683$ and did not occur for cases with larger L^+. This collapse criterion has been found later to be consistent with results from the constant-Δb channel flows and also the Ekman boundary layer studies in Section 3.6.2. Spatial intermittency with co-existing laminar/turbulent patches was observed during the turbulence collapse in the case with $L^+ = 683$, followed by recovery to a fully turbulent state.

3.6.2 Ekman boundary layer

The Ekman boundary layer (EkBL) is the boundary layer that forms when a wind (velocity U_∞ in the x direction) blows over a flat horizontal plane (ground) in a frame that rotates around the vertical axis. As noted in Chapter 1, the wind aloft is in geostrophic balance, i.e., the sideways Coriolis force is balanced by a lateral pressure gradient. A stratified EkBL with a stabilizing temperature (density) difference across it is a simplified model of the stable ABL that ignores complications introduced by terrain roughness, vertical subsidence, lateral buoyancy gradients, etc. By including rotation and momentum/ scalar exchange with the ambient in a top-down coupling, the stratified EkBL introduces greater realism relative to channel flow in representing the stable ABL.

Constant-temperature stability, wherein a potential temperature difference ($\Delta\theta_0$) across the boundary layer is fixed, has been employed to simulate EkBLs with weak to strong stability (Ansorge and Mellado, 2014; Shah and Bou-Zeid, 2014; Deusebio et al., 2014). Here the buoyancy-related control parameter is the bulk Richardson number (Ri_b defined with U_∞ and not the bulk velocity U_b, as in channel flow), and the other control parameter is Re_*:

$$Ri_b = \alpha g \delta_N \frac{\langle \Delta\theta_0 \rangle}{U_\infty^2} \quad , \quad \delta_N = u_{*N}/f \quad , \quad Re_* = \frac{u_{*N}\delta_N}{\nu} \tag{3.28}$$

where u_{*N} is the friction velocity of the neutral EkBL and δ_N is the scale for the neutral EkBL height. (The actual neutral EkBL height is $\approx 0.5\delta_N$ and reduces with increasing stratification.) The target Re_* is achieved by specifying U_∞ appropriately, see Coleman et al. (1990). For the alternative boundary condition of constant-flux stability investigated by Gohari and Sarkar (2017, 2018), the surface buoyancy flux (B_0) and thus the Obukhov length (L) is specified, and $L^+ = Lu_{*N}/\nu$ (equivalently, $L^- = L/\delta_N$) is parametrically varied. Here, $Ri_b(t)$ defined by Equation (3.28) is no longer constant but is proportional to the time-evolving temperature (buoyancy) decrease at the ground.

Ansorge and Mellado (2014) conducted DNS of the EkBL at $Re_* = 478-1399$ and a wide range of $Ri_b = 0.015-0.62$ in large computational boxes. Based on the change in the vertically integrated TKE (say, ΔE) relative to the neutral case, they proposed the following classification: (i) Weakly stable with $\Delta E < 10-20\%$ for $Ri_b = 0.015$, 0.031; (ii) Intermediately stable with significant ΔE of 50% for $Ri_b = 0.15$ and 0.31; and (iii) Very stable with almost complete suppression of ΔE, followed by recovery for $Ri_b = 0.62$. The intermediately stable cases exhibited hodographs ($\langle v \rangle$ plotted against $\langle u \rangle$ for various heights) with Ekman spirals that were between the neutral and laminar cases, and the surface veering angle of about 22° increased towards the laminar value of 45°. MOST was found to apply for $\zeta = z/L < 0.1$, corresponding to cases with $Ri_b \leq 0.15$, and the stability function was found to be $\phi_m(\zeta) - 1 = 5.7\zeta - 0.04$, close to that in Eq. (3.24) that was deduced from ABL data. The very stable case with $Ri_b = 0.62$ exhibited spatial intermittency—laminar/turbulent patches—similar to very stable channel flow.

Shah and Bou-Zeid (2014) also covered a wide range of $Re_* = 338 - 1189$ and $Ri_b = 0.052 - 1.934$. In common with Ansorge and Mellado (2014), the moderately stratified cases showed continuous turbulence, e.g. $Ri_b \leq 0.39$ at $Re_* = 1189$, MOST scaling for the mean velocity, and deviations from the turbulent Ekman spiral including a super-geostrophic velocity maximum. When Ri_b exceeded a threshold, a very stable regime was reached, and turbulence statistics showed large oscillations. The value of the threshold Ri_b increased with increasing Re_*, similar to what has been seen for the threshold Ri_τ in stratified channels. Shah and Bou-Zeid (2014) examined the effect of buoyancy on the budgets of velocity and density variances as well as turbulent fluxes. The damping of vertical motion initiates buoyancy-induced changes in these budgets, but the stability-induced reduction of TKE was primarily due to the reduction of shear production and not the direct destruction term (buoyancy flux), similar to homogeneous shear flow.

Gohari and Sarkar (2017) investigated the stable EkBL at $Re_* = 1121$ under an applied B_0 corresponding to $L^- = L/\delta_N = 0.5$, 1 and 2. There were some differences in the very stable cases with $L^- = 1$ and 0.5 with respect to the constant-temperature stability flows: the super-geostrophic peak of the velocity akin to a low-level jet was stronger, there were *multiple episodes of collapse/rebirth* of turbulence during the transient, and there was a twin-peaked profile of TKE as in the observations of the very stable ABL. The rebirth of turbulence was associated with coupling between a subcritical ($Ri_g < 0.25$) region near the nose of the jet and residual turbulence in the outer layer through pressure transport, followed by energization of near-wall turbulence by shear instability at the jet. After a long time of $ft = 30$, the flow exhibited turbulence with spatial intermittency, there were strong inertial oscillations, and there was no final quasi-steady state of the statistics. Application of B_0 over a finite time of $fT = 6$ by Gohari and Sarkar (2018) allowed a quasi-steady state to be reached during the simulation that extended up to $fT = 30$. Three cases with the same $L^+ \approx 700$ but different Re_* and therefore different $L^- \approx 1, 0.62, 0.4$ were simulated and the final Ri_b values (≈ 0.5, 1 and 2) were different. For all cases with $L^+ \approx 700$, there was an initial collapse of turbulence with a time scale $\propto L/u_*$, as in the channel flow with $L^+ \approx 700$ (Flores and Riley, 2011). During the collapse, the flow exhibited spatial intermittency with laminar/turbulent patches as shown in Fig. 3.5, top row. Temporal intermittency with episodic cessation of turbulence is illustrated by Fig. 3.5 (d) that shows two episodes of collapse/rebirth of turbulence. It can be seen that the rebirth originates with an increase in fluctuation energy away from the wall. The energy is transported towards the wall and ignites wall turbulence. It is Ri_b and not L^+ that determines the final state of the flow. The case with $Ri_b = 2$ remained laminar while the other cases showed temporally continuous turbulence at late time. The case with $Ri_b = 2$ exhibited super-geostrophic velocity and large shear, but, because of the large N in the surface layer, the flanks of the jet were supercritical ($Ri_g > 0.25$) and did not support fluctuations that could be communicated to the near-wall layer.

The oceanic BBL is a turbulent boundary layer on a nominally adiabatic seafloor that is capped by an overlying stratified environment, say, with buoyancy frequency N_∞. Buoyancy effects are introduced by the overlying stratification rather than by wall cooling, as in the examples discussed so far. These effects are governed by the environmental parameter, N_∞/f. A simple surrogate for the BBL on a flat seafloor is a bottom Ekman layer under an outer stratified geostrophic flow, which has been studied using DNS by Taylor and Sarkar (2007) and LES with a near-wall model by Taylor and Sarkar (2008). The flow develops a three-layer structure: a bottom mixed layer, a capping pycnocline, and a uniformly stratified external layer with propagating internal gravity waves. The height, h, of the turbulent boundary layer is found to approximately follow the following scaling law (Zilitinkevich and Esau, 2002) that is based on numerical simulations and observations of a stable ABL: $h = C(u_*/f)(1 + \alpha N_\infty/f)^{-1/2}$ with $C = 0.5$ and $\alpha = 0.15$. Although the turbulent boundary layer thickness decreases substantially, the Ekman velocity component increases, so that the effect on net Ekman transport is insignificant.

The external stratification (N_∞) introduces a departure from the log-law that increases the mean shear relative to the log-law (Perlin et al., 2005). Taylor and Sarkar (2008) modify the length scale, $l = \kappa z$, in the expression for mean shear leading to the following *log-linear* behavior:

$$\frac{U(z)}{u_*} = \frac{1}{\kappa} \log(z/z_0) + z \frac{N_\infty}{C_b u_*} \tag{3.29}$$

It is noted that the authors took $N(z)$ as a constant in the stratified, sheared region—a considerable simplification—to obtain the simple analytical expression. The mean profile given by Eq. (3.29) can be viewed as the oceanic boundary layer analog to the log-linear profile for the stratified ABL.

3.7 Conclusions

An essential aspect of stratified shear flows is the reduction of turbulence intensities and mixing by buoyancy. The reduction is anisotropic so that vertical intensity and vertical length scales are preferentially damped. Such reduction is larger at high Richardson number (Ri_g, Ri_b) for vertical shear flow. The critical value of $Ri_{g,cr} = 0.25$, obtained from linear, inviscid modal stability analysis, is useful since local shear production of turbulence is increasingly inhibited when Ri_g approaches and increases beyond 0.25. However, Ri_g that is locally smaller or larger than $Ri_{g,cr}$ cannot be used as a rigorous criterion for the presence or absence of turbulence.

The flow Reynolds number plays a role, since increasing the buoyancy Reynolds number (Re_b) expands the range of length scales that is locally unaffected by buoyancy. It was also shown that flows with horizontal mean shear exhibit important differences with respect to vertical mean shear of the same magnitude, e.g., the ability to force turbulence at higher stratification and with higher mixing efficiency. The regime of SST (strongly stratified turbulence) with $Fr_h << 1$ and $Re_b > 1$ (or $Re_b >> 1$ for a well-defined inertial range) was introduced as a distinguished limit of high-Re stratified turbulence.

Several distinctive characteristics of stratified shear flows were identified. *First*, the stabilization is not homogeneous in time or space. Free shear flows with vertical shear thicken by turbulent entrainment and therefore, notwithstanding being initially $<< 1$, the values of Ri_g and Ri_b increase until buoyancy effects set in and eventually the vertical spread is suppressed. For instance, the final $Ri_b \approx 0.32 \pm 0.06$ in a simple mixing layer. In flows with vertical or horizontal shear, Fr_h continuously decreases and buoyancy comes into play when $Fr_h = O(1)$. Furthermore, stabilization in wall-bounded flow is strongly inhomogeneous. Although the outer layer with weak mean shear is stabilized, the larger shear and lower Ri_g of the near-wall and log-layer regions can prevent any significant changes from canonical boundary-layer turbulence.

Second, strong stratification introduces space-time intermittency in a sense that is different from the usual internal or external intermittency of turbulence. In particular, within the flow at any given time, distinct localized patches of turbulence may co-exist with laminar-like regions. These patches are germane to the mixing accomplished in the regime of strongly stratified turbulence (SST). The patch geometry is anisotropic, often forming *layers* of vertical thickness $l_v \propto u/N << l_h$, the horizontal scale. Similarly, as a function of time, the entire field of turbulent fluctuations may collapse and then re-emerge at a later time as has been seen in turbulent wall-bounded flows and the ABL when $Ri_b \geq O(1)$.

Third, stratified environmental flows are rich with coherent vortices and internal waves that co-exist with broadband turbulence. On the one hand, vortices and waves support a broad range of flow instabilities that break down to turbulence. On the other hand, vortices and waves emerge as organized structures during the evolution of a turbulent flow in a stable background. Thus the coupling between vortices, waves, and stratified turbulence is integral to a complete analysis of turbulent flows in a stratified environment.

Some simple examples, with and without solid boundaries, were chosen to illustrate buoyancy effects on shear-driven turbulence in a stable background. Environmental shear flows often involve one or more additional effects: earth's rotation; turbulent convection; oscillation, as in tides and internal waves; unsteady mechanical and buoyancy forcing on multiple time scales; multiscale surface roughness from bumps to mountains on land or on the seafloor; the deformable air-sea interface; multiphase flow, as in sediment-laden flow, breaking surface waves, clouds and the wet atmospheric boundary layer; complex geometry in nature, as in overland and underwater topography; and complex geometry in engineered systems such as wind or marine turbines, aerial or marine vehicles, and the urban landscape. Following chapters in this book deal with some of the aforementioned effects.

There has been much progress in our understanding of the fundamentals of stratified turbulence. Nevertheless, the properties of SST, the mixing associated with it, and its coupling with vortices and internal waves remain an outstanding issue, especially at high Reynolds numbers. One or more effects of rotation, horizontal buoyancy contrast, oscillation, and multiscale forcing acting in conjunction with a stable background are also important topics for further research in environmental turbulent flows.

Acknowledgment

I am pleased to acknowledge the assistance of Dr. Hieu Pham who provided comments on a draft of this chapter.

References

Abdilghanie, A.M., Diamessis, P.J., 2013. The internal gravity wave field emitted by a stably stratified turbulent wake. Journal of Fluid Mechanics 720, 104–139.

Ansorge, C., Mellado, J.P., 2014. Global intermittency and collapsing turbulence in the stratified planetary boundary layer. Boundary-Layer Meteorology 153 (1), 89–116.

Armenio, V., Sarkar, S., 2002. An investigation of stably-stratified turbulent channel flow using large eddy simulation. Journal of Fluid Mechanics 459, 1–42.

Arobone, E., Sarkar, S., 2012. Evolution of a stratified rotating shear layer with horizontal shear. Part I. Linear stability. Journal of Fluid Mechanics 703, 29–48.

Banta, R., Newsom, R., Lundquist, J., Pichugina, Y., Coulter, R., Mahrt, L., 2002. Nocturnal low-level jet characteristics over Kansas during cases-99. Boundary-Layer Meteorology 105 (2), 221–252.

Banta, R.M., Mahrt, L., Vickers, D., Sun, J., Balsley, B.B., Pichugina, Y.L., Williams, E.J., 2007. The very stable boundary layer on nights with weak low-level jets. Journal of the Atmospheric Sciences 64, 3068–3090.

Basak, S., Sarkar, S., 2006. Dynamics of a stratified shear layer with horizontal shear. Journal of Fluid Mechanics 568, 19–54.

Billant, P., Chomaz, J.-M., 2000a. Experimental evidence for a new instability of a vertical columnar vortex pair in a strongly stratified fluid. Journal of Fluid Mechanics 418, 167–188.

Billant, P., Chomaz, J.-M., 2000b. Theoretical analysis of the zigzag instability of a vertical columnar vortex pair in a strongly stratified fluid. Journal of Fluid Mechanics 419, 29–63.

Billant, P., Chomaz, J.-M., 2001. Self-similarity of strongly stratified inviscid flows. Physics of Fluids 13 (6), 1645–1651.

Brethouwer, G., Billant, P., Lindborg, E., Chomaz, J.-M., 2007. Scaling analysis and simulation of strongly stratified turbulent flows. Journal of Fluid Mechanics 585, 343–368.

Brucker, K., Sarkar, S., 2007. Evolution of an initially turbulent stratified shear layer. Physics of Fluids 19, 101105.

Brucker, K.A., Sarkar, S., 2010. A comparative study of self-propelled and towed wakes in a stratified fluid. Journal of Fluid Mechanics 652, 373–404.

Carnevale, G., Kloosterziel, R., Orlandi, P., 2013. Inertial and barotropic instabilities of a free current in three-dimensional rotating flow. Journal of Fluid Mechanics 725, 117–151.

Carpenter, J.R., Lawrence, G.A., Smyth, W.D., 2007. Evolution and mixing of asymmetric holmboe instabilities. Journal of Fluid Mechanics 582, 103–132.

Caulfield, C.P., 2021. Layering, instabilities, and mixing in turbulent stratified flows. Annual Review of Fluid Mechanics 53, 113–145.

Caulfield, C.P., Peltier, W.R., 2000. The anatomy of the mixing transition in homogeneous and stratified free shear layers. Journal of Fluid Mechanics 413, 1–47.

Chomaz, J.M., Bonneton, P., Butet, A., Perrier, M., Hopfinger, E.J., 1992. Froude number dependence of the flow separation line on a sphere towed in a stratified fluid. Physics of Fluids 4 (2), 254–258.

Chongsiripinyo, K., Sarkar, S., 2020. Decay of turbulent wakes behind a disk in homogeneous and stratified fluids. Journal of Fluid Mechanics 885, A31.

Coleman, G.N., Ferziger, J.H., Spalart, P.R., 1990. A numerical study of the turbulent Ekman layer. Journal of Fluid Mechanics 213, 313–348.

Cuxart, J., Jiménez, M.A., 2007. Mixing processes in a nocturnal low-level jet: an LES study. Journal of the Atmospheric Sciences 64, 1666–1679.

Cuxart, J., Yagüe, C., Morales, G., Terradellas, E., Orbe, J., Calvo, J., Fernández, A., Soler, M., Infante, C., Buenestado, P., Espinalt, A., Joergensen, H., Rees, J., Vilá, J., Redondo, J., Cantalapiedra, I., Conangla, L., 2000. Stable atmospheric boundary-layer experiment in Spain (SABLES 98): a report. Boundary-Layer Meteorology 96 (3), 337–370.

de Bruyn Kops, S.M., Riley, J.J., 2019. The effects of stable stratification on the decay of initially isotropic homogeneous turbulence. Journal of Fluid Mechanics 860, 787–821.

De Stadler, M.B., Sarkar, S., 2012. Simulation of a propelled wake with moderate excess momentum in a stratified fluid. Journal of Fluid Mechanics 692, 28–52.

Deusebio, E., Brethouwer, G., Schlatter, P., Lindborg, E., 2014. A numerical study of the unstratified and stratified Ekman layer. Journal of Fluid Mechanics 755, 672–704.

Diamessis, P.J., Spedding, G.R., Domaradzki, J.A., 2011. Similarity scaling and vorticity structure in high-Reynolds-number stably stratified turbulent wakes. Journal of Fluid Mechanics 671, 52–95.

Dommermuth, D.G., Rottman, J.W., Innis, G.E., Novikov, E.A., 2002. Numerical simulation of the wake of a towed sphere in a weakly stratified fluid. Journal of Fluid Mechanics 473, 83–101.

Dong, C., McWilliams, J.C., Shchepetkin, A.F., 2007. Island wakes in deep water. Journal of Physical Oceanography 37 (4), 962–981.

Fincham, M., Maxworthy, T., Spedding, G.R., 1996. Energy dissipation and vortex structure in freely-decaying, stratified grid turbulence. Dynamics of Atmospheres and Oceans 23, 155.

Flores, O., Riley, J.J., 2011. Analysis of turbulence collapse in the stably stratified surface layer using direct numerical simulation. Boundary-Layer Meteorology 139 (2), 241–259.

García-Villalba, M., del Álamo, J.C., 2011. Turbulence modification by stable stratification in channel flow. Physics of Fluids 23 (4), 045104.

Garg, R.P., Ferziger, J.H., Monismith, S.G., Koseff, J.R., 2000. Stably stratified turbulent channel flows. I. stratification regimes and turbulence suppression mechanism. Physics of Fluids 12 (10), 2569–2594.

Gerz, T., Schumann, U., Elghobashi, S.E., 1989. Direct numerical simulation of stratified homogeneous turbulent shear flows. Journal of Fluid Mechanics 200, 563–594.

Gohari, S.M.I., Sarkar, S., 2017. Direct numerical simulation of turbulence collapse and rebirth in stably stratified Ekman flow. Boundary-Layer Meteorology 162 (3), 401–426.

Gohari, S.M.I., Sarkar, S., 2018. Stratified Ekman layers evolving under a finite-time stabilizing buoyancy flux. Journal of Fluid Mechanics 840, 266–290.

Gourlay, M.J., Arendt, S.C., Fritts, D.C., Werne, J., 2001. Numerical modeling of initially turbulent wakes with net momentum. Physics of Fluids 13 (12), 3783–3802.

Hanazaki, H., 1988. A numerical study of three-dimensional stratified flow past a sphere. Journal of Fluid Mechanics 192, 393–419.

Hazel, P., 1972. Numerical studies of the stability of inviscid stratified shear flows. Journal of Fluid Mechanics 51, 39–61.

Högström, U., 1996. Review of some basic characteristics of the atmospheric surface layer. Boundary-Layer Meteorology 78 (3), 215–246.

Holliday, D., McIntyre, M., 1981. On potential energy density in an incompressible stratified fluid. Journal of Fluid Mechanics 107, 221–225.

Holt, S.E., Koseff, J.R., Ferziger, J.H., 1992. A numerical study of the evolution and structure of homogeneous stably stratified sheared turbulence. Journal of Fluid Mechanics 237, 499–539.

Issaev, V., Williamson, N., Armfield, S., Norris, S., 2022. Parameterization of mixing in stratified open channel flow. Journal of Fluid Mechanics 935, A17.

Jacobitz, F.G., Sarkar, S., 1998. The effect of nonvertical shear on turbulence in a stably stratified medium. Physics of Fluids 10 (5), 1158–1168.

Jacobitz, F.G., Sarkar, S., VanAtta, C.W., 1997. Direct numerical simulations of the turbulence evolution in a uniformly sheared and stably stratified flow. Journal of Fluid Mechanics 342, 231–261.

Kaminski, A.K., Smyth, W.D., 2019. Stratified shear instability in a field of pre-existing turbulence. Journal of Fluid Mechanics 862, 639–658.

Kang, D., Fringer, O., 2010. On the calculation of available potential energy in internal wave fields. Journal of Physical Oceanography, 2539–2545.

Kimura, Y., Herring, J.R., 1996. Diffusion in stably stratified turbulence. Journal of Fluid Mechanics 328, 253–269.

Kloosterziel, R., Carnevale, G., Orlandi, P., 2007. Inertial instability in rotating and stratified fluids: barotropic vortices. Journal of Fluid Mechanics 583, 379–412.

Komori, S., Ueda, H., Ogino, F., Mizushina, T., 1983. Turbulence structure in stably stratified open-channel flow. Journal of Fluid Mechanics 130, 13–26.

Lamb, K.G., 2014. Internal wave breaking and dissipation mechanisms on the continental slope/shelf. Annual Review of Fluid Mechanics 46, 231–254.

Lewin, S.F., Caulfield, C.P., 2024. Evidence for layered anisotropic stratified turbulence in a freely evolving horizontal shear flow. Journal of Fluid Mechanics 983, A20.

Lin, J.T., Pao, Y.H., 1979. Wakes in stratified fluids. Annual Review of Fluid Mechanics 11, 317–338.

Liu, J., Puthan, P., Sarkar, S., 2025. Turbulence in stratified rotating topographic wakes. Journal of Physical Oceanography 55 (11), 2159–2177.

Mahrt, L., 2014. Stably stratified atmospheric boundary layers. Annual Review of Fluid Mechanics 46, 23–45.

Mashayek, A., Peltier, W.R., 2012. The 'zoo' of secondary instabilities precursory to stratified shear flow transition. Part 2: The influence of stratification. Journal of Fluid Mechanics 708, 45–70.

McWilliams, J.C., 2016. Submesoscale currents in the ocean. Proceedings of the Royal Society A 472, 20160117. https://doi.org/10.1098/rspa.2016.0117.

Métais, O., Herring, J.R., 1989. Numerical simulations of freely evolving turbulence in stably stratified fluids. Journal of Fluid Mechanics 202, 117–148.

Nidhan, S., Schmidt, O.T., Sarkar, S., 2022. Analysis of coherence in turbulent stratified wakes using spectral proper orthogonal decomposition. Journal of Fluid Mechanics 934, A12.

Olsthoorn, J., Kaminski, A.K., Robb, D.M., 2023. Dynamics of asymmetric stratified shear instabilities. Physical Review Fluids 8 (2), 024501.

Orr, T.S., Domaradzki, J.A., Spedding, G.R., Constantinescu, G.S., 2015. Numerical simulations of the near wake of a sphere moving in a steady, horizontal motion through a linearly stratified fluid at Re = 1000. Physics of Fluids 27 (3), 035113.

Ortiz-Tarin, J.L., Nidhan, S., Sarkar, S., 2023. The high-Reynolds-number stratified wake of a slender body and its comparison with a bluff-body wake. Journal of Fluid Mechanics 957, A7.

Osborn, T., 1980. Estimates of the local rate of vertical diffusion from dissipation measurements. Journal of Physical Oceanography 10, 80–89.

Otheguy, P., Chomaz, J.-M., Billant, P., 2006. Elliptic and zigzag instabilities on co-rotating vertical vortices in a stratified fluid. Journal of Fluid Mechanics 553, 253–272.

Pal, A., Sarkar, S., Posa, A., Balaras, E., 2017. Direct numerical simulation of stratified flow past a sphere at a subcritical Reynolds number of 3700 and moderate Froude number. Journal of Fluid Mechanics 826, 5–31.

Perfect, B., Kumar, N., Riley, J., 2018. Vortex structures in the wake of an idealized seamount in rotating, stratified flow. Geophysical Research Letters 45 (17), 9098–9105.

Perlin, A., Moum, J.N., Klymak, J., Levine, M.D., Boyd, T., Kosro, P.M., 2005. A modified law-of-the-wall applied to Oceanic bottom boundary layers. Journal of Geophysical Research: Oceans 110 (C10).

Pham, H.T., Sarkar, S., Brucker, K.A., 2009. Dynamics of a stratified shear layer above a region of uniform stratification. Journal of Fluid Mechanics 630, 191–223.

Pham, H.T., Sarkar, S., Winters, K.B., 2013. Large-eddy simulation of deep-cycle turbulence in an equatorial undercurrent model. Journal of Physical Oceanography 43, 2490–2502.

Pham, H.T., Sarkar, S., Winters, K.B., 2014. Evolution of an asymmetric turbulent shear layer in a thermocline. Journal of Turbulence 15, 449–471.

Portwood, G.D., de Bruyn Kops, S., Caulfield, C., 2019. Asymptotic dynamics of high dynamic range stratified turbulence. Physical Review Letters 122 (19), 194504.

Poulos, G.S., Blumen, W., Fritts, D.C., Lundquist, J.K., et al., 2002. Cases-99: a comprehensive investigation of the stable nocturnal boundary layer. Bulletin of the American Meteorological Society 83 (4), 555.

Praud, O., Fincham, A.M., Sommeria, J., 2005. Decaying grid turbulence in a strongly stratified fluid. Journal of Fluid Mechanics 522, 1–33.

Puthan, P., Jalali, M., Chalamalla, V.K., Sarkar, S., 2019. Energetics and mixing in buoyancy-driven near-bottom stratified flow. Journal of Fluid Mechanics 869, 214–237.

Puthan, P., Pawlak, G., Sarkar, S., 2022. Wake vortices and dissipation in a tidally modulated flow past a three-dimensional topography. Journal of Geophysical Research: Oceans 127 (8), e2022JC018470.

Puthan, P., Sarkar, S., Pawlak, G., 2021. Tidal synchronization of Lee vortices in geophysical wakes. Geophysical Research Letters 48 (4), e2020GL090905.

Riley, J.J., deBruynKops, S.M., 2003. Dynamics of turbulence strongly influenced by buoyancy. Physics of Fluids 15, 2047–2059.

Riley, J.J., Lelong, M.-P., 2000. Fluid motions in the presence of strong stable stratification. Annual Review of Fluid Mechanics 32 (1), 613–657.

Salehipour, H., Caulfield, C.P., Peltier, W.R., 2016. Turbulent mixing due to the holmboe wave instability at high Reynolds number. Journal of Fluid Mechanics 803, 591–621.

Salehipour, H., Peltier, W.R., Mashayek, A., 2015. Turbulent diapycnal mixing in stratified shear flows: the influence of Prandtl number on mixing efficiency and transition at high Reynolds number. Journal of Fluid Mechanics 773, 178–223.

Sarkar, S., Pham, H.T., Ramachandran, S., Nash, J.D., Tandon, A., Buckley, J.M., Lotliker, A.A., Omand, M.M., 2016. The interplay between submesoscale instabilities and turbulence in the surface layer of the Bay of Bengal. Oceanography 29, 146–157.

Sarkar, S., Scotti, A., 2017. From topographic internal gravity waves to turbulence. Annual Review of Fluid Mechanics 49, 195–220.

Scotti, A., White, B., 2014. Diagnosing mixing in stratified turbulent flows with a locally defined available potential energy. Journal of Fluid Mechanics 740, 114–135.

Shah, S.K., Bou-Zeid, E., 2014. Direct numerical simulations of turbulent Ekman layers with increasing static stability: modifications to the bulk structure and second-order statistics. Journal of Fluid Mechanics 760, 494–539.

Shih, L.H., Koseff, J.R., Ferziger, J.H., Rehmann, C.R., 2000. Scaling and parameterization of stratified homogeneous turbulent shear flow. Journal of Fluid Mechanics 412, 1–20.

Skyllingstad, E.D., Samelson, R.M., 2012. Baroclinic frontal instabilities and turbulent mixing in the surface boundary layer. Part I: Unforced simulations. Journal of Physical Oceanography 42, 1701–1716. https://doi.org/10.1175/JPO--D--10--05016.1.

Smedman, A.-S., Tjernström, M., Högström, U., 1993. Analysis of the turbulence structure of a marine low-level jet. Boundary-Layer Meteorology 66 (1–2), 105–126.

Smyth, W.D., Moum, J.N., 2000. Length scales of turbulence in stably stratified mixing layers. Physics of Fluids 12 (6), 1327–1342.

Smyth, W.D., Moum, J.N., 2013. Marginal instability and deep cycle turbulence in the eastern equatorial Pacific Ocean. Geophysical Research Letters 40, 6181–6185.

Smyth, W.D., Moum, J.N., Li, Z., Thorpe, S., 2013. Diurnal shear instability, the descent of the surface shear layer, and the deep cycle of equatorial turbulence. Journal of Physical Oceanography 43, 2432–2455.

Smyth, W.D., Nash, J.D., Moum, J.N., 2019. Self-organized criticality in geophysical turbulence. Scientific Reports 9 (1), 3747.

Smyth, W.D., Winters, K.B., 2003. Turbulence and mixing in Holmboe waves. Journal of Physical Oceanography 33, 694–711.

Spedding, G.R., 1997. The evolution of initially turbulent bluff-body wakes at high internal Froude number. Journal of Fluid Mechanics 337, 283–301.

Spedding, G.R., 2002. Vertical structure in stratified wakes with high initial Froude number. Journal of Fluid Mechanics 454, 71–112.

Spedding, G.R., 2014. Wake signature detection. Annual Review of Fluid Mechanics 46 (1), 273–302.

Srinivasan, K., McWilliams, J.C., Jagannathan, A., 2021. High vertical shear and dissipation in equatorial topographic wakes. Journal of Physical Oceanography 51 (6), 1985–2001.

Sullivan, P.P., McWilliams, J.C., 2018. Frontogenesis and frontal arrest of a dense filament in the Oceanic surface boundary layer. Journal of Fluid Mechanics 837, 341–380.

Sun, J., Nappo, C.J., Mahrt, L., Belušić, D., Grisogono, B., Stauffer, D.R., Pulido, M., Staquet, C., Jiang, Q., Pouquet, A., Yagüe, C., Galperin, B., Smith, R.B., Finnigan, J.J.,

Mayor, S.D., Svensson, G., Grachev, A.A., Neff, W.D., 2015. Review of wave-turbulence interactions in the stable atmospheric boundary layer. Reviews of Geophysics 53.
Taylor, J.R., Ferrari, R., 2010. Buoyancy and wind-driven convection at mixed layer density fronts. Journal of Physical Oceanography 40, 1222–1242.
Taylor, J.R., Sarkar, S., 2007. Internal gravity waves generated by a turbulent bottom Ekman layer. Journal of Fluid Mechanics 590, 331–354.
Taylor, J.R., Sarkar, S., 2008. Stratification effects in a bottom Ekman layer. Journal of Physical Oceanography 38, 2535–2555.
Taylor, J.R., Sarkar, S., Armenio, V., 2005. Large eddy simulation of stably stratified open channel flow. Physics of Fluids 17, 116602.
VanDine, A., Pham, H.T., Sarkar, S., 2020. Investigation of LES models for a stratified shear layer. Computers and Fluids 198, 104405.
Watanabe, T., Riley, J.J., Nagata, K., Onishi, R., Matsuda, K., 2018. A localized turbulent mixing layer in a uniformly stratified environment. Journal of Fluid Mechanics 849, 245–276.
Winters, K., Lombard, P., Riley, J., D'Asaro, E., 1995. Available potential energy and mixing in density-stratified fluids. Journal of Fluid Mechanics 289, 115–128.
Yanase, S., Flores, C., Métais, O., Riley, J.J., 1993. Rotating free-shear flows. I. Linear stability analysis. Physics of Fluids. A, Fluid Dynamics 5 (11), 2725–2737.
Zhou, Q., Diamessis, P.J., 2019. Large-scale characteristics of stratified wake turbulence at varying Reynolds number. Physical Review Fluids 4 (8), 1–30.
Zilitinkevich, S., Esau, I., 2002. On integral measures of the neutral barotropic planetary boundary layer. Boundary-Layer Meteorology 104 (3), 371–379.

Rotating turbulence in the environment

4

Peter A. Davidson
University of Cambridge, Cambridge, United Kingdom

4.1 Definitions, observations, and questions

Flows in a strong background rotation are commonplace in geophysics and astrophysics, including tornadoes, hurricanes, large-scale ocean gyres, and planetary cores. They are characterized by the presence of *inertial waves*, which are a form of incompressible, internal wave motion in which the restoring force is supplied by the Coriolis force. The most important consequence of these inertial waves is the formation of long-lived columnar vortices aligned with the rotation axis (see Fig. 4.1), vortices that eventually dominate the large-scale dynamics of the turbulence. Indeed, the central questions in rotating turbulence are to understand how and when these columnar vortices form, and the influence those columnar vortices have on the large-scale dynamics.

We consider turbulence in a background rotation, $\mathbf{\Omega} = \Omega\hat{\mathbf{e}}_z$, viewed in a frame of reference that rotates with $\mathbf{\Omega}$. Such turbulence is governed by the Navier–Stokes equation in the form

$$\frac{\partial \mathbf{u}}{\partial t} + \mathbf{u} \cdot \nabla \mathbf{u} = 2\mathbf{u} \times \mathbf{\Omega} - \nabla (p/\rho) + \nu \nabla^2 \mathbf{u} \tag{4.1}$$

where $2\mathbf{u} \times \mathbf{\Omega}$ is the Coriolis force, and the irrotational centrifugal force has been absorbed into the pressure gradient. Our discussion builds on that in Davidson (2013, 2022, 2024), and also Davidson et al. (2013). We are concerned particularly with strongly rotating turbulence in which $|\mathbf{u} \cdot \nabla \mathbf{u}| < |\mathbf{u} \times \mathbf{\Omega}|$, i.e., the fluctuating velocity in the rotating frame, $\mathbf{u}$, is smaller than $\Omega \ell_\perp$, where $\ell_\perp$ is the horizontal integral scale of the turbulence (the transverse size of the large eddies). Evidently, the *Rossby number*, $\mathrm{Ro} = u/\Omega \ell_\perp$, measures the relative strengths of the inertial and Coriolis forces in Eq. (4.1), and when $\mathrm{Ro} << 1$, the turbulence is said to be *rapidly rotating*.

The growth and subsequent dominance of columnar vortices in strongly rotating turbulence was evident in the early experiments on rotating turbulence. Perhaps the first was Ibbetson and Tritton (1975), who reported *free-decaying* turbulence in a rotating annulus at $\mathrm{Ro} \sim 1$. In line with most later experiments, these authors observed that columnar vortices aligned with the rotation axis emerge rapidly on a timescale of Ω^{-1}. They also observed that inertial waves are important for transporting kinetic energy across the flow. By contrast, Hopfinger et al. (1982) *continually forced* turbulence in a rotating tank using a vertically oscillating grid. Near the grid, Ro was large, and the turbulence showed few unusual properties, but far from the grid, Ro fell below unity and, once again, columnar vortices formed, those vortices being

Environmental Turbulence. https://doi.org/10.1016/B978-0-32-395873-8.00011-5
Copyright © 2026 Elsevier Inc. and Peter Alan Davidson. All rights are reserved, including those for text and data mining, AI training, and similar technologies.

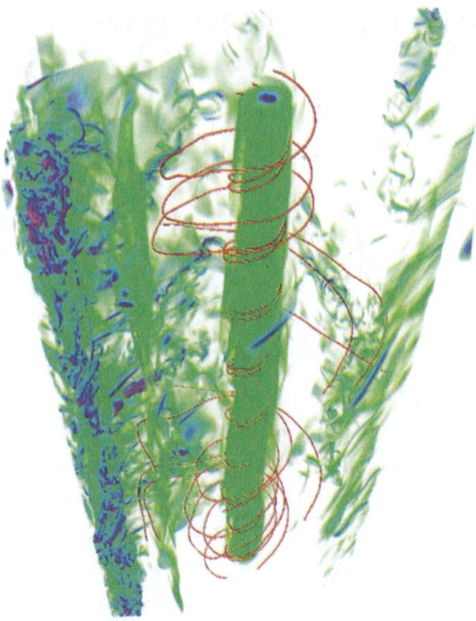

Figure 4.1 A simulation of rapidly rotating turbulence at Ro ~ 0.06, visualized using vorticity. (Courtesy of Pablo Mininni and Annick Pouquet.)

predominantly cyclonic, i.e., $\mathbf{\Omega} \cdot \nabla \times \mathbf{u} > 0$. Dickinson and Long (1983) also continually forced turbulence using a vertically oscillating grid, but they were more interested in how the turbulence created by the grid spreads across a large tank. They observed that, for large Ro, the turbulence spread vertically along the rotation axis by conventional turbulent diffusion, at the rate $(\nu_t t)^{1/2}$, where ν_t is an eddy viscosity. However, for Ro < 1, the turbulence spread along the rotation axis at a rate proportional to time, consistent with turbulent kinetic energy being spread by waves. Subsequently, Davidson et al. (2006) studied *freely-decaying* turbulence in a similar geometry and found that, once Ro drops below $\sim$0.4, columnar vortices spontaneously emerge, whose axial length grows as $\ell_z \sim \ell_\perp \Omega t$. As we shall see, $\ell_z \sim \ell_\perp \Omega t$ is the hallmark of energy dispersion by low-frequency inertial waves.

Given these various observations, it is natural to inquire as to the precise mechanism(s) by which the columnar vortices grow, why this occurs only when Ro < 1, and why some experiments display a statistical bias towards cyclonic columnar vortices.

We explore all of these issues in this chapter.

4.2 The dynamics of inertial waves

Clearly, inertial waves play a central role in rotating turbulence. So, before discussing turbulence, we shall establish the properties of these waves, which are not at all intuitive.

4.2.1 The group velocity and dispersive properties of inertial waves

Consider an inviscid, rotating fluid in the limit of Ro→0, whose vorticity equation is

$$\frac{\partial \boldsymbol{\omega}}{\partial t} = 2(\boldsymbol{\Omega} \cdot \nabla)\mathbf{u}, \quad \boldsymbol{\omega} = \nabla \times \mathbf{u} \tag{4.2}$$

Applying the operator $\nabla \times (\partial/\partial t)$ to (4.2) yields the wave-like equation

$$\frac{\partial^2}{\partial t^2}\nabla^2\mathbf{u} + (2\boldsymbol{\Omega} \cdot \nabla)^2\mathbf{u} = 0 \tag{4.3}$$

which supports plane waves of the form

$$\mathbf{u}(\mathbf{x}, t) = \hat{\mathbf{u}} \exp\left[j\left(\mathbf{k} \cdot \mathbf{x} - \varpi t\right)\right], j = \sqrt{-1} \tag{4.4}$$

where $\mathbf{k}$ is the wavevector and ϖ the frequency. We adopt the sign convention that $\varpi > 0$, and note that $\nabla \cdot \mathbf{u} = 0$ requires $\mathbf{k} \cdot \hat{\mathbf{u}} = 0$. Eq. (4.3) then yields the dispersion relationship

$$\varpi = \pm 2\frac{\mathbf{k} \cdot \boldsymbol{\Omega}}{k} = 2\Omega\left|\sin\theta\right|, 0 < \varpi < 2\Omega \tag{4.5}$$

where θ is the angle that $\mathbf{k}$ makes to the horizontal, with $-\pi/2 < \theta < \pi/2$.

The phase speed, i.e., the speed of the wave crests, is ϖ/k, and so the *phase velocity* is aligned with $\mathbf{k}$ and equal to

$$\mathbf{c}_p = \pm 2\frac{(\mathbf{k} \cdot \boldsymbol{\Omega})\,\mathbf{k}}{k^3} \tag{4.6}$$

The *group velocity*, on the other hand, gives the velocity of propagation of wave energy in the form of wave packets. Since, for any wave, the group velocity is given by $\mathbf{c}_{g,i} = \partial\varpi/\partial k_i$, Eq. (4.5) yields

$$\mathbf{c}_g = \pm 2\frac{\mathbf{k} \times (\boldsymbol{\Omega} \times \mathbf{k})}{k^3} = \pm 2\frac{k^2\boldsymbol{\Omega} - (\mathbf{k} \cdot \boldsymbol{\Omega})\mathbf{k}}{k^3} \tag{4.7}$$

where $c_g = (2\Omega/k)\cos\theta$. Note that, contrary to intuition, the phase and group velocities are orthogonal, with $\mathbf{c}_g$ normal to $\mathbf{k}$. Thus wave crests travel in a direction that is perpendicular to the direction of propagation of the wave packet within which they sit! In short, wave energy propagates *along* the wave crests. Moreover,

$$\mathbf{c}_g + \mathbf{c}_p = \pm 2\boldsymbol{\Omega}/k \tag{4.8}$$

so that $\mathbf{c}_g$ and $\mathbf{c}_p$ have equal and opposite horizontal components. Finally, we have

$$\mathbf{c_g} \cdot \boldsymbol{\Omega} = \pm 2k^{-3}\left[k^2\Omega^2 - (\mathbf{k} \cdot \boldsymbol{\Omega})^2\right] \tag{4.9}$$

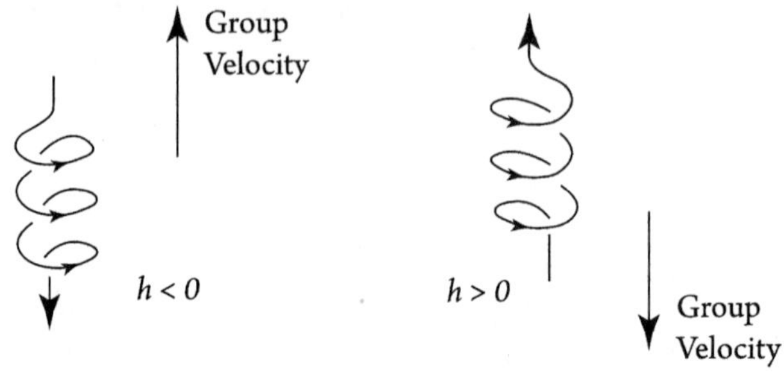

Figure 4.2 A wave packet with $h < 0$ travels upward, while one with $h > 0$ travels downward. (From Davidson (2013).)

and so the upper signs in (4.5)→(4.8) correspond to wave packets propagating upward ($\mathbf{c_g} \cdot \mathbf{\Omega} > 0$), and the lower sign to wave packets travelling downward ($\mathbf{c_g} \cdot \mathbf{\Omega} < 0$).

It is clear from these expressions that low-frequency waves ($\varpi << \Omega$) have $\mathbf{k} \cdot \mathbf{\Omega} \approx 0$ and $\mathbf{c}_g = \pm 2\mathbf{\Omega}/k$ so that $\mathbf{k}$ is horizontal and wave packets travel up and down the rotation axis. For a given k, these waves have the largest group speed. Conversely, high-frequency waves ($\varpi \to 2\Omega$) have $\mathbf{k}$ vertical and a group velocity which is horizontal but vanishingly small. Of course, wave packets of intermediate frequency travel obliquely.

An informative experiment is to suspend a disc horizontally at the center of a rotating tank of water, oscillate the disc, and observe inertial waves radiate from the disc. For low oscillation frequencies, the wave energy disperses up and down the rotation axis to fill a circular cylinder that circumscribes the disc, while at high frequencies ($\varpi \to 2\Omega$), the energy slowly spreads radially outward from the disc. At an intermediate frequency, say $\varpi = \Omega$, the energy radiates to fill two conical annuli, one located above the disc and one below. The orientation of $\mathbf{k}$ is set by Eq. (4.5), which gives $k_z/k = \pm\varpi/2\Omega = \pm 1/2$ for $\varpi = \Omega$, and the orientation of $\mathbf{c}_g$ is then fixed by the fact that $\mathbf{c}_g$ is normal to $\mathbf{k}$.

Note that monochromatic inertial waves have maximal helicity, $h = \mathbf{u} \cdot \boldsymbol{\omega}$, i.e., they have $h = \mp |\mathbf{u}|\,|\boldsymbol{\omega}|$. To see why, first note that Eq. (4.2) requires $2(\mathbf{\Omega} \cdot \mathbf{k})\hat{\mathbf{u}} = -\varpi\hat{\boldsymbol{\omega}}$, where $\hat{\boldsymbol{\omega}} = j\mathbf{k} \times \hat{\mathbf{u}}$ is the amplitude of the vorticity. This, combined with Eq. (4.5), yields $\hat{\boldsymbol{\omega}} = j\mathbf{k} \times \hat{\mathbf{u}} = \mp k\hat{\mathbf{u}}$, so that the velocity and vorticity fields are parallel and in phase. Note that upward propagating wave packets carry negative helicity and downward propagating wave packets carry positive helicity, as shown in Fig. 4.2.

Finally, let us determine the structure of a monochromatic inertial wave. We adopt coordinates (X, Y, Z) with Z parallel to $\mathbf{k}$ and $\hat{\mathbf{u}} = \left(\hat{u}_X, \hat{u}_Y, 0\right)$, which ensures $\mathbf{k} \cdot \hat{\mathbf{u}} = 0$. Then $j\mathbf{k} \times \hat{\mathbf{u}} = \mp k\hat{\mathbf{u}}$ demands $\hat{u}_Y = \mp j\hat{u}_X$, and it follows that $\hat{\mathbf{u}} = \hat{u}_X\,(1, \mp j, 0)$. Taking $\hat{u}_X$ to be real, we have

$$\mathbf{u} = \hat{u}_X\,(\cos(kZ - \varpi t), \pm\sin(kZ - \varpi t), 0)\,. \tag{4.10}$$

This is a circularly polarized wave whose velocity rotates about the Z-axis as the wave propagates. The sense of rotation is determined by the helicity; if $h < 0$, then

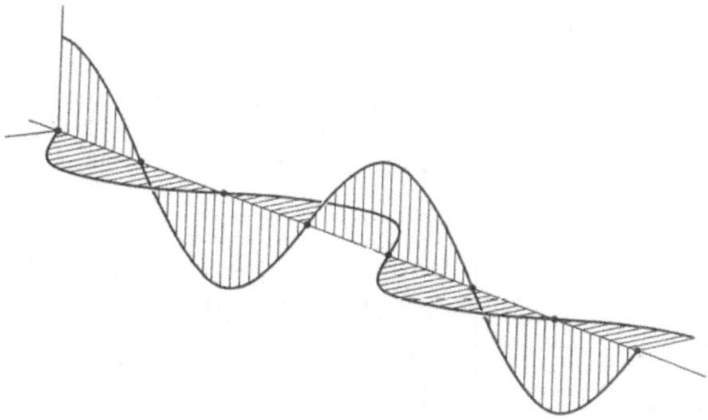

Figure 4.3 The velocity field for an inertial wave with $h < 0$. (From Davidson (2022).)

the upper sign in Eq. (4.10) applies, and the velocity distribution at $t = 0$ is as shown in Fig. 4.3.

4.2.2 The relationship between Taylor columns and low-frequency inertial waves

According to Eq. (4.2), an inviscid, steady flow satisfies $(\mathbf{\Omega} \cdot \nabla)\mathbf{u} = 0$, and so a steady velocity field in a rapidly-rotating fluid is two-dimensional $\mathbf{u} = \mathbf{u}(x, y)$. In particular, the flow must satisfy $\partial u_z/\partial z = 0$, so that columns of fluid oriented with the z-axis cannot change their length. This is known as the Taylor–Proudman theorem, and its power was clearly demonstrated by Taylor (1923) in the experiment shown in Fig. 4.4.

A tank of water is placed on a turntable, and a small object is *slowly* towed across the base of the tank. One might have expected to see the fluid ahead of the object move up and over it. However, the flow is quasi-steady and so must satisfy $\partial u_z/\partial z = 0$, which excludes any such motion. Instead, what happens is that the water sitting above the object drifts across the tank, keeping pace with the object below. It is as if the fluid sitting inside an imaginary cylinder that circumscribes the object is rigidly attached to that object. This cylinder of drifting fluid is known as a Taylor column, and the fluid outside the column flows around it as if the column were rigid. This odd behavior can be confirmed by placing dye at the points A and B shown in Fig. 4.4. The dye at A is seen to drift across the tank, centered above the object, while that at B splits into two as the object passes below it, making way for the seemingly rigid Taylor column.

Although the Taylor–Proudman theorem provides a convenient means of rationalizing these observations, it does not explain how the Taylor column forms in the first place. For example, when the object moves 1 mm to the right, how does the fluid at point A know that it too must move by 1 mm? The answer to this question is surprising: the information is transmitted upward by inertial waves. In fact, Taylor columns are first established, and then subsequently maintained, by low-frequency inertial waves travelling along the rotation axis. Thus it is not correct to think of the flow in Fig. 4.4 as steady. Rather, the time derivative in Eq. (4.2) is important, and for as long as the

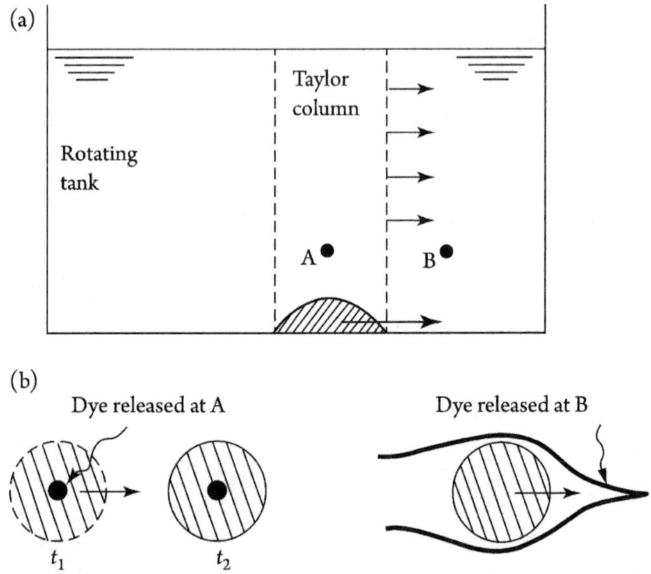

Figure 4.4 A small object is slowly dragged across the base of a rotating tank of water, producing a Taylor column. (a) Side view. (b) Plan view. (From Davidson (2015).)

object at the base of the tank moves, it acts like a radio antenna, continually emitting low-frequency inertial waves which travel upward. These waves carry the information that the object below is moving, and if the object stops, the waves cease, and the Taylor column vanishes.

There is a simple way to confirm that Taylor columns are maintained by the continual radiation of low-frequency inertial waves. Suppose the object in Fig. 4.4 has a diameter d and moves to the right at the speed V, with $\text{Ro} = V/\Omega d << 1$. We adopt a rotating frame of reference that moves slowly across the tank at speed V, so the object appears stationary, while the fluid flows to the left. If the Taylor column is indeed maintained by low-frequency inertial waves streaming upward with a group velocity of $\mathbf{c}_g = 2\boldsymbol{\Omega}/k \sim \boldsymbol{\Omega}d/\pi$ (relative to *stationary* fluid), then the fact that the fluid moves to the left at the speed V means that the column should not be vertical, but rather tilted to the left at a small angle $\theta = V/c_g \sim \pi V/\Omega d$. This is indeed observed, confirming that the Taylor column really does consist of a stream of waves.

4.2.3 The spontaneous formation of columnar vortices by inertial waves

So far, we have considered waves generated by a moving boundary. In such a boundary-value problem, one can control the orientation of $\mathbf{k}$, and hence $\mathbf{c}_g$, through the choice of frequency. However, what happens when the wave source does not impose a timescale?

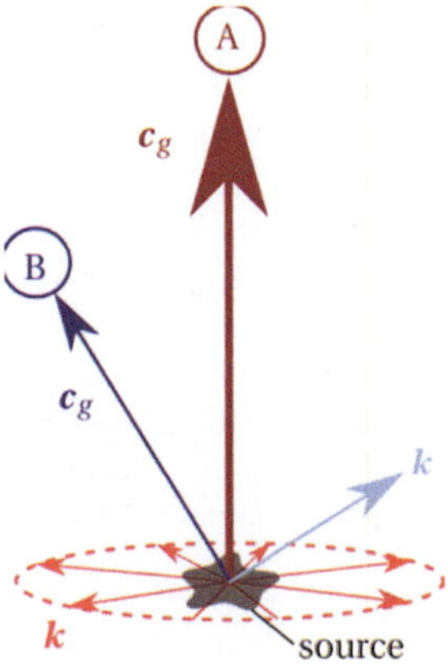

Figure 4.5 Directions of $\mathbf{c}_g$ and $\mathbf{k}$ for on-axis and off-axis waves. (From Davidson (2022).)

Consider inertial waves, which are excited by some initial perturbation in the velocity field. For example, the initial condition might be a localized vortex in the rotating frame of reference. To solve such an initial-value problem, one normally takes a three-dimensional Fourier transform of the initial velocity field. This decomposes $\mathbf{u}_0(\mathbf{x})$ into a sum of Fourier modes of amplitude $\hat{\mathbf{u}}_0(\mathbf{k})$. The transform also attributes a certain amount of kinetic energy to each mode. These modes then propagate as plane waves, carrying their energy with them in accordance with Eq. (4.7). The radiation field at a later time is then found by summing the modes, i.e., by performing an inverse Fourier transform.

Of course, one can vary the way in which energy is distributed across the Fourier modes through the choice of $\mathbf{u}_0\,(\mathbf{x})$. So, it is natural to assume that one can control the dispersion pattern through our choice of the modes excited at $t = 0$. However, it turns out that this is not the case. Rather, for almost any localized disturbance, the wave energy density is largest on the rotation axis. This ability to spontaneously focus the radiated energy onto the rotation axis is a robust phenomenon that explains why inertial-wave dispersion patterns are often dominated by columnar structures aligned with $\boldsymbol{\Omega}$.

The simplest way to understand this spontaneous focusing of the radiated energy onto the rotation axis is to consider the geometric properties of the dispersion relationship, as discussed in Davidson (2013). Consider a wave packet with a dominant wavevector $\mathbf{k}$ that leaves the wave source at $\mathbf{x} = 0$ at time $t = 0$. At time t, the packet reaches the point

$$\mathbf{x} = \mathbf{c}_g t = \frac{2\Omega t}{k^3}\left[k^2\hat{\mathbf{e}}_z - k_z\mathbf{k}\right] \tag{4.11}$$

Now $\mathbf{x}$ is normal to $\mathbf{k}$ and coplanar with $\mathbf{k}$ and $\boldsymbol{\Omega}$. Hence, for any $\mathbf{x}$ that lies off the axis, such as point B in Fig. 4.5, there is one and only one orientation of $\mathbf{k}$ that takes energy from the source to that point. However, if $\mathbf{x}$ is located on the rotation axis, such as point A, then *any* $\mathbf{k}$ that lies in the horizontal plane transports energy to that location.

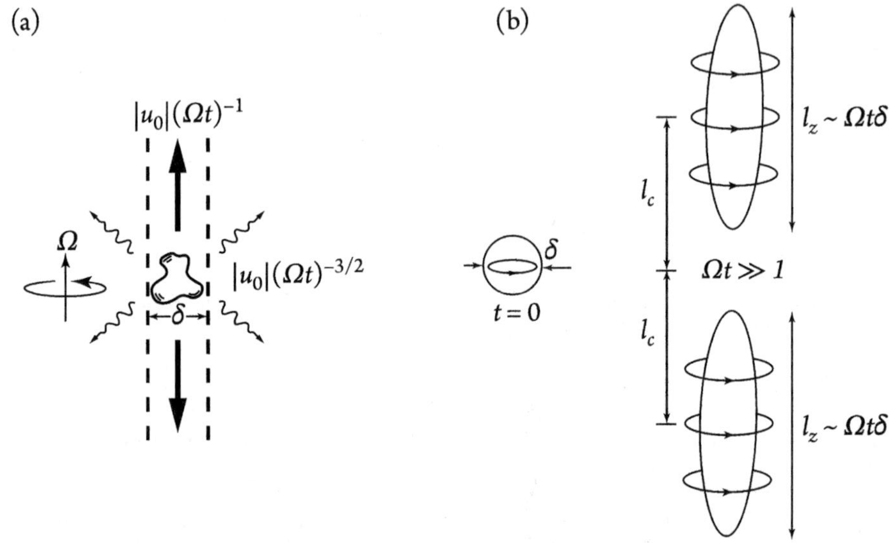

Figure 4.6 The radiation of inertial waves from a localized vortex. (a) The wiggly arrows represent the dispersion of energy, and solid arrows the dispersion of angular momentum. (b) The dispersion pattern is dominated by two columnar vortices. (From Davidson (2022).)

In summary, all the energy that lies within a horizontal disc in **k**-space is folded up onto a line in real space. It is this channelling of energy from two dimensions in **k**-space to one dimension in real space that focuses the wave energy onto the rotation axis.

An alternative explanation for the self-focusing of inertial waves, which yields considerably more information, is given in Davidson et al. (2006). Consider an initial condition that consists of a vortex of scale δ. This is centered at the origin and is circumscribed by a cylindrical control volume, V_c, also of diameter δ. The control volume is oriented parallel to the rotation axis and is infinitely long, as shown in Fig. 4.6(a).

If the fluid is inviscid, its angular momentum density is governed by,

$$\frac{\partial}{\partial t}(\mathbf{x} \times \mathbf{u}) = 2\mathbf{x} \times (\mathbf{u} \times \boldsymbol{\Omega}) + \nabla \times (p\mathbf{x}/\rho) \tag{4.12}$$

the z component of which may be rewritten as

$$\frac{\partial}{\partial t}(\mathbf{x} \times \mathbf{u})_z = -\nabla \cdot \left[r^2 \Omega \mathbf{u}\right] + [\nabla \times (p\mathbf{x}/\rho)]_z \tag{4.13}$$

in cylindrical polar coordinates, (r, θ, z). We now integrate (4.13) over the cylinder V_c and note that the pressure term converts to a surface integral, which is zero. The end result is

$$\frac{d}{dt}\int_{V_c} (\mathbf{x} \times \mathbf{u})_z \, dV = -(\delta/2)^2 \Omega \oint \mathbf{u} \cdot d\mathbf{S} = 0 \tag{4.14}$$

Evidently, the axial component of angular momentum within V_c is conserved, despite the radiation of energy by inertial waves, and it is easy to show that this biases the dispersion pattern. For example, after a time t, inertial waves will have travelled a distance of order $c_g t \sim \Omega\delta t$, and so the kinetic energy fills a volume of order $V_{3D} \sim (\Omega\delta t)^3$. Conservation of energy now requires that a characteristic velocity at time t satisfies $u^2 (\Omega\delta t)^3 \sim u_0^2\delta^3$, where u_0 is a typical velocity at $t = 0$. It follows that, as the energy disperses, the velocity falls off as $u \sim u_0 (\Omega t)^{-3/2}$. However, this is not true near the rotation axis, because angular momentum conservation limits the rate of fall of u within V_c. That is, angular momentum disperses in the axial direction only, and so, after a time t, it fills a cylindrical volume of size $V_{1D} \sim (\Omega\delta t)\delta^2$. Conservation of angular momentum now requires that $(u\delta)(\Omega\delta t)\,\delta^2 \sim (u_0\delta)\delta^3$. In short, a typical velocity near the rotation axis declines no faster than $u \sim u_0 (\Omega t)^{-1}$, as shown in Fig. 4.6(a). These two scaling laws, $u \sim u_0 (\Omega t)^{-1}$ for on-axis radiation and $u \sim u_0 (\Omega t)^{-3/2}$ for off-axis waves, are readily confirmed through an asymptotic analysis of the dispersion integral for inertial waves.

A typical dispersion pattern is shown schematically in Fig. 4.6(b). Any off-axis radiation decays rapidly, and the emergent pattern is dominated by two wave packets propagating along the rotation axis, one upward and the other downward. The centers of the two wave packets propagate at the group velocity corresponding to the dominant horizontal wavenumber, $c_g = 2\Omega/k_{\text{dom}} \sim \Omega\delta$, and so are located at $|z| = \ell_c \sim \Omega\delta t$. Crucially, however, there is a spread of transverse wavenumbers in the initial condition, of the order of $\Delta k_\perp \sim 1/\delta$, and for each transverse wavenumber we have $c_{g,z} = 2\Omega/k_\perp$. Consequently, the wave packets elongate as they propagate, growing in length $\ell_z \sim \Omega\delta t$. In short, the radiation of energy from a localized disturbance generates a pair of columnar vortices, as shown in Fig. 4.6(b).

It is important to note, however, that the net kinetic energy held within the tangent cylinder is of order $\left(u_0^2\delta^3\right)/\Omega t$, and so despite the high energy density within this cylindrical control volume, the bulk of the kinetic energy ultimately disperses in the form of off-axis radiation. Curiously, though, the energy held by columnar structures is often observed to grow with time in rotating turbulence (see, e.g., Smith and Lee (2005)). This tentatively suggests that some form of nonlinear process shadows the linear growth of columnar wave packets, embedding their energy in the turbulence. This will prove to be important when we discuss the maintenance of columnar vortices in rotating turbulence.

It is instructive to consider a specific example of linear dispersion. Consider the case of axisymmetric dispersion from the Gaussian vortex

$$\mathbf{u}_0 = \Lambda r \exp\left[-\left(r^2 + z^2\right)/\delta^2\right]\hat{\mathbf{e}}_\theta \tag{4.15}$$

in (r, θ, z) coordinates. For axisymmetric motion, the θ component of $\nabla^2\mathbf{u}$ satisfies

$$r\left(\nabla^2\mathbf{u}\right)_\theta = r\frac{\partial}{\partial r}\frac{1}{r}\frac{\partial\Gamma}{\partial r} + \frac{\partial^2\Gamma}{\partial z^2} \tag{4.16}$$

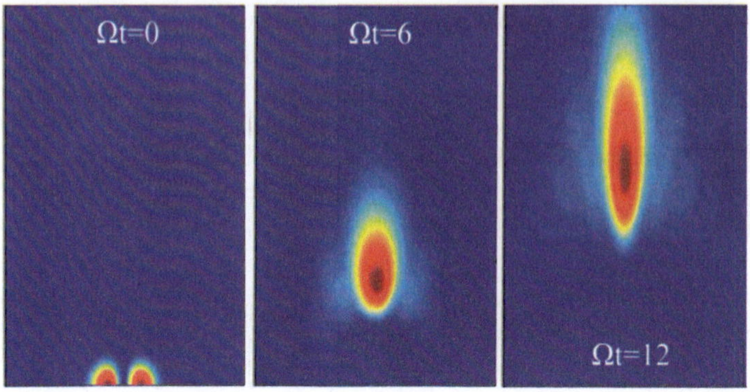

Figure 4.7 Inertial waves dispersing from a Gaussian vortex. (From Davidson (2013).)

where $\Gamma = ru_\theta$, and so the azimuthal component of the wave Eq. (4.3) takes the form

$$\frac{\partial^2}{\partial t^2}\left(r\frac{\partial}{\partial r}\frac{1}{r}\frac{\partial}{\partial r}+\frac{\partial^2}{\partial z^2}\right)\Gamma+(2\Omega)^2\frac{\partial^2\Gamma}{\partial z^2}=0 \tag{4.17}$$

This may be solved using a Hankel-cosine transform, as described in Davidson (2022), and the inverse transform then yields

$$u_\theta \approx \Lambda\delta\int_0^\infty \kappa^2 e^{-\kappa^2} J_1\left(\frac{2\kappa r}{\delta}\right)\left\{\exp\left[-\left(\frac{z}{\delta}-\frac{\Omega t}{\kappa}\right)^2\right]+\exp\left[-\left(\frac{z}{\delta}+\frac{\Omega t}{\kappa}\right)^2\right]\right\}d\kappa \tag{4.18}$$

where J_1 is the usual Bessel function and $\kappa = k_r\delta/2$. Now the integrand in Eq. (4.18) is dominated by $\kappa = O(1)$. Consequently, to keep the argument in the exponentials of order unity for $\Omega t >> 1$, we require $z \sim \pm\Omega\delta t$. This shows that the wave energy is centered at $|z| \sim \Omega\delta t$, as predicted above and shown in Fig. 4.7.

Fig. 4.7 shows the kinetic energy distribution for $z > 0$ at various times. Not only do we observe a wave packet propagating along the rotation axis, but it is also clear that the packet elongates as it propagates. As noted above, this is because there is a range of transverse wavenumbers in the packet, and for each wavenumber we have a different $c_{g,z} = 2\Omega/k_\perp$.

The distribution of u_θ for large Ωt may be found by demanding that the argument in one of the exponentials in Eq. (4.18) remain of order unity as $\Omega t \to \infty$. At locations $z = \pm\delta\Omega t$, this requires that $\kappa \to 1$, and it is readily confirmed that Eq. (4.18) then yields $u_\theta \sim \Lambda\delta J_1(2r/\delta)(\Omega t)^{-1}$. This, in turn, demands that $u_\theta \sim \Lambda r(\Omega t)^{-1}$ near the axis, while $u_\theta \sim \Lambda\delta(\Omega t)^{-3/2}(z/r)^{1/2}$ away from the axis, consistent with the predictions above.

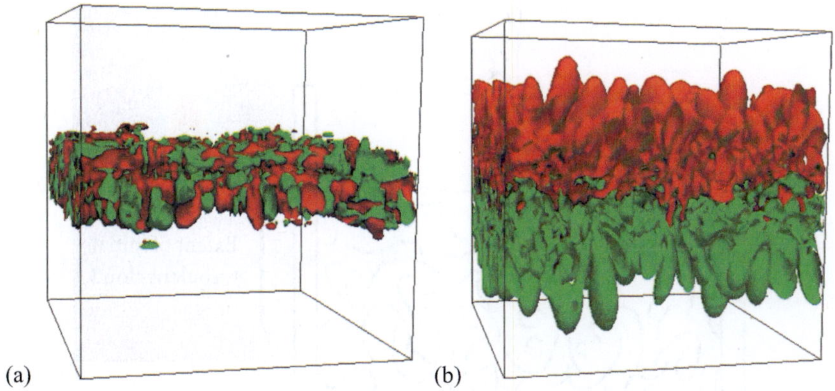

Figure 4.8 Rotating turbulence. (a) The initial condition consists of eddies located near the mid-plane, with Ro ≈ 0.1. (b) The flow at $\Omega t = 6$, colored by helicity. From Davidson (2013).

4.2.4 The spatial segregation of helicity by inertial wave packets

The relationship $\hat{\boldsymbol{\omega}} = \mp k\hat{\mathbf{u}}$ tells us that *monochromatic* inertial waves have maximal helicity. Moreover, (4.9) ensures that $h < 0$ whenever wave energy propagates upward ($\mathbf{c}_g \cdot \boldsymbol{\Omega} > 0$), while $h > 0$ for the downward propagation of energy ($\mathbf{c}_g \cdot \boldsymbol{\Omega} < 0$). It does not necessarily follow that a *wave packet* also has maximal helicity, because wave packets contain a range of overlapping wavenumbers. However, it turns out that, typically, they do have h close to maximal, with $h < 0$ for upward-propagating wave packets and $h > 0$ for downward-propagating wave packets (see, for example, the discussion in Davidson (2022).)

This behavior is illustrated in Fig. 4.8, which shows the computed evolution of a layer of rotating turbulence. The initial condition consists of a collection of eddies which are randomly positioned and randomly orientated, but all located close to the central plane (see Fig. 4.8(a)). Also, the initial Rossby number based on the mean eddy size and rms velocity is Ro ≈ 0.1. For $t > 0$, the turbulent eddies radiate inertial waves, and the resulting motion is shown in Fig. 4.8(b) for $\Omega t = 6$. The wave packets resemble columnar vortices aligned with the rotation axis, and these form on the *fast* timescale of Ω^{-1}, and *not* on the slow timescale of $\ell_\perp/u$, the slow timescale being the hallmark of evolution through weakly nonlinear processes. However, the key point is that the wave packets in Fig. 4.8 are colored according to their helicity, with green for $h > 0$ and red for $h < 0$. As expected, the upward propagating wave packets carry negative helicity and the downward propagating packets positive helicity, so the inertial waves spatially segregate the helicity.

4.3 Three experiments on rapidly-rotating turbulence

We now examine the more recent experimental evidence for rapidly-rotating turbulence. We start with those experiments, either homogeneous or inhomogeneous,

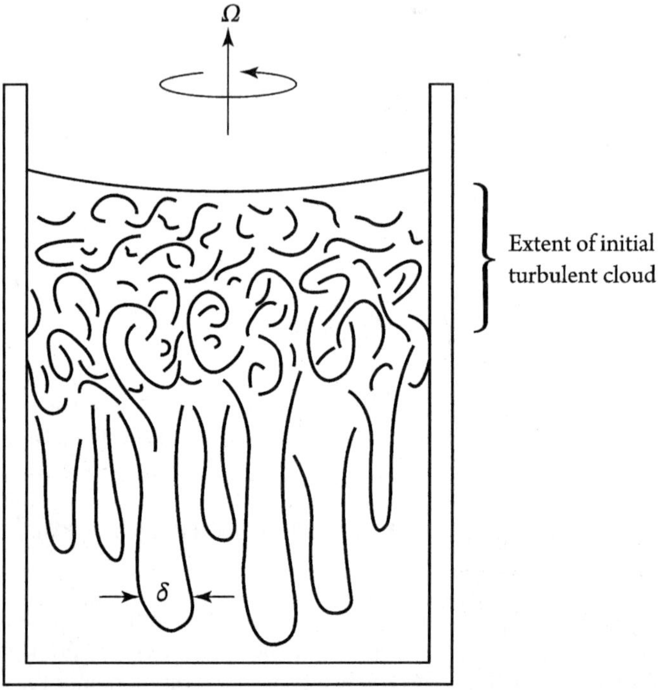

Figure 4.9 Schematic of columnar vortex formation in the experiment of Davidson et al. (2006)). Note that the wider vortices elongate fastest. (From Staplehurst et al. (2008).)

which are most readily interpreted in terms of linear wave packets. We then turn to nonlinear processes in Section 4.3.3.

4.3.1 Columnar vortex formation by linear waves in inhomogeneous turbulence

In the laboratory experiment of Davidson et al. (2006), a flow is generated that is similar to that shown in the lower half of Fig. 4.8. The setup is as shown in Fig. 4.9. Here turbulence is generated in the upper part of a rotating tank of water by dragging a planar mesh partway through the tank, and then removing the mesh. The rotation rate is chosen to ensure that, initially, we have $\text{Ro} = u/\Omega\delta >> 1$, u being the rms velocity and δ the size of the large eddies (the integral scale). Since $\text{Ro} >> 1$, no inertial waves are generated during the formation of the turbulence, which subsequently evolves in a conventional way, at least for a while. However, as the kinetic energy of the turbulence decays, Ro falls, and eventually we enter a regime in which $\text{Ro} \sim O(1)$. The turbulence now feels the background rotation, and it is observed that, when Ro reaches a value of $\text{Ro} \approx 0.4$, columnar vortices aligned with $\mathbf{\Omega}$ emerge from the turbulent cloud, propagating in the axial direction. The overall structure of the flow is as shown in Fig. 4.9.

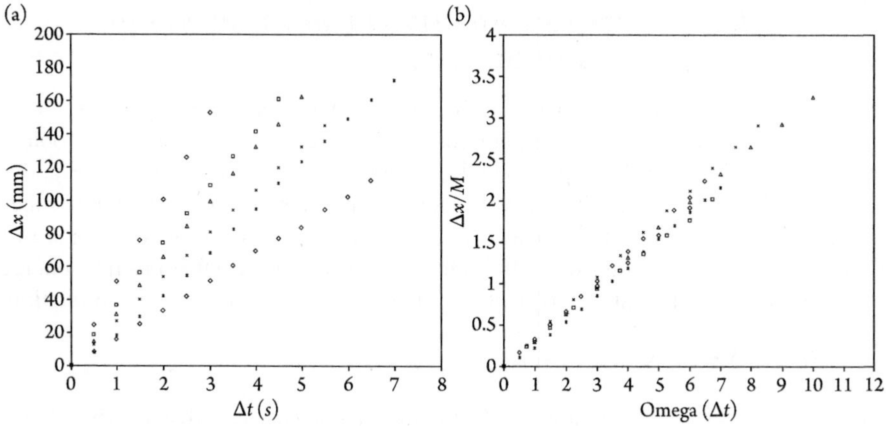

Figure 4.10 The variation of length, Δx, with time of the dominant columnar vortex in six experiments that have different mesh sizes, M, and rotation rates. (a) Δx versus t shows the columnar vortices growing at a constant rate. (b) $\Delta x/M$ versus Ωt confirms that $\Delta x \sim M\Omega t$, consistent with $c_g \sim \Omega\delta$. (From Davidson et al. (2006).)

It is observed that the thick vortices propagate faster than the thin ones, as indicated in Fig. 4.9. This is consistent with these vortices being inertial wave packets, since the group velocity of low-frequency inertial waves is $\mathbf{c}_g = 2\mathbf{\Omega}/k \approx \mathbf{\Omega}\delta/\pi$. Given this observation and the similarities between Figs. 4.9 and the lower half of Fig. 4.8, it seems probable that the columnar structures in this laboratory experiment are also inertial wave packets. This is readily confirmed. First, measurements show that the columnar vortices elongate at a constant rate, as shown in Fig. 4.10(a). This shows the axial length of the dominant columnar vortex, $\Delta x(t)$, in six experiments that have different mesh sizes, M, and different rotation rates. In each case, we find Δx proportional to t. Second, the growth rate is proportional to Ω and to the mesh size, M, which acts as a proxy for the transverse scale of the vortices, δ. This is clear from Fig. 4.10(b), which shows the same data replotted as $\Delta x/M$ versus Ωt. Since the data now all lies on one curve, we have $\Delta x \sim M\Omega t$. This is consistent with $c_g \sim \Omega\delta$, and so confirms that the columnar vortices are indeed inertial wave packets. This crucial role of quasilinear inertial wave packets in columnar vortex formation has been rediscovered several times since 2006.

One curious feature of this experiment is that inertial wave packets exhibiting quasilinear properties are seen to propagate with amplitudes of up to $\text{Ro} = \hat{u}/\Omega\lambda \sim 0.4$, where λ is the dominant wavelength and $\hat{u}$ the wave amplitude. Similar observations have been made in numerical simulations. In particular, the transition from quasilinear wave propagation to the nonlinear suppression of waves starts at around $\text{Ro} \approx 1.4$ for cyclones and $\text{Ro} \approx 0.4$ for anticyclones, although the exact numerical values depend on the precise definition of Ro in a given situation. In any event, this transition turns out to be surprisingly abrupt, requiring only a doubling of Ro to completely suppress the inertial waves. (See, for example, the discussion in Sreenivasan and Davidson (2008).)

4.3.2 Columnar vortex formation by linear waves in homogeneous turbulence

We now turn from inhomogeneous turbulence to statistically homogeneous turbulence, which is often the favorite playground of the turbulence theorist. A problem then arises, at least for the experimentalist. While the columnar vortices in Fig. 4.9 are relatively easy to track, it is hard to identify and then follow individual vortices in statistically homogeneous turbulence, because everything is immersed in a sea of turbulent vortices of varying scale. In such cases, the experimentalist has little choice but to measure various statistical quantities, such as the two-point velocity correlation,

$$Q_{ij}(\mathbf{r}, t) = \left\langle u_i(\mathbf{x}, t) u_j(\mathbf{x} + \mathbf{r}, t) \right\rangle$$

and then try to infer structural information about the turbulence from those correlations. (Here $\langle \sim \rangle$ indicates an ensemble average.) However, as we shall see, the problem of unpacking geometric information from statistical correlations is a nontrivial one, fraught with dangers, especially when there is an interplay between turbulence and waves. The central problem is the following: in forming a two-point correlation, one loses a great deal of phase information, as discussed by Davidson (2015). However, it is precisely the subtle control of phase information that lies at the root of wave dispersion, since the instantaneous location of a wave packet is the point where the phases of the various underlying Fourier modes are such that those modes can add constructively. Before discussing the experimental evidence for columnar vortex formation in homogeneous turbulence, we need to determine exactly which information is retained by the two-point velocity correlations.

It is instructive to return to the inviscid, low Rossby number problem of Section 4.2.3, in which the Gaussian initial condition,

$$u_\theta = \Lambda r \exp\left[-\left(r^2 + z^2\right)/\delta^2\right] \tag{4.19}$$

(expressed in cylindrical polar coordinates) disperses to create a pair of columnar vortices with velocity components

$$u_\theta \approx \Lambda \delta \int_0^\infty \kappa^2 e^{-\kappa^2} J_1\left(\frac{2\kappa r}{\delta}\right) \times \left\{\exp\left[-\left(\frac{z}{\delta} - \frac{\Omega t}{\kappa}\right)^2\right] + \exp\left[-\left(\frac{z}{\delta} + \frac{\Omega t}{\kappa}\right)^2\right]\right\} d\kappa \tag{4.20}$$

$$u_z \approx \Lambda \delta \int_0^\infty \kappa^2 \mathrm{e}^{-\kappa^2} \mathrm{J}_0\left(\frac{2\kappa r}{\delta}\right) \times \left\{-\exp\left[-\left(\frac{z}{\delta} - \frac{\Omega t}{\kappa}\right)^2\right] + \exp\left[-\left(\frac{z}{\delta} + \frac{\Omega t}{\kappa}\right)^2\right]\right\} d\kappa \tag{4.21}$$

(Here J_0 and J_1 are the usual Bessel functions, $\kappa = k_r \delta/2$, and the expressions above are derived by Davidson et al. (2006).) Now consider an initial condition consisting of a sea of such vortices, uniformly but randomly distributed in space. For $t > 0$, this creates a statistically homogeneous (but anisotropic) velocity field whose statistical properties can be determined using superposition and the solution for a single vortex. The resulting two-point velocity correlations

$$Q_{\perp}(\mathbf{r}) = \langle u_x(\mathbf{x})u_x(\mathbf{x}+\mathbf{r}) + u_y(\mathbf{x})u_y(\mathbf{x}+\mathbf{r})\rangle$$

and

$$Q_{\parallel}(\mathbf{r}) = \langle u_z(\mathbf{x})u_z(\mathbf{x}+\mathbf{r})\rangle$$

were calculated by Staplehurst et al. (2008). These turn out to be

$$Q_{\perp} \approx Q_{\perp}^{S}(\mathbf{r}) + \frac{1}{2}\langle \mathbf{u}^2\rangle \int_0^{\infty} \kappa^3 e^{-\kappa^2} J_0\left(\frac{2\kappa r_{\perp}}{\delta}\right) \times \left\{\exp\left[-\left(\frac{r_{\parallel}}{\delta} - \frac{2\Omega t}{\kappa}\right)^2\right] + \exp\left[-\left(\frac{r_{\parallel}}{\delta} + \frac{2\Omega t}{\kappa}\right)^2\right]\right\} d\kappa$$

$$Q_{\parallel} \approx Q_{\parallel}^{S}(\mathbf{r}) - \frac{1}{2}\langle \mathbf{u}^2\rangle \int_0^{\infty} \kappa^3 e^{-\kappa^2} J_0\left(\frac{2\kappa r_{\perp}}{\delta}\right) \times \left\{\exp\left[-\left(\frac{r_{\parallel}}{\delta} - \frac{2\Omega t}{\kappa}\right)^2\right] + \exp\left[-\left(\frac{r_{\parallel}}{\delta} + \frac{2\Omega t}{\kappa}\right)^2\right]\right\} d\kappa$$

where $r_{\parallel} = r_z$, $\mathbf{r}_{\perp} = \mathbf{r} - \mathbf{r}_{\parallel}$, $\kappa = k_{\perp}\delta/2$, and the terms $Q_{\perp}^{S}(\mathbf{r})$ and $Q_{\parallel}^{S}(\mathbf{r})$ are steady contributions to the velocity correlations which remain localized around $|\mathbf{r}| < \delta$.

Evidently, $Q_{\perp} - Q_{\perp}^{S}$ and $Q_{\parallel} - Q_{\parallel}^{S}$ mimic the behavior of individual vortices. In particular, the integrands in these expressions are dominated by $\kappa = O(1)$. Consequently, to keep the argument in the exponentials of order unity for $\Omega t >> 1$, we require $r_{\parallel} \sim \pm\Omega\delta t$, which locates the peaks in the time-dependent part of the correlations. Thus the axial distance over which the velocities are appreciably correlated grows as $\ell_{\parallel} \sim \Omega\delta t$.

Note that, for $\Omega t >> 1$, these expressions simplify considerably. In particular, as shown by Staplehurst et al. (2008), the time-dependent parts of the correlations, $Q_{\perp} - Q_{\perp}^{S}$ and $Q_{\parallel} - Q_{\parallel}^{S}$, become self-similar for large Ωt, provided that $r_{\parallel}$ is normalised by $\ell_{\parallel} = \Omega\delta t$. Evidently, sufficient phase information is retained by the two-point velocity correlations to capture the axial elongation of the vortices by linear wave propagation. Moreover, given the discussion in Section 4.2.3, this behavior is unlikely to be a mere artifact of the Gaussian vortex Eq. (4.19), but rather extends to most homogeneous initial conditions (though see the caveats below).

In summary, if the large eddies elongate by linear wave propagation, then we expect the two-point velocity correlations to become self-similar at large times, except in the region $|\mathbf{r}| < \delta$, where $Q_{\perp}^{S}$ and $Q_{\parallel}^{S}$ dominate. It is necessary only to normalise $r_{\parallel}$ by $\ell_{\parallel} = \Omega\delta t$, where δ is the transverse length scale of the large eddies. Crucially, this provides us with a test for columnar vortex formation by linear wave propagation.

However, the loss of phase information can still be problematic. For example, it is evident from the expressions above that $Q_{ii} = Q_\perp + Q_\parallel$ is constant, despite the elongation of individual eddies. Worse still, it is shown by Staplehurst et al. (2008) that inviscid, linearized dynamics requires

$$\int_0^\infty Q_\perp dr_\parallel = \text{constant}$$

This is problematic because, in axisymmetric turbulence theory, the usual measure of the axial length of the large eddies is

$$\ell_z = \frac{1}{\langle \mathbf{u}_\perp^2 \rangle} \int_0^\infty Q_\perp(r_\perp = 0, r_\parallel) dr_\parallel. \tag{4.22}$$

Evidently, such a definition is entirely inappropriate whenever linear wave propagation plays an important role, as (4.22) is blind to the axial growth of inertial wave packets.

Let us now consider the experimental data of Staplehurst et al. (2008), who generated homogeneous, freely-decaying turbulence at moderate-to-small Ro. Fig. 4.11(a) shows normalized two-point correlations, based on luminosity measurements, as a function of $r_\parallel$ and for one particular value of Ω. It is clear that the axial correlation length, $\ell_\parallel$, increases with time, as the curves systematically shift to the right. The same data is replotted in Fig. 4.11(b) as a function of $r_\parallel / (t - t_0)$, where t_0 represents the notional point at which waves first appear. Evidently, the curves collapse for all but the earliest of times. This confirms that, after an initial transient, the axial correlation length grows linearly with time, which is consistent with the growth of the large eddies by linear wave propagation.

Next, Fig. 4.12 shows normalized two-point correlations (excluding any initial transient) for two experiments performed at different rotation rates. In Fig. 4.12(a), the curves are plotted as a function of $r_\parallel / (t - t_0)$ and the correlations for each experiment collapse, but to different curves. However, when replotted against $r_\parallel / 2\Omega (t - t_0)$, as in Fig. 4.12(b), the two-point correlations from both experiments collapse to a single curve. We conclude that the axial correlation length grows as $\ell_\parallel \sim \Omega t$, which is consistent with the large eddies elongating by linear wave propagation.

4.3.3 Nonlinear effects in approximately homogeneous turbulence

Let us now turn to the experiments of Morize et al. (2005); Morize and Moisy (2006). Their setup was similar to that of Staplehurst et al. (2008), although the depth of water was shallower and so axial confinement effects were more pronounced. As in previous experiments, the large scales were found to be dominated by long-lived columnar vortices. However, these authors focused less on the initial formation of the columnar vortices and more on longer-term, nonlinear effects, such as the spectral exponent in the energy spectrum, $E(k)$, and the rate of decay of kinetic energy.

Regarding the energy spectrum, the inertial-range scaling exponent in $E \sim k_\perp^{-m}$ was found to increase from the Kolmogorov value $m = 5/3$ at short times to around

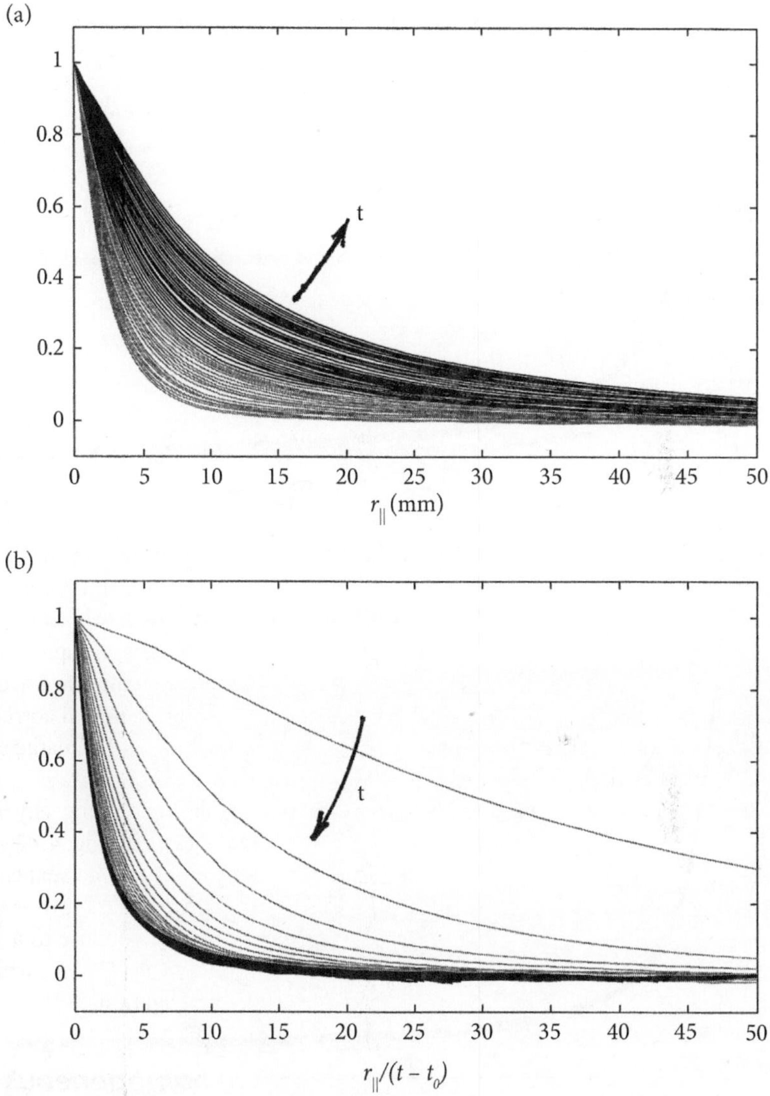

Figure 4.11 Normalized two-point correlations in one of the homogeneous experiments of Staplehurst et al. (2008). The correlations are plotted against: (a) $r_{\|}$; and (b) $r_{\|} / (t - t_0)$. (From Staplehurst et al. (2008).)

$m = 2.3$ at large Ωt, this steepening of the spectrum being consistent with certain phenomenological theories of rotating turbulence. However, at no time did m reach a value of 3, which is characteristic of two-dimensional turbulence, indicating that the columnar vortices are restricted to the large scales. Morize and Moisy (2006) also found that the energy per unit mass decays as $u^2 \sim t^{-n}$ where the exponent n varied from 2 at low rotation rates to 1 for fast rotation. Morize and Moisy (2006)

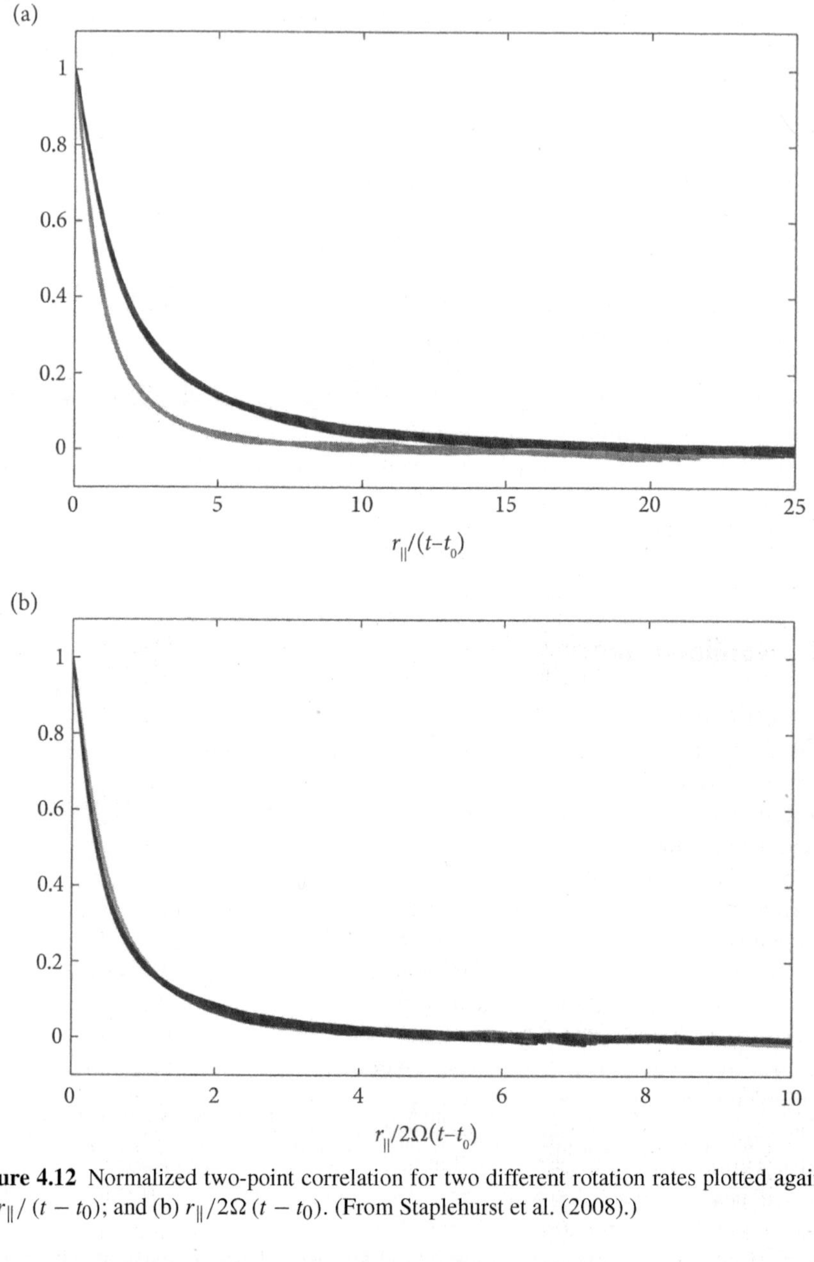

Figure 4.12 Normalized two-point correlation for two different rotation rates plotted against: (a) $r_{\|}/(t-t_0)$; and (b) $r_{\|}/2\Omega(t-t_0)$. (From Staplehurst et al. (2008).)

speculate that the higher values of n are due to the confinement of the turbulence, and so the measured value of $n = 1$ is of most interest from the point of view of homogeneous turbulence. The same exponent (of $n = 1$) was also seen in the experiments of Staplehurst et al. (2008), and indeed, there are tentative theoretical reasons for believing in $n = 1$, as discussed in Davidson et al. (2013). Finally, Morize et al. (2005), like Hopfinger et al. before them, observed a dominance of cyclonic columnar vortices over anticyclonic vortices. They also offered a simple explanation for this statistical bias, which we shall discuss in §4.5.

In summary, then, laboratory experiments at moderate-to-low Ro confirm that, immediately after the turbulence is initiated, the *large eddies* elongate on the fast timescale of Ω^{-1} by quasilinear wave propagation, and this is true in both homogeneous and inhomogeneous turbulence. They also show that these columnar eddies are predominantly cyclonic. This is not the end of the story, however, because at large Reynolds number there is always a wide range of eddy sizes, and while the large scales might be dominated by the background rotation, this is rarely the case for the small scales, as indicated by the spectral exponent of $m = 2.3$ in the experiment of Morize et al. (2005). That is, the small scales are usually too small to feel the background rotation directly, and so they evolve through nonlinear dynamics. There is then the possibility of an interplay between linear and nonlinear processes. Moreover, there is a well-established nonlinear theory that predicts a slow accumulation of energy in the columnar modes on the timescale of ℓ/u, as we now discuss.

4.4 Nonlinear mechanisms of embedding the energy of the columnar modes

Let Re be the Reynolds number based on the large-scale properties of the turbulence. Then, in conventional turbulence, the rms vorticity at the small and large scales are, according to Kolmogorov's classical scaling, related by $\omega_{\text{small}} \sim \text{Re}^{1/2}\,\omega_{\text{large}}$. Thus, for example, if $\text{Re} \sim 10^4$, then $\omega_{\text{small}} \sim 100\omega_{\text{large}}$. Now, we can redefine Ro along the lines of $\text{Ro} = \omega_{\text{large}}/2\Omega$ and, based on this definition, Ro in a rotating turbulence experiment is typically around 0.1 or larger. It follows that the value of Ro based on the small-scale vorticity might be of the order of 10. This means that the small scales in a typical laboratory experiment rarely feel the background rotation directly. Rather, the large-scale columnar eddies are immersed in a sea of smaller, nonlinear vortices. In such a flow, there could exist an intermediate range of scales where both linear and nonlinear dynamical processes co-exist. In this respect, it is interesting to recall that the transition from quasilinear inertial wave propagation to fully nonlinear dynamics can be surprisingly abrupt (see Section 4.3.1), occurring over a relatively narrow range of Ro. This opens up the possibility that a columnar vortex of intermediate scale, which is perhaps undergoing elongation by quasilinear wave propagation, might be immersed in a field of *slightly smaller* vortices that evolve nonlinearly.

Let us now return to Fig. 4.1 and consider eddies of intermediate size sitting in the shadow of a slightly larger, columnar vortex and being subject to the strain field of that vortex. In such a situation, it is natural to ask if some of the characteristics of the columnar vortex could be imprinted on the eddies of intermediate scale through vortex stretching. Certainly, there is some hint of this process in Fig. 4.1, where the smaller eddies are being spiralled around the columnar vortex. If this were a common occurrence in rotating turbulence, it would constitute a simple, systematic, nonlinear mechanism of embedding the axial growth of the large scales on the smaller vortices.

Whether or not this picture is physically reasonable (and there is currently little evidence either way), the nonlinear transfer of energy towards columnar modes has been consistently observed in numerical simulations of rotating turbulence, and so there is some incentive to characterize the interplay between inertial waves and nonlinear turbulence. There are then at least two classes of interactions that need to be considered: (i) *strongly nonlinear* wave-vortex interactions at Ro $\sim$ 1 and; (ii) the *weakly nonlinear* coupling of waves for Ro $<<$ 1. The mathematical apparatus for both regimes has been developed for turbulence in a number of wave-bearing systems, including stratified, magnetohydrodynamic (MHD), and Rossby-wave turbulence.

The strongly nonlinear wave-vortex interactions are often captured using *critical-balance theory*, which applies to the intermediate (i.e., inertial) range of scales. This theory asserts that, *at each scale within the inertial range*, the rms vorticity of the eddies is of the order of the frequency of the appropriate wave type (at that scale). Although critical balance theory is phenomenological, it is well established for stratified turbulence (see Brethouwer et al. (2007), MHD turbulence (see Goldreich and Sridhar (1995)), and Rossby-wave turbulence (see Rhines (1975)). In each case, it makes testable predictions as to the scale-by-scale distribution of energy, predictions which are often consistent with observations. However, critical balance theory is not well developed in the context of inertial-wave turbulence, and so we shall focus here on the weakly nonlinear theory.

Perhaps the most commonly adopted framework for describing the weakly nonlinear coupling of waves is the theory of resonant triad interactions. This framework has been used to describe a number of weakly nonlinear wave-turbulence systems, most famously Rossby-wave turbulence (Rhines, 1975). In the case of weakly nonlinear Rossby waves, two spatially overlapping wave packets with dominant wavevectors $\mathbf{p}$ and $\mathbf{q}$, and corresponding frequencies $\varpi_p = \varpi(\mathbf{p})$ and $\varpi_q = \varpi(\mathbf{q})$, interact nonlinearly to generate a third wave with a wavevector of $\mathbf{k} = \mathbf{p} \pm \mathbf{q}$, and a frequency of $\varpi = \varpi_p \pm \varpi_q$. The response is weak, however, unless the excited frequency is close to the natural frequency of waves of wavevector $\mathbf{k}$. So, to get a significant nonlinear interaction, the dispersion relationship must admit solutions of the form

$$\varpi(\mathbf{p} \pm \mathbf{q}) \approx \varpi(\mathbf{p}) \pm \varpi(\mathbf{q}) \tag{4.23}$$

(See, for example, the discussion in Pedlosky (1979). When this resonance condition is satisfied, a significant nonlinear transfer of energy between wave packets can occur, and this happens on the *slow* timescale of ℓ/u. Now an isolated triad of Rossby waves conserves energy and enstrophy, and its evolution is reversible, exhibiting no arrow of time. However, according to Rhines (1975), the most probable interactions tend to nudge the wave energy towards waves of low frequency, and these have a zonal (east-west) group velocity. At one time, it was hoped that such a cascade of resonant triad interactions could explain the observed east-west winds in the atmospheres of the giant planets, and perhaps similar currents in the terrestrial oceans. It is now thought, however, that the true explanation for the zonal winds on the giant planets is somewhat different.

Let us now return to inertial waves and to rotating turbulence. A similar resonant triad theory has been developed by Waleffe (1993) for inertial waves, which embodies energy and helicity conservation. However, the dispersion relationship for inertial waves makes it harder to find solutions to Eq. (4.23), which is sometimes relaxed to the weaker requirement that

$$\varpi(\mathbf{p} \pm \mathbf{q}) = \varpi(\mathbf{p}) \pm \varpi(\mathbf{q}) + O(\mathrm{Ro}) \tag{4.24}$$

for a small but finite Ro. As with Rossby waves, the resonant triads possess no arrow of time. That is to say, for every initial condition that pushes the triad energy towards a columnar mode, there is another that does the opposite. To resolve this problem, Waleffe (1993) and Smith and Waleffe (1999) argue that certain triadic interactions are statistically more likely to occur than others and, as with Rossby waves, these tend to favour energy passing to low frequencies. (It is probably no accident that the energy is expected to pass to low frequencies for both Rossby and inertial waves. There is a general result for conservative resonant triads, due to Hasselmann (1967), which states that a finite-amplitude wave of non-zero frequency is unstable to two infinitesimal waves of lower frequency. This suggests, but does not prove, that there is a statistical bias for energy to pass to lower frequencies through triadic interactions.) Since low-frequency inertial waves are columnar, Waleffe's theory suggests that a cascade of near-resonant triads could nudge the kinetic energy of the inertial waves towards columnar modes.

Certainly, the weakly nonlinear transfer of wave energy towards columnar modes has been well documented in numerical simulations of rotating turbulence at low Ro (see, for example, Smith and Lee (2005), with the energy transfer occurring on the slow timescale of ℓ/u. It should be noted, however, that these simulations are nearly always carried out in periodic cubes. Crucially, these cubes have unphysical boundary conditions (a wave packet leaving the domain at the top, say, immediately re-enters at the bottom), and it is probable that the long-term evolution of rotating turbulence in a periodic cube is quite different to confined rotating turbulence, say bounded above and below by no-slip boundaries. (For vertically confined turbulence, Ekman pumping at the upper and lower boundaries soon controls the evolution of the columnar eddies, as discussed in Godeferd and Lollini (1999).) Moreover, the precise *physical* mechanism behind the weakly nonlinear shift in energy remains obscure. Given that our understanding of strongly nonlinear wave-vortex interactions at Ro ~ 1 is also poor, it seems that there is still much to learn about nonlinear energy transfer in rotating turbulence.

4.5 The cyclone, anticyclone asymmetry

We end by returning to the observation of Hopfinger et al. (1982) that there is a statistical bias towards cyclonic vortices, that is, columnar vortices in which the axial vorticity is positive in the rotating frame of reference, $\boldsymbol{\omega} \cdot \boldsymbol{\Omega} > 0$. It should be emphasized that this observation is not merely an artefact of the experimental setup in Hopfinger et al. (1982). Rather, it has been consistently observed in many other experiments, such as Morize et al. (2005) and Staplehurst et al. (2008).

There are several tentative explanations for the dominance of cyclones, but we shall focus on just two. Perhaps the most common explanation goes back to Bartello et al. (1994), who restrict attention to two-dimensional, axisymmetric vortices with a monotonic (Gaussian-like) velocity profile and $\mathrm{Ro} \sim 1$. For this subclass of columnar vortices, it may be shown that, on returning to an inertial frame of reference, the anticyclones are more likely to be unstable than cyclones to Rayleigh's centrifugal instability. So the suggestion is that, if anticyclones were to form, they are less likely to survive than cyclones, at least when $\mathrm{Ro} \sim 1$. However, it seems improbable that anticyclones form and then go unstable, whilst cyclones form and persist. It is much more probable that, when $\mathrm{Ro} \sim 1$, cyclones form more readily than anticyclones. This is consistent with the observations of Sreenivasan and Davidson (2008), who note that, for an axisymmetric flow, columnar cyclones can form by quasilinear wave propagation whenever $\mathrm{Ro} < 1.4$, whereas columnar anticyclones require the lower value of $\mathrm{Ro} < 0.4$. In short, when $\mathrm{Ro} \sim 1$, the production of cyclones outstrips that of anticyclones.

The second explanation goes back to Morize et al. (2005), who focus on the vortex stretching term, $\boldsymbol{\omega} \cdot \nabla \mathbf{u}$, in the vorticity equation. Again, they return to an inertial frame of reference and note that, in that frame, the vortex stretching term becomes $(\boldsymbol{\omega} + 2\boldsymbol{\Omega}) \cdot \nabla \mathbf{u}$. The suggestion, then, is that the cyclonic vorticity will be subject to more intensive stretching than the anticyclonic vorticity, and hence will dominate in the long term. Such an explanation is not restricted to $\mathrm{Ro} \sim 1$, nor to axisymmetric flow, and so has the appeal of generality.

4.6 Unanswered questions and future prospects

Despite significant advances over the last forty years, there remain many questions about rotating turbulence. For example, the relative roles played by linear and nonlinear dynamics in establishing the columnar vortices remain controversial. It seems likely that linear dynamics dominate in the early stages, but, depending on the boundary conditions, weakly nonlinear dynamics may dominate later development. Also, nearly all numerical experiments are carried out in periodic cubes, which have idiosyncratic boundary conditions of a type never encountered in the laboratory or in geophysics. In particular, a fast-moving wave packet can traverse the cube and re-enter at the opposite face in a relatively short time, so a single wave packet may be recycled through the cube many dozens of times during a given numerical experiment. This raises the question as to how to relate periodic cube simulations to observations made in real turbulent flows, such as those occurring in rotating machinery, tropical cyclones, and planetary cores, all of which are bounded by impermeable surfaces.

So, perhaps there is a need to move beyond the simple turbulent flows that are the focus of many current studies, such as quasi-homogeneous laboratory experiments and periodic cube simulations, and embrace the fact that naturally occurring turbulence is invariably inhomogeneous and confined. In the case of laboratory experiments, perhaps more emphasis could be placed on the types of inhomogeneous rotating flows observed in nature, such as those in tornadoes or hurricanes. The simulations, on the other hand, might follow Godeferd and Lollini (1999) and impose more physically relevant boundary conditions, even if that limits the ability to explore certain asymptotic regimes.

Then there is the unresolved question of strongly nonlinear wave-vortex interactions in rotating turbulence at Ro $\sim$ 1. While critical balance theory has provided a unified theoretical framework for understanding strongly nonlinear interactions in stratified and MHD turbulence, it has yet to be fully explored in the context of rotating turbulence. This might provide a more physically motivated means of exploring strong nonlinearity than the traditional spectral closure schemes, such as EDQNM.

Finally, perhaps we must guard against the simplistic notion that there is a single, all-encompassing theory of rotating turbulence. It seems probable that the dominant behaviour will depend strongly on whether the waves and turbulence are confined by impermeable boundaries, are continually recycled through a periodic cube, or take the form of a localised cloud evolving in an unbounded fluid. One is reminded of Saffman's observation:

> *Finally, we should not altogether neglect the possibility that there is no such thing as 'turbulence'. That is to say, it is not meaningful to talk of the properties of a turbulent flow independently of the physical situation in which it arises. In searching for a theory of turbulence, perhaps we are looking for a chimera... Perhaps there is no real 'turbulence problem', but a large number of turbulent flows and our problem is the self-imposed and possibly impossible task of fitting many phenomena into the Procrustean bed of a universal turbulence theory. Individual flows should then be treated on their merits and it should not necessarily be assumed that ideas valid for one situation will transfer to others.*
> *P.G. Saffman (1977)*

Perhaps we should leave the last word to Philip Saffman.

References

Bartello, P., Metais, O., Lesieur, M., 1994. Coherent structures in rotating three-dimensional turbulence. Journal of Fluid Mechanics 273, 1–29.

Brethouwer, G., Billant, P., Lindborg, E., Chomaz, J.-M., 2007. Scaling analysis and simulations of strongly stratified flows. Journal of Fluid Mechanics 585, 343–368.

Davidson, P.A., 2013. Turbulence in Rotating, Stratified and Electrically Conducting Fluids. Cambridge University Press.

Davidson, P.A., 2015. Turbulence: an Introduction for Scientists and Engineers, 2nd ed. Oxford University Press.

Davidson, P.A., 2022. Incompressible Fluid Dynamics. Oxford University Press.

Davidson, P.A., 2024. The Dynamics of Rotating Fluids. Oxford University Press.

Davidson, P.A., Staplehurst, P.J., Dalziel, S.B., 2006. On the evolution of eddies in a rapidly rotating system. Journal of Fluid Mechanics 557, 135–144.

Davidson, P.A., et al., 2013. Ten Chapters in Turbulence. Cambridge University Press.

Dickinson, S.C., Long, R.R., 1983. Oscillating-grid turbulence including effects of rotation. Journal of Fluid Mechanics 126, 315–333.

Godeferd, F.S., Lollini, L., 1999. Direct numerical simulations of turbulence with rotation and confinement. Journal of Fluid Mechanics 393, 257–308.

Goldreich, P., Sridhar, S., 1995. Towards a theory of interstellar turbulence. II: Strong Alfvénic turbulence. The Astrophysical Journal 438, 763–775.

Hasselmann, K., 1967. A criterion for nonlinear wave stability. Journal of Fluid Mechanics 30 (4), 737.

Hopfinger, E.J., Browand, F.K., Gagne, Y., 1982. Turbulence and waves in a rotating tank. Journal of Fluid Mechanics 125, 505–534.

Ibbetson, A., Tritton, D.J., 1975. Experiments on turbulence in a rotating fluid. Journal of Fluid Mechanics 68 (4), 639–672.

Morize, C., Moisy, F., Rabaud, M., 2005. Decaying grid-generated turbulence in a rotating tank. Physics of Fluids 17, 095105.

Morize, C., Moisy, F., 2006. Energy decay of rotating turbulence with confinement effects. Physics of Fluids 18, 065107.

Pedlosky, J., 1979. Geophysical Fluid Dynamics, 1st ed. Springer-Verlag.

Rhines, P.B., 1975. Waves and turbulence on a β-plane. Journal of Fluid Mechanics 69 (3), 417–443.

Smith, L.M., Lee, Y., 2005. On near resonance and symmetry breaking in forced rotating flows at moderate Rossby number. Journal of Fluid Mechanics 535, 111–142.

Smith, L.M., Waleffe, F., 1999. Transfer of energy to two-dimensional large scales in forced, rotating three-dimensional turbulence. Physics of Fluids 11 (6), 1608–1622.

Sreenivasan, B., Davidson, P.A., 2008. On the formation of cyclones and anticyclones in a rotating fluid. Physics of Fluids 20 (8), 085104.

Staplehurst, P.J., Davidson, P.A., Dalziel, S.B., 2008. Structure formation in homogeneous freely decaying rotating turbulence. Journal of Fluid Mechanics 598, 81–105.

Taylor, G.I., 1923. Experiments on the motion of solid bodies in rotating fluids. Proceedings of the Royal Society of London. Series A 104, 213–218.

Waleffe, F., 1993. Inertial transfers in the helical decomposition. Physics of Fluids. A, Fluid Dynamics 5 (3), 677–685.

Quasi-geostrophic flows and two-dimensional turbulence

5

Guido Boffetta and Stefano Musacchio
Department of Physics and INFN, University of Torino, Torino, Italy

5.1 The motion of a fluid in a rotating reference frame

In general, the motion of a fluid is described in terms of evolution equations for the mass density field $\rho(\mathbf{x}, t)$ and the velocity field $\mathbf{v}(\mathbf{x}, t)$

$$\frac{D\rho}{Dt} + \rho \nabla \cdot \mathbf{v} = 0 \tag{5.1}$$

$$\frac{D\mathbf{v}}{Dt} = -\frac{1}{\rho}\nabla p - \nabla \phi_e + \nu \nabla^2 \mathbf{v} \tag{5.2}$$

where D/Dt represents the convective (or substantial) derivative $D/Dt \equiv \partial/\partial t + \mathbf{v} \cdot \nabla$ (i.e., the derivative following the local velocity), ν is the kinematic viscosity of the fluid and the density is related to the pressure (and possibly also to the temperature field) by an equation of state $\rho = \rho(p)$. ϕ_e represents the potential of an external conservative force, which, in many applications, is gravity and $\phi_e = -GM/r$ (G is the gravitational constant, M the mass of the Earth, and r the distance from the center). On the surface of the Earth, the external acceleration entering in Eq. (5.2) reduces to $GM/R^2 = g \simeq 9.81\ m\,s^{-2}$ in the radial direction ($R \simeq 6.3 \times 10^6\ m$ is the Earth mean radius and $M \simeq 5.98 \times 10^{24} kg$ is its mass).

In many environmental flows, one can safely consider the density constant in space and time and therefore Eq. (5.2) reduces to the Navier–Stokes equation for an incompressible flow with $\rho = \rho_0$ ($\rho_0 \simeq 10^3\ kg\,m^{-3}$ for water) together with the incompressibility condition which is obtained from Eq. (5.1) assuming ρ constant

$$\nabla \cdot \mathbf{v} = 0 \tag{5.3}$$

Eq. (5.2) is valid in an inertial reference frame. In geophysical applications, one has to take into account the fact that the Earth is rotating around its axis with angular velocity $\Omega \simeq 7.3 \times 10^{-5}\ rad\,s^{-1}$ (i.e., one rotation per day) and therefore the reference frame is not inertial. In the reference frame of the Earth, the momentum Eq. (5.2) has two additional terms

$$\frac{D\mathbf{v}}{Dt} + 2\mathbf{\Omega} \times \mathbf{v} + \mathbf{\Omega} \times (\mathbf{\Omega} \times \mathbf{r}) = -\frac{1}{\rho}\nabla p - \nabla \phi_e \tag{5.4}$$

the first, $2\mathbf{\Omega} \times \mathbf{v}$ is the Coriolis term while the second $\mathbf{\Omega} \times (\mathbf{\Omega} \times \mathbf{r})$ represents the centrifugal acceleration. In Eq. (5.4) we neglect, for simplicity, the viscous term (which is not relevant for large-scale motions) and we assume that the rotation axis

Environmental Turbulence. https://doi.org/10.1016/B978-0-32-395873-8.00012-7
Copyright © 2026 Elsevier Inc. All rights are reserved, including those for text and data mining, AI training, and similar technologies.

is the z axis and thus $\mathbf{\Omega} = (0, 0, \Omega)$. The centrifugal acceleration can be written as the gradient of a centrifugal potential $\phi_c = -\frac{1}{2}r_\perp^2\Omega^2$ (where $r_\perp$ is the perpendicular distance to the Earth's rotation axis), and Eq. (5.4) becomes

$$\frac{D\mathbf{v}}{Dt} + 2\mathbf{\Omega} \times \mathbf{v} = -\frac{1}{\rho}\nabla p - \nabla\phi \tag{5.5}$$

where $\phi = \phi_e + \phi_c$ is called the *geopotential*.

One simple consequence of the centrifugal acceleration is that the Earth at equilibrium under gravity and rotation is not a sphere but a *geoid*, approximately an ellipsoid of constant geopotential. Since $\phi = -GM/r - 1/2r_\perp^2\Omega^2$ the difference of the Earth's radius $\Delta R = R_E - R_P$ from the equator (where $\phi_c = -\frac{1}{2}R_E^2\Omega^2$) and the pole (where $\phi_c = 0$) gives $\Delta R \simeq 11\ km$. We note that the actual deformation of the Earth is about twice this number, $\Delta R \simeq 21\ km$, which can be obtained by going beyond the simple point-mass treatment.

Another important consequence of Eq. (5.5) is obtained for strong rotation, i.e., when the fluid acceleration $D\mathbf{v}/Dt$ can be neglected with respect to the Coriolis term. By taking the curl of Eq. (5.5) in this limit one obtains

$$\nabla \times (\mathbf{\Omega} \times \mathbf{v}) = \mathbf{v} \cdot \nabla\mathbf{\Omega} + \mathbf{\Omega}\nabla \cdot \mathbf{v} - \mathbf{\Omega} \cdot \nabla\mathbf{v} - \mathbf{v}\nabla \cdot \mathbf{\Omega} = 0 \tag{5.6}$$

where we have made use of the identity (5.95). Since we have assumed constant rotation along the z axis (and incompressibility), Eq. (5.6) reduces to

$$\frac{\partial\mathbf{v}}{\partial z} = 0 \tag{5.7}$$

which is the celebrated *Taylor–Proudman theorem*. This theorem states that under the condition of strong rotation the velocity field becomes independent of the coordinate along the rotation axis.

5.2 The geostrophic flow

The large-scale fluid motion in both the atmosphere and the oceans can often be considered almost two-dimensional, since the maximum vertical motion is of a few kms, while the horizontal extension can be of thousands of km. It is therefore convenient to introduce the local reference frame sketched in Fig. 5.1, in which z represents the local vertical (radial) direction at latitude θ. We decompose the Earth's rotation to local vertical and horizontal components, $\Omega_z = \Omega\sin\theta$ and $\Omega_y = \Omega\cos\theta$. Locally, the geopotential is $\phi = gz$ and therefore the momentum Eqs. (5.5), written for the three components of the velocity (u, v, w) in the directions (x, y, z) are

$$\frac{Du}{Dt} - 2\Omega_z v + 2\Omega_y w = -\frac{1}{\rho}\frac{\partial p}{\partial x} \tag{5.8}$$

$$\frac{Dv}{Dt} + 2\Omega_z u = -\frac{1}{\rho}\frac{\partial p}{\partial y} \tag{5.9}$$

$$\frac{Dw}{Dt} - 2\Omega_y u = -\frac{1}{\rho}\frac{\partial p}{\partial z} - g \tag{5.10}$$

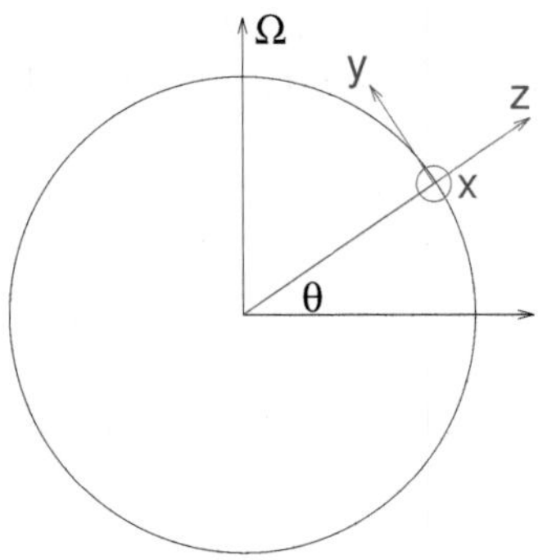

Figure 5.1 The local coordinate system at a point at latitude θ on the Earth.

In several applications, we can neglect the terms containing Ω_y. This can be done at high latitude (where $\Omega_y < \Omega_z$) or when, as is usually the case, the vertical motions are much smaller than the horizontal ones $w \ll (u, v)$. In this case, the term $2\Omega_y w$ in Eq. (5.8) becomes negligible and, for conservation of energy, also the term $2\Omega_y u$ must be removed in Eq. (5.10) Salmon (1998). The vertical component of $\mathbf{\Omega}$, called the Coriolis parameter, is usually written as $f \equiv 2\Omega_z$, which, by definition, depends on the coordinate along the latitude y. For flows of very limited extent, the variation of f can be fully neglected, resulting in the so-called *f-plane approximation* where a constant value is used. For somewhat larger domains horizontally, we can expand the latitude dependence of f to the first order, obtaining the *β-plane approximation* $f = f_0 + \beta y$ where $f_0 = 2\Omega \sin\theta_0$ and $\beta = (2\Omega/R)\cos\theta_0$. For mid-latitude locations, we have $f \simeq 10^{-4}\, rad\, s^{-1}$ and $\beta \simeq 1.6 \times 10^{-11}\, m^{-1}\, s^{-1}$.

The momentum equations in the *β-plane approximation* read

$$\frac{Du}{Dt} - (f_0 + \beta y)v = -\frac{1}{\rho}\frac{\partial p}{\partial x} \tag{5.11}$$

$$\frac{Dv}{Dt} + (f_0 + \beta y)u = -\frac{1}{\rho}\frac{\partial p}{\partial y} \tag{5.12}$$

$$\frac{Dw}{Dt} = -\frac{1}{\rho}\frac{\partial p}{\partial z} - g \tag{5.13}$$

which still contain, to the first order, the dynamically important effect of the dependence of f with the latitude.

One important approximation, the *geostrophic balance*, is obtained from Eqs. (5.11) –(5.12) when the fluid accelerations are small compared to the other terms. In this case, we can put $D/Dt = 0$ in the equations to obtain

$$\rho f v = \frac{\partial p}{\partial x} \tag{5.14}$$

$$\rho f u = -\frac{\partial p}{\partial y} \tag{5.15}$$

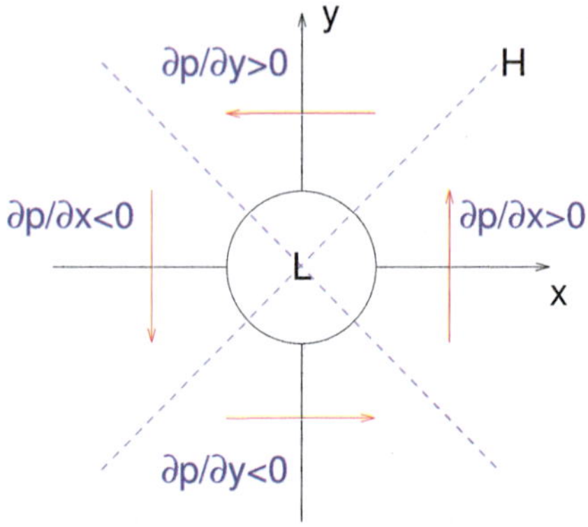

Figure 5.2 The geostrophic wind around a region of low pressure (in the Northern Hemisphere). According to Eqs. (5.14)–(5.15), the vertical (and horizontal) component of the velocity is proportional to $\partial p/\partial x$ (and $-\partial p/\partial y$): the resulting flow is represented by the red arrows.

while Eq. (5.13) gives the hydrostatic balance

$$\frac{\partial p}{\partial z} + \rho g = 0 \tag{5.16}$$

The geostrophic balance is very important in the analysis of the large-scale atmospheric flows, and meteorologists routinely use pressure measurements to estimate the geostrophic velocity. As shown in Fig. 5.2, Eqs. (5.14)–(5.15) predict a counterclockwise-rotating wind (a cyclone) around a region of low pressure (in the Northern Hemisphere where f is positive). Similarly, the geostrophic velocity is a clockwise-rotating wind (an anticyclone) around a region of high pressure.

In principle, the geostrophic balance can also be applied to the ocean. Unfortunately, in this case, it is not of practical use because of the large density of the fluid (remember that water is about 1000 times denser than air). Indeed, a small error in the determination of the horizontal direction produces, because of the large vertical pressure gradient given by Eq. (5.16), a large error in the measurement of horizontal pressure differences and in the corresponding geostrophic currents.

A way to overcome this difficulty is to take the z-derivative of (5.14) and to use Eq. (5.16) to obtain

$$f\frac{\partial}{\partial z}(\rho v) = \frac{\partial^2 p}{\partial x \partial z} = -g\frac{\partial \rho}{\partial x} \tag{5.17}$$

and similarly for Eq. (5.15). Since in the ocean the density changes by only a few percent, we can neglect its z dependence in the first term of Eq. (5.17), replace $\rho \approx \rho_0$, and write the geostrophic balance for the (derivative of) horizontal velocities as

$$\frac{\partial u}{\partial z} = \frac{g}{\rho_0 f}\frac{\partial \rho}{\partial y} \tag{5.18}$$

$$\frac{\partial v}{\partial z} = -\frac{g}{\rho_0 f}\frac{\partial \rho}{\partial x} \tag{5.19}$$

Eqs. (5.18)–(5.19) are important to oceanographers because horizontal differences of the mass density are much easier to measure than horizontal differences of pressure. Unfortunately, the measurements of ρ only reveals the z-derivative of the horizontal velocity and not the velocity itself. These equations are also quite important in atmospheric dynamics, where they are referred to as the *thermal wind equations* (see Chapter 1, Section 1.3.4).

Another general application of the geostrophic balance is obtained by taking the y and x derivatives of Eqs. (5.14) and (5.15), respectively, to get

$$\frac{\partial}{\partial y}(\rho f v) = \frac{\partial^2 p}{\partial x \partial y} \tag{5.20}$$

$$\frac{\partial}{\partial x}(\rho f u) = -\frac{\partial^2 p}{\partial x \partial y} \tag{5.21}$$

and therefore, by adding Eqs. (5.20) and (5.21),

$$\frac{\partial}{\partial x}(\rho f u) + \frac{\partial}{\partial y}(\rho f v) = 0 \tag{5.22}$$

If we assume $\rho \approx \rho_0$ and $f = f_0 + \beta y$, we have

$$\rho_0 f_0 \left(\frac{\partial u}{\partial x} + \frac{\partial v}{\partial y}\right) + \rho_0 \beta v = 0 \tag{5.23}$$

Using the incompressibility condition, $\frac{\partial u}{\partial x} + \frac{\partial v}{\partial y} = -\frac{\partial w}{\partial z}$ (which is consistent with the assumption of constant density, see Eq. (5.1)), we finally obtain

$$\frac{\partial w}{\partial z} = \frac{\beta}{f_0} v \tag{5.24}$$

Since $\frac{\beta}{f_0} \simeq 1/R \simeq 1.6 \times 10^{-7}\, m^{-1}$, Eq. (5.24) indicates that the derivative of the vertical velocity is much smaller (in the usual units) than the horizontal velocity v. Since the vertical velocity itself is zero at the surface of the ocean (or at the bottom of the atmosphere), the above result suggests that, in general, vertical velocities are much smaller than the horizontal ones. From a practical point of view, Eq. (5.24) can be used to estimate the vertical velocity w from the measurements of v.

5.3 The shallow-water equations

The geostrophic balance is a *diagnostic* equation for the horizontal velocity, i.e., it gives the instantaneous value of the velocity given the pressure field, but does not give information on the evolution of the dynamics (in terms of a time-integrable partial differential equation) as the original *prognostic* model, Eqs. (5.1)–(5.2).

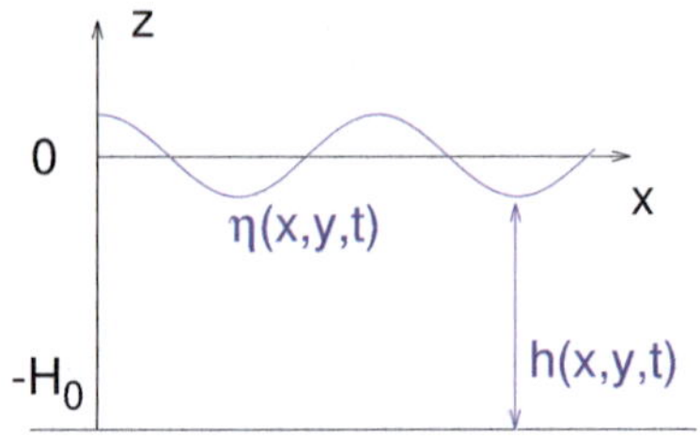

Figure 5.3 The geometry of the shallow water equations.

In order to make the next step in our description of large-scale geophysical flows, we now consider a layer of fluid (e.g., a part of an ocean or atmosphere) of uniform density ρ_0 (for the atmosphere, this would the reference Boussinesq density, see Chapter 1) with a free surface at $z = \eta(x, y, t)$ and a rigid flat bottom at $z = -H_0$ (see Fig. 5.3). At the static equilibrium, the free surface coincides with $z = 0$.

We suppose that the vertical acceleration can be neglected, and therefore we can use the hydrostatic balance. By integrating Eq. (5.16) from a generic depth z to the surface we have

$$p(x, y, z, t) = p_0 + \rho_0 g\,(\eta(x, y, t) - z) \tag{5.25}$$

where p_0 represents the pressure on the surface, which can be taken as $p_0 = 0$. The equation of motion for the horizontal, 2D velocity field $\mathbf{u} = (u, v)$ in the β-plane approximation can be therefore written, from Eqs. (5.11)–(5.12), as

$$\frac{D\mathbf{u}}{Dt} + \mathbf{f} \times \mathbf{u} = -g\nabla\eta \tag{5.26}$$

where we have introduced the Coriolis vector $\mathbf{f} = (0, 0, f)$ and, from now on, $\nabla = (\partial_x, \partial_y)$ represents the horizontal derivative operator. Since the RHS of Eq. (5.26) does not depend on z, we can assume that also $\mathbf{u}(x, y)$ is z-independent.

The incompressibility conditions (assuming constant density) are $\nabla \cdot \mathbf{u} + \partial_z w = 0$. Integrating over z between $-H_0$ and η becomes

$$(\nabla \cdot \mathbf{u})(\eta + H_0) + w(z = \eta) - w(z = -H_0) = 0 \tag{5.27}$$

The boundary conditions on the rigid bottom require $w(z = -H_0) = 0$, while at the free surface, we have the kinematic condition $w(z = \eta) = D\eta/Dt$ (i.e., the surface moves with the fluid velocity). The incompressibility condition (5.27) thus becomes

$$(\nabla \cdot \mathbf{u})(\eta + H_0) + \frac{\partial \eta}{\partial t} + \mathbf{u} \cdot \nabla\eta = 0 \tag{5.28}$$

which can be rewritten in a more familiar form

$$\frac{\partial h}{\partial t} + \nabla \cdot (\mathbf{u}h) = 0 \tag{5.29}$$

where we have introduced the total depth $h \equiv \eta + H_0$. Eqs. (5.26) and (5.29) represent the shallow-water equations for the two-dimensional velocity field $\mathbf{u}(x, y, t)$ and the height of the surface field $h(x, y, t)$ (or η, which is the same). Shallow-water equations find application in several fields of fluid dynamics and, in particular, in the study of large-scale phenomena such as tides and currents in coastal regions. Their one-dimensional formulation, also known as the Saint-Venant equations, is the basis of hydraulic engineering in rivers and channels Whitham (2011).

The shallow-water equations conserve the energy of the flow given by the expression (excluding a constant density factor)

$$E_{sw} = \int d^2x \left[\frac{1}{2}hu^2 + \frac{1}{2}gh^2\right] \tag{5.30}$$

where the integral is defined on the 2D domain (defined with appropriate horizontal boundary conditions). The first term in Eq. (5.30) represents the kinetic energy of the flow (of thickness h), while the second term is the potential energy. To demonstrate the conservation of total energy, in the following we will consider for simplicity periodic boundary conditions (i.e., $\mathbf{u}(x + L_x, y + L_y, t) = \mathbf{u}(x, y, t)$ where L_x and L_y are the sizes of the domain), but the same result is obtained with rigid walls. The time derivative of E_{sw} is

$$\frac{dE_{sw}}{dt} = \int d^2x \left[\frac{1}{2}u^2\frac{\partial h}{\partial t} + h\mathbf{u}\cdot\frac{\partial \mathbf{u}}{\partial t} + gh\frac{\partial h}{\partial t}\right] \tag{5.31}$$

By inserting (5.26) and (5.29) into (5.31), one obtains

$$\frac{dE_{sw}}{dt} = \int d^2x \left[\left(\frac{1}{2}u^2 + gh\right)(-\nabla\cdot(h\mathbf{u})) + h\mathbf{u}\cdot(-\mathbf{u}\cdot\nabla\mathbf{u} - \mathbf{f}\times\mathbf{u} - g\nabla h)\right] \tag{5.32}$$

The term $\mathbf{u}\cdot\mathbf{f}\times\mathbf{u} = 0$ identically. By using the following identities

$$\frac{1}{2}u^2\nabla\cdot(h\mathbf{u}) + h\mathbf{u}\cdot\mathbf{u}\cdot\nabla\mathbf{u} = \nabla\cdot\left[\frac{1}{2}hu^2\mathbf{u}\right] \tag{5.33}$$

and

$$gh\nabla\cdot(h\mathbf{u}) + gh\mathbf{u}\cdot\nabla h = g\nabla\cdot(\mathbf{u}h^2) \tag{5.34}$$

the time derivative of the energy becomes

$$\frac{dE_{sw}}{dt} = -\int d^2x\nabla\cdot\left[\frac{1}{2}hu^2\mathbf{u} + g\mathbf{u}h^2\right] \tag{5.35}$$

By the divergence theorem, the integral is equal to the difference of the function inside the square brackets evaluated on the opposite boundaries of the domain. Both with periodic and with rigid (no-slip) BC, this difference vanishes, and the energy is therefore conserved.

5.4 Potential vorticity

The shallow-water equations are closely related to the concept of *potential vorticity*, which is a scalar quantity conserved by the motion. We introduce the relative vorticity $\omega \equiv \nabla \times \mathbf{u}$. Note that, since the 2D velocity has the (x, y) components only, the vorticity is a (pseudo) scalar field in the z-direction.

In order to derive the equation for the potential vorticity, it is convenient to rewrite the transport term in the time derivative of Eq. (5.26) as $\mathbf{u} \cdot \nabla \mathbf{u} = \nabla u^2/2 - \mathbf{u} \times \omega$ (see Appendix 5.9). By taking the curl of Eq. (5.26), we obtain

$$\nabla \times \frac{\partial \mathbf{u}}{\partial t} + \nabla \times \frac{1}{2} \nabla u^2 - \nabla \times (\mathbf{u} \times \omega) + \nabla \times (\mathbf{f} \times \mathbf{u}) = -g \nabla \times \nabla \eta \tag{5.36}$$

We have $\nabla \times \nabla \eta = 0$ identically and, making use of the identity $\nabla \times (\mathbf{a} \times \mathbf{b}) = \mathbf{b} \cdot \nabla \mathbf{a} + \mathbf{a} \nabla \cdot \mathbf{b} - \mathbf{a} \cdot \nabla \mathbf{b} - \mathbf{b} \nabla \cdot \mathbf{a}$, we can rewrite Eq. (5.36) as the vorticity budget equation

$$\frac{\partial \omega}{\partial t} + \mathbf{u} \cdot \nabla \omega + \omega \nabla \cdot \mathbf{u} + \mathbf{u} \cdot \nabla f + f \nabla \cdot \mathbf{u} = 0 \tag{5.37}$$

which can be rewritten in a simpler form (note that $\partial f / \partial t = 0$)

$$\frac{D}{Dt}(\omega + f) + (\omega + f)\delta = 0 \tag{5.38}$$

where we have introduced the horizontal divergence field $\delta \equiv \nabla \cdot \mathbf{u}$. Similarly, Eq. (5.29) can be rewritten as

$$\frac{Dh}{Dt} + h\delta = 0 \tag{5.39}$$

from which we have an expression for the divergence

$$\delta = -\frac{1}{h} \frac{Dh}{Dt} \tag{5.40}$$

Using now (5.40) in (5.38), we finally obtain the *potential vorticity equation*

$$\frac{D}{Dt} \left(\frac{\omega + f}{h} \right) = 0 \tag{5.41}$$

which simply states that potential vorticity is conserved along the flow or, equivalently, along the Lagrangian trajectories.

The conservation of potential vorticity (in the case $f = 0$) is analogous to the conservation of angular momentum in mechanics. Consider a column of fluid of radius R and height h in solid body rotation with angular velocity ω: if, for any reason, the column increases its radius to $R' > R$, by the conservation of mass (volume) it reduces its height to $h' = h(R/R')^2$. The new moment of inertia increases as $I' = I(R'/R)^2 = I(h/h')$ and therefore the conservation of angular momentum $I\omega = I'\omega'$ is equivalent to the conservation of ω'/h'.

5.5 The quasi-geostrophic equation

The potential vorticity Eq. (5.41) is very compact, but it is of little practical use since it is a single scalar equation in the three variables u, v, and h (or η). In the limit of nearly geostrophic flow, it is, however, possible to express the horizontal velocity in terms of a stream function and to link η to this velocity. In this case, one therefore reduces to a single scalar field Salmon (1998).

Together with the assumption of nearly geostrophic motion, to derive the quasi-geostrophic (QG) equation, we need to further assume that the variation of the Coriolis parameter is small on the scale of the flow (i.e., $\beta y/f_0 \ll 1$ in the β-plane approximation) and that the relative variation of the depth is small (i.e., $\eta/H_0 \ll 1$).

We start by evaluating Eq. (5.25) at $z = -H_0$, i.e.,

$$p = p_0 + \rho_0 g(\eta + H_0) \tag{5.42}$$

Using (5.42) in the geostrophic balance (5.14)–(5.15), we obtain a zero-order approximation for the velocity field

$$\mathbf{u} = (u, v) = \frac{g}{f_0}\left(-\frac{\partial \eta}{\partial y}, \frac{\partial \eta}{\partial x}\right) = \left(-\frac{\partial \psi}{\partial y}, \frac{\partial \psi}{\partial x}\right) \tag{5.43}$$

where we have introduced the *stream function* ψ for the horizontal velocity

$$\psi \equiv \frac{g\eta}{f_0}. \tag{5.44}$$

By definition of vorticity, we have $\omega = \nabla \times \mathbf{u} = \nabla^2 \psi$. We now expand the potential vorticity expression at the first order in the small parameter $f_0\psi/(gH_0)$

$$\frac{\omega + f}{\eta + H_0} = \frac{\nabla^2\psi + f}{f_0\psi/g + H_0} = \frac{f_0}{H_0}\frac{1 + \beta y/f_0 + \nabla^2\psi/f_0}{1 + f_0\psi/(gH_0)} \simeq \frac{f_0}{H_0}\left(1 + \frac{\beta y}{f_0} + \frac{\nabla^2\psi}{f_0} - \frac{f_0\psi}{gH_0})\right) \tag{5.45}$$

By inserting this expression in Eq. (5.41), we get

$$\left(\frac{\partial}{\partial t} - \frac{\partial \psi}{\partial y}\frac{\partial}{\partial x} + \frac{\partial \psi}{\partial x}\frac{\partial}{\partial y}\right)\left(\nabla^2\psi + f - \frac{f_0^2}{gH_0}\psi\right) = 0 \tag{5.46}$$

This is the QG dynamical equation for the evolution of the stream function. It can be written in more compact form by introducing the *quasi-geostrophic potential vorticity*

$$q \equiv \nabla^2\psi + f - \frac{\psi}{\lambda^2} \tag{5.47}$$

where we have also defined the *Rossby deformation radius* $\lambda \equiv \sqrt{gH_0}/f_0$. Introducing the Jacobian operator

$$J(a, b) \equiv \frac{\partial a}{\partial x}\frac{\partial b}{\partial y} - \frac{\partial a}{\partial y}\frac{\partial b}{\partial x} \tag{5.48}$$

the quasi-geostrophic equation is finally written as

$$\frac{\partial q}{\partial t} + J(\psi, q) = 0 \tag{5.49}$$

At variance with the potential vorticity Eq. (5.41), the QG equation is a single PDE for a single field q (since q and ψ are related by Eq. (5.47)) Salmon (1998); Vallis (2017).

The QG equation can be written in the equivalent form

$$\frac{\partial}{\partial t}\left(\nabla^2\psi - \psi/\lambda^2\right) + J(\psi, \nabla^2\psi - \psi/\lambda^2) + \beta\frac{\partial\psi}{\partial x} = 0 \tag{5.50}$$

which is more convenient for obtaining the energy balance. Indeed, by multiplying (5.50) by $-\psi$ and integrating over the 2D domain (x, y) (assuming, as usual, periodic boundary conditions), we obtain

$$\begin{aligned}\int\Big[&-\psi\partial_t\left(\nabla^2\psi - \psi/\lambda^2\right) - \psi(\partial_x\psi)\partial_y(\nabla^2\psi - \psi/\lambda^2) + \\ &\psi(\partial_y\psi)\partial_x(\nabla^2\psi - \psi/\lambda^2) - \beta\psi\partial_x\psi\Big]d^2x = 0\end{aligned} \tag{5.51}$$

By integrating by parts, we can rewrite the above expression as (implicit summation on repeated indices)

$$\int\left[\partial_t\frac{1}{2}\left((\boldsymbol{\nabla}\psi)^2 + \psi^2/\lambda^2\right) + (\partial_i\partial_y\psi)\partial_i(\psi\partial_x\psi) - (\partial_i\partial_x\psi)\partial_i(\psi\partial_y\psi)\right]d^2x = 0 \tag{5.52}$$

while the other terms are identically zero. We now have

$$\begin{aligned}&\int\left[(\partial_i\partial_y\psi)\partial_i(\psi\partial_x\psi) - (\partial_i\partial_x\psi)\partial_i(\psi\partial_y\psi)\right]d^2x = \\ &\int\left[(\partial_i\partial_y\psi)(\partial_i\psi)(\partial_x\psi) - (\partial_i\partial_x\psi)(\partial_i\psi)(\partial_y\psi)\right]d^2x = \\ &\int\left[-\psi\partial_x\left((\partial_i\partial_y\psi)(\partial_i\psi)\right) + \psi\partial_y\left((\partial_i\partial_x\psi)(\partial_i\psi)\right)\right]d^2x = 0\end{aligned} \tag{5.53}$$

where we used integration by parts. We have therefore that the QG equation conserves the total energy of the flow, $dE/dt = 0$, with

$$E \equiv \frac{1}{2}\int\left[(\boldsymbol{\nabla}\psi)^2 + \frac{\psi^2}{\lambda^2}\right]d^2x \tag{5.54}$$

where the first term represents the kinetic energy of the flow, and the second term is the available potential energy.

Since in the QG, Eq. (5.49), the potential vorticity is transported by an incompressible velocity, i.e., $\boldsymbol{\nabla}\cdot\mathbf{u} = 0$, it is straightforward to verify that the integral of any function of q is conserved. In particular, the *enstrophy*

$$Z \equiv \frac{1}{2}\int q^2 d^2x \tag{5.55}$$

is a second quadratic conserved quantity.

One important limit of Eq. (5.49) is obtained for the motion on scales much smaller than the deformation radius λ. In this limit, the term ψ/λ^2 in the potential vorticity is small with respect to the first term $\nabla^2\psi$ and, consistently, also the variation of f with respect to f_0 is negligible, and therefore Eq. (5.49) becomes the two-dimensional vorticity equation for $\omega = \nabla^2\psi$. This equation will be discussed in detail in the following sections.

5.6 Two-dimensional Navier-Stokes equation

The incompressible, two-dimensional Navier–Stokes equation governs the motion of a two-dimensional viscous fluid with constant density ρ:

$$\frac{\partial \mathbf{u}}{\partial t} + (\mathbf{u}\cdot\nabla)\mathbf{u} = -\frac{1}{\rho}\nabla p + \nu\nabla^2\mathbf{u} \tag{5.56}$$

where ν is the kinematic viscosity of the fluid. The pressure p is determined by the incompressibility condition:

$$\nabla\cdot\mathbf{u} = 0 \tag{5.57}$$

which follows from the mass conservation.

The main difference with respect to the three-dimensional case (Eqs. (5.1) and (5.2) is that here the velocity field is two-dimensional, that is, it has two components u and v which depend on two spatial variables x and y:

$$\mathbf{u} = (u(x, y, t), v(x, y, t), 0) \tag{5.58}$$

Because of the incompressibility, the two components of the velocity are not independent. They can be expressed in terms of the stream function $\psi(x, y, t)$ as in Eq. (5.43)

$$u = -\frac{\partial\psi}{\partial y}, \quad v = \frac{\partial\psi}{\partial x}, \tag{5.59}$$

and the vorticity field is $\boldsymbol{\omega} = \nabla\times\mathbf{u} = (0, 0, \omega)$ with $\omega = \nabla^2\psi$. We remark that both the stream function ψ and the vorticity ω are pseudoscalar fields, since they change sign under parity transformation $\mathbf{x} \mapsto -\mathbf{x}$.

It is convenient to rewrite the two-dimensional Navier–Stokes Eqs. (5.56) in terms of a single equation for the vorticity ω, The equation for the vorticity is obtained with the same procedure adopted in Section 5.4 for the potential vorticity. First, we rewrite the transport term in Eq. (5.56) as $\mathbf{u}\cdot\nabla\mathbf{u} = \nabla u^2/2 - \mathbf{u}\times\boldsymbol{\omega}$ (see Appendix 5.9). By taking the curl of (5.56), we obtain

$$\nabla\times\frac{\partial\mathbf{u}}{\partial t} - \nabla\times(\mathbf{u}\times\boldsymbol{\omega}) = -\nabla\times\nabla\left(\frac{p}{\rho}+\frac{1}{2}\nabla u^2\right) + \nabla\times(\nu\nabla^2\mathbf{u}) \tag{5.60}$$

Using the identity $\nabla\times(\mathbf{u}\times\boldsymbol{\omega}) = \boldsymbol{\omega}\cdot\nabla\mathbf{u}+\mathbf{u}\nabla\cdot\boldsymbol{\omega}-\mathbf{u}\cdot\nabla\boldsymbol{\omega}-\boldsymbol{\omega}\nabla\cdot\mathbf{u} = \boldsymbol{\omega}\cdot\nabla\mathbf{u}-\mathbf{u}\cdot\nabla\boldsymbol{\omega}$ (see Appendix 5.9), and recalling that the curl of the pressure gradient vanishes, we have

$$\frac{\partial\boldsymbol{\omega}}{\partial t} + (\mathbf{u}\cdot\nabla)\boldsymbol{\omega} = (\boldsymbol{\omega}\cdot\nabla)\mathbf{u} + \nu\nabla^2\boldsymbol{\omega} \tag{5.61}$$

In two dimensions, the term $(\boldsymbol{\omega}\cdot\nabla)\mathbf{u}$ identically vanishes because the vorticity is orthogonal to the velocity gradients. This is a crucial difference with respect to three-dimensional flows, in which this term is responsible for the phenomenon of *vortex stretching*. As a consequence, in a two-dimensional flow, the (pseudo)scalar vorticity field ω obeys the equation

$$\frac{\partial\omega}{\partial t} + J(\psi,\omega) = \nu\nabla^2\omega \tag{5.62}$$

which shows that, in the absence of viscosity ($\nu = 0$) vorticity is conserved along the streamlines. Comparing Eqs. (5.62) and (5.49), we find that the dynamics described by the two-dimensional Navier-Stokes equation are equivalent to a quasi-geostrophic flow at scales much smaller than the Rossby deformation radius λ, such that the variations of the Coriolis parameter f can be neglected and the potential vorticity reduces to the relative vorticity $q = \nabla^2\psi$.

Despite its relation with the quasi-geostrophic equation, the two-dimensional Navier–Stokes equation is too simplistic a model for geophysical flows, because the stratification and the beta-plane effects are important in the dynamics of oceans and atmosphere. Nonetheless, the relevance of two-dimensional turbulence relies on the fact that it represents the prototype of a fluid system that is able to self-organize into large-scale coherent structures. Moreover, two-dimensional turbulence can be realized, with some degree of approximation, in several laboratory experiments Tabeling (2002); Kellay and Goldburg (2002).

5.7 Energy and enstrophy balance

In the ideal case of an inviscid flow ($\nu = 0$), the Navier–Stokes equation in two dimensions conserves the kinetic energy:

$$E \equiv \frac{1}{2}\int|\mathbf{u}|^2 d^2x = \frac{1}{2}\int\nabla\psi\cdot\nabla\psi d^2x \tag{5.63}$$

and the enstrophy of the flow:

$$Z \equiv \frac{1}{2}\int\omega^2 d^2x = \frac{1}{2}\int(\nabla^2\psi)^2 d^2x \tag{5.64}$$

The conservation of the energy and enstrophy in the inviscid case can be demonstrated as follows:

$$\frac{dE}{dt} = \int u_i \cdot \partial_t u_i d^2x = \int -u_i \left(u_j \partial_j u_i + \frac{1}{\rho} \partial_i p \right) d^2x =$$
$$\int -\partial_j \left(u_j \frac{u_i^2}{2} + \frac{p}{\rho} u_j \right) d^2x = 0 \tag{5.65}$$

$$\frac{dZ}{dt} = \int \omega \cdot \partial_t \omega d^2x = \int -\omega (u_j \partial_j \omega) d^2x =$$
$$\int -\partial_j \left(u_j \frac{\omega^2}{2} \right) d^2x = 0 \tag{5.66}$$

where we assumed that the velocity and vorticity fields satisfy periodic boundary conditions, or that they vanish at the boundary of the domain.

The conservation of enstrophy is a peculiar property of the Navier-Stokes equation in two dimensions. In the 3D case, the enstrophy is not conserved, because the enstrophy balance has the additional source term, which originates from the vortex stretching term $(\boldsymbol{\omega} \cdot \nabla)\mathbf{u}$ in Eq. (5.61).

In a fluid with finite viscosity $\nu \neq 0$, the viscous forces cause a dissipation of energy and enstrophy, which are converted into heat at small scales. As a consequence, the energy and enstrophy balance are modified as follows:

$$\frac{dE}{dt} = \int u_i \cdot \nu \nabla^2 u_i d^2x = \nu \int \left[(\partial_y \psi)(\nabla^2 \partial_y \psi) + (\partial_x \psi)(\nabla^2 \partial_x \psi) \right] d^2x =$$
$$-\nu \int (\nabla^2 \psi)^2 d^2x = -2\nu Z \tag{5.67}$$

$$\frac{dZ}{dt} = \int \omega \nu \nabla^2 \omega d^2x = -\nu \int \nabla \omega \cdot \nabla \omega d^2x = -2\nu P \tag{5.68}$$

where we introduced the *palinstrophy*:

$$P \equiv \frac{1}{2} \int (\nabla \omega \cdot \nabla \omega) d^2x \tag{5.69}$$

In two dimensions, Eq. (5.68) imposes an upper bound for the enstrophy Z, since $P > 0$. Therefore, with a finite Z, the energy dissipation rate vanishes in the limit $\nu \to 0$.

$$\lim_{\nu \to 0} 2\nu Z = 0 \tag{5.70}$$

This is at variance with three-dimensional case, which is characterized by the phenomenon of the *dissipative anomaly*: in the limit $\nu \to 0$, the enstrophy diverges as $Z \sim \nu^{-1}$ such that the energy dissipation rate tends to a finite value $\lim_{\nu \to 0} 2\nu Z = \varepsilon > 0$. In two-dimensional flows, the dissipative anomaly manifests for the enstrophy dissipation rate. Since the vorticity gradients are not bounded, the palinstrophy P diverges in the inviscid limit $\nu \to 0$ such that

$$\lim_{\nu \to 0} 2\nu P = \eta > 0 \tag{5.71}$$

Even if realistic flows have a finite viscosity $\nu \neq 0$, the inviscid invariants, such as the energy and enstrophy, play a crucial role in the turbulent regime, which is discussed in the next section.

5.8 Phenomenology of two-dimensional turbulence

Turbulent flows are produced by external forces that sustain the fluid motion at large scales. As an example, geophysical flows are usually driven by large-scale variations of pressure and/or temperature. The scale separation between the scale ℓ_f at which the kinetic energy is injected in the flow by the external force and the viscous scale ℓ_ν, where it is converted into heat, is in general very large, because the viscosity is small. In three-dimensional turbulent flows, it depends on the Reynolds number as $\ell_f/\ell_\nu \propto Re^{3/4}$.

The process that transfers energy within the fluid from the injection scale to the dissipation scale is described by Richardson as a *turbulent cascade*. The forcing generates large eddies, which are distorted by the turbulent flow and hence transfer their energy to smaller eddies. This process is repeated for smaller and smaller eddies until it reaches the dissipative scales. The turbulent cascade dominates the dynamics in the interval of scales $\ell_\nu \ll \ell \ll \ell_f$, which defines the *inertial range*. Within the inertial range, the external force and viscous forces can be neglected, and the cascading kinetic energy is almost conserved by the dynamics.

In two-dimensional flows, this process is dramatically changed by the constraint that the dynamics in the inertial range conserves both the energy and the enstrophy. In order to illustrate the effects of the inviscid invariants on the turbulent cascade, it is useful to introduce the Fourier transform of the velocity and vorticity fields:

$$\hat{\mathbf{u}}(\mathbf{k}, t) = \frac{1}{2\pi} \int \mathbf{u}(\mathbf{x}, t) e^{-i\mathbf{k}\cdot\mathbf{x}} d^2x \tag{5.72}$$

$$\hat{\omega}(\mathbf{k}, t) = \frac{1}{2\pi} \int \omega(\mathbf{x}, t) e^{-i\mathbf{k}\cdot\mathbf{x}} d^2x \tag{5.73}$$

For simplicity, here we consider the case of an unbounded flow in an infinite domain. The case of a finite domain with periodic boundary conditions is analogous. From the properties of the Fourier transform it follows that $\hat{\omega}(\mathbf{k}, t) = i\mathbf{k} \times \hat{\mathbf{u}}(\mathbf{k}, t)$. The kinetic energy can be written as:

$$E = \frac{1}{2} \int |\hat{\mathbf{u}}(\mathbf{k}, t)|^2 d^2k = \int_0^\infty E(k, t) dk \tag{5.74}$$

where we have introduced the *energy spectrum* $E(k,t)$. Similarly, the enstrophy is given by:

$$Z = \frac{1}{2}\int |\hat{\omega}(\mathbf{k},t)|^2 d^2k = \int_0^\infty k^2 E(k,t)dk \tag{5.75}$$

Fjørtoft (1953) proposed a simple argument to understand the consequences of the conservation of E and Z in the inertial range of two-dimensional turbulence. Let us suppose that the energy and enstrophy are initially contained in a single mode k_I, and that the turbulent dynamics redistribute them among two modes in the inertial range, $k_L \ll k_I$ and $k_S \gg k_I$, which are representative of the large and small eddies, respectively. The energy contained in the initial mode is denoted by E_I, and the enstrophy is $Z_I = k_I^2 E_I^2$. Similarly, the amount of energy and enstrophy, which is transferred to large scales, are denoted by E_L and $Z_L = k_L^2 E_L$, while the energy and enstrophy transferred to small scales are E_S and $Z_S = k_S^2 E_S$.

In two dimensions, the simultaneous conservation of energy and enstrophy in the inertial range implies the following relations:

$$E_I = E_L + E_S \tag{5.76}$$

$$k_I^2 E_I = k_L^2 E_L + k_S^2 E_S \tag{5.77}$$

which allow determining the fraction of energy transferred to large and small scales as a function of the scale separation:

$$\frac{E_L}{E_I} = \frac{k_S^2 - k_I^2}{k_S^2 - k_L^2}, \quad \frac{E_S}{E_I} = \frac{k_I^2 - k_L^2}{k_S^2 - k_L^2} \tag{5.78}$$

In the limit of large scale separation $(k_S/k_I) \gg 1$, one has that $E_L/E_I \to 1$ and $E_S/E_I \to 0$, which shows that the energy is preferentially transferred toward large eddies.

For the enstrophy, one has

$$\frac{Z_L}{Z_I} = \frac{k_L^2(k_S^2 - k_I^2)}{k_I^2(k_S^2 - k_L^2)}, \quad \frac{Z_S}{Z_I} = \frac{k_S^2(k_I^2 - k_L^2)}{k_I^2(k_S^2 - k_L^2)} \tag{5.79}$$

which in the limit $(k_L/k_I) \ll 1$ gives $Z_L/Z_I \to 0$ and $Z_S/Z_I \to 1$, that is, all the enstrophy is transferred to small eddies.

This simple idea is at the basis of the theory of two-dimensional turbulence. Kraichnan (1967, 1971) showed that two-dimensional turbulent flows are characterized by a *double cascade* scenario, which is schematically represented in Fig. 5.4. The energy injected by the external force at the scale ℓ_f is transferred preferentially toward large scales, giving rise to an *inverse energy cascade*, while the enstrophy is transferred toward small scales by a *direct enstrophy cascade* (Fig. 5.4).

The attribute *inverse* for the energy cascade is used to emphasize the difference with respect to the small-scale energy transfer in three dimensions. The large-scale

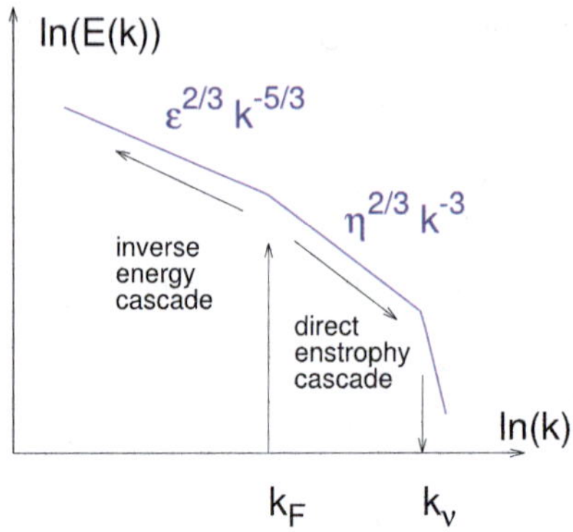

Figure 5.4 Schematic of the double cascade scenario. Energy and enstrophy are injected at the intermediate wavenumber k_F at rates ε and $\eta = k_F^2\varepsilon$. Most of the energy flows to smaller wavenumbers (larger scales), producing the inverse energy cascade with scaling law $k^{-5/3}$, while most of the enstrophy flows to larger wavenumbers and creates the direct enstrophy cascade with scaling k^{-3} (with possible logarithmic corrections). The direct cascade is arrested by viscosity at the wavenumber k_ν.

energy transfer in 2D turbulence represents an instance of self-organization of the flow McWilliams (1990) because the accumulation of energy at large scales causes the spontaneous emergence of coherent structures. This phenomenon has been observed both in experiments with thin fluid layers and in numerical simulations of the 2D Navier–Stokes equation Xia et al. (2008); Chertkov et al. (2007).

The theory of Kraichnan allows us to derive a prediction for the energy spectrum $E(k)$, which describes how the energy is distributed among the different modes k. The rate of variation of the energy spectrum is given by:

$$\frac{\partial E(k)}{\partial t} = T(k) - D(k) + F(K) \tag{5.80}$$

where $T(k)$ represents the rate of energy transfer across the scale $1/k$ due to the nonlinear term of the Navier–Stokes equation, $D(k) = 2\nu k^2 E(k)$ is the viscous dissipation rate and $F(k)$ is the energy injection due to the external force, which is localized at the scale $\ell_f \sim 1/k_f$.

The net spectral fluxes of energy $\Pi_E(K)$ and enstrophy $\Pi_Z(k)$ across scale $1/k$ are defined from $T(k)$ as follows:

$$\Pi_E(k) \equiv \int_k^\infty T(k')dk' \tag{5.81}$$

$$\Pi_Z(k) \equiv \int_k^\infty k'^2 T(k')dk' \tag{5.82}$$

The conservation of energy in the inverse energy cascade implies a constant energy flux $\Pi_E(k) \sim \varepsilon$ in the range of wavenumber $k_{min} \ll k \ll k_f$, where $k_{min} \sim 1/L$ represents the lowest turbulent wavenumber. Similarly, the enstrophy conservation in the direct cascade implies a constant enstrophy flux $\Pi_Z(k) \sim \eta$ in the range of wavenumber $k_f \ll k \ll k_\nu$, where $k_\nu \sim 1/\ell_\nu$ is the ultraviolet (UV) viscous cutoff of the inertial range.

Using a dimensional closure to express the energy flux in terms of the energy spectrum $\Pi_E(k) \sim k^{5/2}E(k)^{3/2}$, the requirement of a constant flux of energy in the inertial range $\Pi_E(k) \sim k^{5/2}E(k)^{3/2} \sim \varepsilon$ gives the prediction for the energy spectrum in the inverse cascade:

$$E(k) = C\varepsilon^{2/3}k^{-5/3}\,, \quad k \ll k_f \tag{5.83}$$

where C is a dimensionless constant. Spectrum (5.83) coincides with the Kolmogorov spectrum for three-dimensional flows with an important difference: In two dimensions, the energy flux is directed toward large scales, and therefore the prediction (5.83) holds for $k \ll k_f$.

According to the picture of Richardson's energy cascade, the distortion of eddies in a turbulent flow causes the transfer of kinetic energy among different active scales. The characteristic distortion time τ_k of the eddies of size $1/k$ can be written in terms of the energy spectrum as

$$\tau_k \sim \left[\int_{k_{min}}^{k} k'^2 E(k')dk'\right]^{-1/2}\,. \tag{5.84}$$

Inserting expression (5.83) into integral (5.84), we obtain $\tau_k \sim \varepsilon^{-1/3}k^{-2/3} + o(k_{min}/k)$. This means that the distortion of an eddy of size $1/k$ is dominated by the contribution of eddies of the same size. This peculiar feature of the energy cascade is called *locality*.

Repeating the dimensional argument for an enstrophy cascade with constant flux $\Pi_Z(k) \sim k^{9/2}E(k)^{3/2} \sim \eta$ gives the prediction:

$$E(k) = C'\eta^{2/3}k^{-3}\,, \quad k \gg k_f \tag{5.85}$$

This result is not fully self-consistent, because using spectrum (5.85) to estimate the distortion time for the eddies in the enstrophy cascade one obtains $\tau_k \sim [\ln(k/k_{min})]^{-1/2}$, which gives a logarithmic dependence on k for the enstrophy flux $\Pi_Z(k) \sim k^3E(k)/\tau_k$. Kraichnan Kraichnan (1971) showed that the constant enstrophy flux can be restored by introducing a logarithmic correction to the spectrum:

$$E(k) = C'\eta^{2/3}k^{-3}[\ln(k/k_{min})]^{-1/3}\,, \quad k \gg k_f \tag{5.86}$$

A consequence of this correction is that in the enstrophy cascade the distortion time τ_k is dominated by the lowest wavenumber k_{min}, i.e., by the large eddies. At variance with the case of the inverse energy cascade, the direct enstrophy cascade is nonlocal, because the stirring of the vorticity at small scales is mostly due to the large-scale flow.

The double cascade scenario predicted by Kraichnan has been observed in numerical simulations of the Navier–Stokes equation forced at intermediate scales Boffetta

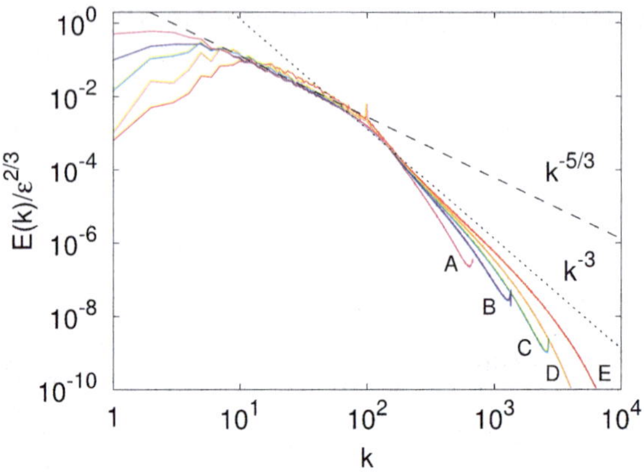

Figure 5.5 The double cascade observed in numerical simulations. The two-dimensional Navier–Stokes Eq. (5.62) is integrated by a pseudo-spectral method on a regular square grid at resolution N^2 with periodic boundary conditions. An additional forcing term, short-correlated in time and active only at the forcing wavenumber $k_F = 100$, is added to (5.62) to sustain the turbulent flow. The approach to the Kraichnan prediction of two different scaling laws for $k < k_F$ (inverse cascade with slope $-5/3$) and $k > k_F$ (direct cascade with slope -3) is shown by decreasing the viscosity ν while increasing the resolution N to resolve the smallest scales. A) (pink) $N = 2048$, $\nu = 2 \times 10^{-5}$; B) (blue) $N = 4096$, $\nu = 5 \times 10^{-6}$; C) (green) $N = 8192$, $\nu = 2 \times 10^{-6}$; D) (orange) $N = 16384$, $\nu = 1 \times 10^{-6}$; E) (red) $N = 32768$, $nu = 2.5 \times 10^{-7}$. Details on the simulations in Boffetta and Musacchio (2010).

and Musacchio (2010), as shown in Fig. 5.5. Experimental studies performed in thin fluid layers and in soap films demonstrated that the phenomenology of two-dimensional turbulence can also be observed in real flows (for a review see, e.g., Boffetta and Ecke (2012)).

The properties of turbulent flows can be characterized in physical space by the statistics of longitudinal velocity increments

$$\delta u_{||}(r) \equiv [\mathbf{u}(\mathbf{x}+\mathbf{r}) - \mathbf{u}(\mathbf{x})] \cdot \hat{\mathbf{r}} \tag{5.87}$$

which is described by the longitudinal velocity structure functions

$$S_{||}^{n}(r) \equiv \langle [\delta u_{||}(r)]^n \rangle \tag{5.88}$$

where the brackets $\langle \cdots \rangle$ represent the average over the whole spatial domain.

Under the hypothesis that the flow is statistically homogeneous and isotropic, the statistics of $\delta u_{||}(r)$ is independent of $\mathbf{x}$ and of the orientation of $\mathbf{r}$; therefore the structure functions depend only on the distance r. The further hypothesis of a constant energy flux in the inverse cascade allows us to derive an exact relation for the third-order structure function:

$$S_{||}^{3}(r) \equiv \langle [\delta u_{||}(r)]^3 \rangle = \frac{3}{2}\varepsilon r \quad r \gg \ell_f \tag{5.89}$$

which is the analog of the Kolmogorov 4/5 law in 3D $S_{||}^3(r) = -(4/5)\varepsilon r$ (see Chapter 1). The opposite sign of the third-order longitudinal velocity structure function in 2D (positive) and 3D (negative) is a direct consequence of the opposite direction of the energy transfer.

Assuming that the statistics at different scales are self-similar, the relation (5.89) for the scaling of $S_{||}^3(r)$ gives the scaling exponent $h = 1/3$ for the typical velocity fluctuations in the inverse energy cascade. This leads to the prediction for the scaling of the velocity structure functions $S^n(r) \sim r^{\zeta_n}$ with $\zeta_n = n/3$, which is confirmed by the results of experiments and numerical simulations of 2D turbulence Boffetta and Ecke (2012). Conversely, the statistics of velocity fluctuations in three-dimensional flows are not self-similar, and the scaling exponent of the structure functions is anomalous $\zeta_n^{(3D)} \neq n/3$.

In the range of scales of the direct enstrophy cascade, the hypothesis of a constant enstrophy flux gives:

$$S_{||}^3(r) \equiv \langle[\delta u_{||}(r)]^3\rangle = \frac{1}{8}\eta r^3 \quad r \ll \ell_f \tag{5.90}$$

which implies a scaling exponent $h = 1$ for the velocity fluctuations at small scales $r \ll \ell_f$.

For further reading about the phenomenology of two-dimensional turbulence and its relevance for geophysical flows, we suggest the reviews Kraichnan and Montgomery (1980); Boffetta and Ecke (2012); Lindborg and Mohanan (2017).

5.9 Appendix: Mathematical formulas

Most of the results of vector calculus used in the chapter can be easily obtained by applying the standard Leibniz product rule to the different components of the expression. Let **a** (and **b** etc.) represents a generic field (which we assume differentiable) of components a_i with $i = 1, 2, 3$, i.e., $\mathbf{a} = (a_1, a_2, a_3)$. We will use the so-called Einstein notation, which implies summation on repeated indices, e.g., $a_i a_i \equiv \sum_{i=1}^{3} a_i^2 = a^2$.

For computing external product (and curl), it is useful to introduce the fully antisymmetric Levi-Civita symbol ε_{ijk}, which assumes nonzero values only if the three indices are all different. In particular, $\varepsilon_{ijk} = +1$ if $(i, j, k) = (1, 2, 3)$ or cyclic permutation, $\varepsilon_{ijk} = -1$ if $(i, j, k) = (1, 3, 2)$ (and cyclic permutation) and $\varepsilon_{ijk} = 0$ in all the other cases. If $\mathbf{c} = \mathbf{a} \times \mathbf{b}$ then $c_i = \varepsilon_{ijk} a_j b_k$. We will make use of the following identity (which can be easily proven directly)

$$\varepsilon_{ijk}\varepsilon_{ipq} = \delta_{jp}\delta_{kq} - \delta_{jq}\delta kp \tag{5.91}$$

where the Kronecker delta is $\delta_{ij} = 1$ if $i = j$ and $\delta_{ij} = 0$ otherwise.

One important relation used in the fluid mechanics is relative to the external product of one vector with its curl:

$$\mathbf{a} \times (\nabla \times \mathbf{a}) = \frac{1}{2}\nabla a^2 - \mathbf{a} \cdot \nabla \mathbf{a} \tag{5.92}$$

in particular when $\mathbf{a}$ is the velocity $\mathbf{a} = \mathbf{u}$, and $\nabla \times \mathbf{a} = \boldsymbol{\omega}$ is the vorticity.

In order to demonstrate (5.92), we write in components

$$(\mathbf{a} \times \nabla \times \mathbf{a})_i = \varepsilon_{ijk} a_j \varepsilon_{kpq} \partial_p a_q \tag{5.93}$$

We can cyclically permutate the indices in the first ε in (5.93) to use (5.91)

$$(\mathbf{a} \times \nabla \times \mathbf{a})_i = \varepsilon_{kij} \varepsilon_{kpq} a_j \partial_p a_q = (\delta_{ip}\delta_{jq} - \delta_{iq}\delta_{jp}) a_j \partial_p a_q = \\ a_j \partial_i a_j - a_j \partial_j a_i = \frac{1}{2} \partial_i a^2 - a_j \partial_j a_i \tag{5.94}$$

which is the i-component of the RHS of (5.92).

Another important expression is the curl of an external product:

$$\nabla \times (\mathbf{a} \times \mathbf{b}) = \mathbf{b} \cdot \nabla \mathbf{a} + \mathbf{a} \nabla \cdot \mathbf{b} - \mathbf{a} \cdot \nabla \mathbf{b} - \mathbf{b} \nabla \cdot \mathbf{a} \tag{5.95}$$

which is one generalization of the Leibniz rule. Again, writing the generic component and using (5.91), we have

$$(\nabla \times (\mathbf{a} \times \mathbf{b}))_i = \varepsilon_{ijk} \varepsilon_{kpq} \partial_j a_p b_q = \varepsilon_{kij} \varepsilon_{kpq} (a_p \partial_j b_q + b_q \partial_j a_p) = \\ b_j \partial_j a_i + a_i \partial_j b_j - a_j \partial_j b_i - b_i \partial_j a_j \tag{5.96}$$

which is nothing but the RHS of (5.95).

Similarly, one can prove the important identity for the curl of a curl:

$$\nabla \times (\nabla \times \mathbf{a}) = \nabla (\nabla \cdot \mathbf{a}) - \nabla^2 \mathbf{a} \tag{5.97}$$

References

Salmon, R., 1998. Lectures on Geophysical Fluid Dynamics. Oxford University Press.

Whitham, G.B., 2011. Linear and Nonlinear Waves. John Wiley & Sons.

Vallis, G.K., 2017. Atmospheric and Oceanic Fluid Dynamics. Cambridge University Press.

Tabeling, P., 2002. Two-dimensional turbulence: a physicist approach. Physics Reports 362, 1–62.

Kellay, H., Goldburg, W.I., 2002. Two-dimensional turbulence: a review of some recent experiments. Reports on Progress in Physics 65, 845.

Fjørtoft, R., 1953. On the changes in the spectral distribution of kinetic energy for twodimensional, nondivergent flow. Tellus 5, 225–230.

Kraichnan, R.H., 1967. Inertial ranges in two-dimensional turbulence. Physics of Fluids 10, 1417.

Kraichnan, R.H., 1971. Inertial-range transfer in two-and three-dimensional turbulence. Journal of Fluid Mechanics 47, 525–535.

McWilliams, J.C., 1990. The vortices of two-dimensional turbulence. Journal of Fluid Mechanics 219, 361–385.

Xia, H., Punzmann, H., Falkovich, G., Shats, M.G., 2008. Turbulence-condensate interaction in two dimensions. Physical Review Letters 101, 194504.

Chertkov, M., Connaughton, C., Kolokolov, I., Lebedev, V., 2007. Dynamics of energy condensation in two-dimensional turbulence. Physical Review Letters 99, 084501.

Boffetta, G., Musacchio, S., 2010. Evidence for the double cascade scenario in two-dimensional turbulence. Physical Review. E 82, 016307.

Boffetta, G., Ecke, R.E., 2012. Two-dimensional turbulence. Annual Review of Fluid Mechanics 44, 427–451.

Kraichnan, R.H., Montgomery, D., 1980. Two-dimensional turbulence. Reports on Progress in Physics 43, 547.

Lindborg, E., Mohanan, A.V., 2017. A two-dimensional toy model for geophysical turbulence. Physics of Fluids 29, 111114.

Turbulent mixing

6

James J. Riley
University of Washington, Seattle, WA, United States

6.1 Introduction

It is difficult to give the meaning of turbulence in a one- or two-sentence definition, so that it is often described by its characteristics (see, e.g., Tennekes and Lumley (1972)), such as being chaotic but deterministic, strongly nonlinear, strongly vortical, three-dimensional, and highly dissipative. One of the most important characteristics of turbulence, and its implications in both nature and technology, is that it irreversibly mixes; it irreversibly mixes mass, e.g., species in chemical reactions, it irreversibly mixes momentum, heat, kinetic energy, and other properties of a turbulent flow. In understanding turbulent mixing, it is important to first distinguish between stirring, an advective process, and mixing (see, e.g., Eckart (1948)). Consider the blob of dye introduced into a homogeneous turbulent field in the first panel of Fig. 6.1. As the flow evolves in time, as shown in the latter two panels, three things are happening. First, the dye is advected by the flow. Second, it is strongly strained by turbulence arising from spatial variations in the velocity field. In general, this straining stretches out the dye and increases its concentration gradients. Third, the dye and the surrounding fluid irreversibly mix, decreasing the maximum concentration of the dye. It is this latter process that constitutes turbulent mixing.

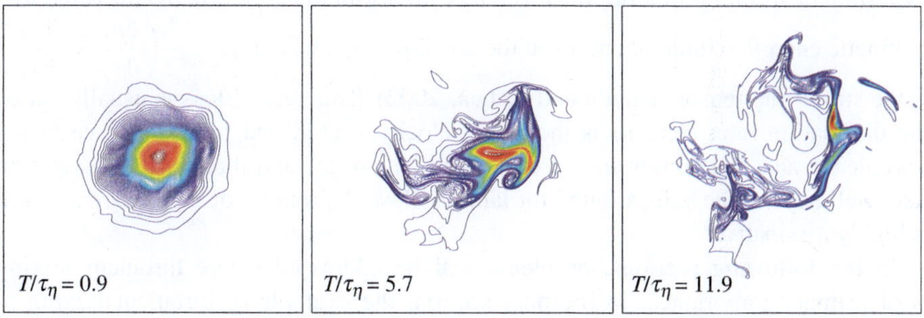

Figure 6.1 Mixing of a passive scalar (Sreenivasan and Schumacher, 2010). The color contours give the concentration of the scalar. T is the characteristic time of the large-scale motions, while τ_η is the Kolmogorov time scale.

There have been numerous excellent reviews of various aspects of turbulent mixing. For example, Dimotakis (2005) provides a broad overview of turbulent mixing, along with attendant experimental, theoretical, and computational challenges. Warhaft (2000) provides an in-depth discussion of the mixing of passive scalars in turbulence, and due to its importance in geophysical flows, especially in oceanographic flows, there have been a series of reviews addressing various issues related to turbulent

Environmental Turbulence. https://doi.org/10.1016/B978-0-32-395873-8.00013-9
Copyright © 2026 Elsevier Inc. All rights are reserved, including those for text and data mining, AI training, and similar technologies.

mixing in stably stratified flows, including Linden (1979), Ivey and Imberger (1991), Fernando (1991), Peltier and Caulfield (2003), Wunsch and Ferrari (2004), Ivey et al. (2008), and Caulfield (2021).

Mixing is of primary importance in many natural and technological processes. For example, in nonpremixed chemical reactions, it is the rate of turbulent mixing that controls the overall reaction rate and the production of product and heat. At the edge of a turbulent shear flow, such as a jet or a boundary layer, it is the rate at which the turbulent fluid, characterized by its high vorticity, mixes with the surrounding, often irrotational flow, that controls the entrainment of ambient fluid and the growth of the turbulent region. Similar processes occur at the top of the turbulent atmospheric boundary layer and at the bottom of the ocean mixed layer, among many examples. At the ocean surface, turbulent mixing controls the flux of heat, mass, and chemical species into the ocean. Moreover, in the ocean interior, it controls the flux of mass across isopycnals, and so the ambient density field, which in turn affects the overall ocean dynamics.

The details of the turbulent mixing process depend strongly on the type of flow and its environment. For example, in a stably stratified ocean, mixing is strongly influenced by stable stratification, whereas in a chemical reaction, it can be strongly influenced by the heat release in the reaction zone. However, in all cases, an essential feature is the straining of the mixed feature by strong velocity gradients in turbulent flow, producing large gradients in the feature and greatly enhancing irreversible mixing processes. We will find that the turbulent mixing is related to the resulting gradients of the feature being mixed.

The most commonly addressed form of turbulent mixing is momentum mixing induced by turbulence. As molecular diffusion reduces momentum gradients caused by turbulent straining motion, it irreversibly mixes momentum. The result is the loss of kinetic energy to internal energy at the rate $2\,\mu s_{ij}s_{ij}$, where $s_{ij} = \frac{1}{2}\left(\frac{\partial u_i}{\partial x_j} + \frac{\partial u_j}{\partial x_i}\right)$ is the strain-rate tensor (Kundu and Cohen, 2002). This rate of loss is usually called the dissipation rate. Here u_i is the fluid velocity vector, and μ the fluid viscosity. Turbulence acts to strongly increase the gradients of u_i, and thereby the dissipation rate, well beyond the values found for laminar flow. A property of turbulence is that it is highly dissipative.

In the following sections, problems will be addressed where turbulent mixing is of primary importance. In the next section, the example of turbulent mixing in chemical reactions will be considered. In the following section, turbulent mixing at the turbulent/nonturbulent interface in shear flows will be discussed. This will be followed by a more lengthy section on mixing in a stratified fluid, with special implications for ocean mixing. In the final section, some summary comments will be made.

6.2 Turbulent mixing in chemical reactions

Turbulent mixing in chemical reactions occurs in numerous environmental situations, from chemical plumes dispersing in the atmosphere and oceans to the mixing of heat and moisture at the edges of clouds, and to the chemical reactions that take place in automobiles, airplanes, and power plants, whose output can lead to significant environmental issues. At the center of these is the flame zone, where the overall reaction rate is controlled by irreversible turbulent mixing. It is usual to consider two distinct types of flames, nonpremixed, or diffusion flames, and premixed flames. Of course, both types of flames may exist in a particular situation. In nonpremixed flames, the reactant and oxidant are initially separated and brought together by the flow field into the reaction zone, where they mix and react. On the other hand, in premixed reactions, the reactant and oxidant are premixed and require a heat source to react. In the flame zone, this is accomplished by the heat of reaction mixing into the premixed fluid as the flame propagates.

In a nonpremixed chemical reaction, the reaction occurs when the reactant and oxidant mix irreversibly at the molecular scale in the flame zone. In turbulent combustion processes it is the turbulent mixing of the chemical species, irreversibly at the molecular scale, that enhances the overall production rate of product and heat essential to combustion. For example, consider the nonpremixed case of a fuel jet issuing into an ambient oxidant environment. An example is given in Fig. 6.2, which is a photograph of a piloted natural gas jet flame (van Veen and Roekaerts, 2003). There are several regions to be noted in the flame. Natural gas, the fuel, is blue, issuing from the nozzle into air, which is black. The region downstream of the fuel, which is white (high temperature) and red, is a combination of heat and the products of reaction. At the outer edges of the fuel is the flame zone, where the fuel and air mix and react.

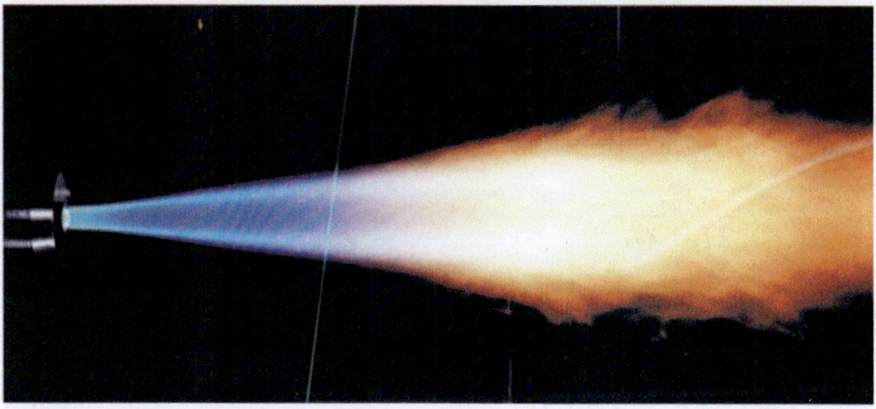

Figure 6.2 Photograph, with limited time exposure, of a natural gas jet flame (van Veen and Roekaerts, 2003).

Following Bilger (1989), consider the chemical species conservation equation,

$$\rho\frac{\partial Y_i}{\partial t}+\rho u_k\frac{\partial Y_i}{\partial x_k}=w_i+\frac{\partial}{\partial x_i}\left(\rho\mathcal{D}_i\frac{\partial Y_i}{\partial x_i}\right) \tag{6.1}$$

where Y_i is the mass fraction of chemical species i that has a net chemical production rate w_i and a mass diffusivity $\mathcal{D}_i$, ρ is the fluid density, and u_k is the component of the fluid velocity in the direction of the coordinate x_k. The fluid temperature and enthalpy also satisfy similar equations. For simplicity of argument, consider the natural gas jet flame as a two species reaction, with $Y_1 = Y_f$ the fuel (F), natural gas, and $Y_2 = Y_o$ the ambient oxidant (O) in a one-step, irreversible reaction. Fuel and oxidant react according to

$$F + r \cdot O \rightarrow (r+1)P \tag{6.2}$$

Here r, the stoichiometric ratio, is the mass of oxidant required to react with a unit mass of fuel, producing the mass of $(r+1)P$ units of the product P. It can be useful to define the mixture fraction Z as

$$Z = \frac{rY_f - Y_o + Y_{oa}}{rY_{f_i} + Y_{oa}} \tag{6.3}$$

where Y_{oa} and Y_{f_i} are the mass fractions of the ambient oxidant and the inlet fuel, respectively. Note that $Z = 1$ in the fuel stream, and $Z = 0$ in the oxidant region. Furthermore, assume for simplicity that the fuel and oxidant have equal diffusivities, i.e., that $\mathcal{D}_f = \mathcal{D}_o = \mathcal{D}$. In addition, from Eq. (6.2), $w_f = w_o/r$, where w_f and w_o are the reaction rates of fuel and oxidant, respectively. Using this in the chemical species conservation equations for the fuel and oxidant, Eq. (6.1), the resulting equation for the mixture fraction Z can be obtained:

$$\rho\frac{\partial Z}{\partial t} + \rho u_k \frac{\partial Z}{\partial x_k} = \frac{\partial}{\partial x_i}\left(\rho\mathcal{D}\frac{\partial Z}{\partial x_i}\right) \tag{6.4}$$

Note that Z does not participate directly in the chemical reaction. Furthermore, if the time scale of the chemical reaction τ_c, is much smaller than the time scale for molecular mixing, τ_m, which is often the case in practice, the reaction occurs in the vicinity of the stoichiometric value of the mixture fraction,

$$Z_{st} = \frac{Y_{oa}}{rY_{f_i} + Y_{oa})} \tag{6.5}$$

The ratio of these two time scales is often referred to as the Damköhler number, $Da = \tau_m/\tau_c$. When Da is much greater than unity (Bilger, 1989), then the thermochemical state of the mixture at any point in space and time is fully determined by Z, i.e.,

$$Y_f = Y_f^e(Z), \quad Y_o = Y_o^e(Z) \tag{6.6}$$

where the functions $Y_f^e(Z)$ and $Y_o^e(Z)$ are determined by the thermochemistry of the problem.

Introducing Eq. (6.6) into Eqs. (6.1) and noting Eq. (6.4) results in:

$$w_f = -\frac{1}{2}\rho\chi\frac{d^2Y_f^e}{dZ^2}, \quad w_o = -\frac{1}{2}\rho\chi\frac{d^2Y_o^e}{dZ^2} \tag{6.7}$$

where $\chi = 2\mathcal{D}\frac{\partial Z}{\partial x_i}\frac{\partial Z}{\partial x_i}$, the scalar dissipation rate of Z. Here χ determines the rate at which turbulent mixing of the scalar occurs, which in turn determines the overall chemical reaction rate. It is the rate that turbulence irreversibly mixes the scalar, and hence from Eq. (6.7) the chemical species, that determines the chemical reaction rate. It will be seen below that quantities analogous to χ determine the mixing rates in other turbulence mixing processes.

Note that the mixing zone itself will depend on the diffusion coefficient $\mathcal{D}$ as well as the turbulent mixing (Kolmogorov) time scale $\tau_\eta = (\nu/\epsilon)^{1/2}$, where ϵ is the turbulent kinetic energy dissipation rate per unit mass, and ν the fluid viscosity. Dimensional analysis gives the width of the mixing zone,

$$\ell_m = \left(\frac{\nu}{\mathcal{D}}\right)^{-1/2} \eta = Pr^{-1/2}\eta \tag{6.8}$$

where $\eta = (\nu^3/\epsilon)^{1/4}$ is the Kolmogorov-length scale, and $Pr = \nu/\mathcal{D}$ is the Prandtl number. Of course, the width of the reaction zone, ℓ_r, also depends on the speed of reaction, decreasing as the reaction rate increases.

Flames are usually more complicated than the example presented. Strong heat release can affect the overall flow and turbulence, raising local temperature and lowering the local Reynolds number. In addition, thermochemistry is usually much more complicated, involving multiple secondary chemical species and reactions. As shown in more complex models, such as laminar flamelet modeling, the overall chemical reaction will still depend on the scalar mixing rate χ.

6.3 Turbulent/nonturbulent interface in turbulent shear flows

In his PhD thesis on the circular turbulent jet in an ambient flow at rest, Corrsin (1947) observed from the oscillograms of the axial velocity signal that 'the turbulent jet' is completely turbulent only from the axis out to approximately $r = r_0$. For $r > r_0$, there exists first an annular transition region, in which the flow at a point alternates between the turbulent and laminar regimes". (Here r_0 is the radial location where the mean axial velocity drops to half of its peak value.) Corrsin had discovered the intermittent layer between a laminar and a turbulent flow, which is now known to be characteristic of any turbulent flow with a free-stream boundary, such as turbulent jets, wakes, mixing layers, and turbulent boundary layers. This discovery was followed by the first definitive studies of the intermittent regions between laminar and turbulent flows (Corrsin and Kistler, 1955).

Fig. 6.3 gives an image of a round turbulent jet (Dimotakis et al., 1983). The experiment was conducted by introducing a jet of water containing fluorescein dye into a tank of still water. The jet was visualized by illuminating it with an argon-ion laser sheet, which caused the dye in the plane of the sheet to fluoresce. The region occupied by the dye ultimately overlaps approximately with the turbulence. The intermittent nature of the turbulent region, especially away from the jet centerline, is clearly visible.

A principal issue in such a study is how to distinguish the turbulent and nonturbulent regimes and, in particular, what measurable characteristic of the turbulence can be used to distinguish these regimes. It was clear that the stochastic

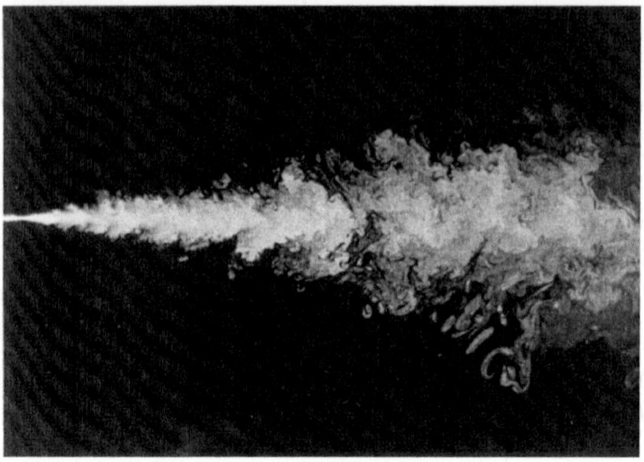

Figure 6.3 Turbulent jet (Dimotakis et al., 1983) visualized by planar laser-induced fluorescence.

nature of the flow could not be used, since the turbulent flow and the 'laminar' flow outside the turbulent region are both stochastic, the latter being driven by the turbulence. Corrsin and Kistler concluded that the characteristic feature of the turbulence to be used for its detection was its high vorticity, compared to the essentially irrotational flow in the laminar region. They ultimately used the streamwise derivative of velocity as a proxy for vorticity to determine the turbulent region. In addition, they surmised that the rotational turbulent region must propagate into the nonturbulent, irrotational region by viscous diffusion, much as "a flame front propagates through a combustion mixture". Consider the vorticity equation for an incompressible, Newtonian fluid,

$$\frac{\partial}{\partial t}\omega_i + u_j \frac{\partial}{\partial x_j}\omega_i = \omega_j s_{ij} + \nu \frac{\partial^2 \omega_i}{\partial x_k^2} \tag{6.9}$$

where ω_i is the vorticity vector, s_{ij} is the strain-rate tensor, and ν the kinematic viscosity. The left-hand side of this equation is the Lagrangian derivative of the vorticity, written in Eulerian form. The equation states that, if the vorticity of the fluid element is initially 0, as it would be in an ambient fluid at rest, then it can only obtain vorticity, i.e., it can only become turbulent, through viscous diffusion. Therefore Corrsin and Kistler reasoned that "the random vorticity field ... can propagate only by direct contact, as opposed to action at a distance, because rotation can be transmitted to irrotational flow only through direct viscous shearing action". They further reasoned that a very thin layer separated the turbulent and nonturbulent regions, which they called the laminar, or viscous superlayer (δ_v). In this region, through turbulent mixing, vorticity in the turbulent region is irreversibly mixed into the originally irrotational laminar fluid, causing this fluid to become rotational and ultimately turbulent, a process called turbulent entrainment. Analogous to the mixing of heat into the reaction zone in a turbulent premixed reacting flow, here vorticity is mixed into the irrotational laminar flow, ultimately causing it to become turbulent. Again, turbulent mixing is at the heart of the process.

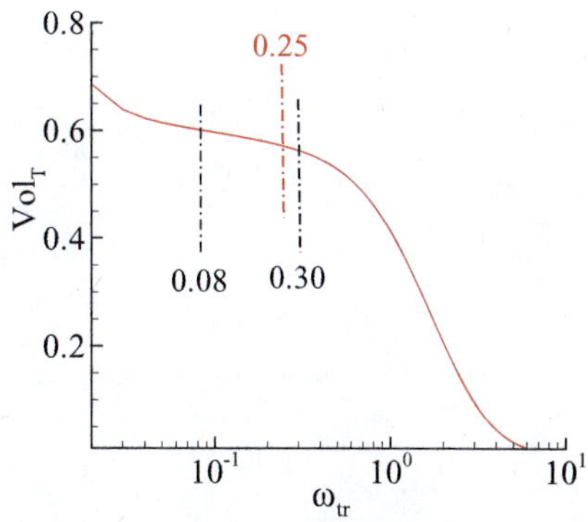

Figure 6.4 Volume of the vorticity region as a function of the vorticity magnitude for an instantaneous vorticity field (Taveira et al., 2013).

Corrsin and Kistler argued that the thickness of the viscous superlayer should depend on the local viscosity ν, and the local strain-rate given by $s \sim (\epsilon/\nu)^{1/2}$, where ϵ is the kinetic energy dissipation rate per unit mass. From dimensional reasoning, the viscous superlayer can be estimated as $\delta_\upsilon \sim (\nu^3/\epsilon)^{1/4}$, the Kolmogorov scale.

It is found, as shown below, that immediately adjacent to the viscous superlayer, there is a thin region where the vorticity goes from the smaller values of vorticity in the superlayer to the vorticity in the internal flow through a thin turbulent sublayer δ_ω (see Fig. 6.6). The scale of the sublayer is found to be related to the thickness of the largest vortices near the turbulent/nonturbulent interface, which scales as λ, the local Taylor microscale (da Silva and Taveira, 2010).

Since the work of Corrsin and Kistler, there has been considerable development, especially in the past 10 years, in the understanding of the interface between the turbulent and nonturbulent flow, now referred to as the turbulent/nonturbulent (T/NT) interface. This development has occurred through more sophisticated, and mainly optical experimental methods, but especially from high-resolution direct numerical simulations. The latter allows a very detailed investigation of the interface. With high-resolution information from numerical simulations, it has become possible to fairly accurately detect the location of the interface, for example, using a very low value of the measured vorticity magnitude. To give an illustration of this, consider the direct numerical simulations of Taveira et al. (2013) for a plane, turbulent jet. Fig. 6.4 shows a plot from this study of the volume of the turbulent (vortical) flow versus a vorticity magnitude threshold. It is observed that there is a significant region for low values of the vorticity magnitude where the turbulent volume changes very little, which indicates that a threshold value in this range can be used to determine the T/NT interface.

An application of this threshold method by Watanabe et al. (2019) is shown in Fig. 6.5, which is a plot of a slice of the vorticity magnitude field at the edge of a plane, turbulent jet. The white lines show the T/NT interface found from the methodology, along with some relevant scales of the motion. This approach allows extensive analysis

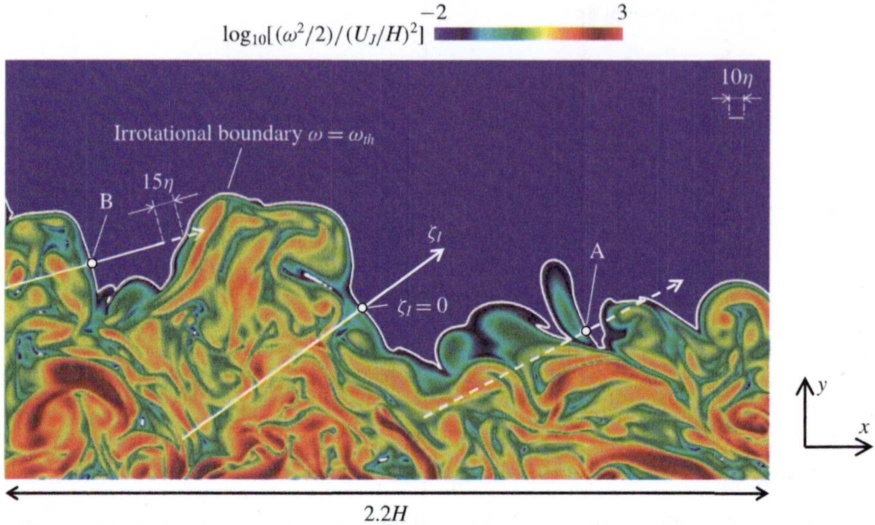

Figure 6.5 Color contour plot of enstrophy at the edge of a plane, turbulent jet showing the T/NT interface as a white line (Watanabe et al., 2019).

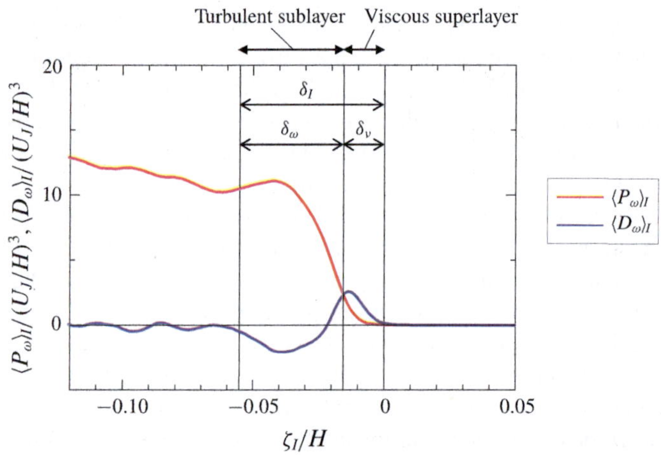

Figure 6.6 Conditional averages of the enstrophy production term $\langle P_\omega \rangle$ and diffusion term $\langle D_\omega \rangle$ through the T/NT interface (Watanabe et al., 2019).

of the properties of the flow in the vicinity of the interface and a better understanding of the behavior of the flow in this region.

For example, given accurate knowledge of the location of the T/NT interface, a method often used to explore the behavior of the flow near the T/NT interface is to use condition sampling, conditioned on the location of the interface. Of importance in understanding this is the behavior of the enstrophy, $\omega^2/2 = \omega_i^2/2$, where ω is the magnitude of the vorticity vector ω_i. For incompressible flow, the equation for the

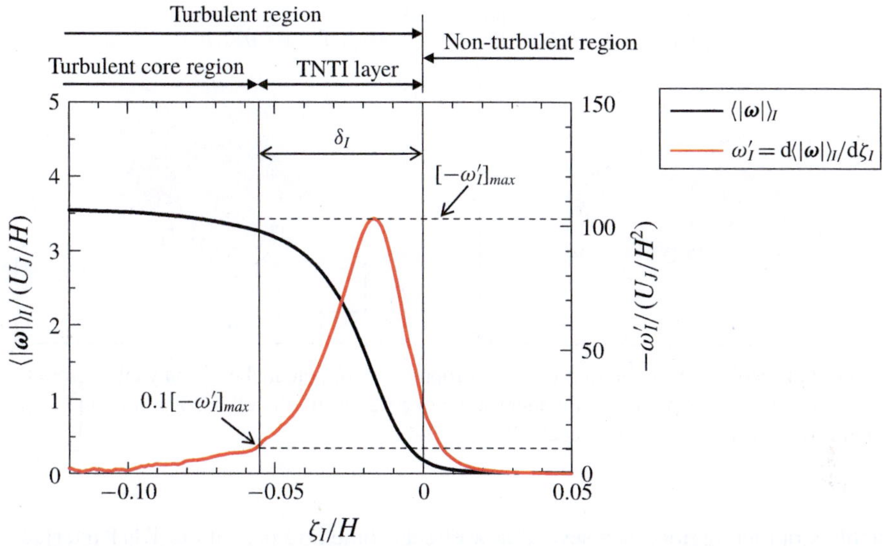

Figure 6.7 Conditional average of the mean vorticity magnitude $\langle|\omega|\rangle$ and its derivative in the vicinity of the T/NT interface (Watanabe et al., 2019).

enstrophy is

$$\left\{\frac{\partial}{\partial t} + u_j \frac{\partial}{\partial x_j}\right\}(\omega^2/2) = \underbrace{\omega_i s_{ij} \omega_j}_{P_\omega} + \underbrace{\nu \nabla^2 (\omega^2/2)}_{D_\omega} - \underbrace{\nu \nabla \omega_i \cdot \nabla \omega_i}_{\epsilon_\omega}, \tag{6.10}$$

where P_ω, D_ω, and ϵ_ω are the production, viscous diffusion, and viscous dissipation terms, respectively. In Fig. 6.6 are given plots of the conditional averages of the production term, $\langle P_\omega \rangle$, and diffusion term, $\langle D_\omega \rangle$, in the vicinity of the T/NT interface. Note that ζ_I/H gives the location relative to the T/NT interface, with positive ζ_I going away from the turbulent region. Both the viscous superlayer and the turbulent sublayer are evident in the figure. There is clearly a diffusion of enstrophy from the turbulent sublayer into the viscous superlayer, where the turbulent mixing of vorticity into the irrotational flow occurs, the basic mechanism for entrainment. Note that there is almost no enstrophy production in the viscous sublayer, but then the production dominates viscous diffusion in the turbulent sublayer. In Fig. 6.7 is plotted the conditional average of the vorticity magnitude $|\omega|$ along with its spatial derivative in the normal direction $\omega_I{}' = d\langle|\omega|\rangle_I / d\zeta_I$. The vorticity is seen to grow from approximately 0 near the interface to the value of the internal flow over the length of the combined interfaces. Note that, for this planar jet case, it was found that $\delta_\nu/\eta = 3.8$ and $\delta_\omega/\eta = 9.6$, which were similar to values obtained for a turbulent mixing layer.

Consider an ambient region which is stably stratified, i.e., defined by an ambient density field $\rho_a(z)$ which decreases with height z. Examples of this condition are in regions of the ocean below the mixed layer, and in regions of the atmosphere such as the stratosphere. When a local turbulent region is contained in a quiescent,

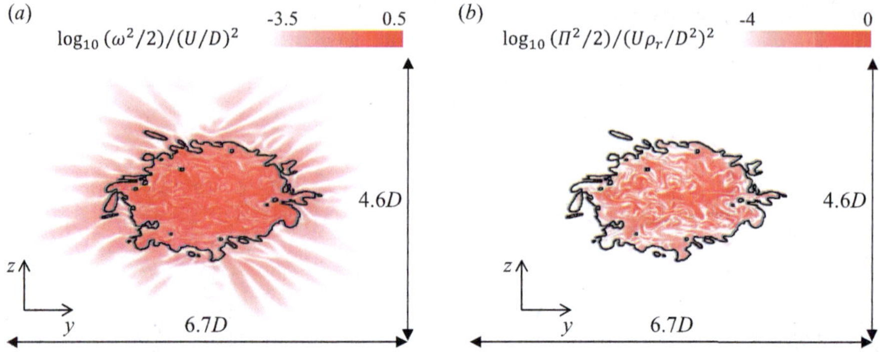

Figure 6.8 Vertical cross-sections of (a) enstrophy and (b) potential enstrophy taken from the direct numerical simulation of a turbulent wake in a stably stratified fluid. The solid black line shows the wake edge (Watanabe et al., 2016).

stably stratified region, then several new effects come into play at the T/NT interface. First of all, internal gravity waves are generated and propagate away from the T/NT interface and the turbulent region (see, e.g., Abdilghanie and Diamessis (2013)). The resultant wave energy flux can be considerable, sometimes the same order as the kinetic energy dissipation rate in the turbulent region (see, e.g., Watanabe et al. (2016)). In addition, as the waves are generated and propagate away from the turbulent region, since they are rotational, the ambient field is no longer irrotational. Therefore vorticity magnitude can no longer be used alone as a means to distinguish the turbulent and nonturbulent regimes. Neglecting viscous and diffusive effects, potential vorticity, defined by $\Pi = \omega_i \dfrac{\partial \rho}{\partial x_i}$, is conserved following the flow (Davidson, 2013). In particular, the internal waves then contain little or no potential vorticity. Consequently, potential vorticity remains 0 outside the turbulent flow field in the nonturbulent regions. Therefore potential vorticity, together with enstrophy, can be used instead of enstrophy alone to distinguish the turbulent and nonturbulent regions for localized turbulence occurring in an ambient, stratified region (Watanabe et al., 2016). Finally, since potential vorticity includes not only vorticity but also density in its definition, at the T/NT interface, not only is there the turbulent mixing of vorticity, but also of density, which together cause the interface to propagate into the nonturbulent regions.

In Fig. 6.8 is given a vertical cross-section of the flow taken from the direct numerical simulations of a turbulent wake in a stably stratified fluid (Watanabe et al., 2016). The threshold method was used to find the T/NT interface, but in this case, using both enstrophy (vorticity magnitude squared) and potential enstrophy (potential vorticity magnitude squared) to identify the interface, which is denoted by the black line in the figure. In Fig. 6.8(a) is shown the enstrophy field. Both the turbulent wake, within the T/NT interface, and the internal wave field exterior to the wake are clearly visible. On the other hand, Fig. 6.8(b) gives the potential enstrophy field, which identifies only the turbulent wake, and not the internal wave field.

6.4 Turbulent mixing in a stratified fluid

The ambient regions in an ocean below the mixed layer, stable (often nocturnal) atmospheric boundary layers, and regions in the upper atmosphere, such as the stratosphere, are stably stratified, and the stratification often has a strong, order one effect on the turbulence and turbulence mixing. Therefore understanding and predicting its effects are important. For example, in the modeling of the oceans, accurate modeling of the turbulent mixing and resulting turbulent fluxes is important in predicting the overall ocean circulation (Jayne, 2009), but also the effects of horizontal circulation, meridional overturning, and poleward heat transport (Gregg et al., 2018). Since turbulent mixing modifies the density structure, it affects the flow dynamics and, in particular, its energetics.

There are many sources of turbulence in stably stratified regions of the atmosphere, oceans, and other bodies of water. For example, the turbulence can be due to shear instabilities, much as for non-stratified regions. However, it can also result from the breakdown of propagating internal waves, for example, due to waves generated by the flow over complex terrain in the atmosphere, and due to the flow over complex bathymetry, especially seamounts, in the ocean. In the latter case, the oscillatory tidal currents can be especially effective in generating internal waves. Moreover, it is important to note that how the turbulence is generated can significantly affect its properties.

6.4.1 Estimates of turbulent buoyancy flux

Of particular interest in modeling turbulence mixing is the accurate representation of the turbulent buoyancy flux and hence the turbulent diffusivity. This was addressed by Osborn (1980), following early work by Osborn and Cox (1972). Consider the turbulent kinetic energy equation for a flow satisfying the Boussinesq approximation:

$$\begin{aligned}\left\{\frac{\partial}{\partial t}+\bar{u}_j\frac{\partial}{\partial x_j}\right\}\frac{1}{2}\overline{u'_iu'_i} &= -\frac{\partial}{\partial x_j}\left\{\frac{1}{\rho_o}\overline{p'u'_j}+\frac{1}{2}\overline{u'_iu'_iu'_j}+2\nu\overline{u'_is'_{ij}}\right\}\\ &\quad -\frac{g}{\rho_o}\overline{\rho'w'}-2\nu\overline{s'_{ij}s'_{ij}}-\overline{u'_iu'_j}\,\bar{s}_{ij}\end{aligned} \tag{6.11}$$

Here the Reynolds decomposition, e.g., $u_i = \bar{u}_i + u'_i$ is used, with $\overline{(\cdot)}$ denoting an average; also $\bar{s}_{ij}$ is the strain-rate of the mean velocity $\bar{u}_i$ while s'_{ij} is the strain rate of the turbulence u'_i. In addition, ρ_o is a constant reference density, ν is the kinematic viscosity (assumed to be constant), and g is the gravitational acceleration, taken to be in the negative x_3 direction. For convenience the turbulent velocity will sometimes be written as $(u'_1, u'_2, u'_3) = (u', v', w')$, with corresponding spatial coordinates $(x_1, x_2, x_3) = (x, y, z)$. The density ρ can also be written in terms of the buoyancy $b = (g/\rho_0)\rho$. The last three terms on the right-hand side of Equation (6.11) are the buoyancy flux, $J_b = \overline{-(g/\rho_o)\rho'w'} = -\overline{b'w'}$, the (irreversible) kinetic energy dissipation rate $\epsilon = 2\nu\overline{s'_{ij}s'_{ij}}$, and turbulent kinetic energy production rate $\mathcal{P} = -\overline{u'_iu'_j}\,\bar{s}_{ij}$, respectively. The buoyancy flux J_b plays two roles: the flux of mean density in the vertical direction and the exchange of kinetic and available potential energies (Eq. (6.26)).

Osborn essentially assumed that the region of the ocean of interest was both spatially homogeneous and stationary. Therefore the energy Eq. (6.11) reduces to:

$$\frac{g}{\rho_o}\overline{\rho' w'} = \overline{b' w'} = -2\nu\overline{s'_{ij}s'_{ij}} - \overline{u'_i u'_j}\,\bar{s}_{ij} \tag{6.12}$$

Assume that the ambient density field is given by the density gradient $\frac{d\bar{\rho}}{dz}$, and the only nonzero components of $\bar{s}_{ij}$ are $\frac{d\bar{u}}{dz}$ and $\frac{d\bar{v}}{dz}$. Making a diffusion approximation for the buoyancy flux, then

$$J_b = -\frac{g}{\rho_o}\overline{\rho' w'} = \frac{g}{\rho_o}K_\rho\frac{d\bar{\rho}}{dz} \tag{6.13}$$

where of principal interest is to determine the turbulent diffusivity K_ρ. From Eqs. (6.12) and (6.13), the turbulent diffusivity K_ρ can be written as

$$K_\rho = \frac{R_f}{1-R_f}\frac{\epsilon}{N^2} = \gamma_\rho\frac{\epsilon}{N^2} \tag{6.14}$$

where N is the buoyancy frequency given by $N^2 = -\frac{g}{\rho_o}\frac{d\bar{\rho}}{dz}$, and R_f is the flux Richardson number defined by

$$R_f = \frac{J_b}{\overline{u'w'}\left(\frac{d\bar{u}}{dz}\right) + \overline{v'w'}\left(\frac{d\bar{v}}{dz}\right)} \tag{6.15}$$

with $\overline{u'w'}$ and $\overline{v'w'}$ the relevant turbulent Reynolds stresses. What becomes of interest then is estimating the factor $\gamma_\rho = R_f/(1-R_f)$, which, as will be discussed below, is related to the mixing efficiency. Once γ_ρ is known, and ϵ and N can be estimated from measurements or from modeling, then K_ρ can be determined. From data, Osborn (1980) treated $\gamma_\rho \approx 0.2$ as an upper bound on γ_R, while from an examination of a number of sets of field data, Gregg et al. (2018) concluded that "there is no observational basis for not using $\gamma_\rho = 0.2$". Note that $R_f = \gamma_\rho/(1+\gamma_\rho)$, so that $\gamma_\rho = 0.2$ corresponds to $R_f \approx 0.17$. Below it will be shown that R_f is sometimes defined as the mixing efficiency. A slight alternative to Eq. (6.14) is to estimate K_ρ as

$$\frac{K_\rho}{\kappa} = \frac{-J_b}{\kappa N^2} = \frac{\nu}{\kappa}\left(\frac{-J_b}{\epsilon}\right)\frac{\epsilon}{\nu N^2} = Pr\,\Gamma\,Re_b \tag{6.16}$$

Here κ is the molecular diffusivity of density ρ, $Pr = \nu/\kappa$ is the Prandtl number, $\Gamma = -J_b/\epsilon$, related to γ_R, is a turbulent flux coefficient, and $Re_b = \epsilon/\nu N^2$ is the buoyancy Reynolds number (Gibson, 1980; Gargett et al., 1984).

6.4.2 Flow energetics

Also important in turbulent mixing in stably stratified flows are the energetics, both the potential and kinetic. A variation of the turbulent kinetic energy equation is given in Eq. (6.11). The behavior of potential energy is, in some ways, more complex. It can be important to distinguish processes in which the potential energy is changed irreversibly, by turbulent mixing, from those due to reversible processes, such as the propagation of internal waves. This partitioning is made possible using the Lorenz concept of available potential energy (Lorenz, 1955). The total potential energy in a stratified fluid volume $\mathcal{V}$, relative to some reference level, say $z = 0$, is given by

$$E_p = g \int_{\mathcal{V}} \rho z dV \tag{6.17}$$

It is useful to define the background potential energy E_b as the minimum potential energy attainable through a reversible (adiabatic) redistribution of ρ in space. Denoting the redistributed density field as $\rho(z_*)$, where $z_*(\mathbf{x}, t)$ is the vertical position in the reference state of the fluid parcel that was at $\mathbf{x}$ at time t, then

$$E_b = g \int_{\mathcal{V}} \rho z_*(\mathbf{x}, t) dV \tag{6.18}$$

The available potential energy E_a then is the energy released in the reversible transition from $\rho(\mathbf{x}, t)$ to $\rho(z_*)$, and is the amount available to drive further motion. From Eqs. (6.17) and (6.18), we have

$$E_a = g \int_{\mathcal{V}} \rho(z - z_*) dV \tag{6.19}$$

or $E_p = E_b + E_a$.

It is important to note that changes in the background potential energy are due to turbulent mixing, i.e., irreversible changes in the potential energy. Winters et al. (1995) showed that the rate of change of background potential energy is given by

$$\begin{aligned} \frac{d}{dt} E_b &= \kappa \rho_o \int_{\mathcal{V}} \left[\frac{|\nabla \rho|}{|dz_*/d\rho|^{-1}} \right]^2 N_*^2 dV \\ &= g \int_{\mathcal{V}} \frac{1}{|dz_*/d\rho|^{-1}} \kappa |\nabla \rho|^2 dV = g \int_{\mathcal{V}} \frac{\chi}{|dz_*/d\rho|^{-1}} dV \,. \end{aligned} \tag{6.20}$$

Therefore the rate of change of background potential energy is related to the turbulent mixing $\chi = \kappa |\nabla \rho|^2$, and is always positive. Winters et al. (1995) also showed that the background potential energy in a three-dimensional flow field, given the density data on a three-dimensional grid, can be approximately found by a three-dimensional sorting of the discrete fluid elements.

Closely related to the change in background potential energy is the irreversible flux of mass across a constant density (diapycnal) surface. For a diapycnal surface S in a volume whose horizontal surface area is A, the diapycnal flux is then defined by

$$\phi_d = \frac{1}{A}\int_S \kappa \nabla\rho \cdot \mathbf{n}\, dS \tag{6.21}$$

where $\mathbf{n}$ is the unit vector locally normal to S in the direction of increasing ρ. Winters and D'Asaro (1996) showed that the flux across the diapycnal surface is given by

$$\phi_d = -\kappa \frac{dz_*}{d\rho}\langle|\nabla\rho|^2\rangle_{z_*} = -\chi_{z_*}\frac{dz_*}{d\rho} \tag{6.22}$$

where $\langle\cdot\rangle_{z_*}$ is the average over the surface S where z_* maintains a constant value. Therefore the diapycnal flux is also proportional to χ.

The available potential energy is inherently a nonlocal quantity; for example, a local change in the density field in one location can affect the reference position z_*, and therefore its change in available potential energy, of a fluid particle away from this local change in density. For small amplitude disturbances, the available potential energy is approximately given pointwise as (Yih, 1965)

$$E_a(\mathbf{x}, t) \approx -\frac{1}{2}\frac{g}{\rho_0}\frac{\rho^2}{d\bar{\rho}/dz} = \frac{1}{2}\frac{b^2}{N^2} \tag{6.23}$$

where $\bar{\rho}(z)$ is the ambient density field, and $\rho = \rho_T - \bar{\rho}$, with ρ_T the total density field. There are often laboratory experiments, numerical simulations, and field measurements where small amplitudes cannot be assumed. In these cases, the definition given by Eq. (6.23) can sometimes be usefully employed in a pointwise manner as a surrogate for available potential energy. Assume that the total density field ρ_T can be written as $\rho_T = \rho_o + \bar{\rho}(z) + \rho_a$, where the ambient density gradient $\frac{d\bar{\rho}}{dz}$ is assumed to be constant, i.e., $N^2 = -\frac{g}{\rho_o}\frac{d\bar{\rho}}{dz}$ is assumed to be constant. Then the equation for ρ_a is:

$$\frac{\partial}{\partial t}\rho_a + u_j\frac{\partial}{\partial x_j}\rho_a = -w\frac{d\bar{\rho}}{dz} + \kappa\frac{\partial^2\rho_a}{\partial x_j^2} \tag{6.24}$$

Following the usual procedure of averaging Eq. (6.24) to obtain the equation for $\bar{\rho}_a$, using the Reynolds decomposition $\rho_a = \bar{\rho}_a + \rho'$, then subtracting the resulting equation from Eq. (6.24), giving an equation for the fluctuation ρ', and finally multiplying this equation by ρ' and averaging results in:

$$\left(\frac{\partial}{\partial t} + \bar{u}_j\frac{\partial}{\partial x_j}\right)\frac{1}{2}\overline{\rho'^2} = \frac{\partial}{\partial x_j}\left\{\kappa\frac{\partial}{\partial x_j}\left(\frac{1}{2}\overline{\rho'^2}\right) - \frac{1}{2}\overline{u'_j\rho'^2}\right\}$$
$$- \overline{\rho' u'_i}\frac{\partial\bar{\rho}_a}{\partial x_i} - \overline{\rho' w'}\frac{d\bar{\rho}}{dz} - \kappa\overline{\left(\frac{\partial\rho}{\partial x_j}\right)\frac{\partial\rho}{\partial x_j}} \tag{6.25}$$

Multiplying by $-\frac{g}{\rho_o}\left(\frac{d\bar{\rho}}{dz}\right)^{-1}$ and using the definition of the available potential energy given by Eq. (6.23) gives

$$\left(\frac{\partial}{\partial t}+\bar{u}_j\frac{\partial}{\partial x_j}\right)E_a=\frac{\partial}{\partial x_j}\left\{\kappa\frac{\partial}{\partial x_j}E_a+\overline{u'_jE'_a}\right\}$$
$$+\overline{b'w'}+\overline{b'u'_j}\left(\frac{d\bar{\rho}}{dz}\right)^{-1}\frac{\partial}{\partial x_j}\bar{\rho}_a-\frac{1}{N^2}\chi \tag{6.26}$$

where $E'_a=-\frac{1}{2}\frac{g}{\rho_o}\frac{\rho'^2}{d\bar{\rho}/dz}=\frac{1}{2}\frac{b'^2}{N^2}$, and here $\chi=\kappa\langle\nabla b'^2\rangle$. Note that the last term indicates a rate-of-loss of available potential energy proportional to χ, i.e., through turbulent mixing. This irreversible loss is consistent with the gain in background potential energy given by Eq. (6.20). This equation for E_a may only be useful when the penultimate term on the right-hand side can be neglected, i.e., when the coefficient $\frac{\partial}{\partial z}\bar{\rho}_a\Big/\left(\frac{d\bar{\rho}}{dz}\right)$ is small. For statistically homogeneous flows, is can be shown that $\frac{d\bar{\rho}}{dz}$ is a constant and $\frac{\partial}{\partial z}\bar{\rho}_a=0$ (Riley et al., 1981; Shih et al., 2005). See, for example, Scotti and White (2014) for some generalizations of Eq. (6.25), bringing in some of the features of Eq. (6.20).

6.4.3 Mixing efficiency

Mixing efficiency is the ratio of the energy expended to mix the fluid to the overall energy expended. In a stratified fluid, the total energy required would be the energy expended to irreversibly mix the fluid plus the energy lost to heat, i.e., to internal energy. The specific mathematical definition of mixing efficiency depends on the problem. For example, for a single process, like a shear instability, breakdown into turbulence, and ultimate decay, the mixing efficiency might be defined as the energy required to mix over the entire event, compared to the energy required to mix plus the energy lost to heat over the event. For continuous processes, a more useful definition could be the rate at which energy is lost to mixing relative the overall rate at which energy is lost (Peltier and Caulfield, 2003; Venayagamoorthy and Koseff, 2016). In this discussion, the latter definition will be used, although it should be noted that the former may be more useful for some problems.

From Eq. (6.11), it can be seen that turbulent kinetic energy is reversibly exchanged with potential energy through the buoyancy flux, J_b, and is irreversibly lost at the dissipation rate ϵ. On the other hand, available potential energy is irreversibly lost to background potential energy at a rate given by Eq. (6.20). Of interest then is the overall rate that available potential energy is irreversibly lost to the background, compared to the overall rate that energy is irreversibly lost, i.e., the rate of loss of available potential and kinetic energies combined. This is not only of fundamental interest, but is related to the buoyancy flux and the computation of the turbulent diffusivity, for example, given in Eqs. (6.14) and (6.16).

The mixing efficiency η is defined here as the rate of irreversible loss of energy to background potential energy compared to the rate of total irreversible loss of energy, i.e., to background potential energy and heat. Considering a volume $\mathcal{V}$, and defining $\mathcal{B} = dE_b/dt$ (see Eq. (6.20)), and $\mathcal{E} = \int_{\mathcal{V}} \epsilon dV$, then for this volume η can be defined mathematically as:

$$\eta = \frac{\mathcal{B}}{\mathcal{B} + \mathcal{E}} \tag{6.27}$$

where, for example, $\mathcal{B}$ could be computed by a sorting algorithm from a numerical simulation (Winters et al., 1995). Note that several other definitions of mixing efficiency are used. For example, the definition in terms of the net irreversible loss of potential energy, rather than the net total loss of energy is more useful for analyzing a specific event. Also, $\mathcal{B}/\mathcal{E}$ is sometimes used as the definition, although this ratio is often referred to as the mixing coefficient.

Various approximations for η can be obtained. For example, following the analysis of Osborn (1980) applied to a region of the ocean, then, since the loss of available potential energy would be equal to its gain in terms of the buoyancy flux, J_b,

$$\eta_o = \frac{-J_b}{-J_b + \epsilon} \tag{6.28}$$

From Eq. (6.12), using the additional assumptions of Osborn regarding $d\bar{\rho}/dz$ and $\bar{u}_i$, we obtain

$$\epsilon = J_b - \overline{u'w'}\left(\frac{d\bar{u}}{dz}\right) - \overline{v'w'}\left(\frac{d\bar{v}}{dz}\right) \tag{6.29}$$

which gives, when combined with Eq. (6.28),

$$\eta_o = R_f \tag{6.30}$$

where R_f is the flux Richardson number defined in Eq. (6.15). Therefore R_f is sometimes used as the definition of mixing efficiency (Venayagamoorthy and Koseff, 2016; Ivey and Imberger, 1991). But note that a number of simplifying assumptions have been used in its definition. Finally, using Eq. (6.13) for J_b in terms of the turbulent diffusivity K_ρ, then K_ρ can be expressed in terms of η_o as

$$K_\rho = \frac{\eta_o}{1 - \eta_o}\frac{\epsilon}{N^2} \tag{6.31}$$

which is the same as Eq. (6.14), given Eq. (6.30).

Using direct numerical simulations, a number of researchers have attempted to determine the dependency of mixing efficiency on various key non-dimensional parameters for several different simplified problems. For example, Maffioli et al. (2016) performed simulations of forced, homogeneous stratified turbulence for a range of stratifications. A forcing function $\mathcal{F}$ was used to drive the flow to steady state, and the surrogate form of the potential energy equation, Eq. (6.26), was used. Because of homogeneity, the mean density gradient remains constant, so that the coefficient $\frac{\partial}{\partial x_j}\bar{\rho}_a$ on the right-hand side of Eq. (6.26) remains 0. Therefore the energy Eqs. (6.11) and (6.26) reduce to

$$\frac{d}{dt}E_k = \mathcal{F}_i u_i' - \overline{b'w'} - \epsilon = 0 \tag{6.32}$$

$$\frac{d}{dt}E_a = \overline{b'w'} - \frac{1}{N^2}\chi = 0 \tag{6.33}$$

with $E_k = (1/2)\overline{u_i' u_i'}$. Important non-dimensional parameters in this problem are the Froude number $Fr = \mathcal{U}/NL_i$, where $\mathcal{U}$ is an rms velocity, and L_i is an 'integral scale' defined by $L_i = \mathcal{U}^3/\epsilon$; the Reynolds number $Re = \mathcal{U}L_i/\nu$; and the buoyancy Reynolds number, $Re_b = \epsilon/\nu N^2$. The Froude number can be considered as the ratio of the turbulence time scale $L_i/\mathcal{U}$ compared to the buoyancy time scale N^{-1}; it is large when the effects of density stratification are weak, and small when the effects of stratification are strong. Note that these parameters are related by $Re_b = Fr^2 Re$. An additional parameter of importance is the Ozmidov scale, $L_o = (\epsilon/N^3)^{1/2}$, an estimate of the largest horizontal scale of motion which can overturn (Riley and Lindborg, 2008). In terms of L_o the buoyancy, Reynolds number can be written as $Re_b = (L_o/\eta)^{4/3}$. Gibson (1980) and Gargett et al. (1984) pointed out that when $Re_b \sim 1$, so that $L_o \sim \eta$, then turbulence cannot be sustained, since there are no length scales above the Kolmogorov scale which can overturn. Note that at least two of these parameters can be defined independently, so that, in general, two independent parameters are needed to define this problem, either (Fr, Re) or (Fr, Re_b).

From their simulations Maffioli et al. (2016) computed, in particular, the ratio $\Gamma = (\chi/N^2)/\epsilon$, related to the mixing efficiency as $\eta = \Gamma/(1+\Gamma)$, and examined it as a function of Fr. Assuming that the Reynolds number is very large, they argued that, when Fr is also very large, so that stratification effects are not important, they could use a non-stratified estimate for χ/N^2 and determined that $\Gamma \sim Fr^{-2}$. On the other hand, as Fr goes to 0, again assuming that the Reynolds number is very large, they argued that Γ should asymptote to a constant. Fig. 6.9 is a plot of their simulation results, giving a log-linear plot of Γ versus Fr. It is seen that the Fr^{-2} behavior is obtained for large Fr, while the results appear to asymptote to a constant with a value of about 0.34 as Fr becomes small, corresponding to a value of η of about 0.25. From the expression for the buoyancy Reynolds number written as $Re_b = Fr^2 Re$, it is seen that, for Fr becoming very small, the Reynolds number would need to become increasingly large for turbulence to be sustained (Gibson, 1980; Gargett et al., 1984). Because this is a two-parameter problem, probably the limit for small Fr must be examined as a function of Re or Re_b to be assured of proper results.

Another example of the use of the results of direct numerical simulations to investigate and parameterize the mixing efficiency is given by Venayagamoorthy and Koseff (2016), using principally the results of numerical simulations performed by Shih et al. (2005). These simulations were for cases of homogeneous shear flows, where it can be shown that both the mean density gradient and the mean shear,

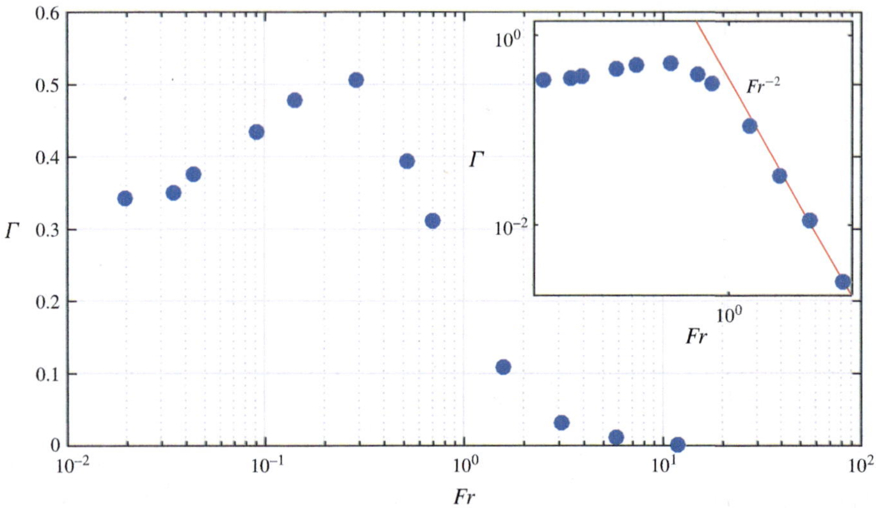

Figure 6.9 Mixing coeffiicient as a function of the Froude number (log-log plot is shown in the inset (Maffioli et al., 2016)).

$S = \dfrac{\partial \bar{u}_h}{\partial z}$, are constant, where $\bar{u}_h$ is the horizontal component of the vector $\bar{\mathbf{u}}$. In this case, the kinetic and potential energy equations, Eqs. (6.11) and (6.26), reduce to

$$\frac{d}{dt}E_k = -\overline{b'w'} - \overline{u'_h w'}S - \epsilon \tag{6.34}$$

$$\frac{d}{dt}E_a = \overline{b'w'} - \frac{1}{N^2}\chi \tag{6.35}$$

where u'_h is the component of the vector $\mathbf{u}'$ in the direction of $\bar{u}_h$. Additional important parameters for this problem are the gradient Richardson number, $R_g = N^2/S^2$, which governs the stability of stratified shear flows, and the flux Richardson number (Eq. (6.15)), which is used as a parameter to govern mixing efficiency. By comparing various definitions of the mixing efficiency with the results from the simulations, Venayagamoorthy and Koseff (2016) conclude that the most useful definition of mixing efficiency is

$$\eta_v = \frac{\chi/N^2}{\chi/N^2 + \epsilon} \tag{6.36}$$

Fig. 6.10 is a plot from the results of Shih et al. (2005) of the mixing efficiency η_v versus the gradient Richardson R_g for a range of values of R_g. In the plot $R_f^* = \eta_v$. The agreement is rather good when comparing the simulation results with the following curve fit:

$$\eta_v = \eta_{v\infty}[1 - \exp(-\beta Ri_g)]\,, \tag{6.37}$$

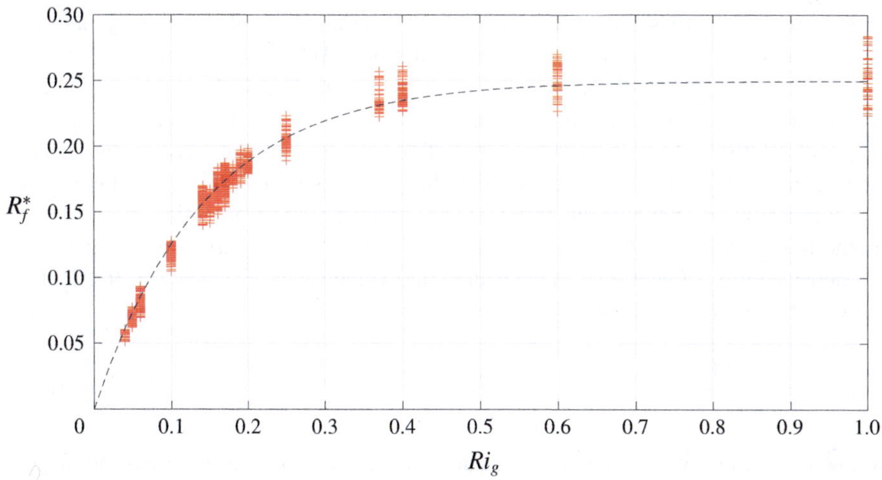

Figure 6.10 Mixing coefficient as a function of the gradient Richardson number (Venayagamoorthy and Koseff, 2016).

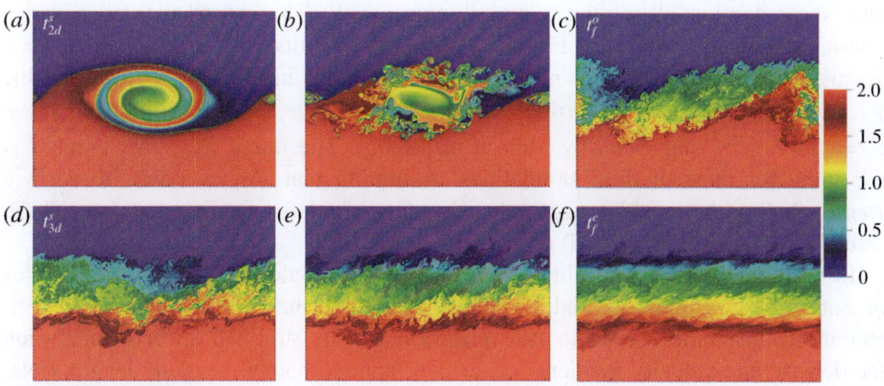

Figure 6.11 Evolution of the density field in a vertical plane, from Salehipour and Peltier (2015).

with $\eta_{v\infty} = 0.25$, the large Ri_g asymptote of the curve, and $\beta = 7$. Note that their asymptote of $R_f = 0.25$ for large R_g, i.e., for strong stratification, is consistent with the asymptote obtained by (Maffioli et al., 2016) for small Fr, i.e., strong stratification. In addition, both of the curves of Maffioli et al. (2016) and Venayagamoorthy and Koseff (2016) asymptote to 0 as the stratification weakens. In addition, note that this problem is also expected to be a two-parameter problem with some dependence, for example, on Re or Re_b. So the results might break down, for example, at moderate values of Re and larger values of Ri_g (corresponding roughly to smaller values of Fr and moderate values of Re in the previous example).

Salehipour and Peltier (2015) addressed the mixing efficiency of turbulent flows using the direct numerical simulation of unstable shear layers susceptible to Kelvin–

Helmholtz instabilities. Fig. 6.11 shows color contour plots of the density field in a vertical plane taken from one of their simulations. These plots show the evolution of the flow, from a two-dimensional instability in panel a, to the three-dimensional breakdown of the flow in panels b and c, and finally to the decay of the flow in panels d through f, leaving a steppy fine-structure in the density field. They performed simulations at buoyancy Reynolds numbers up to $\mathcal{O}(10^3)$, and computed the conversion of available potential energy to background potential energy by the methods suggested by Winters et al. (1995). Determining the irreversible mixing of the density field $\mathcal{M}$ as the vertically integrated value of ϕ_d given by Eq. (6.22), they defined the mixing efficiency as

$$\mathcal{E} = \frac{\mathcal{M}}{\mathcal{M} + \epsilon_k} \tag{6.38}$$

where ϵ_k is the volume integrated value of ϵ. Then, considering Re_b and Ri_g as the vertically integrated values of the buoyancy Reynolds number and the gradient Richardson number, respectively, they categorized the intermediate (i.e., $Re_b = \mathcal{O}(10^1 - 10^2)$) and high (i.e., $Re_b > \mathcal{O}(10^2)$) ranges of Re_b as 'buoyancy-dominated' and 'shear dominated' mixing regimes", respectively. They found that the $\mathcal{E}$ varies non-monotonically with Re_b. Fig. 6.12 gives their results for the mixing efficiency versus the buoyancy Reynolds number. The dashed line in the figure is for the value $\gamma_\rho = 0.2$ suggested by Osborn (1980) (see Eq. (6.14)). The solid line is the curve $\mathcal{E} = 1.5Re_b^{-1/2}$ suggested by Shih et al. (2005). The initial Reynolds number is the same for most of their simulations. Assuming that $Re = \text{const}$, then since $Re_b = ReFr^2$, their results suggest that $\mathcal{E} \sim Fr^{-1}$ for large Fr, inconsistent with the results of (Maffioli et al., 2016) given above.

In treating isolated instability and transition to turbulence, a number of studies of both Kelvin–Helmholtz and Holmboe instabilities have been pursued. Kelvin-Helmholtz instabilities tend to take place on density-stratified shear layers when the density and velocity gradients occur over approximately the same length scale, whereas Holmboe instabilities occur when the scale of the density gradient is much smaller than that of the velocity gradient. The instabilities and turbulence develop very differently, the Kelvin–Helmholtz instability leading to turbulence with a strong vortex rollup and mixing, as in Fig. 6.11, whereas the Holmboe instability leads to weaker, but longer-lasting turbulence which 'scours' the density interface (Salehipour et al., 2016). Both methods tend to leave a somewhat well-mixed layer with high-density gradient regions above and below. In the end, however, the mixing due to the two mechanisms is comparable. The fact that the Holmboe instability occurs with the density interface much thinner than the velocity interface suggests that it might be more typical of flows in fluids with high Prandtl number, since with higher Pr, mass diffuses much more slowly than momentum. Such conditions with high Pr are common in many regions of the oceans and estuaries.

A key assumption in most of the numerical simulations is that the Prandtl number ν/D is $\mathcal{O}(1)$. Based upon the dependence of Pr upon temperature, however, its value is around 7, whereas based upon salinity, it can be as high as several hundred.

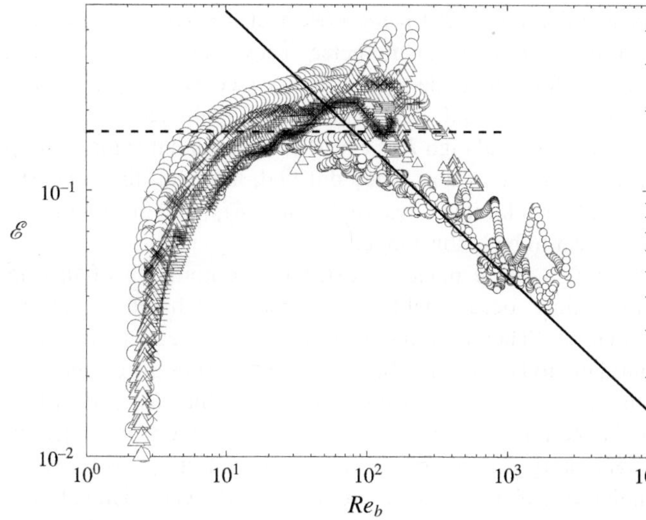

Figure 6.12 Mixing efficiency versus the buoyancy Reynolds number from Salehipour and Peltier (2015).

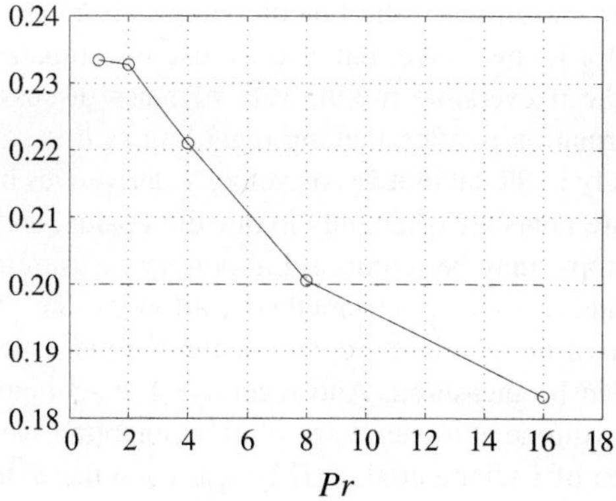

Figure 6.13 Mixing efficiency versus the Prandtl number, from Salehipour et al. (2015).

Therefore it is important to determine the sensitivity of the various results to the variation in the Prandtl number. This issue has been addressed by Salehipour and Peltier (2015), who performed turbulent shear layer simulations for a range of Prandtl numbers up to 16, using the same analysis of mixing efficiency as Salehipour et al. (2015). They found that, indeed, the effects of the Prandtl number were significant. As Fig. 6.13 shows, the mixing efficiency decreased from a value of over 0.23 down to near 0.18 as the Prandtl number was increased from 1 to 16. Using a very detailed stability analysis of the evolving flow, they suggested that this change in mixing efficiency was due to subtle changes in the transitional properties of the flow. Furthermore, they suggested that the effect may be much stronger as the Prandtl number is further decreased.

Lozovatsky and Fernando (2013) reported measurements in a stable atmospheric boundary layer using a sonic anemometer. They were able to measure K_ρ and ϵ directly and therefore infer the efficiency γ_ρ (see Eq. (6.14)). They found that γ_ρ depended on both Ri_g and Re_b, and varied significantly depending on these parameters. For very low and high Re_b and Ri_g they found that the assumptions made by Osborn (1980) were untenable. They did find, however, that a constant value of $\gamma_\rho \approx 0.17$ is tenable in a limited range of Re_b and Ri_g, and that many oceanographic measurements appear to be in this range.

Gregg et al. (2018) have made an extensive compilation of mixing efficiency obtained from mainly ocean field measurements, laboratory experiments, and numerical simulations. They conclude that "it has long been understood that mixing efficiency is unlikely to be constant, but, because K_ρ from tracers and microstructure evaluated with $\gamma_\rho = 0.2$ agree better than do coefficients and efficiencies from simulations and experiments, observations should continue to be scaled with 0.2 until observations, laboratory experiments, and numerical simulations converge on a more accurate formulation". Note that with the mixing efficiency defined in terms of the flux Richardson number R_f, then $\gamma_\rho = 0.2$ corresponds to $\eta = 0.17$.

Finally, the discussion here has been devoted to the irreversible turbulent mixing of mass. Of equal and possibly of more importance, however, is the irreversible turbulent mixing of momentum, resulting in the kinetic energy dissipation rate ϵ. This information is important for its own sake, but also for use in estimates of other important quantities, such as the irreversible mixing rate of mass discussed above. Unfortunately, accurate information is often lacking regarding ϵ. It is sometimes difficult to be measured accurately in the laboratory, as velocity derivatives have to be accurately measured; the measurements are often only in one dimension, and additional assumptions, e.g., local isotropy, may be required. Laboratory measurements of ϵ are being improved with the addition of optical methods, allowing the almost instantaneous measurements of one or more velocity components in a plane, so that more terms in the definition of ϵ can be measured. And recently, three-component measurements have been made in a volume at moderate resolution, enabling measurement of all the terms in the definition of ϵ (Jiang et al., 2022). It is even more difficult to measure ϵ in the field, as the measurements are usually of one or at most two components along a line, and often with limited resolution. So, again, assumptions such as local isotropy (Oakley, 1982; Wesson and Gregg, 1994) or curve fitting, e.g., Luketina and Imberger (2001), are required. Thus significant errors are expected in the measurement of ϵ in the field. In numerical simulations, ϵ can be a difficult quantity to accurately treat. Reynolds-averaged Navier–Stokes (RANS) modeling often requires an equation for ϵ. The generation and dissipation terms in this equation are modeled with very little information and confidence, rendering them very suspect. In large eddy simulation (LES) modeling, ϵ is sometimes modeled with its own equation, giving the same problems as with RANS. In other LES approaches, ϵ results from the type of sub-grid scale model used, which again can lead to inaccurate results.

To obtain ϵ more accurately in the field, there is clearly a need for more theoretical and modeling support for how to determine ϵ more accurately, given one-dimensional measurements of the velocity gradient and other properties of the flow, such as local ambient shear and stratification. Very high resolution direct numerical simulations are beginning to be used for this purpose, using *a priori* assessment of the modeling or approximations, e.g., Lewin et al. (2021); Almalkie and de Bruyn Kops (2012).

6.5 Summary

Turbulent mixing is one of the most important features of turbulence. It can cause high kinetic energy dissipation rates, which results in more power requirements. On the other hand turbulent mixing can enhance overall chemical reaction rates, and can lead to enhanced dilution as contaminants are released into the environment. And turbulent mixing affects the evolution of the ocean density field, and thus strongly influences the overall ocean dynamics. There have been significant advances in determining how mixing occurs in the ocean, and how to model mixing events, although much remains to be done. What is probably as important now is to learn the principal sources of turbulence in the ocean, and to determine how the overall forcing of the motion in the ocean, through mainly wind stress and tidal motions, ultimately leads to turbulence (Wunsch and Ferrari, 2004). Furthermore, since much has been learned about turbulent mixing from a variety of problem areas, ranging from turbulent combustion to turbulence in stratified flows, there is probably much to be gained from cross-fertilization of results in these areas.

Acknowledgments

The author would like to thank S. M. de Bruyn Kops for many discussions of the issues related in this chapter. This work was partially funded by the U.S. Office of Naval Research, grant N00014-19-1-2154.

References

Abdilghanie, A.M., Diamessis, P.J., 2013. The internal gravity wave field emitted by a stably stratified turbulent wake. Journal of Fluid Mechanics 720, 104–139.

Almalkie, S., de Bruyn Kops, S.M., 2012. Energy dissipation rate surrogates in incompressible Navier-Stokes turbulence. Journal of Fluid Mechanics 697, 204–236.

Bilger, R.W., 1989. Turbulent diffusion flames. Annual Review of Fluid Mechanics 21, 101–135.

Caulfield, C.P., 2021. Layering, instabilities, and mixing in turbulent stratified flows. Annual Review of Fluid Mechanics 53, 113–145.

Corrsin, S., 1947. I. Extended applications of the hot-wire anemometer – II. Investigations of the flow in round turbulent jets. Ph.D. Thesis. Caltech.

Corrsin, S., Kistler, A.L., 1955. Free-stream boundaries of turbulent flows. NACA Report 1244, 1033–1064.

Davidson, P.A., 2013. Turbulence in Rotating, Stratified, and Electrically Conducting Fluids. Cambridge University Press.

Dimotakis, P.E., 2005. Turbulent mixing. Annual Review of Fluid Mechanics 37, 329–356.

Dimotakis, P.E., Miake-Lye, R.C., Papantoniou, D.A., 1983. Structure and dynamics of round turbulent jets. Physics of Fluids 26, 3185–3192.

Eckart, C., 1948. An analysis of the stirring and mixing processes in incompressible fluids. Journal of Marine Research 7 (3), 265–275.

Fernando, H.J.S., 1991. Turbulent mixing in stratified fluids. Annual Review of Fluid Mechanics 23, 455–493.

Gargett, A., Osborn, T., Nasmyth, P., 1984. Local isotropy and the decay of turbulence in a stratified fluid. Journal of Fluid Mechanics 144, 231–280.

Gibson, C.H., 1980. Fossil turbulence, salinity, and vorticity turbulence in the ocean. In: Nihous, J.C.J. (Ed.), Marine Turbulence. Elsevier, pp. 221–257.

Gregg, M.C., D'Asaro, E.A., Riley, J.J., Kunze, E., 2018. Mixing efficiency in the ocean. Annual Review of Marine Science 10, 443–473.

Ivey, G.N., Imberger, J., 1991. On the nature of turbulence in a stratified fluid. Part 1: The energetics of mixing. Journal of Physical Oceanography 21, 650–658.

Ivey, G.N., Winters, K.B., Koseff, J.R., 2008. Density stratification, turbulence but how much mixing? Annual Review of Fluid Mechanics 40, 169–184.

Jayne, S., 2009. The impact of abyssal mixing parameterizations in an ocean general circulation model. Journal of Physical Oceanography 39 (7), 1756–1775.

Jiang, X., Lefauve, A., Dalziel, S., Linden, P., 2022. The evolution of coherent vortical structures in increasingly turbulent stratified shear layers. Journal of Fluid Mechanics 947, A30.

Kundu, P.K., Cohen, I.M., 2002. Fluid Mechanics, 2nd edn. Academic Press.

Lewin, S.E., de Bruyn Kops, S.M., Portwood, C.P., Caulfield, C.P., 2021. Probabilistic neural networks for predicting energy dissipation rates in geophysical flows. In: Fourth Workshop on Machine Learning and the Physical Sciences.

Linden, P.F., 1979. Mixing in stratified fluids. Geophysical and Astrophysical Fluid Dynamics 13, 431–464.

Lorenz, E.N., 1955. Available potential energy and the maintenance of the general circulation. Tellus 7, 157–167.

Lozovatsky, I.D., Fernando, H.J.S., 2013. Mixing efficiency in natural flows. Philosophical Transactions of the Royal Society of London A: Mathematical, Physical and Engineering Sciences 371 (1982), 20120213.

Luketina, D.A., Imberger, J., 2001. Determining turbulent kinetic energy dissipation from batchelor curve fitting. Journal of Atmospheric and Oceanic Technology 18, 100–113.

Maffioli, A., Brethouwer, G., Lindborg, E., 2016. Mixing efficiency in stratified turbulence. Journal of Fluid Mechanics 794, R3.

Oakley, N.S., 1982. Determination of the rate of dissipation of turbulent energy from simultaneous temperature and velocity shear microstructure measurements. Journal of Physical Oceanography 12, 256–271.

Osborn, T.R., 1980. Estimates of the local-rate of vertical diffusion from dissipation measurements. Journal of Physical Oceanography 10, 83–89.

Osborn, T.R., Cox, C.S., 1972. Oceanic fine structure. Geophysical Fluid Dynamics 3 (1), 321–345.

Peltier, W.R., Caulfield, C.P., 2003. Mixing efficiency in stratified shear flows. Annual Review of Fluid Mechanics 35, 135–167.

Riley, J.J., Lindborg, E., 2008. Stratified turbulence: a possible interpretation of some geophysical turbulence measurements. Journal of the Atmospheric Sciences 65, 2416.

Riley, J.J., Metcalfe, R.W., Weissman, M.A., 1981. Direct numerical simulations of homogeneous turbulence in density stratified flows. In: West, B.J. (Ed.), Proc. AIP Conf. Nonlinear Properties of Internal Waves. American Institute of Physics, New York, pp. 79–112.

Salehipour, H., Caulfield, C.P., Peltier, W.R., 2016. Turbulent mixing due to the holmboe wave instability at high Reynolds number. Journal of Fluid Mechanics 803, 591–621.

Salehipour, H., Peltier, W.R., 2015. Diapycnal diffusivity, turbulent Prandtl number and mixing efficiency in Boussinesq stratified turbulence. Journal of Fluid Mechanics 775, 464–500.

Salehipour, H., Peltier, W.R., Mashayek, A., 2015. Turbulent diapycnal mixing in stratified shear flows: the influence of Prandtl number on mixing efficiency and transition at high Reynolds number. Journal of Fluid Mechanics 773, 178–223.

Scotti, A., White, B., 2014. Diagnosing mixing in stratified turbulent flows with a locally defined available potential energy. Journal of Fluid Mechanics 740, 114.

Shih, L.H., Koseff, J.R., Ivey, G.N., Ferziger, J.H., 2005. Parameterization of turbulent fluxes and scales using homogeneous sheared stably stratified turbulence simulations. Journal of Fluid Mechanics 525, 193–214.

da Silva, C.B., Taveira, R.R., 2010. The thickness of the turbulent/nonturbulent interface is equal to the radius of the large vorticity structures near the edge of the shear layer. Physics of Fluids 22, 121702.

Sreenivasan, K.R., Schumacher, J., 2010. Lagrangian views on turbulent mixing of passive scalars. Philosophical Transactions of the Royal Society of London. A 368, 1561–1577.

Taveira, R.R., Diogo, J.S., Lopes, D.C., da Silva, C.B., 2013. Lagrangian statistics across the turbulent-nonturbulent interface in a turbulent plane jet. Physical Review E 88, 043001.

Tennekes, H., Lumley, J.L., 1972. A First Course in Turbulence. MIT Press, Cambridge.

van Veen, E.H., Roekaerts, D., 2003. On the accuracy of temperature measurements in turbulent jet diffusion flames by coherent anti-Stokes-Raman spectroscopy. Combustion Science and Technology 175, 1893–1914.

Venayagamoorthy, S.K., Koseff, J.R., 2016. On the flux Richardson number in stably stratified turbulence. Journal of Fluid Mechanics 798, R1.

Warhaft, Z., 2000. Passive scalar in turbulent flows. Annual Review of Fluid Mechanics 32, 203–240.

Watanabe, T., Riley, J.J., de Bruyn Kops, S.M., Diamessis, P.J., Zhou, Q., 2016. Turbulent/non-turbulent interfaces in wakes in stably stratified fluids. Journal of Fluid Mechanics 797, R1.

Watanabe, T., da Silva, C.B., Nagata, K., 2019. Non-dimensional energy dissipation rate near the turbulent/nonturbulent interfacial layer in free shear flows and shear free turbulence. Journal of Fluid Mechanics 875, 321–344.

Wesson, J.J., Gregg, M.C., 1994. Mixing at camarinal sill in the Strait of Gibraltar. Journal of Geophysical Research 99 (C5), 9847–9878.

Winters, K.B., D'Asaro, E.A., 1996. Diascalar flux and the rate of fluid mixing. Journal of Fluid Mechanics 317, 179–193.

Winters, K.B., Lombard, P.N., Riley, J.J., D'Asaro, E.A., 1995. Available potential energy and mixing in density-stratified fluids. Journal of Fluid Mechanics 289, 115–128.

Wunsch, C., Ferrari, R., 2004. Vertical mixing, energy and the general circulation of the oceans. Annual Review of Fluid Mechanics 36, 281–314.

Yih, C.-S., 1965. Dynamics of Nonhomogeneous Fluids. Collier-Macmillan, New York-London.

Turbulent transport in the environment

7

Benoit Cushman-Roisin
Thayer School of Engineering, Dartmouth College, Hanover, NH, United States

7.1 Introduction

Whenever a substance is carried by a turbulent flow, ambient air or surface water in the outdoors, it is not merely transported by the averaged velocity of the flow; it is also actively dispersed by its turbulent fluctuations. This turbulent dispersion causes dilution, which is beneficial, and also multiplies encounters with other substances carried by the same fluid, thus promoting chemical reactions and/or biological processes that may degrade the substance or generate new ones. This latter part may or may not be beneficial.

For example, turbulence in a river facilitates the contact between its water and the surface where oxygen is taken up from the air to become dissolved oxygen (DO) in the water, and this dissolved oxygen is effectively distributed across the water column to be made available to the fish, which depend on its presence for their survival. Should the stream become calmer, oxygen is not only taken up from the air more slowly but is also less effectively conveyed to the fish. It is thus no wonder that active fish like salmon and trout prefer highly turbulent waters to satisfy their high oxygen demand.

In contrast, the convective turbulence inside a cloud promotes encounters between chemical species; so, should there be some sulfur dioxide (SO_2), the higher the intensity of the turbulence, the greater the contact rate of SO_2 with water droplets (H_2O), and the more effective the conversion of sulfur dioxide into sulfuric acid (H_2SO_4), which will eventually fall to the ground in the form of acid precipitation. In the river case, the stronger the turbulence, the better; in the cloud case, the stronger the turbulence, the worse.

Environmental examples of turbulent dispersion abound. In the lower atmosphere, specifically in the so-called atmospheric boundary layer (Lyons and Scott, 1990, Chapter 2), where humans and animals breathe, and plants perform photosynthesis, exhaust fumes typically enter through smokestacks and vehicle tailpipes. These plumes grow to a much wider size initially by the shear flow caused by their own exhaust speed and eventually by turbulent whirls of the ambient air caused by building or topographic obstructions, solar-induced convection, or larger-scale winds. The aggregation effect of these individual emission points demands, at times, consideration as a distributed source, and the question is no longer how one particular plume grows but how a distribution of sources contributes to a contaminant's budget in a large volume subjected to stirring by turbulence.

Examples are likewise common in aquatic systems. With their submerged or emerging vegetation, wetlands exhibit peculiar types of turbulent flows (Nepf, 1999)

Environmental Turbulence. https://doi.org/10.1016/B978-0-32-395873-8.00014-0
Copyright © 2026 Elsevier Inc. All rights are reserved, including those for text and data mining, AI training, and similar technologies.

that disperse both healthy nutrients and harmful contaminants (Kadlec, 1994). In rivers, the gravitational flow down the slope of the channel causes friction-induced shear near the bottom and banks, which in turn generates turbulence in all three possible directions, vertical, transverse, and downstream (Chanson, 2004), with a different intensity in each of these directions (anisotropy). Lakes are subjected to wind-driven vertical mixing, typically on a daily basis, as well as to vertical convective mixing, on both diurnal and seasonal time scales (Fischer et al., 1979). Phases of restratification and turbulence suppression alternate with those episodes of mixing. In estuaries, tidal flow adds its energy to the river's own turbulence, while buoyancy of the freshwater in contact with seawater tends to put a damper on turbulence by consuming kinetic energy for the lifting of denser waters and the downward pushing of lighter waters (Lewis, 1997). Once into the sea, the partially diluted river waters spread and are diluted under the turbulent action of coastal currents.

Needless to say, a reliable formulation of turbulent motions is crucial for the correct representation of the dispersion and fate aspects in all these types of turbulent flows. The problem, as we know all too well, is that turbulence is not easily described, and all of its models fall short in one way or another. So, what has to be done but to seek a yet better model?

7.1.1 The nature of mixing

Because mixing is an everyday word, most readily associated with our daily food preparations, it tends to lack a sharp definition. It is nonetheless accompanied by an extensive body of knowledge, mostly stemming from chemical engineering. Indeed, a frequent preoccupation among chemical engineers is the design of swirling flows that maximize mixing, followed by the determination of the energy required to stir the contents of a chemical reactor in a certain amount of time. The study of mixing also pertains to environmental fluid mechanics because it is the process by which fluid parcels with different origins, and possibly different contaminant loadings, come into contact, dilute their properties into one another, and promote transformative chemical reactions.

A first approach to the study of mixing is the consideration of its geometrical aspects, that is, the topology of streamlines in the flow (Ottino, 1989). The questions then concern the folding of fluid trajectories and rates of deformation of a cloud of tracked fluid particles. Mixing can be viewed as a three-dimensional "shuffling process" (Brothman et al., 1945) during which there is an increasing probability over time that a fluid parcel of a given size will pass through a certain region. Following this line of thought and carefully distinguishing between stirring and mixing,[1] Eckart (1948) calculated "mixing times" of temperature and salt anomalies in the ocean without resorting to any dynamical description of the underlying processes.

[1] According to Eckart (1948), *stirring* should be used only to refer to the kinematic stretching and folding of lines of fluid particles, whereas *mixing* is the process that erases differences (gradients) under the subsequent action of molecular diffusion.

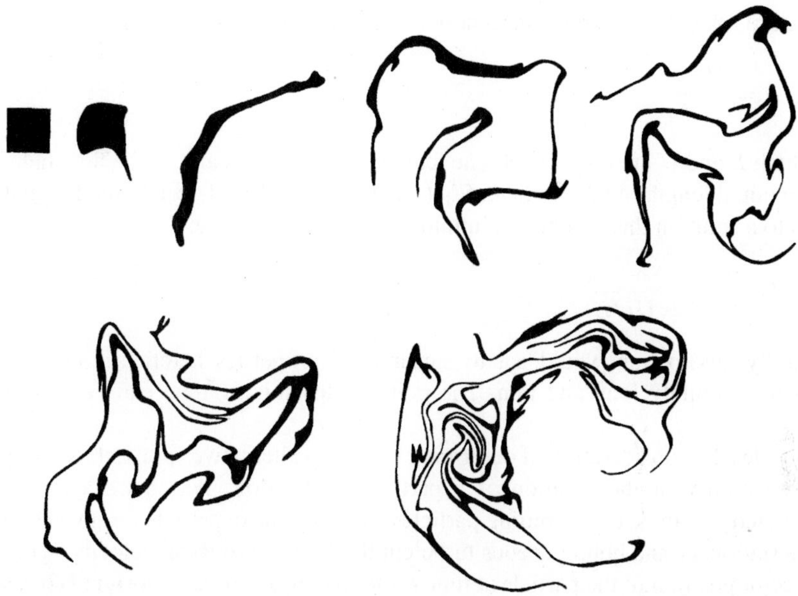

Figure 7.1 Sequence of snapshots showing the temporal deformation of a patch of butanol floating on water in a rotating vessel. Note how the extent of the patch grows faster initially than in later stages when bending causes a tendency to return toward the center of the patch. [*Source*: Welander, 1955, Figure 3].

Welander (1955) offers the following description for stirring in two dimensions. An initial square element with a side short compared to the length scale of the deforming flow field is subjected to translation, rotation, and shearing. As time passes, the deformation of the element becomes significant, and the element gradually acquires a highly irregular shape. Increasingly long and thin bands are formed (Fig. 7.1). Whenever the band lies at an angle to the flow, folds develop at points of maximum velocity. Folds also appear where the velocity gradient is nearly opposite to the velocity direction. As the band folds, it becomes increasingly elongated and thin while not occupying much more overall space. Within a finite amount of time, the situation appears chaotic and hopeless to a modeler.

One reasonable assumption to begin to describe the situation is to consider that the rate of stretching of a segment of length ℓ may grow proportionally to ℓ itself,

$$\frac{d\ell}{dt} = \alpha\,\ell \tag{7.1}$$

with the coefficient of proportionality α depending on the features of the flow field. Considering a stationary turbulence field, Batchelor (1952) took α as a constant over time, with ℓ then increasing exponentially but never reaching an infinite value in a finite time. Welander (1955), in contrast, argued that an infinite length could be reached in a finite time. His argument, based on properties of stretching in high-curvature folds, leads to the expression

$$\alpha = \frac{U}{L} \frac{\ell}{\ell_0} \tag{7.2}$$

in which L and U are respectively the length and velocity scales of the flow field, and ℓ_0 the initial length of the segment. Eq. (7.1) then includes a factor ℓ^2 on the right and leads to a solution that reaches an infinite value in a finite time:

$$\ell(t) = \frac{\ell_0}{1 - Ut/L} \tag{7.3}$$

Naturally, dissipation would act to smear fluid properties before a band of fluid becomes infinitely long and thin. The point is that mixing time can be as short as L/U.

Besides the consideration of deforming bands, an alternative approach is to explore the probability of the position of a particle in the flow. For this, Taylor (1921) considered a point source emitting particles, such as smoke particles from a chimney, into a stationary and homogeneous turbulent fluid with zero mean velocity. With $\vec{x}(t)$ the vector position of the particle at time t after its release from the origin $[\vec{x}(0) = 0]$, the following variance is defined:

$$R^2(t) = <\vec{x}(t) \cdot \vec{x}(t)> \tag{7.4}$$

in which the brackets $< \ldots >$ indicate an average over turbulent fluctuations and the dot represents the scalar product of the two vectors. Although the particle may occasionally move in a direction that takes it back toward its origin, it is unlikely that it will ever return there, and the more time elapses, the further away from the origin it is likely to be. The correlation $R^2(t)$ measures the square of the distance from the origin by time t, regardless of direction, and the vigor of mixing is measured by the rate at which $R^2(t)$ grows over time. So, we take its time derivative:

$$\frac{dR^2}{dt} = 2 < \vec{x} \cdot \frac{d\vec{x}}{dt} > \tag{7.5}$$

Since the time derivative of position is velocity, the preceding equation may also be successively expressed as

$$\begin{aligned} \frac{dR^2}{dt} &= 2 < \left(\int_0^t \vec{u}(t')dt' \right) \cdot \vec{u}(t) > \\ &= 2 \int_0^t < \vec{u}(t') \cdot \vec{u}(t) > dt' \\ &= 2 \int_0^t < \vec{u}(t-\tau) \cdot \vec{u}(t) > d\tau \end{aligned} \tag{7.6}$$

in which $\vec{u} = d\vec{x}/dt$ is the vector velocity in the turbulent flow, which is also a random variable. The second expression was obtained by switching the order of operation between time integration and averaging over turbulent fluctuations, and the change of variable $\tau = t - t'$ was made to arrive at the third expression.

If the mixing fluid is in a state of stationary, homogeneous and isotropic turbulence, the dimensionless velocity auto-correlation over time,

$$S^2(\tau) = \frac{< \vec{u}(t) \cdot \vec{u}(t-\tau) >}{< \vec{u}(t) \cdot \vec{u}(t) >} = \frac{< \vec{u}(t) \cdot \vec{u}(t-\tau) >}{u_*^2} \tag{7.7}$$

is independent of time t, position, and direction and can only be a function of the time separation τ. The velocity variance u_*^2 measures the turbulence intensity, which is constant in stationary conditions. Eq. (7.6) can then be cast as

$$\frac{dR^2}{dt} = 2u_*^2 \int_0^t S^2(\tau)\, d\tau \tag{7.8}$$

Two limiting cases are of interest. For short separation times ($\tau \to 0$), the velocity cannot be significantly decorrelated because acceleration must remain finite. Hence, $S^2(\tau) \simeq 1$ for short τ, and

$$\frac{dR^2}{dt} = 2u_*^2\, t \tag{7.9}$$

of which the time integration yields:

$$R(t) = u_*\, t \tag{7.10}$$

Thus the spread of the particle position increases as the first power of time. In essence, the particle is propelled in the direction of its initial velocity. This initial behavior can be considered as a ballistic regime.

But this behavior will not last long because velocity fluctuations will soon redirect the particle and force it to meander and at times reverse direction. At long separation times, the velocity auto-correlation vanishes (Fig. 7.2), and the relevant quantity becomes the integration of $S^2(t)$ from $t = 0$ to $t = \infty$, which defines a correlation time T_{corr} of the turbulent flow,

$$T_{corr} = \int_0^\infty S^2(t)\, dt, \tag{7.11}$$

interpreted as the memory time of the turbulent flow and based on the assumption that $S^2(t)$ is integrable to infinity (i.e., finite area under the curve). In that limit, the variance of the particle position becomes

$$R^2(t) = 2u_*^2\, T_{corr}\, t \tag{7.12}$$

of which the square root is

$$R(t) = u_* \sqrt{2T_{corr} t} \tag{7.13}$$

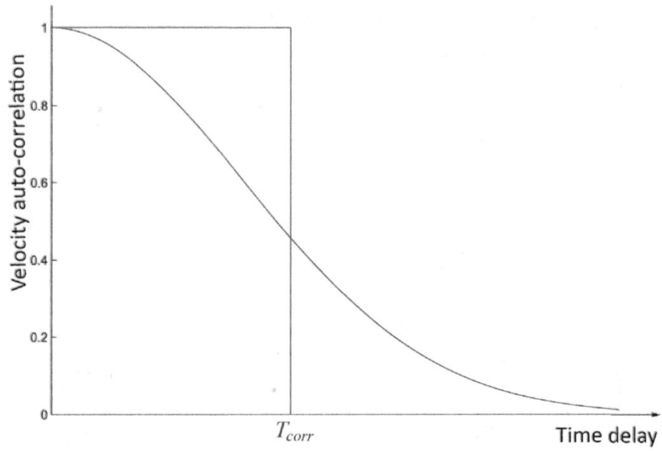

Figure 7.2 Typical velocity autocorrelation in a turbulent flow. With zero time delay, the velocity is perfectly correlated to itself, and the autocorrelation value is 1. With increasing time difference, the auto-correlation diminishes until it vanishes. The time T_{corr} such that the area of the rectangle is equal to the area under the curve may be considered as the time of memory allowed by the turbulence.

Thus, over long periods, the standard deviation increases as the square root of time. The situation is similar to the process of molecular diffusion. Indeed, with molecular diffusion, spread occurs at the rate of $\sigma = \sqrt{2Dt}$ in which σ is the standard deviation of the patch undergoing dispersion, D is the molecular diffusivity, and t is time counted from the time of a punctual release. The equivalence with Eq. (7.13) is that the turbulent diffusivity is $K = u_*^2 T_{corr} >> D$.

7.1.2 The eddy diffusivity model

Inspired by the preceding equivalence, a very common and elementary approach for the representation of turbulent dispersion is the eddy diffusivity model, whereby turbulent dispersion is viewed merely as a form of molecular diffusion, albeit greatly enhanced. With this approach, the molecular diffusivity is simply replaced by a much larger eddy diffusivity—denoted here as K—the value of which must somehow be tied to the vigor of the turbulence in the flow field. For a brief historical account, see Sutton (1932, p. 143).

Among many others, before and after them (e.g., Csanady, 1973; Pacanowski and Philander, 1981; Roberts and Webster, 2002), Fischer et al. (1979) proposed a number of these expressions for rivers and estuaries. For example, the eddy diffusivity for vertical mixing in a river is given as

$$K = 0.067 H u_* \tag{7.14}$$

which increases with both the water depth H and the turbulent velocity u_* related to the friction stress along the bed. The underlying physics are that deeper water allows for the existence of larger top-to-bottom eddies and that stronger bed friction causes more active shear instability and thus faster rotating eddies. The larger and the stronger the eddies, the more efficient the stirring. The numerical factor $0.067 = \kappa/6$ stems from the physical rationale used in deriving the expression (with κ being the von Kármán constant).

More sophisticated models, such as the family of k turbulence closure models (k, k-ϵ, k-ω; for a summary, see Cushman-Roisin and Beckers, 2011, Sections 14.3 & 14.4) solve an evolution equation for the turbulent kinetic energy k and make the eddy diffusivity dependent on it. For example (Umlauf and Burchard, 2005), the eddy diffusivity may be expressed as:

$$K = C\,\frac{\mathrm{k}^2}{\epsilon} \tag{7.15}$$

in which the factor C depends on the velocity shear, and ϵ is the turbulent energy dissipation rate, which like k needs its own evolution equation. The function C and several terms in the evolution equations for k and ϵ are formulated based more on inferred phenomenology and empirical evidence rather than on basic physics.

These formulations have some physical basis complemented by some heuristics. They are ultimately accepted based more on post-model validation than on pre-model physics. Some values have also been inferred from measurements of dispersion (e.g., Yanagi et al., 1982), by a procedure that might be called reverse engineering, without recourse to physics beyond the arguable statement that an eddy diffusivity exists.

7.1.3 The problem with eddy diffusivity

Aside from the fact that existing formulations for the eddy diffusivity K involve some heuristics and much parameter tuning, the eddy diffusivity approach itself suffers from a major deficiency, namely that it predicts a growth proportional to the square root of time, with $\sigma = \sqrt{2Kt}$ in which σ is the standard deviation of the patch undergoing dispersion and t is time counted from the time of the punctual release. A multiplicity of observations (Cushman-Roisin, 2013) reveal beyond a doubt that patch size in a turbulent flow environment grows rather like the first power of time or even faster (like $t^{\frac{3}{2}}$). Thus entirely different models are necessary to predict either $\sigma \simeq u_* t$ in which u_* is a turbulent velocity (that may be related either to a bottom stress or to the turbulent kinetic energy k) or $\sigma \simeq \epsilon^{\frac{1}{2}} t^{\frac{3}{2}}$ in which ϵ is the energy dissipation rate in the turbulent inertial cascade.

Pertinent examples are shown in Figs. 7.3 and 7.4. The unmistakable triangular shape of the plumes indicates linear growth over distance, which in a larger-scale and therefore locally uniform wind corresponds to spreading proportional to the first power of time. It should be noted that, for cases like these, a spatially dependent eddy diffusivity as that proposed in Eq. (7.15) is of no use. The straight path of the plume indicates a uniform wind field from which we can presume homogeneous turbulence conditions and thus a uniform eddy diffusivity. The fact that the plume does not widen

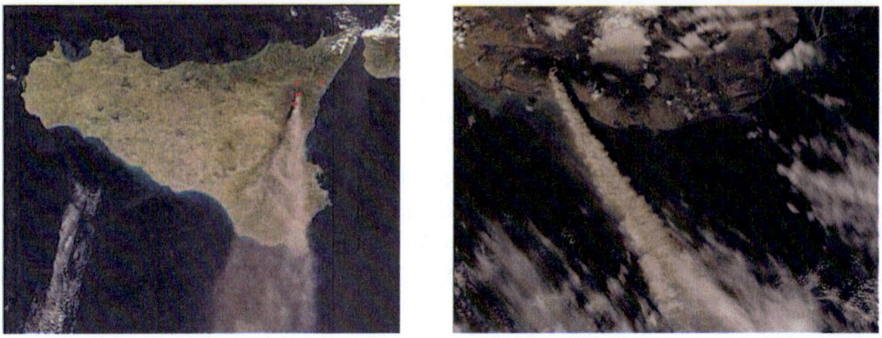

Figure 7.3 Satellite images of the ash and smoke plumes emanating from the Etna Volcano in Sicily on 27 October 2002 (left) and the Eyjafjallajokull Volcano in Iceland on 10 May 2010 (right), captured by the MODIS sensor onboard NASA's Terra and Aqua satellites, respectively. Note the conical shape of both plumes as indicated by their linear growth over distance (proxy for time). [*Photo credits*: NASA Goddard / MODIS Rapid Response Team].

Figure 7.4 Smokestack plumes during a morning neutral atmosphere. As in the previous figure, note the conical shape of the plumes. [Photos by the author].

like the square root of distance but faster points to a fundamental flaw with the eddy diffusivity approach.

Put another way, the mathematics of turbulent dispersion should be governed by an operator other than a second derivative, as in molecular diffusion. This argument was already voiced by Sutton (1932), who remarked in the context of atmospheric dispersion that "the difference between the eddy structure of a turbulent fluid and the molecular structure of a fluid at rest is more than one of scale, and it is now clear that there is a need for an extended theory to express this difference." Similar observations in the context of marine dispersion led Richardson and Stommel (1948) to state that "The variation of K depends on a geometrical quantity σ, and Fick's equation is also geometrical in so far as it contains $\partial^2/\partial x^2$. For this reason, it is difficult to regard the variation of K as an outer circumstance detached from Fick's equation. There appears to be a fault in the equation itself." (*Note*: In this latter quote the variables K and σ stand for the same quantities as here.) Notwithstanding their own admonitions, Sutton (1932) and Richardson and Stommel (1948) retained the eddy diffusivity concept and Fick's equation, each preferring to rely on an eddy diffusivity that grows over time, as t^{m-1} with $1.75 \leq m \leq 2$ (Sutton, 1932) or as σ^α with $1.0 \leq \alpha \leq 4/3$ (Richardson and Stommel, 1948). We note that the superdiffusive growth rates mentioned above, $\sigma = u_* t$ and $\sigma = \epsilon^{\frac{1}{2}} t^{\frac{3}{2}}$, correspond to ($m = 2$, $\alpha = 1$) and ($m = 3$, $\alpha = 4/3$), respectively. In a follow-up study of turbulent dispersion in the ocean, Stommel (1949) reiterated the conclusion that the Fickian model fails to describe horizontal diffusion in the sea.

Richardson and Stommel (1948) were not the only ones to suggest a patch-size-dependent diffusivity; others followed, including Okubo (1971) and Clark et al. (1996), among others. But this fix is wholly inadequate in modeling. Indeed, how could a model be so constructed when the system includes multiple overlapping patches at various stages of development? What should be the local value of K where an older, wider patch overtakes a more recent, smaller patch? To be effective, a model of turbulent dispersion ought to be able to adapt to the local and instantaneous distribution of the substance's concentration without a particular time or point of departure. The answer lies in the formulation of a different operator than the second derivative, preceded by an eddy diffusivity. Put another way, turbulent dispersion in the environment proceeds in a qualitatively different manner than enhanced molecular diffusion, and one ought to seek a mathematical formulation that is not based on Fickian diffusion.

7.2 A better model

7.2.1 *Model requirements*

Here we are not interested in developing arguments about the statistics of dispersion, such as considering the evolution of two-particle separation or moments of displacements (e.g., Richardson and Stommel, 1948; Hunt, 1985; Chatwin and Sullivan, 2006; Ferrari, 2007). Instead, we seek the development of a mathematical tool that can readily be implemented in existing models of environmental transport and fate in which unresolved sub-grid-scale processes can be represented on the resolving grid by means of automated calculations that provide the spatio-temporal evolution of concentration fields.

The arguments made in the preceding section lead us to formulate a model of turbulent dispersion that possesses the following three properties:

1. It must be based on physical principles rather than being validated after trials.
2. It must be such that the rate of dispersion automatically adapts to the spatial distribution of the substance's concentration.

3. It should lead to patch growth proportional to the first or higher power of time; for example, in wall turbulence characterized by an eddy velocity $u_* = \sqrt{\tau_{wall}/\rho}$, it should lead to patch growth proportional to the first power of time ($\sigma \simeq u_* t$).

7.2.2 Model development

To establish a working model, it is helpful to recall the essence of turbulent dispersion. The cause of spreading is the eddy field inside which the tracer patch resides, with eddies moving, distorting, and stirring the patch (Fig. 7.5). Eddies that are significantly larger than the patch sweep the latter around their orbits, merely changing its location without affecting its size significantly. At the opposite end of the spectrum, eddies much smaller than the patch size only stir the inside of the patch, while marginally increasing its extent along the edges. In contrast, eddies of diameter comparable to the patch size greatly distort it, pull its edges in multiple directions, enlarge its overall extent, and cause the most dispersion. While the eddy population may remain statistically unchanged over time, eddies take their turn in affecting dispersion, with the smaller eddies acting first when the patch is small and increasingly larger eddies acting sequentially as the patch widens.

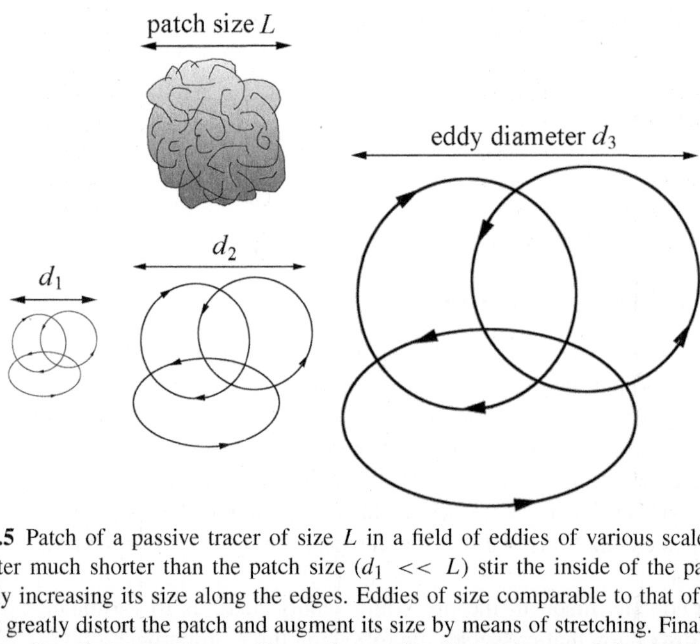

Figure 7.5 Patch of a passive tracer of size L in a field of eddies of various scales. Eddies of diameter much shorter than the patch size ($d_1 << L$) stir the inside of the patch while marginally increasing its size along the edges. Eddies of size comparable to that of the patch ($d_2 \sim L$) greatly distort the patch and augment its size by means of stretching. Finally, eddies much larger than the patch ($d_3 >> L$) merely translate the patch around their orbits, causing large displacements but little dispersion. The conclusion is that, of all eddies, those that most effectively contribute to the enlargement of the patch are those of size comparable to the patch size. As the patch grows in time, eddies take their turn at being those of the most effective kind.

In environmental situations, the large domain size ensures that there are always turbulent eddies larger than the patch, unless, of course, the patch size reaches the size of the domain, which also sets the largest eddy diameter, and ceases to grow for lack of room. In other words, in environmental situations, as long as the patch grows, its growth is in the presence of eddies larger than itself. These larger eddies contribute to a non-zero velocity auto-correlation on a scale larger than the patch size, and this implies that, in environmental situations, the tail end of the velocity auto-correlation invoked by Taylor (1921) (Fig. 7.2) is never reached, invalidating his result for long asymptotic time, namely Eqs. (7.11) and (7.13). Since the latter equation has served as a justification for the use of an eddy diffusivity as an enhanced form of molecular diffusivity, its jeopardy is yet another argument against the eddy diffusivity approach for turbulent dispersion.

The phenomenology based on Fig. 7.5 means that, should there be an eddy diffusivity, its value would somehow vary with the dominant scale of the patch. Thus, if the patch is characterized by the wavenumber[2] $k \simeq \sigma^{-1}$, the diffusivity should be a function of k, and as k evolves because the patch grows, so should the eddy diffusivity value adapt locally and instantaneously without the need to record the time elapsed or distance covered since the start.

It follows from this phenomenology that the instantaneous rate of growth of the patch depends on the orbital velocity of the eddies that have a diameter comparable to the patch size, and the dispersion of the patch over time depends on how the orbital velocity varies with eddy size. Growth proceeds at a constant rate when all eddies share the same orbital velocity scale regardless of diameter. This is the case of shear turbulence in the proximity of a wall, characterized by a single orbital velocity $u_* = \sqrt{\tau_{wall}/\rho}$, the so-called friction velocity related to the unique wall stress. In the case of inertial turbulence, the Kolmogorov cascade prescribes that larger eddies have larger orbital velocities, proportional to $(\epsilon d)^{1/3}$ with d being their diameter. Patch growth proportional to the square root of time, such as that generated by a diffusivity model, would imply eddy velocities that decrease with increasing eddy diameter, a situation never encountered in environmental forms of turbulence.

To put these elements in mathematical terms, let us assume with Kolmogorov (1941) that the eddy field is a collection of vortices of various sizes d with orbital velocity $u_*(d)$, i.e., with all eddies of comparable size having similar orbital velocities. The corresponding half-turnover time is $\Delta t = \pi d / u_*(d)$ over which the net displacement is one diameter d. In terms of an eddy diffusivity, we would have

$$K = \frac{1}{2}\frac{d\sigma^2}{dt} = \sigma\frac{d\sigma}{dt} = \sigma\frac{d}{\Delta t} = \frac{1}{\pi}\sigma u_*(d) \tag{7.16}$$

As remarked earlier, the eddies contributing most to the dispersion are those with diameters d comparable to the patch size σ. Thus the dominant contribution to K is

[2] Note the subtlety in notation: straight k for the earlier turbulent kinetic energy and presently italic k for the dominant wavenumber magnitude.

due to $d = \mathcal{O}(\sigma)$ and

$$K = \mathcal{O}(\sigma u_*(\sigma)) \tag{7.17}$$

or, in terms of the patch wavenumber $k \simeq \sigma^{-1}$,

$$K = \mathcal{O}\left(\frac{u_*(k^{-1})}{k}\right) \tag{7.18}$$

In Fourier space, the diffusion term $K\nabla^2 c$ of a concentration field $c(\mathbf{x}, t)$ is $-k^2 K\, \hat{c}(\mathbf{k}, t)$, and the preceding considerations suggest that we replace this term by something of which the Fourier transform ought to be

$$-k^2 K\, \hat{c} \quad \rightarrow \quad -C\, k u_*(k^{-1})\hat{c} \tag{7.19}$$

with a constant dimensionless coefficient C and where k stands for the magnitude of the three-dimensional wavenumber $\mathbf{k}$. The inverse Fourier transform provides the expression in physical space that should supersede the second derivative preceded by an eddy diffusivity coefficient:

$$-C\, \frac{1}{(2\pi)^{\frac{3}{2}}} \iiint_{-\infty}^{\infty} k u_*(k^{-1})\, \hat{c}(\mathbf{k}, t)\, e^{i\mathbf{k}\cdot\mathbf{x}}\, d\mathbf{k} \tag{7.20}$$

For wall turbulence with uniform orbital velocity u_* across scales, the substituted dispersion operator reduces to

$$K\, \nabla^2 c \quad \rightarrow \quad C_3 u_* \iiint \frac{c(\mathbf{x}', t) - c(\mathbf{x}, t)}{|\mathbf{x}' - \mathbf{x}|^4}\, d\mathbf{x}' \tag{7.21}$$

with the new coefficient C_3 related to the previous constant by $C_3 = 2C/\pi^2$. The index 3 is used to indicate that it corresponds to the case of three dimensions.

Should the concentration field depend on fewer than three spatial dimensions, the integral can be calculated over the missing dimensions. For the case of a two-dimensional distribution of a pollutant with $c = c(x, y, t)$, we have:

$$K\left(\frac{\partial^2 c}{\partial x^2} + \frac{\partial^2 c}{\partial y^2}\right) \quad \rightarrow \quad C_2 u_* \iint \frac{c(x', y', t) - c(x, y, t)}{|(x' - x)^2 + (y' - y)^2|^{3/2}}\, dx' dy' \tag{7.22}$$

with $C_2 = \pi C_3/2$ and index 2 to indicate two dimensions.

In the case of a single dimension with $c = c(z, t)$ (with z chosen as the single dimension because those cases usually involve dispersion in the vertical direction), we have:

$$K \frac{\partial^2 c}{\partial z^2} \quad \rightarrow \quad C_1 u_* \int \frac{c(z', t) - c(z, t)}{(z' - z)^2}\, dz' \tag{7.23}$$

with $C_1 = \pi C_3 = 2C_2$ and index 1 to indicate one dimension. The full 1D equation with added terms for accumulation, advection (with velocity w), and source/sink terms is:

$$\frac{\partial c}{\partial t} + w\,\frac{\partial c}{\partial z} = C_1 u_* \int_{-\infty}^{\infty} \frac{c(z',t) - c(z,t)}{(z'-z)^2}\,dz' + \text{Source} - \text{Sink} \tag{7.24}$$

7.2.3 Fractional derivatives

We recognize in the integrals of Eqs. (7.21)–(7.23) expressions for fractional derivatives. Their use in the modeling of dispersion processes is not new, especially in the context of advection-dispersion modeling in groundwater hydrology and porous media (Schumer et al., 2001; Meerschaert and Tadjeran, 2004; Meerschaert et al., 2006). Application more broadly to fluid turbulence is slightly more recent (Cushman-Roisin, 2013). In particular, Epps and Cushman-Roisin (2018) and Cushman-Roisin (2022) proposed derivations and applications for cases involving a variety of fractional derivatives.

Note that the integrals of Eqs. (7.21)–(7.23) extend over an infinite domain (in 3D, 2D, or 1D, respectively). In finite domains, simple truncation of the integrals is generally the efficient way to proceed as long as close attention is paid to the scheme that sets the boundary value. This can be a complicated matter, and according to Baeumer et al. (2018), in the context of fractional derivatives, "the specification, implementation, and interpretation of meaningful boundary conditions remains an open problem."

7.2.4 Flux

In one dimension, the flux J_z of the substance can be inferred from the mass balance (Fick's second equation). It must be such that the diffusion/dispersion term is equal to minus its divergence:

$$-\frac{\partial J_z}{\partial z} = C_1 u_* \int_{-\infty}^{\infty} \frac{c(z',t) - c(z,t)}{(z'-z)^2}\,dz'$$

Some algebra leads to

$$J_z = -C_1 u_* \int_{-\infty}^{z} dz' \int_{z}^{\infty} \frac{c(z'',t) - c(z',t)}{(z''-z')^2}\,dz'' \tag{7.25}$$

The mathematical interpretation of this expression is that the flux of the substance at level z is the result of all the possible concentration differences between any point on one side of z and any point on the other side. The numerator of the integrand may be interpreted as the result of an exchange between two concentration values, at z' and z'' on opposite sides of z, and the integral as a weighted average of all such exchanges across level z with weight decreasing as the square of the distance between the two levels (the denominator).

By adding and subtracting $c(z,t)$ to the numerator of the integrand, integrating the first half with respect to z' and the second half with respect to z'', and then combining the two pieces, we can reduce the expression to a single integral over the entire domain:

$$J_z = -C_1 u_* \int_{-\infty}^{\infty} \frac{c(z', t) - c(z, t)}{z' - z} \, dz' \tag{7.26}$$

It must be noted that the latter (7.26) expression is equivalent to the former (7.25) only in a doubly infinite domain. In a finite domain, Eq. (7.25) must be used.

7.2.5 Determination of the front coefficient

We now show that the front coefficient C_1 in the one-dimensional Eq. (7.23) can be related to the von Kármán constant and thereby determined unequivocally. For this, we apply the dispersive operator not to a passive scalar but to momentum, and choose the case of a semi-infinite domain extending from a boundary at $z = 0$. For a uniform, steady, sheared flow $u(z)$ parallel to the boundary with no pressure gradient, the x-momentum equation has all vanishing terms, requiring the remaining frictional (dispersive) term also to vanish:

$$C_1 u_* \int_{0}^{\infty} \frac{u(z') - u(z)}{(z' - z)^2} \, dz' = 0 \tag{7.27}$$

The general solution to this integral equation is

$$u(z) = A \ln z + B$$

with free multiplicative and additive constants A and B.

We recognize here the well-known logarithmic velocity profile of a turbulent flow along a flat boundary. Using conventional notation for the constants, we use the friction velocity u_* and the roughness height z_0 of the asperities along the boundary and write instead

$$u(z) = \frac{u_*}{\kappa} \ln \frac{z}{z_0} \tag{7.28}$$

in which $\kappa = 0.40$ is the von Kármán constant. The relation between friction velocity u_* and wall stress τ_{wall} is

$$u_* = \sqrt{\frac{\tau_{wall}}{\rho}} \tag{7.29}$$

in which ρ is the fluid density (water or air).

Now, the stress τ across the flow is the momentum flux. Since it is directed toward the wall, opposite to the z-direction, it is negative. To work with a positive quantity, we reverse the sign in Eq. (7.25) and use:

$$\tau = C_1 u_* \int_0^z dz' \int_z^{\infty} \frac{\rho u(z'') - \rho u(z')}{(z'' - z')^2} \, dz''$$

Use of the logarithmic velocity profile (7.28) in this expression yields after some algebra:

$$\tau = -\frac{C_1}{\kappa}\rho u_*^2 \int_0^1 \int_0^1 \frac{\ln(ab)}{(1-ab)^2}\, da\, db = \frac{\pi^2 C_1}{3\kappa}\rho u_*^2 \tag{7.30}$$

We note that this is a constant. Indeed, the stress must be uniform in the absence of other forces, and its value is none other than the wall stress value $\tau_{wall} = \rho u_*^2$. Eq. (7.30) then becomes

$$\rho u_*^2 = \frac{\pi^2 C_1}{3\kappa}\rho u_*^2$$

out of which we extract the value of the coefficient C_1:

$$C_1 = \frac{3\kappa}{\pi^2} = 0.1216 \tag{7.31}$$

The values of the coefficients at two and three dimensions follow:

$$C_2 = \frac{1}{2}C_1 = \frac{3\kappa}{2\pi^2} = 0.06079 \tag{7.32}$$

$$C_3 = \frac{2}{\pi}C_2 = \frac{3\kappa}{\pi^3} = 0.03870 \tag{7.33}$$

We consider these values as universal and adopt them in the rest of this chapter irrespective of the application.

7.3 One-dimensional applications

7.3.1 Punctual release

Our first application of the formalism is to the spreading in an infinite 1D domain of a pollutant patch $c(x, t)$ following an instantaneous and punctual release, say of mass M at time $t = 0$ and at location $x = 0$. The equation to be solved is:

$$\frac{\partial c}{\partial t} = C_1 u_* \int_{-\infty}^{+\infty} \frac{c(x', t) - c(x, t)}{(x' - x)^2}\, dx' \tag{7.34}$$

with the initial condition $c(x, t = 0) = M\delta(x - 0)$ and subject to the boundary conditions $c(x, t) \to 0$ for $x \to \pm\infty$. The solution is

$$c(x, t) = \frac{M}{\pi} \frac{(\pi C_1 u_* t)}{x^2 + (\pi C_1 u_* t)^2} \tag{7.35}$$

with $\pi C_1 = 3\kappa/\pi = 0.3820$.

We recognize here the Cauchy statistical distribution with algebraic tails rather than exponential tails. The so-called 'fat' tails may be interpreted as caused by the ballistic behavior of the larger eddies, which are able to move the edges of the patch outward at speed $\pi C_1 u_* = 0.38 u_*$.

Although the standard deviation of the Cauchy distribution is infinite, a patch width may still be defined by analogy with the Gaussian distribution. When a Gaussian distribution is fitted to observations, tails tend to be ignored because values there are small and noisy, and the curve fit proceeds from the hump of the distribution. This is easily done by considering the value of the width at half peak, which is $W_{50\%} = 2x_{50\%} = 2\sqrt{2 \ln 2}\, \sigma$ for the bell curve $\exp(-x^2/2\sigma^2)$. Applying this method to the present Cauchy distribution for which $W_{50\%} = 2x_{50\%} = 2\pi C_1 u_* t$, we have

$$\sigma = \frac{\pi C_1}{\sqrt{2 \ln 2}}\, u_* t = 0.3244\, u_* t \tag{7.36}$$

If the 4-σ rule is used to estimate the patch width (2σ on each side of the center), the width of the Cauchy patch can be taken as $1.30\, u_* t$.

7.3.2 Rivers

Water flow in rivers is almost always in a state of turbulence. This is because their dimensions and flow speed correspond to very high Reynolds numbers. For example, a 1-m deep river (to take its shortest dimension to be conservative) with a flow speed of 1 m/s has a Reynolds number on the order of 10^6. What characterizes river turbulence is a high degree of anisotropy because the spatial dimensions of the river channel are vastly different in all three directions. Generally, the depth H is much smaller than the width W, and both are much shorter than the length L of the river under consideration, called the reach:

$$H << W << L \tag{7.37}$$

Thus one expects vertical mixing (accomplished by vertical eddies extending from bottom to surface) to be vastly different from transverse mixing (accomplished by eddies extending from bank to opposite bank), itself vastly different from longitudinal mixing (accomplished chiefly by shear dispersion).

Vertical dispersion

We start with vertical dispersion because, in order of processes, it is the first one to affect the release of a pollutant. Indeed, vertical mixing of a tracer concentration takes much less time than its transverse mixing, which itself takes much less time than longitudinal spread.

The eddies responsible for vertical stirring are those that rotate in the vertical around horizontal axes directed from bank to bank, as one would imagine water proceeding forward by rolls not unlike bicycle wheels rolling down the sloping channel. The physical cause of these vortices is shear instability, with the shear caused by bottom friction. The vertical velocity profile of the downstream velocity is given to a good degree of accuracy by the logarithmic profile over a rough surface, already stated in Eq. (7.28), which we found in Section 7.2.5 to be eminently compatible with the present dispersion model for which the logarithmic profile is a solution.

In uniform flow,[3] the bottom stress is obtained from a simple balance between the force of gravity down the slope S (expressed in meter of drop per meter of reach, thus dimensionless) of the channel and the retarding force of friction along the wetted bottom, per unit length in the downstream direction (Chanson, 2004, Page 29):

$$P\tau_b = \rho g S A \tag{7.38}$$

in which P is the wetted perimeter along the bottom ($P \simeq W$ for a stream much wider than deep) and A the cross-sectional area occupied by water. This implies that the friction velocity can be determined directly from the bottom slope of the channel and how much of its cross-section is occupied by water:

$$u_* = \sqrt{g R_h S} \tag{7.39}$$

in which the so-called hydraulic radius $R_h = A/P$ is nearly the water depth H for a channel that is much wider than deep.

We now apply the dispersion formalism of Section 7.2.2 in one dimension to the vertical transport of dissolved oxygen from the surface into the water column. For this, we begin by defining the saturated value of dissolved oxygen (DO_s, expressed in mg/L), which is the amount of oxygen in the water in equilibrium with the amount in the air. Although the amount of oxygen in the air barely varies, the solubility of oxygen in water varies with temperature, and DO_s is a function of temperature.[4]

Many water quality models for rivers assume efficient mixing across the water column and are formulated in terms of vertically averaged quantities. At the basis of this model is an equation with the following terms:

$$\frac{\partial \overline{DO}}{\partial t} + \overline{u}\frac{\partial \overline{DO}}{\partial x} = K_r\left(DO_s - \overline{DO}\right) - \overline{\mathscr{D}} \tag{7.40}$$

in which $\overline{DO}(x,t)$ is the vertically averaged value of dissolved oxygen, $\overline{u}$ is the vertically averaged water velocity in the downstream direction x, K_r is a re-aeration coefficient (with dimension of 1/time), and $\overline{\mathscr{D}}$ is the vertically averaged oxygen demand (expressed in mg/L per time) caused by oxygen-depleting organisms, such as fish, algae, or organic matter. It is important to track the oxygen level in the water as it is a convenient, catch-all measure of water quality.

O'Connor and Dobbins (1958), hereafter denoted OCD, theorized and verified with observations that a practical formulation [OCD Eq. (27)[5]] for the re-aeration coefficient of rivers is, in the present notation:

$$K_r = \frac{\sqrt{D_L r}}{H} \tag{7.41}$$

[3] Usually a good approximation at low Froude number.

[4] Tabulated values are easy to find online.

[5] There is an unfortunate typographical error in that equation in the article. The pair of parentheses should only include the numerator. The present K_r is equal to their $2.31k_2$.

in which D_L is the molecular diffusivity of oxygen in water, equal to 2.01×10^{-9} m^2/s at 20°C and varying significantly with temperature.[6] The factor r is "a mean rate of surface renewal" by "turbulent sweeps" equal to [OCD Eq. (37)]

$$r = \frac{u_*}{\kappa H} \tag{7.42}$$

providing

$$K_r = \sqrt{\frac{D_L u_*}{\kappa H^3}} \tag{7.43}$$

as a practical expression for the re-aeration coefficient of rivers.

The appearance of the molecular diffusivity of oxygen in this formula is due to the crucial role of a sub-millimeter-thin "liquid-film" layer below the surface through which oxygen must first diffuse before participating in the turbulent mixing that occurs below in the water column (OCD page 643).

At steady state and in the absence of streamwise variations, Eq. (7.40) tells that the oxygen deficit $DO_s - \overline{DO}$ is directly related to the oxygen demand via the re-aeration coefficient:

$$DO_s - \overline{DO} = \frac{\overline{\mathscr{D}}}{K_r} \tag{7.44}$$

Reversing this relation allows an estimation of the re-aeration coefficient in terms of measurable quantities:

$$K_r = \frac{\overline{\mathscr{D}}}{DO_s - \overline{DO}} \tag{7.45}$$

With this posited, we now turn our attention to the manner in which dissolved oxygen makes its way down from the surface through the water column. In other words, our interest now lies in the profile $DO(z)$. First, it can be reasonably assumed that at any given time, the dissolved oxygen at the very surface of the water is in equilibrium with the air and therefore equal to the saturated value DO_s. In steady state under sustained oxygen demand, the surface flux must compensate for the total demand $\int_0^H \mathscr{D}(z)dz = \overline{\mathscr{D}}H$ along the water column. This sets the value of the surface flux. Immediately below the surface is the thin liquid film (the diffusive layer) through which dissolved oxygen must first diffuse before reaching the interior of the water column. There is no appreciable oxygen demand in such a thin layer, and in a steady state, the dissolved oxygen must decrease linearly with depth to accommodate the surface intake flux. These considerations allow us to determine the level of dissolved oxygen at the base of the liquid-film layer. If we denote by δ the thickness of the thin

[6] With multiplying factor 1.0378^{T-20} for T in degree Celsius. See: Engineering Toolbox – Gases Solved in Water – Diffusion Coefficients.

film, the dissolved oxygen value at its base must be

$$DO(z=\delta) = DO_s - \frac{\overline{\mathscr{D}} H\delta}{D_L} \tag{7.46}$$

O'Connor and Dobbins (OCD page 652) argue that the thickness of the thin film is proportional to the square root of the molecular diffusivity D_L of oxygen:

$$\delta \simeq \sqrt{\frac{D_L}{r}} = \sqrt{\frac{\kappa D_L H}{u_*}} \tag{7.47}$$

At common ambient temperatures, the thin-film thickness is on the order of 4×10^{-4} m [OCD page 652].

Condition (7.46) with δ given by Eq. (7.47) serves as the top boundary condition for the dispersion equation applied through the bulk of the water column where turbulence occurs. The bottom boundary condition is that of zero flux.

The dispersion equation to be solved in the bulk of the water column has the following form:

$$\frac{\partial DO}{\partial t} = \text{Dispersion} - \text{Demand} \tag{7.48}$$

The demand term, denoted $\mathscr{D}$, is taken uniform over depth (fish and oxygen-depleting substances may be located at any level), thus $\overline{\mathscr{D}} = \mathscr{D}$, and constant over time for simplicity. For the dispersion term, we employ the 1D formulation (7.23) derived in Section 7.2.2:

$$\text{Dispersion} = C_1 u_* \int_\delta^H \frac{DO(z') - DO(z)}{(z'-z)^2}\, dz' \tag{7.49}$$

with the coefficient C_1 set at its canonical value $3\kappa/\pi^2 = 0.1216$, thus rendering the model free from any tunable parameters. The depth coordinate z runs downward from δ at the base of the thin-film layer (virtually the surface) to H at the bottom.

With constant demand, the solution reaches a steady state with an aeration influx of new oxygen from the surface, balancing the sustained demand along the water column. Close examination of a series of numerical solutions by varying river conditions and numerical resolution reveals that the $DO(z)$ profile at steady state is primarily composed of a logarithmic profile with a singularity at the surface and of a parabolic profile that vanishes at both surface and bottom. Thanks to the existence of the thin-film layer in which molecular diffusivity operates, the logarithmic singularity is neatly avoided at the surface.

Adapting the logarithmic velocity of Section 7.2.5 to the present case of oxygen dispersion, we deduce that the logarithmic profile that accommodates the required surface flux of re-aeration must be:

$$DO(z) = DO_s - \frac{\mathscr{D} H}{\kappa u_*} \ln\left(\frac{z}{z_L}\right) \tag{7.50}$$

The constant z_L, analogous to a viscous layer thickness and to the roughness height z_0 of the turbulent velocity profile, is a free parameter that will be adjusted shortly by matching the rapidly varying logarithm near the surface to the sharp gradient in the liquid film caused by the weak molecular diffusivity. [The gradient at the surface of the parabolic component of the solution is insignificant compared to that of the singular logarithm.]

Now, we attach the boundary layer at the top by matching the value of the interior solution at $z = \delta$ with that at the base of the liquid-film boundary layer, which was given in (7.46)

$$\frac{1}{\kappa u_*} \ln\left(\frac{\delta}{z_L}\right) = \frac{\delta}{D_L} \tag{7.51}$$

which, after some manipulation using (7.47), sets the value of the dangling parameter z_L in the logarithmic profile in terms of the liquid-film thickness δ.

The vertical average of the oxygen profile (7.50) predicted by the turbulent dispersion model is

$$\begin{aligned}\overline{DO} &= \frac{1}{H}\int_\delta^H DO(z)\,dz \\ &= DO_s - \frac{\mathscr{D}H}{\kappa u_*}\left[\ln\left(\frac{H}{z_L}\right) - \frac{\delta}{H}\ln\left(\frac{\delta}{z_L}\right) - 1 + \frac{\delta}{H}\right]\end{aligned} \tag{7.52}$$

With $\delta << H$ and $\ln(\delta/z_L)$ given by (7.51), we note that some terms are much smaller than others. Retaining only the leading term and using (7.47) and (7.51), we have:

$$\overline{DO} \simeq DO_s - \frac{\mathscr{D}H\delta}{D_L} \simeq DO_s - \mathscr{D}\sqrt{\frac{\kappa H^3}{D_L u_*}} \tag{7.53}$$

Using this value in the expression (7.45) giving the re-aeration coefficient K_r, we obtain its leading term:

$$K_r \simeq \sqrt{\frac{D_L u_*}{\kappa H^3}} \tag{7.54}$$

This is exactly the formula proposed and tested by O'Connor and Dobbins (1958) cited earlier as (7.43). We are brought to conclude that the dispersion model employed here provides a vertical profile of dissolved oxygen that is thoroughly consistent with the leading investigation on aeration of rivers.

Longitudinal dispersion

We now turn our attention to another form of dispersion in a river, which is mixing in the downstream direction.[7]

[7] We omit here consideration of transverse, bank-to-bank dispersion as it does not feature any particular process that is not already contained in the 1D-vertical and 1D-longitudinal dispersion.

Clark et al. (1996) reported on a passive tracer release experiment they performed in the Hudson River (northeast USA) for the purpose of determining the longitudinal eddy diffusivity in a major river. Their tracer was a mixture of SF_6 and ^3He, both of which can be precisely measured in water samples. As expected, they saw their patch of tracer expand in size as it drifted downstream. The concentration distributions collected over the course of two weeks displayed a center concentration decreasing over time, flanked by symmetric tails that fattened over time, to which they fitted Gaussian distributions. From the Gaussian fits, they estimated a standard deviation σ for each time of observations, almost daily from Day 3 to Day 13, counted from release.

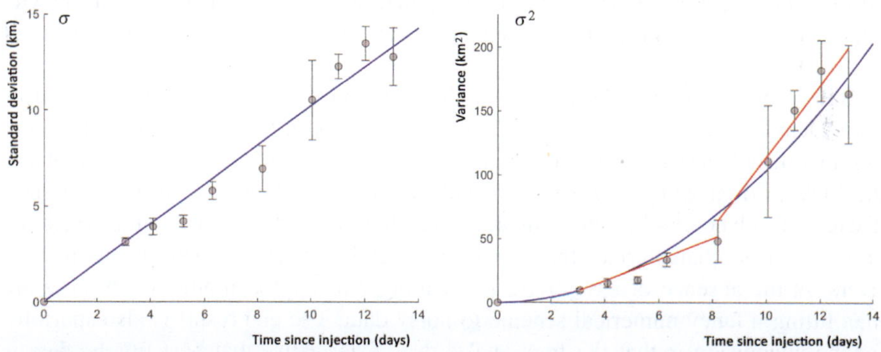

Figure 7.6 Observations and curve fits from a tracer release experiment performed in the Hudson River by Clark et al. (1996). *Left Panel*: Standard deviation σ versus time since release taken from Table 1 of Clark et al. (1996) with linear regression fit passing through the origin (blue line). The line fit has a correlation of 97% with a slope of 1.019 km/day. *Right Panel*: Variance σ^2 of the same data as in Figure 5 of Clark et al. (1996), with their two separate fits to obtain longitudinal dispersion coefficients, 47 ± 6 m^2/s for Days 3–8 and 162 ± 22 m^2/s for Days 8–13 (red segments). The blue line is the parabola that corresponds to the blue straight line in the left panel after squaring the numbers. In both plots, the vertical error bars correspond to the 95% confidence intervals.

Fig. 7.6 shows the data displayed twice, once in the form of the standard deviation σ and again as the variance σ^2, both versus time since release. We immediately note that the variation of σ over time is nearly linear. Forcing the linear fit to pass through the origin, because the patch was nearly punctual at the time of release, we obtain a slope $d\sigma/dt = 1.019$ km/day $= 0.0118$ m/s with a 97% correlation coefficient. This is in keeping with the prediction of the growth of a patch following an instantaneous punctual release as described in Section 7.3.1 when using the turbulent mixing model developed in Section 7.2.2, which predicts patch growth proportional to the first power of time. Applying rule (7.36), we infer that the turbulent velocity causing the observed growth is

$$u_* = \frac{\sqrt{2 \ln 2}}{\pi C_1} \frac{d\sigma}{dt} = 0.0364 \text{ m/s} \tag{7.55}$$

Unfortunately, Clark et al. (1996) did not report sufficient hydraulic data to validate this number in a direct manner, but there is a way to work around the lack of data. The river velocity averaged over the cross-section of the channel in the upper reach of the experiment location can be estimated from the reported discharge at a nearby dam:

$$\overline{u} = \frac{210 \pm 40 \text{ m}^3/\text{s}}{2000 \text{ m}^2} = 0.105 \pm 0.020 \text{ m/s} \tag{7.56}$$

A turbulent velocity $u_* = 0.0364$ m/s, smaller than, but on the same order as, a mean flow velocity $\overline{u} = 0.105$ m/s, is a very reasonable expectation in river flow.

The key result, of course, is the fact that the turbulent model of Section 7.2.2 fits all the data in one scoop (blue line in left panel of Fig. 7.6) whereas the diffusivity approach requires a multiplicity of separate curve fits (red segments in right panel of Fig. 7.6).

Before bringing this subsection to a close, the papers by Deng and collaborators (e.g., Deng et al., 2004; Deng et al., 2006) deserve a brief mention because they, too, consider turbulent river dispersion by means of fractional calculus. This work replaces the Fickian double derivative by *fiat* with a fractional derivative in a numerically discretized advection-diffusion equation and then seeks the value of the fractional order that best characterizes the data from several dye release experiments in U.S. rivers. In the absence of any physical reasoning, this approach amounts to no more than fitting a fancy numerical scheme to noisy data. The end result is also spurious, for the conclusion is that the fractional order of derivative that best fits the data is not unique but ranges between 1.4 and 2.0, even for the same stream, with the most likely value being around 1.65. It would be hard to develop a theory based on physical dimensional quantities that would produce an odd power like 1.65 (even if assimilated to 5/3 = 1.67) because it is not describable in terms of the dimensional quantities at hand. Mostly, the existence of a range of values for the fractional order has no physical interpretation; it appears to exist only because the model fails to capture the observational variation with a single number.

7.3.3 Smokestack plumes

The billowing of smokestack plumes is arguably the most noticeable manifestation of turbulent dispersion in the environment, and because smokestack emissions have a major impact on air quality, there is abundant literature on the subject, starting with the seminal works of Pasquill (1961) and Gifford (1961) based on a Gaussian plume model. This model expresses the crosswind and vertical dimensions of the plume in terms of standard deviations, denoted σ_y and σ_z, respectively. Several refinements have been proposed over the years, and pertinent from the perspective of the present work are the formulas derived by Weil and Brower (1984) in terms of the wind velocity u, the vertical turbulent velocity w_*, and the distance x downwind from the smokestack. These are:

$$\sigma_y = \sqrt{0.08^2 + \left(0.56\frac{w_*}{u}\right)^2}\, x \tag{7.57a}$$

$$\sigma_z = \sqrt{0.06^2 + \left(0.56\frac{w_*}{u}\right)^2}\, x \tag{7.57b}$$

The turbulent velocity w_* is either the convective velocity scale (related to the velocity of rising thermals) or the wind-induced friction velocity otherwise denoted u_*, whichever is greater.

We immediately note that the preceding expressions are linearly growing functions of x, the downwind distance from the smokestack, as the present theory predicts. The first term under each square root is suspicious because it appears to be causing plume growth in the absence of turbulence ($w_* = 0$); this is none other than an artifact of curve fitting; the curve fit should have been forced to start from the origin ($\sigma_y = \sigma_z = 0$ at $x = 0$), but the authors restricted their attention to significant distances away from the smokestacks. Without this first constant term under the square root, which is quite small compared to the second term, the expressions share the same approximation:

$$\sigma_y \simeq \sigma_z \simeq 0.56\frac{w_*}{u}\, x$$

In terms of downwind travel time $t = x/u$, the shared expression becomes:

$$\sigma_y \simeq \sigma_z \simeq 0.56\, w_* t \tag{7.58}$$

This is remarkably close in both form and magnitude to the prediction (7.36) of the present model, given that the one at hand was derived from large sets of atmospheric data without explicit physics, and the former purely from basic physical principles with only the von Kármán constant as experimental input.

So, once again, the Fickian diffusion model, predicting growth proportional to the square root of time, is found to be incorrect in the presence of turbulence, and support is given to a model like the present one that predicts spatial growth proportional to the first power of time or traveled distance.

7.4 Generalization

The theory for turbulent dispersion exposed here is rooted in the existence of a characteristic turbulent velocity, such as the friction velocity u_* of a turbulent shear flow along a boundary. There are cases, however, where the quantity characteristic of the ambient turbulence is some other physical variable (Cushman-Roisin, 2022).

For example, in inertial turbulence, the eddy orbital velocity varies with the eddy diameter d or, equivalently, with the wavenumber k according to the Kolmogorov cascade theory (Pope, 2000, page 187):

$$u_*(k) \propto (\epsilon d)^{1/3} = \mathcal{O}\left(\frac{\epsilon^{1/3}}{k^{1/3}}\right) \tag{7.59}$$

in which ϵ is the energy dissipation rate. Thus there is no single value for u_*. Instead, the unique value characterizing the turbulence field is ϵ, which is the uniform rate at which energy cascades across scales from the longest to the shortest in the turbulent flow field. The earlier expression (7.18) leads to:

$$K = \mathcal{O}\left(\frac{u_*}{k}\right) = \mathcal{O}\left(\frac{\epsilon^{1/3}}{k^{4/3}}\right) \tag{7.60}$$

and expression (7.20) in Fourier space that replaces Fickian diffusion is:

$$-\frac{C}{(2\pi)^{\frac{3}{2}}}\epsilon^{1/3} \iiint_{-\infty}^{\infty} k^{2/3}\, \hat{c}(\mathbf{k}, t)\, e^{i\mathbf{k}\cdot\mathbf{x}}\, d\mathbf{k} \tag{7.61}$$

Expressed in terms of the spatial concentration distribution $c(\mathbf{x}, t)$, the dispersion term in the advection-diffusion equation takes the form:

$$K\, \nabla^2 c \quad \to \quad C_3\, \epsilon^{1/3} \iiint \frac{c(\mathbf{x}', t) - c(\mathbf{x}, t)}{|\mathbf{x}' - \mathbf{x}|^{11/3}}\, d\mathbf{x}' \tag{7.62}$$

which replaces (7.21). This approach was initiated by Chen (2006) in the pursuit of a fractional Laplacian operator to model Richardson's turbulence superdiffusion (Richardson and Stommel, 1948).

More generally, if the physical variable characterizing the intensity of the turbulent flow is, say, q with dimensions L^α/T, then the scale for the eddy orbital velocity is:

$$u_* = \mathcal{O}\left(q k^{\alpha-1}\right) \tag{7.63}$$

and the pertinent dispersion term is

$$C_3\, q \iiint \frac{c(\mathbf{x}', t) - c(\mathbf{x}, t)}{|\mathbf{x}' - \mathbf{x}|^{\alpha+3}}\, d\mathbf{x}' \tag{7.64}$$

An example of such a situation is the turbulent planar jet (Pope, 2000, page 134) in which the conserved quantity along the jet is the momentum flux per unit width, equal to $\rho H U^2$ where ρ is the fluid density, H the thickness of the jet at the discharging slit, and U the exit velocity. The corresponding defining physical characteristic q is $\sqrt{HU^2}$ with dimensions $L^{3/2}/T$, thus with $\alpha = \frac{3}{2}$. This problem was solved by Cushman-Roisin (2022), who found good agreement with laboratory data, showing heavier tails than those predicted by the Gaussian fit.

7.5 Synthesis and challenges

This chapter developed and applied an alternative to the conventional approach of modeling turbulent dispersion by means of an eddy-enhanced diffusivity. The underlying phenomenology is that, among the many different eddies that make up the turbulent flow field, those with size comparable to the size of the contaminant patch are the most effective in dispersing the contaminant (Fig. 7.5), and as the patch grows in size, eddies of different sizes take their turn at being of the most effective kind.

Translated into mathematical language, this conception of turbulent dispersion leads to the replacement of the second-derivative diffusive operator by an integral across the domain that amounts to a weighted average of the difference between the local contaminant concentration and all other concentration values in the domain. In other words, the expression is nonlocal, and this makes sense in the context of turbulence.

The weighing factor used in the integral has a power law, the exponent of which depends on the nature of the turbulence. It is different for shear (wall) turbulence and inertial (Kolmogorov) turbulence.

The method is relatively novel in the sense that it has not yet been applied widely. A great number of environmental situations await application, including, but not limited to, those mentioned in the second half of the Introduction. The main challenge in each application is the computational burden, especially at dimensions higher than one. The reason is that the temporal update of the concentration value at each grid point in the domain necessitates the scanning of the values not just at a few surrounding grid points but at all other grid points across the domain. This is because the mathematical operator does not consist of derivatives but of an integral across the domain. At a fundamental level, there is no avoiding this burden because, by means of its many eddies, some of which are as large as the domain, turbulence is intrinsically non-local and therefore computationally demanding. This is especially true of environmental systems by virtue of their large spatial dimensions. It is hoped that computational scientists will devise ways to reduce the number of calculations, for example, by eliminating contributions deemed negligible in the calculation of the nonlocal integral.

A second challenge, which is not peculiar to turbulent dispersion but common to all applications of fractional derivatives in finite domains, is the proper implementation of boundary conditions. There is still a need for greater clarity on the subject (Baeumer et al., 2018).

Last but not least is the following theoretical challenge. It was shown in Section 7.2.5 that application of the formalism to wall turbulence predicts the well-known logarithmic velocity profile to which the von Kármán constant is closely connected. This raises the hope that with some additional effort, someone may finally determine the value of this constant theoretically instead of experimentally (Bailey et al., 2014) or numerically (Jiménez and Moser, 2007).

References

Baeumer, B., Kovács, M., Meerschaert, M.M., Sankaranarayanan, H., 2018. Boundary conditions for fractional diffusion. Journal of Computational and Applied Mathematics 336, 408–424.

Bailey, S.C.C., Vallikivi, M., Hultmark, M., Smits, A.J., 2014. Estimating the value of von Kármán's constant in turbulent pipe flow. Journal of Fluid Mechanics 749, 79–98. https://doi.org/10.1017/jfm.2014.208.

Batchelor, G.K., 1952. The effect of homogeneous turbulence on material lines and surfaces. Proceedings of the Royal Society A. Mathematical, Physical and Engineering Sciences 213, 349–366.

Brothman, A., Wollan, G.N., Feldman, S.M., 1945. New analysis provides formula to solve mixing problems. Chemical and Metallurgical Engineering 52, 102–106.

Chanson, H., 2004. Environmental Hydraulics of Open Channel Flows. Elsevier Butterworth-Heinemann. 430 pp.
Chatwin, P., Sullivan, P., 2006. Turbulent Diffusion. Entry in Encyclopedia of Environmetrics. John Wiley & Sons. 20 pp.
Chen, W., 2006. A speculative study of 2/3-order fractional Laplacian modeling of turbulence: some thoughts and conjectures. Chaos 16, 023126. https://doi.org/10.1063/1.2208452.
Clark, J.F., Schlosser, P., Stute, M., Simpson, H.J., 1996. $SF^6-{}^3He$ tracer release experiment: a new method of determining longitudinal dispersion coefficients in large rivers. Environmental Science & Technology 30, 1527–1532.
Csanady, G.T., 1973. Turbulent Diffusion in the Environment. Geophysics and Astrophysics Monographs, vol. 3. Springer. 248 pp.
Cushman-Roisin, B., Beckers, J.-M., 2011. Introduction to Geophysical Fluid Dynamics – Physical and Numerical Aspects. Academic Press. 828 pp.
Cushman-Roisin, B., 2013. Turbulent dispersion. In: Fernando, H.J.S. (Ed.), Handbook of Environmental Fluid Dynamics – Volume 1: Overview and Fundamentals. CRC Press, pp. 263–271. Chapter 21.
Cushman-Roisin, B., 2022. Turbulent dispersion. In: Schuttelaars, H., Heemink, A., Deleersnijder, E. (Eds.), The Mathematics of Marine Modelling. Springer, pp. 167–196.
Deng, Z.-Q., Singh, V.P., Bengtsson, L., 2004. Numerical solution of fractional advection-dispersion equation. Journal of Hydraulic Engineering 130 (5), 422–431. https://doi.org/10.1061/(ASCE)0733-9429(2004)130:5(422).
Deng, Z., Bengtsson, L., Singh, V.P., 2006. Parameter estimation for fractional dispersion model for rivers. Environmental Fluid Mechanics 6. https://doi.org/10.1007/s10652-006-9004-5.
Eckart, C., 1948. An analysis of the stirring and mixing processes in incompressible fluids. J. Marine Res. 7, 265–275.
Epps, B.P., Cushman-Roisin, B., 2018. Turbulence modeling via the fractional Laplacian. https://arxiv.org/pdf/1803.05286.pdf.
Ferrari, R., 2007. Statistics of dispersion in flows with coherent structures. In: Proceedings of the 15th ‘Aha Huliko’a Hawaiian Winter Workshop on Extreme Events’, pp. 23–26.
Fischer, H.B., List, E.J., Koh, R.C.Y., Imberger, J., Brooks, N.H., 1979. Mixing in Inland and Coastal Waters. Academic Press. 483 pp.
Gifford Jr., F.A., 1961. Use of routine observations for estimating atmospheric dispersion. Nuclear Safety 2, 47–57.
Hunt, J.C.R., 1985. Turbulent diffusion from sources in complex flows. Annual Review of Fluid Mechanics 17, 447–485.
Jiménez, J., Moser, R.D., 2007. What are we learning from simulating wall turbulence? Philosophical Transactions of the Royal Society of London. A 365, 715–732. https://doi.org/10.1098/rsta.2006.1943.
Kadlec, R.H., 1994. Detention and mixing in free water wetlands. Ecological Engineering 3, 345–380. https://doi.org/10.1016/0925-8574(94)00007-7.
Kolmogorov, A.N., 1941. Dissipation of energy in locally isotropic turbulence. Doklady Akademii Nauk SSSR 32, 19–21. in Russian.
Lewis, R., 1997. Dispersion in Estuaries and Coastal Waters. John Wiley & Sons. 312 pp.
Lyons, T.J., Scott, W.D., 1990. Principles of Air Pollution Meteorology. CRC Press. 224 pp.
Meerschaert, M.M., Tadjeran, C., 2004. Finite difference approximations for fractional advection-dispersion flow equations. Journal of Computational and Applied Mathematics 172 (1), 65–77.
Meerschaert, M.M., Mortensen, J., Wheatcraft, S.W., 2006. Fractional vector calculus for fractional advection-dispersion. Physica. A 367, 181–190.

Nepf, H.M., 1999. Drag, turbulence, and diffusion in flow through emergent vegetation. Water Resources Research 35 (2), 479–489. https://doi.org/10.1029/1998WR900069.

O'Connor, D.J., Dobbins, W.E., 1958. Mechanism of reaeration in natural streams. Transactions of the American Society of Civil Engineers 123 (1), 641–666. https://doi.org/10.1061/TACEAT.0007609.

Okubo, A., 1971. Oceanic diffusion diagrams. Deep-Sea Research 18, 789–802.

Ottino, J.M., 1989. The Kinematics of Mixing: Stretching, Chaos, and Transport. Cambridge University Press. 364 pp.

Pacanowski, R.C., Philander, S.G.H., 1981. Parameterization of vertical mixing in numerical models of tropical oceans. Journal of Physical Oceanography 11, 1443–1451. https://doi.org/10.1175/1520-0485(1981)011<1443:POVMIN>2.0.CO;2.

Pasquill, F., 1961. The estimation of the dispersion of windborne material. Meteorological Magazine 90, 33–49.

Pope, S.B., 2000. Turbulent Flows. Cambridge University Press. 771 pp.

Richardson, L.F., Stommel, H., 1948. A note on eddy diffusion in the sea. Journal of Meteorology 5, 238–240.

Roberts, P.J.W., Webster, D.R., 2002. Turbulent diffusion. In: Shen, H.H., Cheng, A.H.D., Wang, K.-H., Teng, M.H., Liu, C.C.K. (Eds.), Environmental Fluid Mechanics–Theories and Applications. American Society of Civil Engineers, pp. 7–45.

Schumer, R., Benson, D.A., Meerschaert, M.M., Wheatcraft, S.W., 2001. Eulerian derivation of the fractional advection-dispersion equation. Journal of Contaminant Hydrology 48, 69–88.

Stommel, H., 1949. Horizontal diffusion due to Oceanic turbulence. Journal of Marine Research 8, 199–225.

Sutton, O.G., 1932. A theory of eddy diffusion in the atmosphere. Proceedings of the Royal Society A. Mathematical, Physical and Engineering Sciences 135 (826), 143–165. https://doi.org/10.1098/rspa.1932.0025.

Taylor, G.I., 1921. Diffusion by continuous movements. Proceedings of the London Mathematical Society 20, 196–212.

Umlauf, L., Burchard, H., 2005. Second-order turbulence closure models for geophysical boundary layers. A review of recent work. Continental Shelf Research 25, 795–827.

Weil, J.C., Brower, R.P., 1984. An updated Gaussian plume model for tall stacks. Journal of the Air Pollution Control Association 34, 818–827.

Welander, P., 1955. Studies on the general development of motion in a two-dimensional, ideal fluid. Tellus 7, 141–156.

Yanagi, T., Murashita, K., Higuchi, H., 1982. Horizontal turbulent diffusivity in the sea. Deep-Sea Research 29 (2), 217–226.

Turbulent flows in urban terrain

8

Elie Bou-Zeid[a], *Marco Giometto*[b], *and Qi Li*[c]
[a]Department of Civil and Environmental Engineering, Princeton University, Princeton, NJ, United States, [b]Department of Civil Engineering and Engineering Mechanics, Columbia University, New York, NY, United States, [c]Department of Atmospheric and Oceanic Sciences, Peking University, Beijing, China

8.1 Introduction

Environmental flows over hydrodynamically smooth terrain (where the viscous sublayer depth exceeds the height of surface roughness elements k) are the exception, only occurring over flat ice, smooth sand, or quiescent water surfaces. Therefore, environmental turbulence is almost always interacting with a rough wall where the geometric and topographic surface properties exert a strong influence on the structure of the eddies and the statistics of the turbulent flow. In this chapter, we will focus on flows where the roughness elements consist of buildings and similar human-made structures (the following two chapters in this book introduce flows over vegetation and wind farms). While these rough surfaces exhibit various differences, their flows share some general features (see a comparison of these three types of rough surfaces in Appendix A (Table 8.1)).

One common feature of environmental turbulence over such 'very' rough surfaces is that the roughness scale k can exceed 10% ($\approx$ 100 m) of the full turbulent atmospheric boundary layer (ABL) depth δ ($\approx$ 1000 m). In inner units and at a bulk Reynolds number of $Re \sim 10^8$, typical of the ABL, this is equivalent to a $k^+ = ku_*/\nu > 10^6$ (where u_* is the friction velocity that typically ranges from 0.1 to 0.5 ms^{-1}, and ν the air's kinematic viscosity $\approx 1.5 \times 10^{-5}$ m^{2}s^{-1}). This roughness scale is much larger than what is typically considered in the engineering literature of flows over rough walls, where $k^+ \sim 50 - 100$ (see Jiménez, 2004, for a detailed review of that literature), and where the focus is on the flow and turbulence above the roughness layer. In environmental applications, the details of the flow inside the roughness layer and even around individual roughness elements are, in fact, of great importance. In cities, for example, we need to understand flow and turbulence in between buildings and at street/pedestrian level to forecast chemical transport (Tseng et al., 2006; Grylls and van Reeuwijk, 2022), avoid urban designs that cause excessive wind gusts or stagnation zones (Wang et al., 2025; Roth, 2000), predict wind forces on structures (Holmes et al., 2007; Gorlé, 2025) and pedestrians (Blocken et al., 2016), and fly drones safely in cities (Gianfelice et al., 2022; Patil and García-Sánchez, 2025). We thus need a deep understanding of the effect of **urban form** (or morphology) on the flow field (see Kadivar et al. (2021) for a review of urban roughness characterization and geometric parameters).

Environmental Turbulence. https://doi.org/10.1016/B978-0-32-395873-8.00015-2
Copyright © 2026 Elsevier Inc. All rights are reserved, including those for text and data mining, AI training, and similar technologies.

Another key feature of urban flows is the strong influence of buoyancy (as in almost all environmental flows) and the resulting coupling between flow and heat transfer in cities. The sources of buoyancy in cities are linked to (i) the "unnatural" materials we use to build them such as dark asphalt (absorbing most of the incoming radiation), high thermal capacity concrete (storing a lot of heat), and irrigated lawns that can be cooler than surrounding deserts in some cities, and (ii) to the complex insolation and shading patterns resulting from urban form. The properties of urban materials, abstractly referred to as the **urban fabric** (Oke et al., 2017), are thus of importance since they modulate the conversion of potential to kinetic energy, and hence urban flow and turbulence.

The last feature we will focus on in this introduction to urban flows is the human dimension. We are interested in urban flows because of the large number of people they affect, but the activities of these humans also have a significant impact on the airflow beyond the initial urban design decisions. From the anthropogenic heat production in buildings and through combustion of fuels that can create mixing, to the turbulence produced in the wake of cars, to the emissions of gases and aerosols that can modify radiative transfer and cloud formation, anthropogenic activities that are linked to the **urban function**, or activities happening in a given area, modulate flow and turbulence in various direct and indirect ways. There is thus a pronounced interaction between the metabolism of cities (Wolman, 1965) and the flow of air that transports mass and energy in and out of the urban airshed.

The interest in these various fundamentals and applications of urban fluid mechanics (a term first introduced by Fernando et al. (2001)) has significantly increased over the past three decades due to the very rapid urbanization of the world's population. While about 20% of humans lived in cities in 1950, by 2025, this fraction has increased to about 45%, with another 36% living in smaller towns (UN Population Division, 2025). Therefore, about 81% of the world's population now lives in 'built terrain'. With these fractions continuously increasing, the grand challenges facing humanity—from the energy transition to resource overuse to mitigating and adapting to climate change and weather extremes—are, by and large, urban challenges (Masson et al., 2020; González et al., 2021; Gorlé, 2025).

This chapter aims to introduce a reader who is already somewhat versed in the physics of turbulent flows to their unique urban attributes. We start with an overview of the main challenges, focusing on horizontal variability and vertical layers in the urban boundary layer. We then examine the first principles and equations that describe (i) the mean streamwise momentum budget and how its dominant terms change with height, (ii) the intricacies of spatial averaging in complex domains, (iii) the various kinetic energy components of the flow, and (iv) the sources of shear and buoyant turbulence generation in cities. The primary research questions and flow features are subsequently introduced at three scales of analysis: the building-to-street scale, the street-to-neighborhood scale, and the neighborhood-to-metropolitan scale. After that, we delve into two applications of particular interest, examining the role of urban turbulence in their dynamics: urban heat islands and air quality in cities. We conclude with a synthesis and an overview of the open challenges for understanding and predicting turbulent flows in built terrain.

8.2 Spatial variability and other challenges

Research and applications of flow and turbulence in urban terrain have to contend with various hurdles, beyond the canonical complexities of geophysical flows (see Chapter 1). First, no two cities or neighborhoods are exactly alike, so there is constant tension between the appeals of studying idealized geometries that are amenable to generalization, versus the practicality of studying real cities that give more realistic but very particular results (Belcher, 2005; Lu et al., 2023a; Wang et al., 2013). Second, in any real city, there is a broad spectrum of geometric feature scales, the smallest of which (window sills, roof or façade protrusions, pedestrians, etc.) are nearly impossible to reproduce in lab experiments or resolve in simulations, and are often in fact impossible to even map accurately; so any study is in some sense an idealization (Bou-Zeid et al., 2009; Ricci et al., 2017). Third, urban flows and turbulence are forced by time-varying large-scale (meso to synoptic) pressure gradients and a diurnal cycle of buoyancy flux at the surface (Momen and Bou-Zeid, 2017a), which often result in transient turbulence statistics (making time averaging intricate) and even nonequilibrium of the turbulence with the mean flow (Li and Giometto, 2023, 2024). However, the major challenge in studying urban flows is undoubtedly the strong spatial variability of the mean and higher-order statistics at scales ranging from the building to the metropolitan.

8.2.1 Horizontal variability

The strong spatial variability of urban flows implies a lack of a spatial direction in which the statistics are expected to be homogeneous. The flow in real cities, and over other rough geophysical surfaces, is fully three-dimensional (3D), which invalidates spatial ergodicity and the use of spatial averages as surrogates for Reynolds ensemble averages (Katul et al., 2004; Ghannam et al., 2015). The mean flow ('mean', everywhere in this chapter, strictly refers to the average in a Reynolds ensemble sense) may also be varying in time as mentioned above, invalidating temporal ergodicity, but a common (and often unavoidable) assumption in studying atmospheric turbulence is that the mean flow variations are primarily driven by meso-alpha to synoptic scale systems (>200 km, varying over time scales $\geq$ 1 day) that are much slower than the largest eddies. Therefore, a 'spectral gap' exists, allowing us to postulate that, over periods of about 1 hour or less, the mean flow is quasi-stationary (Stull, 1988), and thus temporal averaging can provide a practical and acceptable approximation of the Reynolds mean (see Mahrt and Bou-Zeid (2020) and Momen and Bou-Zeid (2017b) for further discussions on this point). This assumption allows the conventional Reynolds decomposition of any flow variable (e.g., for the velocity) along $u_i = \overline{u}_i + u'_i$, where i is a free index; the overbar denotes time/Reynolds averaging; and the prime denotes turbulent fluctuations. The important point to keep in mind here is that

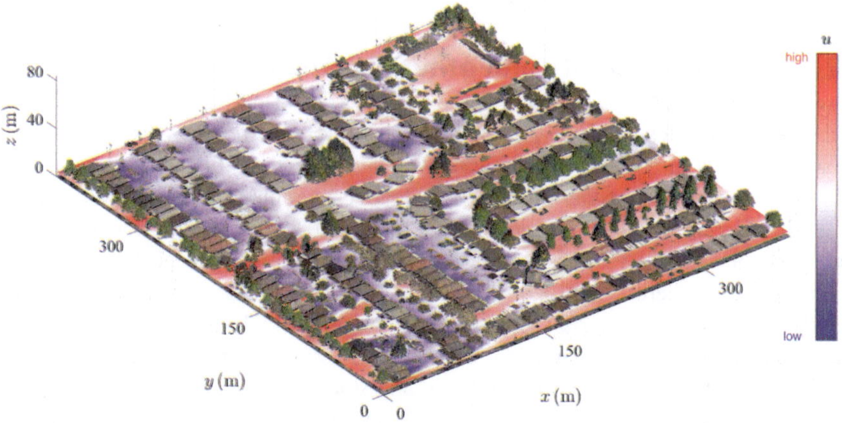

Figure 8.1 Time-averaged streamwise velocity field from a large eddy simulation of flow over a portion of the Sunset neighborhood in Vancouver, BC, Canada. Results are based on simulations reported by Giometto et al. (2017).

the mean velocity is still a field that varies in all directions (x, y, and z, respectively, denoting streamwise, cross-stream, and vertical directions). A simulated sample of such a variable flow in a real city is shown in Fig. 8.1, depicting regions where the flow is slower (blue) or faster (red) than the planar-average (white). The velocity patterns are complex, reflecting the unstructured organization of houses, trees, and streets in that neighborhood.

In most urban applications, this local mean velocity $\overline{u}_i$ and the local statistics of u_i' (e.g., its variance $\overline{u_i'^2}$) are of importance, for example, when analyzing the gustiness or stagnation propensity of the flow at a given location. However, we often also want to obtain information about the turbulent kinetic energy or the carbon, momentum, and heat fluxes over a whole neighborhood, which then requires some method to aggregate the statistics in space. Defining spatial averaging over some (x, y) horizontal slab (or plane) using angled brackets, we can decompose the mean velocity field into its spatial average (white regions in Fig. 8.1) and the spatial deviation from that average (colored regions in Fig. 8.1), denoted by a tilde, which we will henceforth refer to as the dispersive component. At a given location, this decomposition is expressed as $\overline{u}_i = \langle \overline{u}_i \rangle + \widetilde{u}_i$. This spatial averaging operation is defined over the fluid volume only (excluding the solid volume), and simply referred to as "spatial averaging" in this chapter (in the literature, it is often defined as intrinsic averaging), and it will satisfy the property $\langle \widetilde{u}_i \rangle = 0$. In Section 8.3.2, we will distinguish it from spatial averaging over the entire fluid+solid volume, referred to as superficial averaging (Schmid et al., 2019).

Here note that the spatio-temporal average $\langle \overline{u}_i \rangle$ is only a function of the height over the averaging domain z (but it can vary horizontally at larger scales), while both $\overline{u}_i$

and $\widetilde{u}_i$ are stationary 3D fields (also note that $\overline{\widetilde{u}_i} = \widetilde{u}_i$ since $\widetilde{u}_i$ is not a time varying quantity). These Reynolds and spatial decompositions can then be lumped into a triple decomposition expressed as

$$u_i = \langle \overline{u}_i \rangle + \widetilde{u}_i + u_i' \tag{8.1}$$

which can be applied to the instantaneous stress tensor $u_i u_j$ to yield

$$\begin{aligned} u_i u_j = \langle \overline{u}_i \rangle \langle \overline{u}_j \rangle + \langle \overline{u}_i \rangle \widetilde{u}_j + \langle \overline{u}_i \rangle u_j' \\ + \widetilde{u}_i \langle \overline{u}_j \rangle + \widetilde{u}_i \widetilde{u}_j + \widetilde{u}_i u_j' + u_i' \langle \overline{u}_j \rangle + u_i' \widetilde{u}_j + u_i' u_j' \end{aligned} \tag{8.2}$$

Using this decomposed expression and the commutativity of time averaging and summation—and noting that $\overline{\langle \overline{u}_i \rangle u_j'} = 0$ and $\overline{\widetilde{u}_i u_j'} = 0$ and that the spatial average and dispersive components are constant in time—we can compute the time-averaged stress tensor (total momentum flux) as

$$\overline{u_i u_j} = \langle \overline{u}_i \rangle \langle \overline{u}_j \rangle + \widetilde{u}_i \widetilde{u}_j + \overline{u_i' u_j'} + \widetilde{u}_i \langle \overline{u}_j \rangle + \langle \overline{u}_i \rangle \widetilde{u}_j \tag{8.3}$$

Note that in the 3D flow in between and right above the buildings, none of the terms above is necessarily zero. The first term is the stress produced by the spatio-temporally averaged flow; the second is a new addition that results from the dispersive velocities at a local point in space (but averaged in time); and the third is the conventional turbulent (Reynolds) stress resulting from the temporal correlation of the turbulent perturbations. The last two terms are the stresses resulting from the interaction of the mean and dispersive components (physically, the advection of mean momentum by the dispersive velocity field, and vice versa).

If the equation above is then averaged in space—and noting that $\langle \langle \overline{u}_i \rangle \widetilde{u}_j \rangle = \langle \langle \overline{u}_i \rangle \rangle \langle \widetilde{u}_j \rangle = 0$—we obtain

$$\langle \overline{u_i u_j} \rangle = \langle \overline{u_i} \rangle \langle \overline{u_j} \rangle + \langle \widetilde{u}_i \widetilde{u}_j \rangle + \langle \overline{u_i' u_j'} \rangle \tag{8.4}$$

where the first term on the right is the spatio-temporally averaged stress resulting from the spatio-temporally averaged flow, exactly as in the previous equation. If the spatial averaging area is selected such that the vertical $\langle \overline{u}_3 \rangle = \langle \overline{w} \rangle \approx 0$,[1] this first term would not contribute significantly to the vertical flux of horizontal momentum.

[1] Just as averaging in time requires a minimal time span for the statistics to converge and be interpretable, the spatial averaging to define the dispersive components requires a minimal spatial area. Ideally, it would be done over a large enough area that (i) has a characteristic horizontal fetch $\gg$ building height or footprint so that it contains a sufficient number of roughness elements and features to make the spatial averaging representative of that type of urban land use, (ii) yields a vertical velocity averaged in time and space $\langle \overline{w} \rangle \approx 0$, precluding any mean subsidence or uplift, which results in $\partial \langle \overline{w} \rangle / \partial z \approx 0$, and (iii) filters out any small scale, strong horizontal gradients leading to $\partial \langle \overline{\cdot} \rangle / \partial x \approx \partial \langle \overline{\cdot} \rangle / \partial y \approx 0$ (horizontal homogeneity and negligible advection for space and time average quantities). These conditions are difficult to satisfy for real canopies, even in simulations (although periodic simulations enforce them by design), since the area would often become large enough and comparable to the scale at which the spatially averaged means of the canopy properties would vary (see Schmid et al. (2019) and Whitaker (1999) for more details).

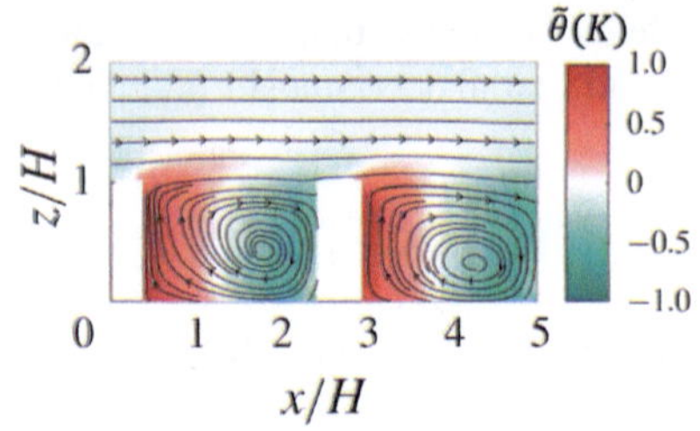

Figure 8.2 Mean flow streamlines overlaid on a pseudocolor plot of the temperature $\widetilde{\theta}$, depicting the dispersive fluxes (mean wind from the left). The building walls and ground are heated; the recirculation occurring in the cavity results in hot air ($\widetilde{\theta} > 0$, heated from its contact with the surface) being advected upwards ($\widetilde{w} > 0$) on the back/lee side, while colder air ($\widetilde{\theta} < 0$, cooled by turbulent mixing with the fresh cold air above the canyon) is advected downward ($\widetilde{w} < 0$) on the front/windward side. When averaged over the street width, these result in $\langle\widetilde{\theta}\widetilde{w}\rangle > 0$, a net upward heat flux.

The last term $\langle\overline{u_i'u_j'}\rangle$ is the conventional turbulent stress, but now it is also averaged in space to get its net spatial contribution. The middle term, $\langle\widetilde{u}_i\widetilde{u}_j\rangle$, is again a new addition; it is commonly referred to as the dispersive stress (Raupach and Shaw, 1982), and that is why we refer to the spatial deviations as the dispersive components (the nomenclature is in analogy to mass dispersion in subsurface flows that is due to the tortuous fluid flow in the pore network). As can be inferred from its expression, it represents the spatial correlation of the spatial deviations from the spatio-temporal averaged flow field. Physically, it can be viewed as the net (over the whole volume or surface of spatial averaging) advective flux by the velocity component $\widetilde{u}_i$ of the momentum component $\widetilde{u}_j$, or alternatively, of heat or any trace gas. In fact, the same triple decomposition can be applied to the flux $u_i\xi$ of any scalar ξ. Fig. 8.2 illustrates the net upward dispersive flux of heat over a hot urban surface (all facets are heated), with the mean flow going from left to right.

8.2.2 Vertical layers of the flow

Flows over urban terrain also display pronounced vertical variability (see Fig. 8.3). The dispersive fluxes are very significant in between the buildings, the so-called **urban canopy sublayer (UCL)**, where they can contribute over 50% of the total fluxes of momentum or scalars, with this fraction increasing for areas with variable building height (Li and Bou-Zeid, 2019; Li et al., 2025). The flow in the UCL below h ($\approx$ height of the tallest building), is fully 3D, and the turbulence statistics display strong spatial variability. The flow, temperature, and concentrations of gases or aerosols in that layer are obviously important, since these are the conditions to which pedestrians and buildings are exposed.

Above the tallest building, the **roughness sublayer (RSL)** begins and extends to about $3h$ above ground level (Roth, 2000) (sometimes, the RSL is defined as including the UCL and extending to the ground). The mean flow and turbulence in that layer are

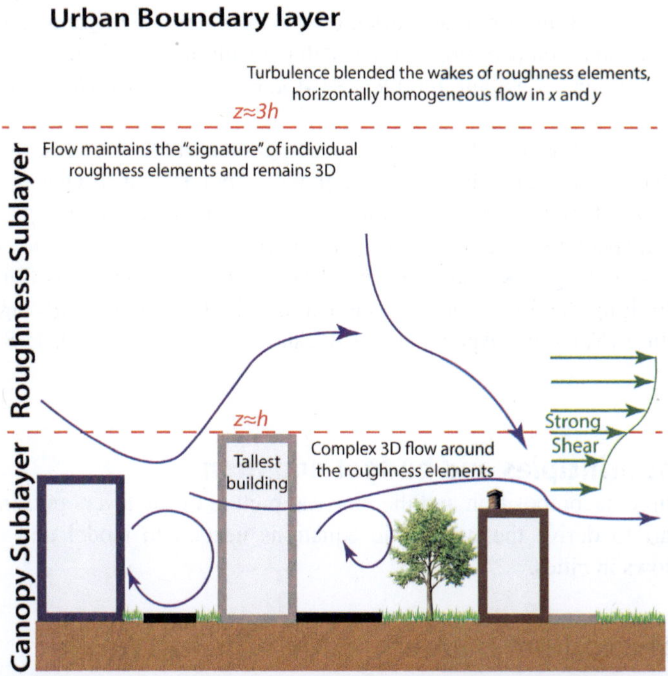

Figure 8.3 Schematic of flow and layers in urban terrain depicting the urban canopy sublayer (UCL) and the roughness sublayer (RSL). The flow is very complex and depends on the morphology of the canopy, but strong shear is invariably present near the top of the buildings.

still strongly influenced by individual roughness elements, and thus the statistics vary in x and y. The horizontal variability and the associated dispersive fluxes gradually decrease with height, as the turbulent fluxes increase to sustain the total vertical transport of mass, momentum, and heat.

Above the roughness sublayer, almost all flow features arising from individual elements have been blended by the turbulence. The flow becomes horizontally homogeneous, recovering the general attributes of wall-bounded flows over rough walls. One then expects an **inertial layer** (with a logarithmic velocity profile if the flow is neutral and the Rossby number is very high) extending till about 15% of the depth of the full atmospheric boundary layer ($\delta \approx 1000$ m), and an **outer (wake) layer** above that. In the outer and inertial urban boundary layers, the general turbulence statistics and large structures are not too dissimilar from those over smooth or lower roughness walls since they result from the bulk shear and dynamics across the whole ABL (Inagaki et al. (2017); Kadivar et al. (2021), see Smits et al. (2011) for a review of large structures in wall-bounded flows). However, imprints of the low-momentum wakes of buildings can often still be detected in the outer layer (Hong et al., 2012). In addition, as with most geophysical flows, the role of the Coriolis force is non-negligible (since the Rossby number is not large enough) and manifests as an Ekman turning of the mean flow (see, e.g., Ghannam and Bou-Zeid (2021)), especially in the urban outer layer. Built terrain also often overheats relative to the rural surroundings, creating horizontal temperature gradients that result in baroclinic conditions that further alter the outer flow and turbulence (Momen et al., 2018; Ghannam and Bou-Zeid, 2021).

This canonical layering, however, may sometimes break down. If the buildings are taller than ≈ 50 m, resulting in a roughness sublayer that reaches ≈ 150 m, an inertial layer might not exist; the transition is then directly to the outer layer. Moreover, under certain special sparse-roughness configurations, recent numerical experiments have shown that roughness-induced secondary circulations can persist far above the canopy, inhibiting the formation of a classical inertial layer and modifying the outer-layer structure (Yang and Anderson, 2018; Sathe and Giometto, 2024; Sathe et al., 2025).

8.3 First principles and budget equations

With the triple decomposition and the physical picture of the layers presented above, we can start to derive the prognostic equations needed to model and understand turbulent flows in cities.

8.3.1 Momentum budgets in the different layers

First, we will examine the spatio-temporally averaged force balance that dictates the correspondingly averaged streamwise velocity at various heights. If the canopy is statistically homogeneous in the horizontal direction (that is, its geometric characteristics are invariant to translation of the spatial averaging surface), we can assume that $\partial\langle\overline{\cdot}\rangle/\partial x = \partial\langle\overline{\cdot}\rangle/\partial y \approx 0$ and $\langle\overline{w}\rangle \approx 0$. We can also neglect viscous forces on account of Re at the building scale being $> 10^6$, even for a 10-m building. Therefore, the pressure/form drag D should dominate over the viscous drag,[2] which is indeed confirmed by direct numerical simulation results (Coceal et al., 2006). This streamwise force balance can then be expressed as

$$\frac{\partial\langle\overline{u}\rangle}{\partial t} = -\frac{1}{\rho_r}\frac{\partial\langle\overline{p}\rangle}{\partial x} - \left\langle\frac{\partial\overline{u'w'}}{\partial z}\right\rangle - \left\langle\frac{\partial\widetilde{u}\,\widetilde{w}}{\partial z}\right\rangle + f_c\langle\overline{v}\rangle + D \tag{8.5}$$

where the tendency term is on the left-hand side (LHS) and is equal to zero for steady-state mean flows. On the right-hand side (RHS), the first term is the streamwise pressure gradient driving the flow; the second is the Reynolds stress gradient (Reynolds force); the third is the equivalent dispersive stress gradient (dispersive force); the fourth is the Coriolis force (typically negligible below 50 m); and the last term is the (predominantly form) drag exerted by the buildings in the canopy sublayer.

[2] Note that this is only correct for the bulk force balance equation outside of the viscous sublayer. Viscous forces, in fact, play a significant role in modulating the details of the flow near the solid-fluid interfaces, and, if these interfaces have complex shapes, these forces can modify the point where the flow streamlines separate and thus modify the drag coefficient. This, for example, has been shown for round and angled roofs (compared to flat roofs) by Llaguno-Munitxa et al. (2017).

In the roughness sublayer above the buildings, we can simplify the equation above by simply omitting D since there are no structures to exert a drag on the flow. Finally, above the roughness sublayer, we expect the dispersive force to become negligible, and hence we can omit the third term on the RHS.

In various applications, it is often convenient to rewrite the second and third terms on the RHS of (8.5) as vertical gradients of the time- and spatially averaged Reynolds and dispersive stresses. An important complication then arises: the spatial-averaging operation over the fluid plane does not, in general, commute with vertical differentiation because the area of the averaging plane varies with height, and the resulting errors increase with solid volume fraction. One practical remedy is to use superficial averaging (over the combined solid and fluid plane) when formulating the force balance in (8.5), since that operation does commute with vertical differentiation (Schmid et al., 2019). This choice, however, introduces additional interpretational challenges. In the next section, we elaborate on the implications of these two averaging choices and the assumptions inherent to each.

8.3.2 The spatial averaging operation

Because urban canopies occupy a significant fraction of space (unlike vegetation or wind farm canopies, see Appendix A), any horizontal spatial averaging operation must account explicitly for the coexistence of fluid and solid volumes. This leads to a clear distinction between intrinsic and superficial averages, which becomes essential for the correct formulation and interpretation of mean profiles, fluxes, and budget equations in cities (e.g., Whitaker, 1999; Schmid et al., 2019). While we have already introduced spatial averages and dispersive components in the previous sections in terms of the intrinsic averaging over the fluid volume, we now formalize the two averaging operators and clarify their different physical meanings and mathematical properties. We define a horizontal averaging volume as a thin slab centered at height z. Within this slab, the total volume V is composed of a fluid volume $V_f(z)$ and a solid volume $V_s(z)$ occupied by buildings and other obstacles. The corresponding fluid volume fraction (or void fraction) is

$$\lambda_v(z) \equiv \frac{V_f(z)}{V} \tag{8.6}$$

In urban terrain, $\lambda_v(z)$ generally decreases from unity above the roughness sublayer to much smaller values within the canopy. For any flow quantity φ defined in the fluid, two spatial averages can be defined:

$$\langle\varphi\rangle(z) \equiv \frac{1}{V_f(z)} \int_{V_f(z)} \varphi(\mathbf{x})\, \mathrm{d}V \qquad \text{(intrinsic average)} \tag{8.7}$$

$$[\varphi]\,(z) \equiv \frac{1}{V} \int_{V_f(z)} \varphi(\mathbf{x})\, \mathrm{d}V \qquad \text{(superficial average)} \tag{8.8}$$

The intrinsic average is normalized by the fluid volume and therefore, represents a typical value within the air phase. It is the appropriate average for comparison with in situ measurements and for defining spatial deviations and dispersive fluxes. The superficial average, by contrast, is normalized by the total volume and is directly related to extensive quantities per unit total volume, such as bulk momentum, heat, or mass (Whitaker, 1999). The two averages are related through the simple identity

$$[\varphi](z) = \lambda_v(z) \langle\varphi\rangle(z) \tag{8.9}$$

Above the canopy, where $\lambda_v \approx 1$, the intrinsic and superficial averages coincide. Within the canopy, intrinsic averages exceed their superficial counterparts by a factor of $1/\lambda_v$.

A key difference between intrinsic and superficial averages is that they exhibit different commutation properties with vertical differentiation. Based on the spatial averaging theorem (Whitaker, 1999), the superficial averaging operation can be written as

$$\left[\frac{d\varphi}{dz}\right] = \frac{d[\varphi]}{dz} + \frac{1}{V}\int_{A_{fs}} \varphi\, n_z\, \mathrm{d}A \tag{8.10}$$

where A_{fs} is the total fluid–solid interface within the averaging volume and n_z is the vertical component of the unit normal oriented such that it points from the fluid region V_f into the solid region V_s. For intrinsic averages, an additional geometric term appears:

$$\left\langle\frac{d\varphi}{dz}\right\rangle = \frac{d\langle\varphi\rangle}{dz} + \frac{\langle\varphi\rangle}{\lambda_v}\frac{d\lambda_v}{dz} + \frac{1}{V_f}\int_{A_{fs}} \varphi\, n_z\, \mathrm{d}A \tag{8.11}$$

The surface integrals in (8.10) and (8.11) represent net interfacial fluxes at solid boundaries, while the $d\lambda_v/dz$ term is purely geometric and reflects the vertical variation of the cross-sectional area available to the flow.

This distinction has important consequences for the definition of horizontally averaged equations. Intrinsic formulations contain explicit $d\lambda_v/dz$ correction terms whenever the fluid fraction varies significantly with height, as is generally the case in urban canopies (this variation is often ignored in vegetation and wind farm canopies since the solid fraction is minimal and $\lambda_v \approx 1$). As emphasized by Schmid et al. (2019), neglecting the correction terms in intrinsic formulations results in incomplete momentum and scalar budgets in dense urban terrain. Even for regular building arrays, it would lead to significant errors near the canopy top where $d\lambda_v/dz$ is the highest. Superficial formulations, on the other hand, absorb these geometric effects into the definition of the averaged variables, leading to formally simpler one-dimensional vertical balances. This is the fundamental reason why superficial averages are often preferred when writing vertically integrated force balances, whereas intrinsic averages are preferred for defining mean profiles, dispersive fluxes, and neighborhood-scale statistics that are directly comparable to observations.

The streamwise momentum budget provides a concrete illustration of how the choice between superficial and intrinsic averaging impacts the structure of the time and volume-averaged equations. We first need to distinguish between the dispersive perturbations relative to the intrinsic average, which we have denoted with a tilde, and the perturbations relative to the superficial average that we will denote with a hat. The time-average velocity is thus now decomposed following: $\overline{u}_i = [\overline{u}_i] + \widehat{u}_i$. Applying the time and superficial-averaging operations to the incompressible Navier–Stokes equations in flow over urban canopies, we then obtain the counterpart to the intrinsically averaged Eq. (8.5):

$$\frac{\partial [\overline{u}]}{\partial t} = -\frac{1}{\rho_r}\frac{\partial [\overline{p}]}{\partial x} - \frac{\partial \left[\overline{u'w'}\right]}{\partial z} - \frac{\partial [\widehat{u}\,\widehat{w}]}{\partial z} + f_c[\overline{v}] + D \tag{8.12}$$

The appeal of the superficial averaging here is clear: in the second and third term on the RHS, averaging and differentiation can now be commuted. Since the velocities are assumed to be zero at fluid–solid interfaces, the vertical gradients of the turbulent and dispersive stresses do not produce surface integrals. The viscous term would produce an integral of the viscous drag at the fluid–solid interface, in addition to the viscous transport along the mean velocity gradient. For high Reynolds number urban canopy flows, however, both of these terms are typically negligible, and we have made this assumption here. In studies that employ superficial averaging, the triple decomposition (8.1) is typically also defined relative to the superficial mean. As discussed in Schmid et al. (2019), this choice is, in general, problematic because it can generate nonzero dispersive fluxes (in Eq. (8.4)) even when the corresponding intrinsic mean fields are spatially uniform, thereby introducing spurious spatial correlations that have no physical origin.

If intrinsic averaging is instead required or preferred, then the additional terms resulting from application of Eq. (8.5) cannot be ignored when the fluid volume fraction λ_v differs substantially from unity and varies rapidly with height, which is almost invariably the case in urban applications. The correct intrinsic form of the momentum equation then becomes (e.g., Nikora et al., 2007):

$$\begin{aligned}\frac{\partial \langle \overline{u}\rangle}{\partial t} = &-\frac{1}{\rho_r}\frac{\partial \langle \overline{p}\rangle}{\partial x} - \frac{\partial}{\partial z}\overline{\langle u'w'\rangle} - \frac{\partial}{\partial z}\langle \widetilde{u}\,\widetilde{w}\rangle \\ &\underbrace{-\frac{\overline{\langle u'w'\rangle}}{\lambda_v}\frac{d\lambda_v}{dz}}_{\text{geometric correction}} \underbrace{-\frac{\langle \widetilde{u}\,\widetilde{w}\rangle}{\lambda_v}\frac{d\lambda_v}{dz}}_{\text{geometric correction}} + f_c\langle \overline{v}\rangle + D\end{aligned} \tag{8.13}$$

Applying Eq. (8.11) introduces two additional "geometric correction" terms proportional to the vertical gradients of the fluid fraction. These terms reflect the fact that vertical variations of intrinsically averaged fluxes arise not only from true sources or sinks of momentum, but also from purely geometric effects. The physical origin of these correction terms can be understood by comparing two horizontal averaging slabs located at heights z and $z + \delta z$. Differences between the total integrated momentum fluxes through the upper and lower surfaces contribute to the momentum budget within the considered layer. However, the intrinsic flux may also change simply because the cross-sectional area available to the flow changes with height as the solid fraction varies. In this case, a decrease in the intrinsic flux does not imply that momentum has been locally extracted from the flow; rather, the same total momentum is redistributed over a larger fluid area. The geometric correction terms in the intrinsic momentum balance compensate for this effect by counteracting flux gradients that are solely due to changes in the geometry, thereby ensuring that only genuine dynamical sources and sinks appear in the budget.

Analogous superficial or intrinsic volume-averaging procedures can be applied to the turbulent kinetic energy, Reynolds stress transport, or various equations for higher-order moments. In those cases, the same commutation properties lead to additional geometric contributions in the intrinsic formulation, while the superficial formulation retains a compact conservative structure suitable for vertically one-dimensional neighborhood-scale budgets (Akinlabi et al., 2023). For this reason, we adopt superficial averaging as the primary framework for the kinetic energy budgets introduced in the next section. However, the physical interpretations and discussions are almost identical for the two averaging procedures, with the caveat that the correction terms must be included if intrinsic averaging is used.

8.3.3 Mean, dispersive, and turbulent kinetic energies

The triple decomposition applied above to the stresses can also be used to decompose all higher-order moments and derive their budget equations, but here we will only cover the balance equations for various components of the kinetic energy. Specifically, we will consider the kinetic energy components that emerge when considering Eq. (8.4) or its superficial counterpart with $i = j$:

$$\langle \overline{u_i^2} \rangle = \langle \overline{u}_i \rangle^2 + \langle \widetilde{u}_i^2 \rangle + \langle \overline{u_i'^2} \rangle \; ; \; \left[\overline{u_i^2} \right] = [\overline{u}_i]^2 + \left[\widehat{u}_i^2 \right] + \left[\overline{u_i'^2} \right] \tag{8.14}$$

We will focus henceforth on the superficial averaging for mathematical simplicity, as noted above. This expression implies that the spatio-temporal average of the total kinetic energy has three components: the averaged kinetic energy of the spatio-temporally averaged flow $E_A = [\overline{u}_i]^2/2$ that primarily informs on the potential for large-scale horizontal advection, the spatial average of the dispersive kinetic energy (which by definition is also averaged in time) $E_D = \left[\widehat{u}_i^2\right]/2$ that represents the activity of steady secondary circulations and their ability to mix the urban layers vertically and horizontally, and the spatial average of the conventional turbulent kinetic energy $E_T = \left[\overline{u_i'^2}\right]/2$ that also results in 3D mixing, but through the chaotic turbulent eddies.

The easiest approach for deriving budget equations for these three components is to start from the Reynolds decomposition into a mean kinetic energy (MKE, denoted as $E = \overline{u_i}^2/2$) and a time-averaged turbulent kinetic energy (TKE, denoted as $\overline{e} = \overline{u_i'^2}/2$), at each point in space. This is simply a Reynolds decomposition and yields the equations that can be found in various textbooks (e.g., Wyngaard, 1992), without any assumptions on horizontal homogeneity. Neglecting viscous transport terms for both budgets since they are only relevant very close to the boundaries, the viscous dissipation term for E but not for $\overline{e}$, as well as the Coriolis terms, which are identically zero (Coriolis is a fictitious pseudo force, so it is conservative and does not produce or extract any net energy from the flow; it simply redistributes it between components Momen et al. (2018)), we can write

$$\frac{\partial E}{\partial t}+\overline{u}_j\frac{\partial E}{\partial x_j}=+\overline{u'_iu'_j}\frac{\partial \overline{u}_i}{\partial x_j}-\frac{1}{\rho_r}\frac{\partial\left(\overline{u}_i\overline{\check{p}}\right)}{\partial x_i}-\frac{\partial\left(\overline{u}_i\overline{u'_iu'_j}\right)}{\partial x_j}+\delta_{i3}\frac{g}{\theta_r}\overline{u}_i\overline{\check{\theta}} \tag{8.15}$$

$$\frac{\partial \overline{e}}{\partial t}+\overline{u}_j\frac{\partial \overline{e}}{\partial x_j}=-\overline{u'_iu'_j}\frac{\partial \overline{u}_i}{\partial x_j}-\frac{1}{\rho_r}\frac{\partial\left(\overline{u'_j\check{p}'}\right)}{\partial x_j}-\frac{\partial\left(\overline{u'_je}\right)}{\partial x_j}+\delta_{i3}\frac{g}{\theta_r}\left(\overline{u'_i\check{\theta}'}\right)-\varepsilon \tag{8.16}$$

In these equations, the inverted hats (checks) over p and θ denote the perturbations of these variables from the Boussinesq hydrostatic reference state (denoted by subscript r), as detailed in Chapter 1. For the rest of this chapter, however, we will omit the inverted hats for notational simplicity, but they are always implied over p and θ. In the first equation on the RHS, the first term is the conversion of MKE to TKE; the second term is the transport (locally, can be a source or sink) of MKE by the mean pressure gradient (can be viewed as a divergence of the work term $\overline{u}_i\overline{p}$); the third is the mean of the turbulent transport of cross (turbulent-mean) kinetic energy (this may be clearer when viewed as the divergence of $(\overline{u'_j\overline{u}_iu'_i})$); and the last term is the production or destruction of MKE by buoyancy associated with the Boussinesq perturbations of the potential temperature θ (or virtual potential temperature if air humidity has a significant impact on buoyancy (Brutsaert, 2005)). The buoyancy term in the MKE equation is thus not identically zero unless the Boussinesq and Reynolds mean states are identical; they need not be so in general, as explained in Chapter 1. In fact, they are definitely not identical in urban flows since the Reynolds mean state varies in all directions, while the Boussinesq state must be viewed as a spatially averaged reference over a much larger scale (Spiegel and Veronis, 1960). In the second equation for the TKE, on the RHS, the first term is the production (though rarely it can be a destruction) of TKE from MKE (equal but opposite in sign to the corresponding term in the MKE equation); the second and third terms are the pressure and turbulent transports; the fourth is the buoyant conversion or destruction (though rarely it can be a destruction) of TKE from or into the turbulent potential energy associated with potential temperature perturbations $\overline{\theta'^2}$ (equivalent to density perturbations $\overline{\rho'^2}$). The last term is the viscous dissipation into thermal energy.

We can now decompose the mean kinetic energy into a spatial average and a perturbation thereof, and then average it in space to get:

$$[E]=\frac{1}{2}\left[\overline{u}_i^2\right]=\frac{1}{2}\left[\overline{u}_i\right]^2+\frac{1}{2}\left[\widehat{\overline{u}}_i^2\right]=E_A+E_D \tag{8.17}$$

Recall that the cross terms are zero after averaging in space, but, as we will see later, their correlation with the dispersive velocity before averaging leads to transport terms. To derive the separate budget equations for E_A and E_D, we would start from the equation of the time-averaged (mean) velocity field, decompose it into an equation for the space-time average and the dispersive perturbation, space-average the equation, to obtain an equation for $[\overline{u_i}]$, and then subtract that equation from the decomposed equation to obtain an equation for $\widehat{u_i}$. Then, we multiply each budget equation by the velocity component it is predicting. Mathematically, the procedure is identical to the Reynolds decomposition and the derivation of the MKE and TKE equations, but with some additional terms arising from the Reynolds stress divergence term in the mean velocity equation (which are not present in the full Navier–Stokes). The resulting budget equations are:

$$\frac{\partial E_A}{\partial t} + [\overline{u}_j]\frac{\partial E_A}{\partial x_j} = +[\widehat{u}_i\widehat{u}_j]\frac{\partial[\overline{u}_i]}{\partial x_j} - \frac{\partial[\overline{u}_i][\widehat{u}_i\widehat{u}_j]}{\partial x_j}$$
$$+\left[\overline{u'_j u'_i}\right]\frac{\partial[\overline{u}_i]}{\partial x_j} - \frac{\partial[\overline{u}_i]\left[\overline{u'_j u'_i}\right]}{\partial x_j} - \frac{1}{\rho_r}\frac{\partial[\overline{u}_i][\overline{p}]}{\partial x_i} + \delta_{i3}\frac{g}{\theta_r}[\overline{u}_i][\overline{\theta}] \quad (8.18)$$

$$\frac{\partial E_D}{\partial t} + [\overline{u}_j]\frac{\partial E_D}{\partial x_j} = -[\widehat{u}_j\widehat{u}_i]\frac{\partial[\overline{u}_i]}{\partial x_j} - \frac{\partial[\widehat{u}_j(0.5\widehat{u}_i^2)]}{\partial x_j}$$
$$-\left[\overline{u'_j u'_i}\frac{\partial\widehat{u}_i}{\partial x_j}\right] + \left[\frac{\partial\widehat{u}_i\overline{u'_j u'_i}}{\partial x_j}\right] - \frac{1}{\rho_r}\frac{\partial[\widehat{u}_i\widehat{p}]}{\partial x_i} + \delta_{i3}\frac{g}{\theta_r}[\widehat{u}_i\widehat{\theta}] \quad (8.19)$$

The terms can be interpreted in parallel to the Reynolds decomposition of MKE and TKE. The first terms in both equations on the RHS are the exchange of KE between E_A and E_D, which tends to produce E_D. The second term in Eq. (8.18) is the spatial-average of the dispersive transport of cross (mean-dispersive) energy by the dispersive velocity (which can also be written as the divergence of $[\widehat{u}_j[\overline{u}_i]\widehat{u}_i]$); the third is the conversion of E_A into TKE; the fourth is the spatial-average of the turbulent transport of cross (mean-turbulent) energy by the turbulent velocity (which can also be written as the divergence of $[[\overline{u}_i]\overline{u'_j u'_i}]$); the fifth is the transport of E_A by the spatially averaged component of the mean pressure gradient; and the last term is buoyancy conversion of E_A into/from spatio-temporally averaged potential energy (which will be zero if $[\overline{w}] = 0$). In Eq. (8.19), the second term on the RHS is the dispersive transport of DKE; the third is the spatially averaged conversion of DKE into TKE; the fourth is the spatial-average of the turbulent transport of cross (dispersive-turbulent) energy by the turbulent velocity (which can also be written as the divergence of $[\overline{u'_j\widehat{u}_i u'_i}]$); the fifth is the transport of E_D by the dispersive component of the mean pressure gradient; and the last term is the buoyancy conversion of E_D into/from spatio-temporally averaged potential energy, which is unlikely to be zero.

Finally, we will examine the spatial average of Eq. (8.16) that yields the budget equation for the spatio-temporally averaged turbulent kinetic energy $E_T = [\overline{e}] = [\overline{u_i'^2}]/2$:

$$\frac{\partial[\overline{e}]}{\partial t} + [\overline{u}_j]\frac{\partial[\overline{e}]}{\partial x_j} = -\left[\overline{u_i'u_j'}\right]\frac{\partial[\overline{u_i}]}{\partial x_j} - \left[\widehat{\overline{u_i'u_j'}}\frac{\partial\widehat{u}_i}{\partial x_j}\right]$$
$$-\frac{1}{\rho_r}\frac{\partial\left[\overline{u_j'p'}\right]}{\partial x_j} - \frac{\partial\left[\overline{u_j'e}\right]}{\partial x_j} + \frac{\partial[\widehat{u}_j\widehat{e}]}{\partial x_j} + \delta_{i3}\frac{g}{\theta_r}\left[\overline{u_i'\theta'}\right] - [\varepsilon] \quad (8.20)$$

All terms are the spatial-averaged equivalents of their spatially local counterparts, except the second and fifth terms on the RHS, which emerge from the decomposition of the spatial averages of the shear production term and advective term, respectively. The second RHS term is simply the shear production of TKE by the dispersive stress and strain, where $\widehat{u_i'u_j'} = \overline{u_i'u_j'} - \left[\overline{u_i'u_j'}\right]$ (compare it to the first RHS term, which is the equivalent production by the spatio-temporally averaged stress and strain). This term can be further subdivided into the component related to $\overline{u_i'u_j'}$ and the one related to $-\left[\overline{u_i'u_j'}\right]$ (see Giometto et al., 2016). The fifth term is the dispersive transport of TKE (compared to the one before it, which is the turbulent transport of TKE). These equations will be used in the next subsection to unravel the sources of dispersive and turbulent mixing and transport in an urban canopy and the structure of the flow and turbulence.

8.3.4 Sources of turbulence and dispersion: shear, buoyancy, and human activities

Mixing in urban terrain is done by turbulent and dispersive motions, with the balance shifting rapidly to a dominance of the turbulent transport towards and above the top of the roughness sublayer. An important question is thus what are the physical mechanisms for the production of turbulent eddies and dispersive motions. Considering the averaged TKE Eq. (8.20), we will note the usual shear production (first term on RHS), but now averaged in space. This tensor contraction will be dominated by the term involving the spatio-temporally averaged vertical gradient of the horizontal velocity and the vertical transport of horizontal momentum $-\left[\overline{u'w'}\right](\partial\langle\overline{u}\rangle/\partial z)$. It will thus peak near the building tops (depends on the building height variability), where the strongest mean shear can be found (Giometto et al. (2016), see Fig. 8.3). This term will thus only produce $\langle\overline{e}\rangle$ in the streamwise component, resulting in turbulence anisotropy. However, as in all wall-bounded flows, this streamwise TKE then gets redistributed to the cross-stream and vertical components through the pressure redistribution term that appears in the velocity variance equations of the three individual components (Bou-Zeid et al., 2018; Yi et al., 2025). The second RHS term in Eq. (8.20) produces turbulence from the dispersive motions, but in all components, resulting in a more isotropic and vigorous turbulent field compared to flat terrain. It is the result of the complex three-dimensionality of the mean (time-averaged) flow field that produces strong dispersive shear and stresses in all directions and that strongly varies in space (before it is averaged). The other new term (fifth on the RHS in Eq. (8.20)) further illustrates that urban TKE undergoes dispersive transport like all scalars in urban terrain.

Considering the buoyancy terms in the averaged TKE equation, we see that there is only a single spatially averaged buoyancy term and that the dispersive heat fluxes do not produce $\langle\overline{e}\rangle$ directly. It is also dominated by a single component, $(g/\theta_r)\left[\overline{w'\theta'}\right]$, that only produces averaged TKE in the vertical component; however, that TKE can also be redistributed via the return-to-isotropy pressure terms. The dispersive heat fluxes show up in the DKE equation (last term on RHS of Eq. (8.19)), $(g/\theta_r)\left[\widehat{w}\widehat{\theta}\right]$, as a source of vertical dispersive kinetic energy. Physically, the reader can picture an example of this term as a buoyant jet rising from a hot wall, creating a local dispersive circulation. The other source of DKE is the first term on the RHS of its equation, where the contraction is again dominated by one term, $-\left[\widehat{u}\,\widehat{w}\right]\left(\partial\left[\overline{u}\right]/\partial z\right)$, that produces DKE (converting it from E_A) only in the horizontal DKE component associated with $\widehat{u}$. Finally, we should reiterate that, while not identically zero, the buoyancy term in the MKE Eq. (8.18) is usually small if the spatial average is done over a sufficiently large area that yields $[\overline{w}] \approx 0$.

The sources of TKE and DKE discussed above are the conventional ones and are mainly related to urban form and fabric; they can be transported by all three components of the velocity (mean, dispersive, and turbulent). The material properties of the city, associated with its fabric, can induce variability in heat transfer and buoyancy fluxes, thereby strengthening the 3D variability of the flow beyond the complexity of the urban form alone. For example, in a street canyon with an asphalt pavement that heats up significantly in the daytime and side concrete pavements that remain cooler, thermal dispersive circulations can be induced by the resulting heterogeneity of the surface temperature. Urban function also exercises a significant influence on urban flows and turbulence. Moving hot cars (Solazzo et al., 2008; Klein et al., 2000; Chen et al., 2021) or hot jets from chimneys, vents, or air conditioning compressors (especially the large numbers of such heat pumps needed to cool modern data centers) can produce TKE and DKE through the same terms and mechanisms discussed above. Urban activities can influence the flow indirectly as well, for example, through the emission of aerosols and gaseous pollutants that modify cloud formation and then the radiation and heat fluxes in the entire urban boundary layer (Wang et al., 2019). The three urban attributes, form, fabric, and function, interact to produce strong daytime and nighttime urban heat islands (UHIs), defined as the difference in temperature between the city and its less urbanized surroundings. UHIs can be measured in terms of the air temperature (usually peaking during nighttime) or surface temperatures (usually peaking during daytime) (Oke et al., 2017), and they invariably boost the buoyant production of TKE (and DKE), reducing the probability of statically stable stratifications and increasing the occurrence of neutral or unstable conditions at night in cities. For example, during a field campaign in Basel, Switzerland, stable conditions occurred 5% of the time inside the city, compared to 60% at rural sites. Buoyancy under unstable conditions produces TKE and mixing, which is helpful for cities since it prevents the onset of strong inversions that trap pollutants.

8.4 Flow and turbulence: from the building to the metropolitan scale

One of the defining characteristics of turbulence is its broad spectrum of scales. ABL flows over smoother Earth surfaces, for example, have eddies ranging from the Kolmogorov or inner scale ($\eta \sim 1$ mm) all the way to the outer scales ($\sim \delta \approx 1$ km) and to the large and very large scale motions that can be $\sim 10\delta$ (Smits et al., 2011;

Shah and Bou-Zeid, 2014). A powerful simplifying framework for such flows is that only two external length scales affect the flow, the outer and inner scales (the ratio of which is, in fact, encoded in the Reynolds number). TKE is produced at the largest of scales (outer scales, with production locally peaking at a scale $\sim z$, the height above the ground according to Townsend's attached eddy hypothesis (see Chapter 1)), cascades down following its own internal dynamics, and then gets dissipated at the smaller viscous (inner) scales.

This picture needs modification for flows over urban surfaces and similar, very rough surfaces. Such walls have their own broad range of geometric scales that can interact with the turbulence spectrum. For example, tree branches can produce wakes $\sim$ 0.01–1 m, individual cars excite eddies $\sim$ 5 m, buildings create wakes and shear layers that inject DKE and TKE at scales $\sim$ 10-100 m, the complex shear at the top of UCL generates mixing layers that also produce eddies of broad sizes, and the outer flow still contains the largest $\sim$ 1–10 km structures. Thus, realistic turbulence in cities may be too complex to fully unravel due to scale interactions; however, for most applications, we can try to focus on a subrange of that large spectrum and aim to develop a sufficiently accurate understanding and useful models for that range. In this chapter, we will not examine turbulence generated by urban vegetation (see Giometto et al. (2017) and Grylls and van Reeuwijk (2022) for overviews of the effect of vegetation on urban flows) or vehicles (see Solazzo et al. (2008); Klein et al. (2000); Chen et al. (2021)). We will instead focus on turbulence and flow structures generated by buildings (akin to a single cuboid); we will then examine the dynamics at the intermediate scale $\sim$ a block or neighborhood and the flow features in a single "street canyon"; and finally, we will overview the largest circulations at the scale of the whole city.

8.4.1 Flow over a single building or street canyon

Flow over a single cuboid is well-understood (relative to its neighborhood and city scales counterparts), with a rich literature exploring the mean structure and turbulence dynamics (see a recent review in Gorlé (2025)). The focus of the studies at this scale is usually on the flow structure around the building, the wake and turbulence characteristics, the influence on pedestrian level wind, the convective heat exchange between the building and the air, as well as the forces that the flow imposes on the structure, which has to be designed accordingly. Here, we discuss the main flow features that are also visualized in Fig. 8.4. For wall-mounted cuboids, representing individual buildings, the incoming velocity increases with height above the wall. Bernoulli's equation thus predicts higher pressures on the upper part of the windward face as the flow decelerates, and a higher drag force. However, near the top of the building, the flow is deflected over the roof and is not brought to a complete halt. So the highest pressure point usually occurs at a height $\approx$ 2/3 of the building height (Gorlé, 2025). Since the air has very low velocity and inertia in the vicinity of that point, its flow is mainly dictated by the pressure gradients, and since pressure decreases away from the highest pressure point, the flow must diverge horizontally and vertically from

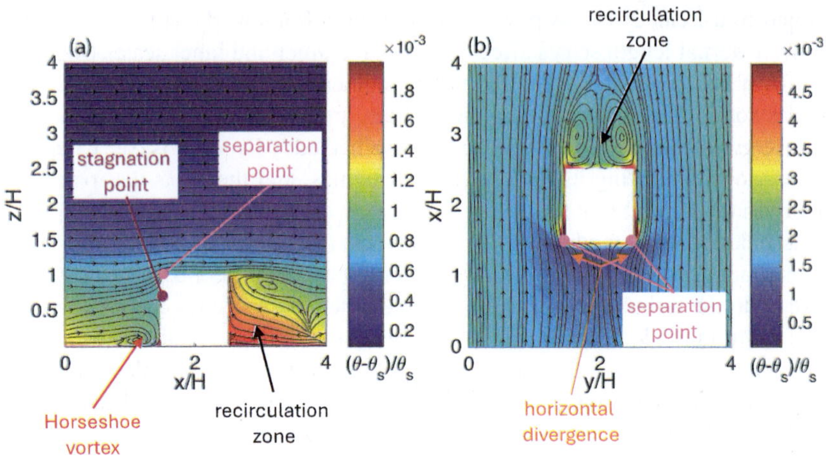

Figure 8.4 Flow features over a single wall-attached cuboid, depicting the stagnation and separation points, the recirculation zone in the lee of the structure, and the horseshoe vortex formed upwind and diverted around the building and into the wake.

that point around the windward face. A diverging flow also involves a stagnation point (zero velocity), which is thus nearly identical to the maximum pressure point (viscous and unsteady effects neglected in Bernoulli's equation could cause a mismatch). It should be noted that, instantaneously, these points move around due to turbulent fluctuations (Castro and Robins, 1977).

The diverging streamlines on the windward face typically separate at the edges of that face for cuboids, if one face is perpendicular to the incoming velocity. Downstream of the separation, small recirculation zones form on the side walls and roof. The flow deflected downward at the stagnation points moves towards the ground, but as the incoming wind is still impinging on that face, the perpendicular velocity components create a circulation zone with high cross-stream vorticity near the base of the building. The flow continuously converging in the (x, z) plane must diverge in the cross-stream y direction, and that divergence carries with it and stretches this vorticity, creating a horseshoe-shaped vortex that wraps around the building base. The diverging streamlines converge back behind the building, but that only occurs at a distance larger than the scale of the cuboid; therefore, in the lee of (behind) the structure, a large recirculation occurs where pollutants and heat can be trapped.

The details of the flow are modified by the aspect ratio of the cuboid and the orientation relative to the mean wind (Gorlé, 2025), and even more so if the building is cylindrical or pyramidal (Martinuzzi et al., 2007), or features a nonflat roof. To the best of our knowledge, there is no comprehensive review that considers multiple building shapes and how the change in form alters the flow and wake. These flow modifications also alter the building-air heat exchange (Meinders et al., 1999), which can also feed back to alter the flow (when some or all the facets are heated) if it generates strong buoyancy forces (Torres et al., 2018). At the building scale and considering heat transfer, three nondimensional parameters can be postulated to be sufficient to describe the flow: the Reynolds number $Re = UH_b/\nu$ (where H_b is the characteristic building scale); the Richardson number, which here we could define as $Ri = (g/\theta_r)(\Delta\theta H_b/U^2)$, where $\Delta\theta$ is temperature difference between the building surface and the air; and the Strouhal number $St = fH_b/U$, where f is the vortex shedding frequency. A challenge that arises is that these three numbers scale differently with H_b and U, and it is categorically impossible to match all three of them at a reduced scale in a wind tunnel or a direct numerical simulation. Even matching Re and Ri, while possible, is challenging due to their opposite scaling with U (see relevant discussion in Williams et al. (2017) on modeling a stable atmospheric boundary layer in a wind tunnel).

8.4.2 Flow over an array of buildings: street to neighborhood scales

Obviously, very few buildings are sufficiently isolated to generate the patterns of flow over a single cuboid overviewed in the previous subsection. Therefore, various studies have examined configurations of aligned or staggered cube arrays, or 2D ribs that are very elongated in the cross-stream direction, as idealizations of regular homogeneous neighborhoods. Nevertheless, an increasing number of studies are now considering real urban neighborhoods that feature the full complexity of urban form to make the findings more translatable to real cities and more relevant to urban design. The focus of studies at this scale is usually on (i) the general structure of the flow and turbulence, and the resulting mixing (dispersive and turbulent) between the UCL and RSL, (ii) the modeling of the velocity or scalar concentration profiles across the various sublayers, and their linkages to fluxes, to serve as flux-profile parametrizations in coarse models that cannot resolve buildings, and the related determination of the roughness lengths and displacement heights (refer to Chapter 1) of the urban canopy, and (iii) linking these flow dynamics to the morphometric characteristics of the canopy, with the ultimate goal of deriving general models that can ingest morphometric data and accurately predict flow information (increasingly using machine learning approaches). These three goals obviously intersect in many ways.

Two parameters that are necessary, but not sufficient, for characterizing the packing density of regular arrays of cuboids are the planar area ratio λ_p and the frontal area ratio λ_f (see Grimmond and Oke (1999) for one of the earlier reviews). The planar area ratio is defined as the ratio of the total built area (A_p, green surfaces in Fig. 8.5) to the ground area (or lot area, A_l, grey surface in Fig. 8.5), hence $\lambda_p = A_p/A_l$. It characterizes the total built fraction of the surface. The frontal area ratio, on the other hand, characterizes the propensity of the buildings to block the incoming flow, and it is thus defined as $\lambda_f = A_f/A_l$, where A_f is the area of the walls facing the wind (brown surfaces in Fig. 8.5), projected onto a plane normal to the wind. The same fractions can be used for 2D ribs that are quasi-infinite in the cross-stream direction. For such a setup, $\lambda_p = W_b/(W_b + W_s)$, where W_b is the streamwise width of the building and W_s is the streamwise width of the streets, and $\lambda_f = H_b/(W_b + W_s)$, with H_b denoting the characteristic height of the buildings (recall that previously we

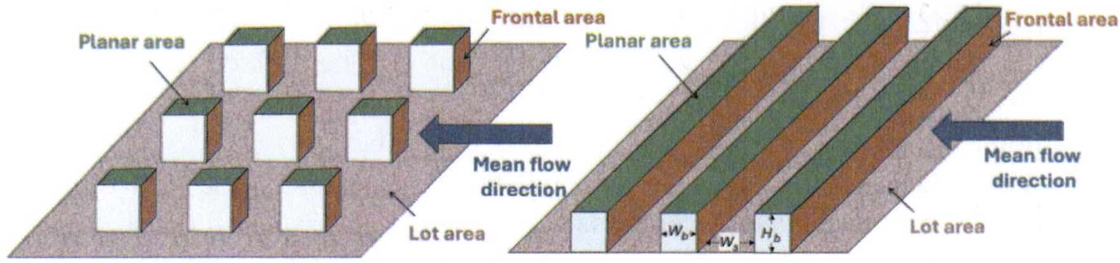

Figure 8.5 The idealized geometries consisting of aligned 3D cuboids (left) or elongated 2D ribs (right), with the planar, frontal, and lot areas depicted.

denoted the height of the tallest building as h, which is not necessarily equal to H_b). For buildings that change cross-section with height, and more generally for canopies with variable building heights, one can (and should) compute the profiles of λ_f and λ_p that now vary with height (e.g., Wang et al., 2025).

When $W_s >> H_b$ (buildings are far apart), the flow separated from the upstream building reattaches to the ground (this may take a separation $W_s > 4H_b$ (see Oke, 1988)) and the wake vortices weaken; therefore, the downstream building receives a nearly wall-parallel flow (albeit with stronger turbulence), and generates a similar wake to the upstream building. This regime is referred to as the isolated roughness flow in the urban meteorology literature due to the weak interactions between the elements (Oke et al., 2017). In the engineering fluid mechanics literature, it is referred to as k-type roughness because the scale of eddies shed by the element and the velocity profile both depend on the height of the roughness elements (Leonardi et al., 2007), denoted by k in that literature (here we use H_b). When $W_s \sim H_b$, the downstream building receives a complex 3D inflow that contains the wake and vortices of its upstream neighbor; the flow over the downstream building is thus very different from a single isolated building and depends on W_s/H_b. However, the flow behind the buildings re-enters the urban canopy layer as it begins to subside, and significant air mixing occurs near the building tops. This is known as the wake interference regime in the meteorological literature (Oke et al., 2017), and a transitional roughness regime (between k- and d-type roughness) in the engineering fluids literature (Leonardi et al., 2007). As the buildings move closer and $W_s << H_b$, the flow deflected from the upstream buildings has no time to re-enter the UCL, and the buildings start to act like a single block. The flow deflected by the first row of buildings skims above the downstream ones, while the air in between the buildings is recirculated, reducing mixing with the flow aloft. This regime is known as the skimming flow regime in urban meteorology (Oke et al., 2017), while in the engineering literature, it is labelled as the d-type roughness regime since the turbulence and the velocity profile both depend on the boundary layer depth, typically denoted as d in that literature (Leonardi et al., 2007). Fig. 8.6 depicts the typical urban geometric proportions where these three regimes are expected to occur.

While idealized, this classification clearly illustrates the importance of the ratio $W_s/H_b \propto (1 - \lambda_p)/\lambda_f$ in setting the mixing between the UCL and the RSL aloft. Poor mixing implies trapping of the pollutants and heat released in the UCL,

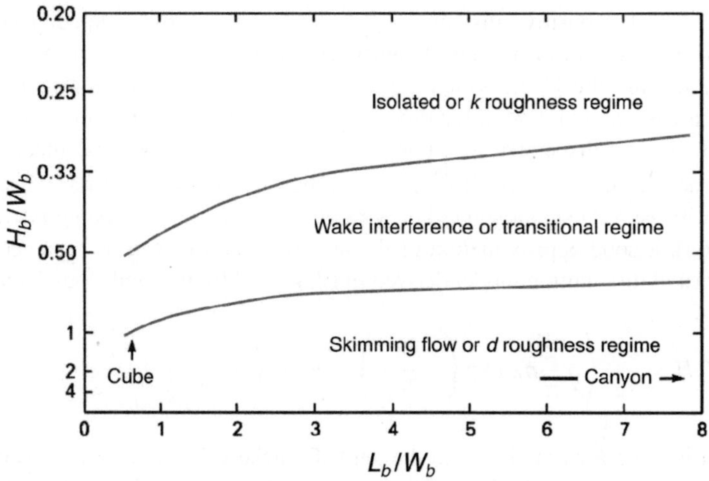

Figure 8.6 The different flow regimes in street canyons between buildings of cross-stream length L_b, adapted from Oke (1988), with the equivalent terms from the engineering literature added.

while enhanced ventilation is associated with improved air quality and environmental conditions (Britter and Hanna, 2003). The two $\lambda's$, though necessary, are clearly not sufficient to fully characterize the flow; a simple illustration is the difference between the flow over a regular and a staggered cube array that have an identical λ (Coceal et al., 2006). In addition, they have almost opposite impacts on the flow if they are increased or decreased independently (Li and Bou-Zeid, 2019). Recent work has investigated flow over real urban canopies and aimed to translate these findings from the idealized versions, potentially adding new geometric parameters to complement the λ parameters (Lu et al., 2023a; Wang et al., 2025). The skimming flow regime, for example, might be seldom observed in real canopies since it is easily disturbed by variability in the building height (Britter and Hanna, 2003), and the three-dimensionality of the flow. The exact details of any urban flow are highly sensitive to the form details, making generalizations nearly unattainable (see excellent reviews of the various flow patterns for different geometries in Zajic et al. (2011) and Oke et al. (2017)). However, general attributes and statistics may be linked to statistical geometric surface characteristics.

One key attribute is the air exchange rate (ACH) between the UCL and RSL, across an interface of area A, that quantifies the ventilation of street air for a given urban morphology. It can be computed exactly from LES or DNS data following:

$$ACH\left(m^3/s\right) = \underbrace{\iint_A \underbrace{\left(\frac{1}{T}\int_0^T w^+ dt\right)}_{\text{temporal averaging}} dA}_{\text{spatial integration}} = A\left\langle\overline{w^+}\right\rangle \tag{8.21}$$

where w^+ are the positive instantaneous values of the vertical velocity at a given vertical level (e.g., mean building height). This expression simply measures the air vertically leaving the UCL, which approximately equals the air vertically entering for large areas (the conditions for this assumption are identical to the ones yielding $\langle \overline{w} \rangle = 0$ for the dispersive averaging). If only the spatially local mean $(\overline{w})$ and standard deviation $\sigma_w = (\overline{w'^2})^{1/2}$ of the vertical velocity are available (e.g., from RANS or measured using a single hot-wire or sonic anemometer being moved along the canyon), a good approximation of the air exchange rate can be obtained using a folded normal distribution model derived in Llaguno-Munitxa and Bou-Zeid (2018):

$$ACH = \frac{A_c}{2} \left\langle \sqrt{\frac{2}{\pi}} \sigma_w \exp\left(-\frac{\overline{w}^2}{2\sigma_w^2}\right) - \overline{w} \operatorname{erf}\left(-\frac{\overline{w}}{\sqrt{2}\sigma_w}\right) \right\rangle \tag{8.22}$$

where erf is the error function, and the angled brackets denote intrinsic averaging as before. The first term on the RHS estimates the turbulent mass flux, while the second quantifies the dispersive one. These air exchange rates are often divided by the canyon (fluid) volume in the UCL to obtain a frequency, the inverse of which denotes the time it takes to exchange a full air volume once. Analyses of air exchange rates show that ventilation is sensitive to various morphometric properties of the urban canopy besides λ_p and λ_f, including roof shape and the presence of balconies (Llaguno-Munitxa and Bou-Zeid, 2018), as well as the variability of building heights and shapes in real cities (Wang et al., 2025).

While the analyses above require flow and turbulence data that resolve individual buildings, in many practical applications, a large domain size is more critical than a finer resolution (e.g., urban weather and climate modeling, or long-range air pollution transport in or out of cities). In the corresponding simulations, the grid cells are often ~ 1 km and cannot resolve individual buildings, calling for succinct parametrizations to couple the urban-land surface to the atmosphere that yield the correct drag, heat flux, and gas exchanges (e.g., H_2O and CO_2) averaged over the model grid cell (Grimmond et al., 2010; Lipson et al., 2024). At such grid scales, one can make the assumption that the averaging area is large enough to satisfy the dispersive averaging conditions detailed in a previous footnote and yield $\langle \overline{w} \rangle \approx 0$ and $\partial \langle \overline{\cdot} \rangle / \partial x \approx \partial \langle \overline{\cdot} \rangle / \partial y \approx 0$, but an implicit assumption is also that the canopy properties are statistically homogeneous over the grid cell, such that the mean flow being parameterized is in equilibrium with the underlying surface. A classic parametrization for the surface drag is to link it to the fluid velocity at the first grid level in the atmosphere using the logarithmic law (log law in short, modified via Monin–Obukhov similarity to account for buoyancy, as detailed in Chapter 1, but here we will ignore this correction in the equations for simplicity). For rough walls, this law links the z-varying, planar-temporally averaged streamwise velocity to the momentum flux encoded in the friction velocity u_* (here evaluated above the UCL) via

$$\frac{\langle \overline{u} \rangle}{u_*} = \frac{1}{\kappa} \ln\left(\frac{z-d}{z_0}\right) \tag{8.23}$$

where d is the displacement height and z_0 the roughness length for momentum, both of which depend on the canopy morphology and have been modeled using urban form parameters (Grimmond and Oke, 1999), though more accurate models are still being pursued. Simplistically, they should depend on the building height alone in k-type roughness conditions (far apart buildings) and on the building spacing alone in d-type roughness flows (very close packing of buildings). In reality, many cities may be in the transitional regime where both are important, and recent literature has also shown that d and z_0 also depend on building height variability (its variance and skewness (Zhu et al., 2016)). More broadly, for rough surfaces, these roughness parameters may also depend on stability. While Monin-Obukhov similarity theory is supposed to encode the effect of buoyancy independently from d and z_0, over rough walls, buoyancy may alter the mixing across the top of the canopy layer and the "effective surface" the flow interacts with (Zilitinkevich et al., 2008).

The log law is strictly only applicable above the RSL where the flow has been homogenized, and the stress (linked to u_*) is nearly constant. Inside the RSL, corrections have been introduced that modify the log law, averaged in space, using empirical functions of the variable $(z - d)/z_{RSL}$, where z_{RSL} is the depth of the roughness sublayer (Garratt, 1980), and they seem to improve the match with observed profiles over real urban canopies (see Theeuwes et al., 2019).

Inside the UCL, the velocity deviates from this logarithmic profile significantly, and is instead often modeled using an exponential profile of the form

$$\langle \overline{u} \rangle = \langle \overline{u} \rangle_{H_b} \exp\left[-a\left(1 - \frac{z}{H_b}\right)\right] \tag{8.24}$$

where $\langle \overline{u} \rangle_{H_b}$ is the averaged velocity at the building height H_b, and a is an empirical coefficient, though recent efforts have succeeded in linking it also to morphometric parameters (Li and Katul, 2022). The log law (or its RSL-modified version) and the exponential law can then be blended smoothly if a continuous velocity profile is needed, and the details of both profiles and their merger depend on canopy form and density. With increasing packing density, the flow transitions from a wall-bounded type flow to a mixing layer type flow (see Ghisalberti (2009); Li and Bou-Zeid (2019)).

For parametrizations in coarse atmospheric models, only the logarithmic part of the profile is needed. If one knows the velocity at the first grid level (from a previous time step), and the roughness parameters from the urban morphology, then the friction velocity can be computed from an inverted form of Eq. (8.23) and used to compute the "surface" stress needed as a boundary condition for the model. A very similar approach can also be used to obtain heat or other scalar fluxes. The mean concentrations of these scalars also obey a logarithmic law that depends on their surface fluxes (Brutsaert, 2005), and their values at the first grid level can be used to model these fluxes. However, the roughness parameters d and z_0 are in fact not identical for scalars and momentum, and they are the subject of active research (see, e.g., Li et al. (2020)). More advanced frameworks have been developed to allow the urban canopy to interact with elevated grid cells, which is especially useful in dense urban cores with tall buildings. These approaches represent the neighborhood buildings as a vertically distributed drag and heat source and blend the urban flow and turbulence closures of the coarse models (Santiago and Martilli, 2010).

8.4.3 Flow over a city: neighborhood to regional scales

Cities' influence is not restricted to flow modification within their boundaries. Their indirect impacts shape our entire interaction with the planet (for example, they are responsible for a disproportionate amount of greenhouse gas emissions, estimated in 2015 to be 62%–70% when their population share was < 55% (Moran et al., 2018; Gurney et al., 2022)). In addition, even their direct influence has substantial repercussions on regional flows, meteorology, and climate. Cities create plumes of pollution and heat that extend far downwind, emit aerosols, and deflect the regional flow to modify clouds and rainfall, and generate large-scale thermal circulations—the urban-rural breezes—that are akin to land-sea breezes and may interact with them in coastal cities (Ramamurthy et al., 2025). The city-scale fluid mechanics and turbulence are thus of importance to a wide set of applications that are no less critical than their building and neighborhood scale counterparts. At the city scale, researchers tend to be interested in the upward and sideways flow deflection due to the bluff roughness, the added buoyancy over the city that can generate large-scale convection and thermal circulations, and how the relative strength of the synoptic advection (regional wind speeds), compared to the differential urban versus rural heating, modifies the structure of the city-scale circulations. There are other interesting questions related to urban areas in complex terrain that we will not delve into here, but that have been reviewed authoritatively in Fernando (2010). We will also not review the interaction of urban flows with hydrometeorology and rainfall, but there is a wide literature on the topic (e.g., see Liu and Niyogi, 2019). Geographic complexity and interaction with rainfall have also been reviewed recently in the context of a changing global climate by Masson et al. (2020).

With similar buoyancy fluxes from the urban and surrounding terrain, a city generally behaves like a single large, porous bluff body. A small fraction of the incoming flow can enter the UCL of a dense city, and it quickly equilibrates with the new rough environment it finds itself in (Belcher et al., 2003). The aspect ratio (height of building over extent of the city) being small, most of the flow is deflected upward, rather than sideways. Downwind, the flow reattaches and merges with the flow exiting from the UCL to advect the urban emissions of heat and pollution downwind. This regime is depicted in the upper panel in Fig. 8.7, which shows the streamlines and streamwise velocity pseudocolors with strong incoming wind and similar urban and rural heat fluxes. With weak synoptic winds and large differences in urban and rural heating, we shift to a regime of thermal secondary circulation where the heated flow rises over the city and subsides over the surrounding rural areas, creating a closed bubble that can trap heat and pollutants in the urban boundary layer. This is depicted in the lower panel of Fig. 8.7. In between these two regimes, we get a range of transitional flows that have features of both the recirculating bubble and the advective urban plume (middle panel of Fig. 8.7). The shifting from one regime to the other can be predicted

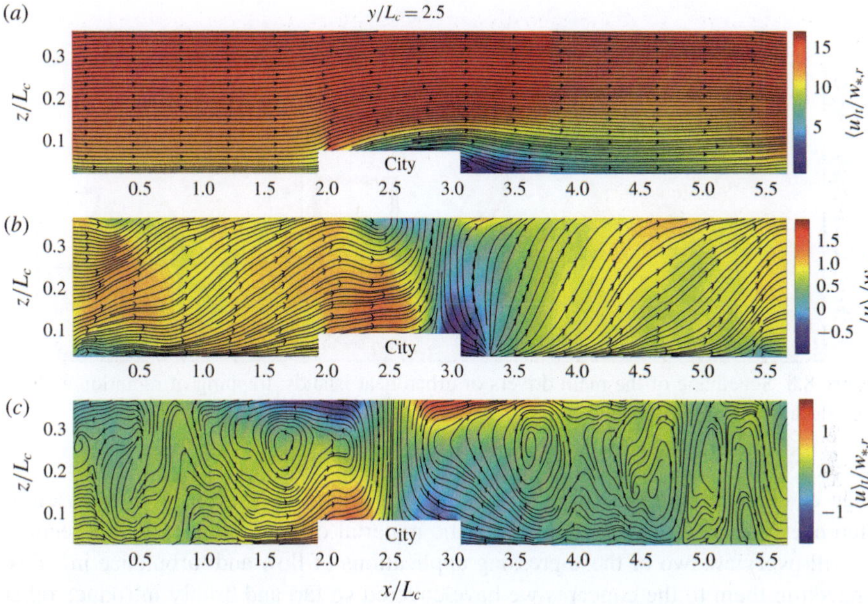

Figure 8.7 Pseudocolor plot of the mean streamwise velocity averaged in time $\langle u \rangle_t$ overlaid by streamlines in a vertical (x, z) plane (dimensions are normalized by the city size L_C) cutting through the center of an idealized city (from LES resolving the city blocks, but not the individual buildings). Top panel shows conditions with a weak heat flux difference and strong advecting wind. Middle panel displays a case where the heat flux contrast and incoming wind are comparable in impact. Bottom panel shows a case with weak wind and strong heating differential. The velocities are normalized by the convective velocity scale computed from the rural heat flux. Figure adapted from Omidvar et al. (2020).

using a dimensionless parameter that compares the large-scale advective velocity (e.g., characterized through the geostrophic wind speed M_g) to the convective velocity that is generated by the heat flux contrast between urban and rural terrain (e.g., estimated as $W_* = \sqrt{(g/\theta_r)\delta\Delta\theta_s}$, where δ is the urban boundary layer depth and $\Delta\theta_s$ is the difference between urban and rural effective surface temperatures (see Omidvar et al. (2020) for the urban-specific simulations and analysis, and Margairaz et al. (2020); Allouche et al. (2023); Fogarty and Bou-Zeid (2023) for more general discussions on the scaling of advection versus buoyancy for thermal secondary circulations). The ratio W_*^2/M_g^2 is equivalent to a "heterogeneity Richardson number".

8.5 Urban applications and implications of turbulent fluid mechanics

The concepts covered in this chapter are the necessary basics for research on the fundamentals of flow in urban terrain and other similar geophysical flows over very rough surfaces. However, they are by no means sufficient to cover the broad

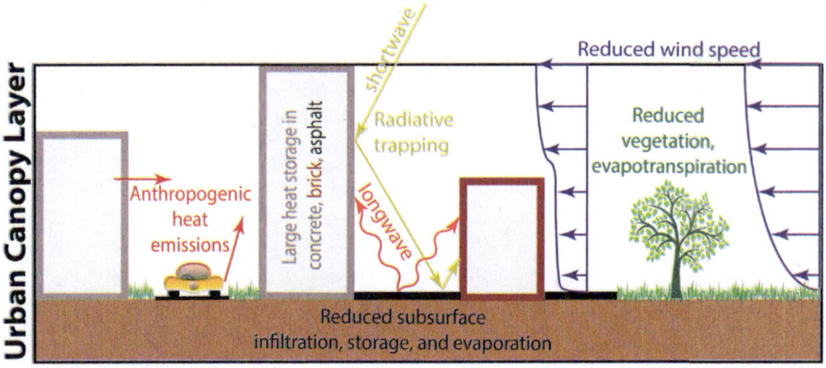

Figure 8.8 Schematic of the main drivers of urban heat islands: trapping of radiation and air, large storage and anthropogenic emissions of heat, and reduced evaporative cooling.

range of classical and emerging applications of urban fluid mechanics (the various references provided above complement the material of the chapter). In this section, we will overview two of these pressing applications of flow and turbulence in cities, connecting them to the concepts we have covered so far, and briefly introduce other pertinent turbulence dynamics that we cannot fully cover in this chapter.

8.5.1 Urban heat islands

As discussed above, cities tend to be hotter than their surroundings due to their (i) form, which traps radiation, sensible heat, and mass, (ii) fabric, which absorbs radiation due to darker surfaces (like asphalt with its low albedo) and stores heat more effectively (like concrete with its large thermal effusivity), and (iii) function, which increases anthropogenic heat emissions from combustion and heat pumps (Fig. 8.8). Fluid mechanics is central to the development of UHIs, as well as to their mitigation.

Starting at the finest urban element, the facet (e.g., a single roof, wall, or ground surface), turbulence plays a central role in controlling the heat and water vapor (evaporative) exchanges with the air aloft. To illustrate this, one can write a surface energy balance equation for an infinitesimally thin layer (so that its heat storage is ≈ 0) of solid material at the building-air interface:

$$Q_R = Q_S + Q_E + Q_G \tag{8.25}$$

where Q_R (defined positive into the interface) is the net radiative energy absorbed by the surface (equal to the sum of the downwelling solar shortwave and terrestrial longwave, minus the sum of the upwelling reflected shortwave and reflected and emitted longwave). All terms on the RHS are defined positive away from the interface, which is the conventional notation in geophysics. Q_S is the sensible heat flux that first diffuses across the diffusive sublayer and then is transported by turbulent eddies further away from the wall; Q_E is the latent heat consumed by evaporating water molecules (to break their hydrogen bonds) that again first diffuse and then are transported by turbulence; and Q_G is the heat flux conducted into the ground surface where it is stored or transmitted to the indoor spaces of buildings.

During the daytime, $Q_R > 0$ is high, leading to higher surface temperatures. The energy input from this radiation term is then redistributed into the terms on the RHS. Q_S is the sensible heating of the air that increases outdoor temperatures, while Q_G heats the solid materials and the interiors of buildings; the heat flux that goes indoors then determines the cooling load of air conditioning systems. If the surface is wet, $Q_E > 0$ dissipates some of the radiative energy into latent heat that does not increase air temperature, cooling the city; but it also raises relative humidity, which offsets some of the cooling benefits for thermal comfort and building cooling demand (Huang et al., 2025). At night, the terms reverse sign and the net radiative flux Q_R becomes negative. The surface is now cooling by longwave emissions, reversing the temperature gradients in the walls and leading to the release of the stored energy; $Q_G < 0$ becomes a main source of interface heating. Over natural materials, the surface often becomes cooler than the air, but that happens less frequently over urban materials due to their high thermal mass (Oke et al., 2017). So often Q_S remains positive, and heat continues to be released from the surface to the air, leading to a strong nighttime UHI, but also to mostly unstable stratifications (as discussed before) that mitigate inversions and the trapping of this heat (and air pollution). Q_E is often negligible during nighttime.

Turbulence directly modulates Q_S and Q_E. If the urban facets are rough (this can be viewed as a micro-roughness, compared to the macro-roughness at the building to city scale, as defined by Li et al. (2020)), they will exchange sensible heat more effectively with the air compared to smoother facets (higher Nusselt number). Rougher surfaces also enhance the upward mass flux of water vapor molecules, which is needed to dry the diffusive sublayer and sustain higher evaporation rates, boosting Q_E. Unlike momentum, for which pressure drag can dominate the exchange between the air and surface (and where viscous drag can be neglected as discussed above), heat and mass transfer from the urban surface are always dependent on molecular transport effectiveness near the wall (no equivalent to pressure drag). This implies that the Nusselt number (for heat) or Sherwood number (for mass) cannot become *Re* independent since viscous and molecular processes always modulate the exchange (Li et al., 2016). This has two important implications. First, not only the roughness, but also the very fine details of the texture of the facet and any protrusions or changes in the shape will have important impacts on heat (Q_S) and mass transfer (and hence Q_E). Second, while for forces and momentum laboratory experiments or lower *Re* DNS can be more easily translated into the real-world design of buildings and cities (the results become *Re*-invariant when the pressure drag dominates, but again one has to be careful with round shapes where the flow separation point can change with *Re* Llaguno-Munitxa et al. (2017)), for heat and mass transfer, lab and DNS results might not be representative of the scaled-up real world versions that have a higher *Re* (or equivalently, a higher Peclet number).

Zooming out now beyond the viscous and diffusive layers at the solid-fluid interface, to the building and UCL scales, one has to consider the complex mean and turbulence dynamics of urban flows to understand how the heat and mass extracted from a facet are going to be transported further aloft by turbulence. Each facet generates a developing boundary layer, while the mean flow features separation and recirculation zones, and is rarely steady and wall-parallel as in textbook versions of wall-bounded flow problems. Heat exchange is also often modulating the flow, and mixed convection regimes (which remain much less studied than their forced and natural counterparts, (Petukhov et al., 1988)) are omnipresent. However, heat flux from horizontal surfaces, the effects of which can be represented through Monin-Obukhov corrections, has a very different impact on the flow compared to heat flux from vertical facets. Buoyancy produces TKE and MKE in the vertical direction; over horizontal surfaces, this enhances the surface air exchanges. Over vertical walls, however, the produced TKE and MKE are parallel to the wall, and they enhance (or dampen) streamwise advection; this effect should not be represented via a Monin-Obukhov correction; more research is required to elucidate what, if any, correction to the log law is needed (Sutherland et al., 2015).

Zooming out even further to the neighborhood and RSL scales, the trapping of heat (and pollutants) in the UCL, or their effective ventilation to the RSL and then to the outer layer, depends on the upward transport by the dispersive and turbulent fluxes that were discussed in previous sections. Advection here also plays a role, with neighborhoods at the upwind edge of the city, or those downstream of large parks or water bodies, benefiting from cooler inflow. The evolution of the flow and turbulence at transitions from flat rural terrain to complex urban terrain, and vice versa, or at transitions between neighborhoods of different morphological characteristics, has been the subject of a few studies (Belcher et al., 2003; Bou-Zeid et al., 2009), but more research is needed as the urban climate community continues to increase the resolution at which it studies and models UHIs. How should models, for example, evolve as their resolution increases beyond the scales that satisfy the dispersive averaging conditions, but remain coarse for resolving a single building (this is a sort of an urban version of the *terra incognita* or gray zone challenge in computational fluid dynamics, where the resolution lies between the RANS- and LES-appropriate scales (Wyngaard, 2004)).

At the even larger scale of the city to metropolitan region, the aforementioned bubble and plume regimes can modulate the intensity of the urban heat island. Advection plays a key role in determining the thermal conditions over a city, and expanding urbanization around urban cores increases their temperatures (Zhang et al., 2009; Moustaoui and Georgescu, 2025). Additionally, urban-rural-sea breezes can modify the timing and intensity of extreme heat in cities and provide much-needed transport of heat away from the city, even when the synoptic flows are weak (Ramamurthy et al., 2017, 2025). The diurnal cycle of solar radiation can then modify the atmospheric stability and the mixing intensity across the entire urban boundary layer, but it can also induce inertial oscillations during nighttime, when weaker turbulence reduces the coupling of the flow to the surface. The resulting low-level jets can create intense turbulence that washes down onto the urban surface and can reduce the intensity of UHI, as shown by Hu et al. (2013). At this large scale, feedback also comes into play. For example, cities can act as topographic features that lift and also heat the flow, with buoyancy then further boosting these upward motions. Pollution from cities can then contribute cloud condensation nuclei, which act with the uplift of the flow to increase clouds (also see Chapter 13 about the role of turbulence in cloud formation and dynamics) and rainfall, which can then have a significant impact on Q_R and Q_E, and on urban hydrology more broadly.

8.5.2 Air quality in cities

Urban air pollution remains one of the most adverse impacts of anthropogenic activities on the health of citizens (WHO, 2024). Emissions of pollutants in cities are linked to urban function, but their fate and transport are strongly modulated by flow and turbulence (and other urban attributes, Fig. 8.9). All the discussions about the mixing and transport of heat from the previous subsection, and earlier in the chapter, apply to air pollutants as well. For example, pollutants tend to have higher concentrations on the lee side of buildings (see Fig. 8.2, also Oke et al. (2017)), and their ventilation from the UCL, where they are predominantly emitted, to the RSL depends on the turbulent and dispersive motions near the top of the buildings (Akinlabi et al., 2022; Wang et al., 2025). However, air exchange rate formulations derived earlier only compute the mass flux across a plane, including parcels that move up and then back down across that horizontal cross-section without much mixing. Such parcel motions do not actually ventilate the canyon, though they are counted in the ACH formulations (Llaguno-Munitxa and Bou-Zeid, 2018). This then requires a ventilation efficiency (as a multiplicative factor in front of the ACH), but there are very few studies that have attempted to quantify that efficiency. One study estimated that anywhere between 30% and 50% of the pollutants removed are actually re-entrained into the canyon (Liu et al., 2005). Another key driver of air pollution is the relative timing of emissions and TKE and DKE generation. All are strongest during the day and weakest at night, but the diurnal pattern of the emissions tends to dominate, often causing higher concentrations during daytime despite the enhanced mixing (Oke et al., 2017).

Turbulence and air pollutants interact in other ways as well. Gaseous pollutants (CO, SO_2, NO_x, VOCs, O_3) are transported in the urban boundary layer as chemically reactive but buoyantly passive tracers, but their reactions are modulated by urban turbulence. For example, NO_x and VOC react to produce O_3, in the presence of sunlight, but they are not all emitted primarily by the same sources (Oke et al., 2017) and need to be brought into contact, and into sunlight, by urban air flow. This is a slow process and causes ozone concentrations to peak well downwind of cities. Modeling this chemistry-turbulence interaction requires a comparison of their time scales. When mixing (or subgrid scale mixing for LES) is much faster than reaction (low Damköhler number), the transport of these gases is conservative, and they can be modeled using the advection-diffusion equation (ignoring reaction). When the reaction is much faster, a reaction term must be added, but is easy to model since one can assume that, at any

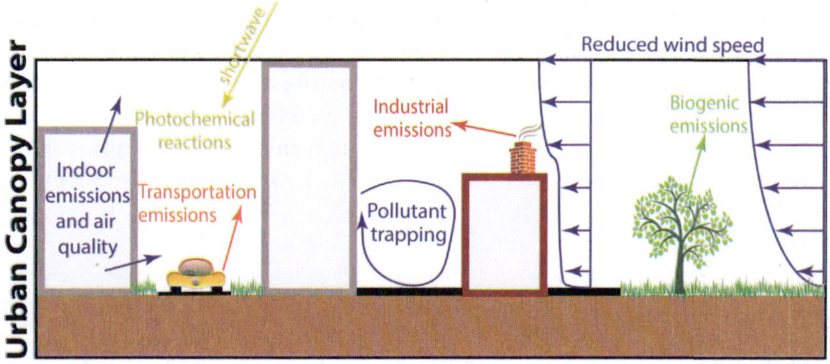

Figure 8.9 Schematic of the processes modulating urban air quality: emissions from anthropogenic and natural sources, trapping and reduced ventilation, interaction with the complex urban thermofluidic and radiative environment.

point in time and space, the reaction kinetics are in equilibrium, which then allows us to compute the concentration of each reactant and product. At intermediate Damköhler numbers, models need to account for partial reactions (Vinuesa et al., 2006).

Liquid or solid phase pollutants (aerosols) are even more sensitive to the turbulent and dispersive velocity fields. Aerosols tend to settle, and only turbulence or dispersive motions can keep them suspended in the air for a long enough time to allow transport over significant distances (Seinfeld and Pandis, 2016; Parolari et al., 2016). The Rouse number, which is the ratio of the settling velocity to the vertical velocity standard deviation (root-mean-square or rms velocity), can be used to characterize the settling efficiency. While in neutral wall-bounded flows the rms is often taken as κu_*, this is not an adequate estimate for complex urban flows with buoyancy, where the combination of turbulence and dispersive upward velocities is needed, and must include the thermal effects. This gravitational settling is also assisted by inertial impaction of the aerosols as they are advected by the wind into urban features, including trees and walls. Aerosols and gaseous pollutants are also removed from urban air by wet deposition, where they are absorbed by falling raindrops and then deposited onto the ground (Seinfeld and Pandis, 2016). This illustrates the complexity of the interactions between urban turbulence and its topographical, chemical, ecohydrological, and thermodynamic environments.

8.6 Challenges and opportunities

Cities are the interface between Earth and human-made systems, and modeling or observing their states or dynamics is undoubtedly complex (Alexander, 2015). However, urban populations (including cities and towns) are booming and will increase from about 6.4 billion in 2025 to nearly 7.8 billion in 2050 (UN Population Division, 2025); the area occupied by urban infrastructure will increase even more rapidly (Angel, 2012); and climate change will stress urban systems beyond historical precedents (Masson et al., 2020; González et al., 2021). Therefore, cities are the nexus of the most consequential challenges facing humanity. Our goal should thus be to make cities more resilient to natural disasters, more sustainable in their use of resources, and simply more pleasant to live in. Fluid mechanics and the turbulent flow of urban air are absolutely central to this endeavor.

As reviewed in the chapter, urban flow and turbulence are complicated by the multiscale imprints of urban form, fabric, and function on the flow and turbulent eddies, ranging from the micro-roughness of a facet to the scale of the whole city. Temporal variability also challenges our turbulence theories; flow dynamics can vary on the timescale of a car passing by, to the diurnal cycle scale of radiative heating, to the synoptic scale of weather variability ($\sim$ 1 week). Turbulence in the city may not be in quasi-equilibrium with the mean flow when the forcing and the mean are changing rapidly, and turbulence retains memory of previous flow states; this "stretches" almost all the simplifying theories presented in Chapter 1 and this chapter. The dual actions of shear, forced by the large-scale pressure gradient, and buoyancy, modulated by the surface energy budget and heat flux, require an understanding of mixed convection, a topic that has received less attention than it deserves in the geophysical and engineering literature alike (Petukhov et al., 1988). There is also an increasingly pressing need to examine urban turbulence under rare, but very consequential extreme conditions (hurricanes, tornadoes, heat waves, intense storms, etc., (Gorlé, 2025)).

In addition to these fundamental fluid mechanics challenges, there are more practical challenges in the way we design and conduct urban flow studies, as well as in the way we apply the knowledge in real cities. We have generated a relatively detailed understanding of the flow and turbulence over idealized geometries, but flows over real urban canopies are very different and can be highly sensitive to minor modifications in the shape or orientation of the canopies relative to the wind. What is required is a meta-analysis to understand what flow features over idealized geometries are persistent enough to be useful for urban design and planning (e.g., nonflat roofs tend to increase canyon ventilation, or building height variability reduces the risk of skimming flow and stagnation zones). Specificities of the flow and turbulence over idealized canopies are not of great value if they never occur in real canopies. Similarly, specificities over a given real canopy that are solely pertinent to that canopy are only useful in cases where, for example, one is planning for action after a hazardous chemical leak or is investigating the impact of a new high-rise building that is being constructed in that specific location. The urban fluid mechanics community needs to advance our predictive capability of flow and transport over real canopies by continuous improvement of the models and observations, and their fusion, because they have critical and valuable real-world applications, but we should avoid overanalyzing the fine details of flow and turbulence dynamics in these case studies since they are specific to that location. Here, too, a meta-analysis might help us understand which features of flow and turbulence over realistic canopies are common and worth investigating to develop a generalized understanding of their impacts.

Fortunately, our ability to model and observe flow in cities has advanced significantly to assist in addressing these many research challenges. Computational Fluid Dynamics (CFD) models for urban flows have made significant advances in their numerics (e.g., the development of GPU-based solvers that are able to attain resolutions that would have seemed to be decades away (Muñoz-Esparza et al., 2020)), as well as in their physics (e.g., the representation of key thermal and hydrological processes in widely used codes (Maronga et al., 2020)). At larger scales, many numerical weather prediction (NWP) models (and in particular the Weather Research and Forecasting (WRF) model) now have increasingly realistic parametrization of unresolved thermo-fluidic urban processes, including radiative trapping of heat and pollution in urban canyons and the vertical distribution of the urban drag force on the atmosphere (Chen et al., 2011; Santiago and Martilli, 2010). These CFD and NWP frameworks are also fusing together, either through GPU codes that can resolve the building to synoptic scales (Muñoz-Esparza et al., 2020) or through online or offline coupling of the models (Vogel et al., 2022). A lot of these advances are also in codes that are open-source or planned to become open source, allowing urban fluids researchers to accelerate their advances. Hurdles remain on this front, however. The long-standing challenge of wall modeling in LES is particularly important for urban flows, and we need advances in allowing these models to handle nonequilibrium and separated flows, flows with buoyancy over horizontal or vertical walls, heat and mass transfer at the full range of realistic Re, and facets with fine-scale unresolved features. Another challenge is to translate the knowledge we continue to gain from CFD on flow and turbulence into parametrization for NWP (and climate) models. Progress has been made on this front (Lu et al., 2024), but most urban canopy models still use simplistic schemes to represent flow, and all models ignore dispersive transport.

Observationally, sensor technologies and platforms have advanced significantly over the past two decades as well, and have also helped us probe urban flows and transport with unprecedented detail. Scanning Doppler lidars can capture the mean flow and turbulence and are now revealing, in real cities, some of the flow features that we have hypothesized or observed only in simulations and wind tunnels (Lamer et al., 2022). Mobile sensors and low-cost sensor networks, while usually not well-suited for measuring flow and turbulence, are increasingly able to map mean temperature and concentrations across the city, which can then indirectly reveal areas of stagnation or trapping. Uncrewed aerial vehicles (UAVs), also known as drones, have a transformative potential (Varentsov et al., 2023), but are now difficult to deploy in dense population centers due to logistical and safety considerations. However, if commercial fleets become more trusted and reliable, they can open the possibility of ubiquitous 3D measurements in cities. Finally, the biggest impacts have been from bringing these novel and more traditional sensors and platforms together in large coordinated field campaigns (a few recent examples include (Fenner et al., 2024; Kollias et al., 2025)) that have allowed us to understand the complex interaction of air flow and turbulence with the thermodynamics, weather dynamics, hydrometeorology, and land use variability. Observationally, challenges also remain. The extreme flow variability makes generalization from any urban measurement at specific locations or periods tenuous. UAVs may help in that regard, but all mobile sensors are moving in time and space (assuming they are not dedicated to urban observations) through a temporally and spatially variable field, making conventional averaging approaches inapplicable. We need novel ideas on how to analyze and interpret these measurements (e.g., two-point statistics with spatial and temporal lags across multiple UAVs could be utilized). Remote sensors (e.g., satellites) have also been widely used, but their view angle biases the facets they see and how they represent the city (Chang et al., 2025). Nevertheless, these many challenges should make urban fluid dynamics researchers excited and optimistic, as they arise from the explosion of new sensing technologies and their applications in cities, and they signal myriad new research possibilities ahead.

There are also opportunities at the interface of modeling and observations related, for example, to the assimilation of observational data into models, the design of experiments using simulations, or the fusion of observational and modeling data to produce more accurate and complete datasets for analysis (Dzambo et al., 2025). Machine learning (ML) can be a very useful tool for such data assimilation and fusion due to the ease with which it can ingest heterogeneous data in its analysis. But ML is also increasingly showing its skill in a whole range of urban fluid dynamics applications including: turbulence closure, fast emulation of physical models (surrogate CFD), downscaling (which is a specific form of emulation), rapid prediction of key flow features given arbitrary urban topographies, coherent structure detection, flow regime classification and switching detection, to name a few (Lu et al., 2023b, 2025; Hora and Giometto, 2024; Peng et al., 2025). One challenge that remains quite limiting for ML is its inability to simulate novel urban designs (form, fabric, and function) that lie beyond its training data. For example, it cannot yet predict how a new material with unusual thermal properties will affect heat flux, or how subtle form features that result in atypical flow separation locations would impact the flow. However, the case remains that artificial intelligence is opening myriad new possibilities and opportunities for urban turbulence researchers, and some of its current limitations may be overcome in the future.

However, maybe the biggest challenge for urban fluid dynamicists is how to translate our advancing knowledge into information that the average citizen can understand and use, and the urban planner and decision maker can build upon in their efforts to create sustainable, resilient, and liveable cities. This increasingly requires us to step out of our laminar zones of comfort and into the turbulent fields of interdisciplinary work, where we need to integrate our knowledge with that of health experts, economists, atmospheric chemists, and transportation planners, to name a few. But this requires us to synthesize the essence of what we know about flow in cities. We need to create a map of our knowledge and field that is legible to researchers in other fields, not only maps whose size and complexity are as large as our collective deep expertise.[3]

[3] On Exactitude in Science, a short story by Jorge Luis Borges, Collected Fictions, translated by Andrew Hurley.

In that Empire, the art of cartography attained such perfection that the map of a single province occupied the entirety of a city, and the map of the empire, the entirety of a province. In time, those unconscionable maps no longer satisfied, and the Cartographers Guilds struck a map of the Empire whose size was that of the Empire, and which coincided point for point with it. The following generations, who were not so fond of the study of cartography as their forebears had been, saw that that vast map was

Appendix A

Table 8.1 Comparison of urban, vegetation, and wind farm roughness.

	Urban Roughness	Vegetation Roughness	Wind Farm Roughness
Similarities	Wake production and spectral short-circuiting very active		
	Excite spatial variability and local non-equilibrium between turbulence and mean		
	Inflection point, mixing layer analogy at top of roughness (for wind farm, also below disks)		
	Canopy sublayer and roughness sublayer are of most interest		
	Require adjustment of log-law and Monin-Obukhov functions		
Differences	Impervious elements	Pervious (at branch scale)	Impervious elements
	Rigid (mostly) and immobile	Flexible and oscillating	Rigid and rotating (for floating offshore, flexible, rotating, and oscillating)
	Height variability can be large	Height variability usually moderate	Height variability very minor (in current designs)
	Organization mostly random, irregular, and difficult to characterize	Organization obeys eco-physics, can be clustered or homogeneous	Organization planned and known, but may irregular
	Variable solid density, highest at the ground	Low and distributed density, may peak away from ground	low density, peaks at hub height
	Drag coefficient perennial, varying minimally with wind speed, significantly with direction	Drag coefficient seasonal, decreasing moderately with wind speed, not varying with direction	Drag coefficient varies significantly with wind speed, at hourly scales, and with direction for irregular farms
	Pressure drag dominant	Pressure and viscous drag both relevant	Pressure drag dominant, linked to energy extraction
	High Bowen ratio	Low Bowen ratio	Bowen ratio varies depending on the underling surface
	Highly heterogeneous surface temperature due to contrasting surface material properties	Temperature varying with height, controlled by biophysical processes and vegetation structure	Depends on underlying surface heterogeneity, influenced by turbine mixing
	Heat flux and shading from buildings very important	Heat flux and shading from leaves important	Heat flux and shading from turbines of minimal importance

References

Akinlabi, E., Maronga, B., Giometto, M., Li, D., 2022. Dispersive fluxes within and over a real urban canopy: a large-eddy simulation study. Boundary-Layer Meteorology 185, 93–128.

Akinlabi, E.O., Giometto, M., Li, D., 2023. Budgets of second-order turbulence moments over a real urban canopy. Boundary-Layer Meteorology 188 (3), 351–387.

Alexander, C., 2015. A City Is Not a Tree, 50th anniversary edition. Sustasis Press, Portland, Oregon.

useless, and not without some pitilessness was it, that they delivered it up to the inclemencies of sun and winters. In the deserts of the west, still today, there are tattered ruins of that map, inhabited by animals and beggars; in all the land, there is no other relic of the disciplines of geography...

Allouche, M., Bou-Zeid, E., Iipponen, J., 2023. The influence of synoptic wind on land–sea breezes. Quarterly Journal of the Royal Meteorological Society 149 (757), 3198–3219.

Angel, S., 2012. Planet of Cities. Lincoln Institute of Land Policy, Cambridge, MA.

Belcher, S.E., 2005. Mixing and transport in urban areas. Philosophical Transactions of the Royal Society A: Mathematical, Physical and Engineering Sciences 363 (1837), 2947–2968.

Belcher, S.E., Jerram, N., Hunt, J.C.R., 2003. Adjustment of a turbulent boundary layer to a canopy of roughness elements. Journal of Fluid Mechanics 488, 369–398.

Blocken, B., Stathopoulos, T., Van Beeck, J., 2016. Pedestrian-level wind conditions around buildings: review of wind-tunnel and CFD techniques and their accuracy for wind comfort assessment. Building and Environment 100, 50–81.

Bou-Zeid, E., Gao, X., Ansorge, C., Katul, G.G., 2018. On the role of return to isotropy in wall-bounded turbulent flows with buoyancy. Journal of Fluid Mechanics 856, 61–78.

Bou-Zeid, E., Overney, J., Rogers, B.D., Parlange, M.B., 2009. The effects of building representation and clustering in large-eddy simulations of flows in urban canopies. Boundary-Layer Meteorology 132, 415–436.

Britter, R.E., Hanna, S.R., 2003. Flow and dispersion in urban areas. Annual Review of Fluid Mechanics 35 (1), 469–496.

Brutsaert, W., 2005. Hydrology. Cambridge University Press.

Castro, I.P., Robins, A.G., 1977. The flow around a surface-mounted cube in uniform and turbulent streams. Journal of Fluid Mechanics 79 (2), 307–335.

Chang, Y., Weng, Q., Voogt, J.A., Xiao, J., 2025. Urban thermal anisotropies by local climate zones: an assessment using multi-angle land surface temperatures from ECOSTRESS. Remote Sensing of Environment 322, 114705.

Chen, F., Kusaka, H., Bornstein, R., Ching, J., Grimmond, C., Grossman-Clarke, S., Loridan, T., Manning, K., Martilli, A., Miao, S., 2011. The integrated WRF/urban modelling system: development, evaluation, and applications to urban environmental problems. International Journal of Climatology 31, 273–288.

Chen, X., Yang, J., Zhu, R., Wong, M.S., Ren, C., 2021. Spatiotemporal impact of vehicle heat on urban thermal environment: a case study in Hong Kong. Building and Environment 205, 108224.

Coceal, O., Thomas, T.G., Castro, I.P., Belcher, S.E., 2006. Mean flow and turbulence statistics over groups of urban-like cubical obstacles. Boundary-Layer Meteorology 121 (3), 491–519.

Dzambo, A.M., Bruning, E., Oue, M., Brunner, K., Singewald, D., Rosky, E., Allwayin, N., Patil, S., Qiao, Y.J., Xia, Z.H., Hu, Y., Lebo, Z.J., Huang, Y., McFarquhar, G.M., Kollias, P., Souza, J., Miller, M., Weiss, S., Ascher, B., 2025. Forecasting for ESCAPE: a multi-institution hybrid forecasting and nowcasting operation for sea-breeze convection supporting a ground-based and airborne field campaign. Bulletin of the American Meteorological Society 106 (3), E456–E472.

Fenner, D., Christen, A., Grimmond, S., Meier, F., Morrison, W., Zeeman, M., Barlow, J., Birkmann, J., Blunn, L., Chrysoulakis, N., Clements, M., Glazer, R., Hertwig, D., Kotthaus, S., König, K., Looschelders, D., Mitraka, Z., Poursanidis, D., Tsirantonakis, D., Bechtel, B., Benjamin, K., Beyrich, F., Briegel, F., Feigel, G., Gertsen, C., Iqbal, N., Kittner, J., Lean, H., Liu, Y., Luo, Z., McGrory, M., Metzger, S., Paskin, M., Ravan, M., Ruhtz, T., Saunders, B., Scherer, D., Smith, S.T., Stretton, M., Trachte, K., Van Hove, M., 2024. Urbisphere-Berlin campaign: investigating multiscale urban impacts on the atmospheric boundary layer. Bulletin of the American Meteorological Society 105 (10), E1929–E1961.

Fernando, H., 2010. Fluid dynamics of urban atmospheres in complex terrain. Annual Review of Fluid Mechanics 42 (1), 365–389.

Fernando, H., Lee, S., Anderson, J., Princevac, M., Pardyjak, E., Grossman-Clarke, S., 2001. Urban fluid mechanics: air circulation and contaminant dispersion in cities. Environmental Fluid Mechanics 1 (1), 107–164.

Fogarty, J., Bou-Zeid, E., 2023. The atmospheric boundary layer above the marginal ice zone: scaling, surface fluxes, and secondary circulations. Boundary-Layer Meteorology 189 (1–3), 53–76.

Garratt, J.R., 1980. Surface influence upon vertical profiles in the atmospheric near-surface layer. Quarterly Journal of the Royal Meteorological Society 106 (450), 803–819.

Ghannam, K., Bou-Zeid, E., 2021. Baroclinicity and directional shear explain departures from the logarithmic wind profile. Quarterly Journal of the Royal Meteorological Society 147 (734), 443–464.

Ghannam, K., Poggi, D., Porporato, A., Katul, G.G., 2015. The spatio-temporal statistical structure and ergodic behaviour of scalar turbulence within a rod canopy. Boundary-Layer Meteorology 157 (3), 447–460.

Ghisalberti, M., 2009. Obstructed shear flows: similarities across systems and scales. Journal of Fluid Mechanics 641, 51–61.

Gianfelice, M., Aboshosha, H., Ghazal, T., 2022. Real-time wind predictions for safe drone flights in Toronto. Results in Engineering 15, 100534.

Giometto, M., Christen, A., Meneveau, C., Fang, J., Krafczyk, M., Parlange, M., 2016. Spatial characteristics of roughness sublayer mean flow and turbulence over a realistic urban surface. Boundary-Layer Meteorology 160, 425–452.

Giometto, M.G., Christen, A., Egli, P.E., Schmid, M.F., Tooke, R., Coops, N., Parlange, M.B., 2017. Effects of trees on mean wind, turbulence and momentum exchange within and above a real urban environment. Advances in Water Resources 106, 154–168.

González, J.E., Ramamurthy, P., Bornstein, R.D., Chen, F., Bou-Zeid, E.R., Ghandehari, M., Luvall, J., Mitra, C., Niyogi, D., 2021. Urban climate and resiliency: a synthesis report of state of the art and future research directions. Urban Climate 38, 100858.

Gorlé, C., 2025. Urban fluid mechanics, resilience, and sustainability. Annual Review of Fluid Mechanics 58, 193–219.

Grimmond, C.S.B., Blackett, M., Best, M.J., Barlow, J., Baik, J.-J., Belcher, S.E., Bohnenstengel, S.I., Calmet, I., Chen, F., Dandou, A., Fortuniak, K., Gouvea, M.L., Hamdi, R., Hendry, M., Kawai, T., Kawamoto, Y., Kondo, H., Krayenhoff, E.S., Lee, S.-H., Loridan, T., Martilli, A., Masson, V., Miao, S., Oleson, K., Pigeon, G., Porson, A., Ryu, Y.-H., Salamanca, F., Shashua-Bar, L., Steeneveld, G.-J., Tombrou, M., Voogt, J., Young, D., Zhang, N., 2010. The international urban energy balance models comparison project: first results from phase 1. Journal of Applied Meteorology and Climatology 49 (6), 1268–1292.

Grimmond, C.S.B., Oke, T.R., 1999. Aerodynamic properties of urban areas derived from analysis of surface form. Journal of Applied Meteorology 38 (9), 1262–1292.

Grylls, T., van Reeuwijk, M., 2022. How trees affect urban air quality: it depends on the source. Atmospheric Environment 290, 119275.

Gurney, K.R., Kılkış, Ş., Seto, K.C., Lwasa, S., Moran, D., Riahi, K., Keller, M., Rayner, P., Luqman, M., 2022. Greenhouse gas emissions from global cities under SSP/RCP scenarios, 1990 to 2100. Global Environmental Change 73, 102478.

Holmes, J.D., Paton, C., Kerwin, R., 2007. Wind Loading of Structures. CRC Press.

Hong, J., Katz, J., Meneveau, C., Schultz, M.P., 2012. Coherent structures and associated subgrid-scale energy transfer in a rough-wall turbulent channel flow. Journal of Fluid Mechanics 712, 92–128.

Hora, G.S., Giometto, M.G., 2024. Surrogate modeling of urban boundary layer flows. Physics of Fluids 36 (7), 076625.

Hu, X.-M., Klein, P.M., Xue, M., Lundquist, J.K., Zhang, F., Qi, Y., 2013. Impact of low-level jets on the nocturnal urban heat island intensity in Oklahoma City. Journal of Applied Meteorology and Climatology 52 (8), 1779–1802.

Huang, X., Bou-Zeid, E., Vanos, J.K., Middel, A., Ramamurthy, P., 2025. Urban heat mitigation through misting, and its role in broader blue infrastructure portfolios. Landscape and Urban Planning 256, 105290.

Inagaki, A., Kanda, M., Ahmad, N.H., Yagi, A., Onodera, N., Aoki, T., 2017. A numerical study of turbulence statistics and the structure of a spatially developing boundary layer over a realistic urban geometry. Boundary-Layer Meteorology 164 (2), 161–181.

Jiménez, J., 2004. Turbulent flows over rough walls. Annual Review of Fluid Mechanics 36 (1), 173–196.

Kadivar, M., Tormey, D., McGranaghan, G., 2021. A review on turbulent flow over rough surfaces: fundamentals and theories. International Journal of Thermofluids 10, 100077.

Katul, G., Cava, D., Poggi, D., Albertson, J., Mahrt, L., 2004. Stationarity, Homogeneity, and Ergodicity in Canopy Turbulence. Kluwer Academic Publishers, pp. 161–180.

Klein, P.K., Berkowicz, R., Plate, E., 2000. Modelling of vehicle-induced turbulence in air pollution studies for streets. International Journal of Environment and Pollution 14 (1-6), 496.

Kollias, P., McFarquhar, G.M., Bruning, E., DeMott, P.J., Kumjian, M.R., Lawson, P., Lebo, Z., Logan, T., Lombardo, K., Oue, M., Roberts, G., Shaw, R.A., van den Heever, S.C., Wolde, M., Barry, K.R., Bodine, D., Bruintjes, R., Chandrasekar, V., Dzambo, A., Hill, T.C.J., Jensen, M., Junyent, F., Kreidenweis, S.M., Lamer, K., Luke, E., Bansemer, A., McCluskey, C., Nichman, L., Nguyen, C., Patnaude, R.J., Perkins, R.J., Powers, H., Ranjbar, K., Roux, E., Snyder, J., Treserras, B.P., Tsai, P., Wales, N.A., Wolf, C., Allwayin, N., Ascher, B., Barr, J., Hu, Y., Huang, Y., Litzmann, M., Mages, Z., McKeown, K., Patil, S., Rosky, E., Tuftedal, K., Tzeng, M.-D., Zhu, Z., 2025. Experiment of sea breeze convection, aerosols, precipitation, and environment (escape). Bulletin of the American Meteorological Society 106 (2), E310–E332.

Lamer, K., Luke, E.P., Walsh, B., Andrade, S., Mages, Z., Zhu, Z., Leghart, E., Treserras, B.P., Emrick, A., Kollias, P., Vogelmann, A., Schoonen, M., 2022. Going mobile to address emerging climate equity needs in the heterogeneous urban environment. Bulletin of the American Meteorological Society 103 (9), E2069–E2080.

Leonardi, S., Orlandi, P., Antonia, R.A., 2007. Properties of d- and k-type roughness in a turbulent channel flow. Physics of Fluids 19 (12), 125101.

Li, F., Wang, R., Chen, G., Mo, Z., Ikegaya, N., Liu, C.-H., 2025. Turbulence scale and strength analysis in the roughness and inertial sublayers over urban areas: a wind tunnel study. Building and Environment 267, 112193.

Li, Q., Bou-Zeid, E., 2019. Contrasts between momentum and scalar transport over very rough surfaces. Journal of Fluid Mechanics 880, 32–58.

Li, Q., Bou-Zeid, E., Anderson, W., Grimmond, S., Hultmark, M., 2016. Quality and reliability of LES of convective scalar transfer at high Reynolds numbers. International Journal of Heat and Mass Transfer 102, 959–970.

Li, Q., Bou-Zeid, E., Grimmond, S., Zilitinkevich, S., Katul, G., 2020. Revisiting the relation between momentum and scalar roughness lengths of urban surfaces. Quarterly Journal of the Royal Meteorological Society 146 (732), 3144–3164.

Li, Q., Katul, G., 2022. Bridging the urban canopy sublayer to aerodynamic parameters of the atmospheric surface layer. Boundary-Layer Meteorology 185 (1), 35–61.

Li, W., Giometto, M.G., 2023. Mean flow and turbulence in unsteady canopy layers. Journal of Fluid Mechanics 974, A33.
Li, W., Giometto, M.G., 2024. The structure of turbulence in unsteady flow over urban canopies. Journal of Fluid Mechanics 985, A5.
Lipson, M.J., Grimmond, S., Best, M., Abramowitz, G., Coutts, A., Tapper, N., Baik, J., Beyers, M., Blunn, L., Boussetta, S., Bou-Zeid, E., De Kauwe, M.G., de Munck, C., Demuzere, M., Fatichi, S., Fortuniak, K., Han, B., Hendry, M.A., Kikegawa, Y., Kondo, H., Lee, D., Lee, S., Lemonsu, A., Machado, T., Manoli, G., Martilli, A., Masson, V., McNorton, J., Meili, N., Meyer, D., Nice, K.A., Oleson, K.W., Park, S., Roth, M., Schoetter, R., Simón-Moral, A., Steeneveld, G., Sun, T., Takane, Y., Thatcher, M., Tsiringakis, A., Varentsov, M., Wang, C., Wang, Z., Pitman, A.J., 2024. Evaluation of 30 urban land surface models in the URBAN-PLUMBER project: phase 1 results. Quarterly Journal of the Royal Meteorological Society 150 (758), 126–169.
Liu, C., Leung, D., Barth, M., 2005. On the prediction of air and pollutant exchange rates in street canyons of different aspect ratios using large-eddy simulation. Atmospheric Environment 39 (9), 1567–1574.
Liu, J., Niyogi, D., 2019. Meta-analysis of urbanization impact on rainfall modification. Scientific Reports 9 (1), 7301.
Llaguno-Munitxa, M., Bou-Zeid, E., 2018. Shaping buildings to promote street ventilation: a large-eddy simulation study. Urban Climate 26, 76–94.
Llaguno-Munitxa, M., Bou-Zeid, E., Hultmark, M., 2017. The influence of building geometry on street canyon air flow: validation of large eddy simulations against wind tunnel experiments. Journal of Wind Engineering and Industrial Aerodynamics 165, 115–130.
Lu, J., Li, W., Hobeichi, S., Azad, S.A., Nazarian, N., 2025. Machine learning predicts pedestrian wind flow from urban morphology and prevailing wind direction. Environmental Research Letters 20 (5), 054006.
Lu, J., Nazarian, N., Hart, M.A., Krayenhoff, E.S., Martilli, A., 2023a. Novel geometric parameters for assessing flow over realistic versus idealized urban arrays. Journal of Advances in Modeling Earth Systems 15 (7), e2022MS003287.
Lu, J., Nazarian, N., Hart, M.A., Krayenhoff, E.S., Martilli, A., 2024. A one-dimensional urban flow model with an eddy-diffusivity mass-flux (EDMF) scheme and refined turbulent transport (MLUCM v3.0). Geoscientific Model Development 17 (7), 2525–2545.
Lu, Y., Zhou, X.-H., Xiao, H., Li, Q., 2023b. Using machine learning to predict urban canopy flows for land surface modeling. Geophysical Research Letters 50 (1), e2022GL102313.
Mahrt, L., Bou-Zeid, E., 2020. Non-stationary boundary layers. Boundary-Layer Meteorology 177 (2–3), 189–204.
Margairaz, F., Pardyjak, E.R., Calaf, M., 2020. Surface thermal heterogeneities and the atmospheric boundary layer: the thermal heterogeneity parameter. Boundary-Layer Meteorology 177 (1), 49–68.
Maronga, B., Banzhaf, S., Burmeister, C., Esch, T., Forkel, R., Fröhlich, D., Fuka, V., Gehrke, K.F., Geletič, J., Giersch, S., Gronemeier, T., Groß, G., Heldens, W., Hellsten, A., Hoffmann, F., Inagaki, A., Kadasch, E., Kanani-Sühring, F., Ketelsen, K., Khan, B.A., Knigge, C., Knoop, H., Krč, P., Kurppa, M., Maamari, H., Matzarakis, A., Mauder, M., Pallasch, M., Pavlik, D., Pfafferott, J., Resler, J., Rissmann, S., Russo, E., Salim, M., Schrempf, M., Schwenkel, J., Seckmeyer, G., Schubert, S., Sühring, M., von Tils, R., Vollmer, L., Ward, S., Witha, B., Wurps, H., Zeidler, J., Raasch, S., 2020. Overview of the PALM model system 6.0. Geoscientific Model Development 13 (3), 1335–1372.
Martinuzzi, R., AbuOmar, M., Savory, E., 2007. Scaling of the wall pressure field around surface-mounted pyramids and other bluff bodies. Journal of Fluids Engineering 129 (9), 1147–1156.

Masson, V., Lemonsu, A., Hidalgo, J., Voogt, J., 2020. Urban climates and climate change. Annual Review of Environment and Resources 45 (1), 411–444.

Meinders, E.R., Hanjalic, K., Martinuzzi, R.J., 1999. Experimental study of the local convection heat transfer from a wall-mounted cube in turbulent channel flow. Journal of Heat Transfer 121 (3), 564–573.

Momen, M., Bou-Zeid, E., 2017a. Analytical reduced models for the non-stationary diabatic atmospheric boundary layer. Boundary-Layer Meteorology 164 (3), 383–399.

Momen, M., Bou-Zeid, E., 2017b. Mean and turbulence dynamics in unsteady Ekman boundary layers. Journal of Fluid Mechanics 816, 209–242.

Momen, M., Bou-Zeid, E., Parlange, M.B., Giometto, M., 2018. Modulation of mean wind and turbulence in the atmospheric boundary layer by baroclinicity. Journal of the Atmospheric Sciences 75 (11), 3797–3821.

Moran, D., Kanemoto, K., Jiborn, M., Wood, R., Többen, J., Seto, K.C., 2018. Carbon footprints of 13,000 cities. Environmental Research Letters 13 (6), 064041.

Moustaoui, M., Georgescu, M., 2025. Winds of change: the role of urban expansion and thermal advection in driving Phoenix's (AZ) warming trends. Journal of Geophysical Research. Atmospheres 130 (16), e2024JD043166.

Muñoz-Esparza, D., Sauer, J.A., Shin, H.H., Sharman, R., Kosović, B., Meech, S., García-Sánchez, C., Steiner, M., Knievel, J., Pinto, J., Swerdlin, S., 2020. Inclusion of building-resolving capabilities into the FASTEDDY® GPU-LES model using an immersed body force method. Journal of Advances in Modeling Earth Systems 12 (11).

Nikora, V., McEwan, I., McLean, S., Coleman, S., Pokrajac, D., Walters, R., 2007. Double-averaging concept for rough-bed open-channel and overland flows: theoretical background. Journal of Hydraulic Engineering 133 (8), 873–883.

Oke, T., 1988. Street design and urban canopy layer climate. Energy and Buildings 11 (1–3), 103–113.

Oke, T.R., Mills, G., Christen, A., Voogt, J.A., 2017. Urban Climates. Cambridge University Press.

Omidvar, H., Bou-Zeid, E., Li, Q., Mellado, J.-P., Klein, P., 2020. Plume or bubble? Mixed-convection flow regimes and city-scale circulations. Journal of Fluid Mechanics 897, A5.

Parolari, A.J., Li, D., Bou-Zeid, E., Katul, G.G., Assouline, S., 2016. Climate, not conflict, explains extreme Middle East dust storm. Environmental Research Letters 11 (11), 114013.

Patil, A., García-Sánchez, C., 2025. Quantifying the impact of urban geometric detail for urban air mobility risk forecasting. Sustainable Cities and Society 132, 106750.

Peng, K., Xin, J., Zhu, X., Cao, X., Wang, Z., Ma, Y., Zhao, D., Ren, X., 2025. A novel 3d physics-integrated swin-transformer model for precise high-resolution urban boundary layer wind speed estimation. Geophysical Research Letters 52 (11), e2025GL115246.

Petukhov, B.S., Poleiakov, A.F., Launder, B.E., 1988. Heat Transfer in Turbulent Mixed Convection. Taylor & Francis, Philadelphia, PA.

Ramamurthy, P., Li, D., Bou-Zeid, E., 2017. High-resolution simulation of heatwave events in New York City. Theoretical and Applied Climatology 128 (1–2), 89–102.

Ramamurthy, P., Pena, J.C., Liu, S., Gonzalez-Cruz, J., 2025. The interaction between urban heat island intensity and sea-breeze effect. Philosophical Transactions of the Royal Society A: Mathematical, Physical and Engineering Sciences 383 (2308).

Raupach, M.R., Shaw, R.H., 1982. Averaging procedures for flow within vegetation canopies. Boundary-Layer Meteorology 22 (1), 79–90.

Ricci, A., Kalkman, I., Blocken, B., Burlando, M., Freda, A., Repetto, M., 2017. Local-scale forcing effects on wind flows in an urban environment: impact of geometrical simplifications. Journal of Wind Engineering and Industrial Aerodynamics 170, 238–255.

Roth, M., 2000. Review of atmospheric turbulence over cities. Quarterly Journal of the Royal Meteorological Society 126 (564), 941–990.

Santiago, J.L., Martilli, A., 2010. A dynamic urban canopy parameterization for mesoscale models based on computational fluid dynamics Reynolds-averaged Navier–Stokes microscale simulations. Boundary-Layer Meteorology 137 (3), 417–439.

Sathe, A., Anderson, W., Calaf, M., Giometto, M., 2025. On the structure and dynamics of secondary flows over multi-column roughness in turbulent boundary layers. arXiv preprint. arXiv:2508.05889.

Sathe, A.S., Giometto, M.G., 2024. Impact of the numerical domain on turbulent flow statistics: scalings and considerations for canopy flows. Journal of Fluid Mechanics 979, A36.

Schmid, M.F., Lawrence, G.A., Parlange, M.B., Giometto, M.G., 2019. Volume averaging for urban canopies. Boundary-Layer Meteorology 173 (3), 349–372.

Seinfeld, J.H., Pandis, S.N., 2016. Atmospheric Chemistry and Physics: From Air Pollution to Climate Change, 3rd edition. John Wiley & Sons.

Shah, S., Bou-Zeid, E., 2014. Very-large-scale motions in the atmospheric boundary layer educed by snapshot proper orthogonal decomposition. Boundary-Layer Meteorology 153 (3), 355–387.

Smits, A.J., McKeon, B.J., Marusic, I., 2011. High–Reynolds number wall turbulence. Annual Review of Fluid Mechanics 43 (1), 353–375.

Solazzo, E., Cai, X., Vardoulakis, S., 2008. Modelling wind flow and vehicle-induced turbulence in urban streets. Atmospheric Environment 42 (20), 4918–4931.

Spiegel, E.A., Veronis, G., 1960. On the Boussinesq approximation for a compressible fluid. The Astrophysical Journal 131, 442.

Stull, R.B., 1988. An Introduction to Boundary Layer Meteorology. Springer, Netherlands.

Sutherland, D., Chung, D., Ooi, A., Bou-Zeid, E., 2015. The law-of-the-wall in mixed convection flow in a vertical channel. In: APS Division of Fluid Dynamics Meeting Abstracts, APS Meeting Abstracts, p. A19.009.

Theeuwes, N.E., Ronda, R.J., Harman, I.N., Christen, A., Grimmond, C.S.B., 2019. Parametrizing horizontally averaged wind and temperature profiles in the urban roughness sublayer. Boundary-Layer Meteorology 173 (3), 321–348.

Torres, J.F., Ghanadi, F., Nock, I., Arjomandi, M., Pye, J., 2018. Mixed convection around a tilted cuboid with an isothermal sidewall at moderate Reynolds numbers. International Journal of Heat and Mass Transfer 119, 418–432.

Tseng, Y.-H., Meneveau, C., Parlange, M.B., 2006. Modeling flow around bluff bodies and predicting urban dispersion using large eddy simulation. Environmental Science & Technology 40 (8), 2653–2662.

UN Population Division, 2025. World Urbanization Prospects 2025. Population Studies. United Nations, Department of Economic and Social Affairs. Population Division, New York, NY.

Varentsov, M., Konstantinov, P., Repina, I., Artamonov, A., Pechkin, A., Soromotin, A., Esau, I., Baklanov, A., 2023. Observations of the urban boundary layer in a cold climate city. Urban Climate 47, 101351.

Vinuesa, J.F., Porté-Agel, F., Basu, S., Stoll, R., 2006. Subgrid-scale modeling of reacting scalar fluxes in large-eddy simulations of atmospheric boundary layers. Environmental Fluid Mechanics 6 (2), 115–131.

Vogel, J., Afshari, A., Chockalingam, G., Stadler, S., 2022. Evaluation of a novel WRF/PALM-4U coupling scheme incorporating a roughness-corrected surface layer representation. Urban Climate 46, 101311.

Wang, J., Llaguno-Munitxa, M., Li, Q., Giometto, M., Bou-Zeid, E., 2025. Wind extremes over built terrain: characterization and geometric determinants. Boundary-Layer Meteorology 191 (2), 11.

Wang, L., Liu, J., Gao, Z., Li, Y., Huang, M., Fan, S., Zhang, X., Yang, Y., Miao, S., Zou, H., Sun, Y., Chen, Y., Yang, T., 2019. Vertical observations of the atmospheric boundary layer structure over Beijing urban area during air pollution episodes. Atmospheric Chemistry and Physics 19 (10), 6949–6967.

Wang, Z., Bou-Zeid, E., Smith, J.A., 2013. A coupled energy transport and hydrological model for urban canopies evaluated using a wireless sensor network. Quarterly Journal of the Royal Meteorological Society 139 (675), 1643–1657.

Whitaker, S., 1999. The Method of Volume Averaging. Springer, Netherlands.

WHO, 2024. Sustainable development goal indicator 3.9.1: mortality attributed to air pollution. Technical report. World Health Organization.

Williams, O., Hohman, T., Van Buren, T., Bou-Zeid, E., Smits, A.J., 2017. The effect of stable thermal stratification on turbulent boundary layer statistics. Journal of Fluid Mechanics 812, 1039–1075.

Wolman, A., 1965. The metabolism of cities. Scientific American 213 (3), 178–190.

Wyngaard, J., 1992. Atmospheric turbulence. Annual Review of Fluid Mechanics 24 (1), 205–233.

Wyngaard, J.C., 2004. Toward numerical modeling in the “terra incognita”. Journal of the Atmospheric Sciences 61 (14), 1816–1826.

Yang, J., Anderson, W., 2018. Numerical study of turbulent channel flow over surfaces with variable spanwise heterogeneities: topographically driven secondary flows affect outer-layer similarity of turbulent length scales. Flow, Turbulence and Combustion 100 (1), 1–17.

Yi, Y.R., Koseff, J.R., Bou-Zeid, E., 2025. Nonlinear and buoyancy pressure correlations in stably stratified turbulence. Journal of Fluid Mechanics 1019, A14.

Zajic, D., Fernando, H.J.S., Calhoun, R., Princevac, M., Brown, M.J., Pardyjak, E.R., 2011. Flow and turbulence in an urban canyon. Journal of Applied Meteorology and Climatology 50 (1), 203–223.

Zhang, D., Shou, Y., Dickerson, R.R., 2009. Upstream urbanization exacerbates urban heat island effects. Geophysical Research Letters 36 (24), L24401.

Zhu, X., Iungo, G.V., Leonardi, S., Anderson, W., 2016. Parametric study of urban-like topographic statistical moments relevant to a priori modelling of bulk aerodynamic parameters. Boundary-Layer Meteorology 162 (2), 231–253.

Zilitinkevich, S.S., Mammarella, I., Baklanov, A.A., Joffre, S.M., 2008. The effect of stratification on the aerodynamic roughness length and displacement height. Boundary-Layer Meteorology 129 (2), 179–190.

Turbulence in plant canopies

9

Sylvain Dupont
INRAE, Bordeaux Sciences Agro, ISPA, Villenave d'Ornon, France

9.1 Introduction

Plants cover a large fraction of the Earth's surface, creating a porous layer between the ground surface and the atmosphere called the canopy. Compared to bare surfaces, vegetated surfaces absorb a larger fraction of incoming solar radiation due to their lower albedo (darker surface); they transpire water vapor during photosynthesis, transferring water from the deep soil to the atmosphere; they sequester a significant amount of carbon through photosynthesis; and they emit volatile organic compounds during abiotic stress. These plant features have strong impacts on the climate and the atmospheric composition (Bonan, 2008).

Tall plant canopies also alter the wind dynamics and turbulence by absorbing momentum and by emitting or absorbing heat throughout the canopy depth. In contrast, bare or low-roughness surfaces only absorb or emit momentum and heat at the ground surface. Such differences with low-roughness surfaces give rise to specific canopy-scale turbulent motions that penetrate the entire canopy and drive all biosphere-atmosphere exchanges, including dispersal of biotic (pollens, seeds, spores, and others) and abiotic (nutrients, pollutants, and others) particles released in plant canopies.

Understanding turbulence in plant canopies is therefore essential to properly represent all biosphere-atmosphere exchanges in weather and climate models. It is also important in many environmental applications where the wind flow around vegetation plays a key role, such as in evaluating new agricultural practices and their consequences on the environment, biodiversity, and natural resources, particularly in terms of dispersion of contaminants, pollutants, genes, and nutrients, in terms of wind stability of forests, or wildfire propagation. The current context of climate change, with the potential for increased extreme weather events such as windstorms, heat-waves, droughts, and forest fires, emphasizes the need to understand turbulence in plant canopies to mitigate their impacts on agriculture, forestry, and biodiversity. Although climate change corresponds to a long-term evolution in the atmosphere's dynamics, its impacts and consequences are felt locally through the occurrence of extreme events with serious environmental, health, and economic consequences.

For these reasons, turbulence in plant canopies has been the subject of significant research over the last few decades. Several reviews have been published on the subject (Raupach, 1981; Finnigan, 2000; Belcher et al., 2012; Patton and Finnigan, 2012; Brunet, 2020). This chapter provides an overview of turbulence in tall plant canopies and how it interacts and evolves with the main turbulent motions of the atmospheric boundary layer, depending on thermal stability. This description of vegetation-canopy flow will be compared with turbulence above surfaces with low roughness elements

Environmental Turbulence. https://doi.org/10.1016/B978-0-32-395873-8.00016-4
Copyright © 2026 Elsevier Inc. All rights are reserved, including those for text and data mining, AI training, and similar technologies.

(bare soils and grassy surfaces), referred to as surface-shear flows, to highlight the differences in turbulence between the two types of surfaces.

In the following, we describe first the main turbulence properties and coherent motions of the atmospheric boundary layer (ABL) and how they evolve with thermal stability (Section 9.2). We address in Section 9.3 the turbulence in plant canopies, focusing first on the wind dynamics according to the thermal stability and its sensitivity to plant motions, then we focus on scalar turbulence, and on the limit of applying similarity theory to near-surface flows. Section 9.4 discusses the influence of large-scale atmospheric motions on the canopy turbulence. A brief description of the main sources of investigation of turbulence in plant canopies is presented in Section 9.5, followed by two examples of applications, one related to wind damage in forests and the other to particle dispersal. Finally, in Section 9.6, we offer a perspective on the major challenges facing research on turbulence in plant canopies.

9.2 Turbulence in the atmospheric boundary layer

9.2.1 ABL diurnal cycle

The ABL is a region of continuous turbulence located at the base of the troposphere (~1–2 km). This turbulent layer contrasts with the stably stratified, nonturbulent free troposphere that extends above it. Within the ABL, the flow near the surface is mainly influenced by the surface friction and heating (or cooling), which produce mechanical and buoyant turbulence, respectively (Wyngaard, 2010, see also Chapter 2 of this book). This region corresponds to the atmospheric surface layer (ASL), extending over about ~50–100 m. Above, the flow is mainly influenced by surface heating and the Earth's rotation, resulting in a dominant buoyancy-driven turbulence. The character of the ABL turbulent motions depends on the predominant turbulence production mechanism, whether it is mechanical or buoyant. The Obukhov length L (Obukhov, 1946, see also Chapter 1) informs about the height above which the buoyant production (or destruction) of turbulent kinetic energy dominates the mechanical production. Hence, the ratio of the main eddy length scale to L is often used as a measure of thermal stratification in the lower atmosphere to group main flow characteristics and turbulent motions.

The ABL is in constant adjustment between different forcing, the surface radiative heating/cooling, the near-surface wind shear, and the entrainment of air from the free troposphere, to which secondary forcings add, such as advection, atmospheric radiation, and others (Dupont and Patton, 2022). The surface radiative forcing leads to a diurnal cycle in the ABL and its thermal stratification, characterized by the development of a mixed layer above the ASL after sunrise, as the ASL instability increases with surface warming (Fig. 9.1). This mixed layer is dominated by large-scale convective motions, spreading throughout the whole ABL, carrying heat, gases, and particles emitted at the surface, and bringing toward the surface larger momentum from the upper ABL. At the end of the day, after sunset, a stable boundary layer develops at the surface (see Chapter 3) beneath the remaining mixed layer, as the surface cools radiatively more quickly than the atmosphere aloft. Meanwhile, the

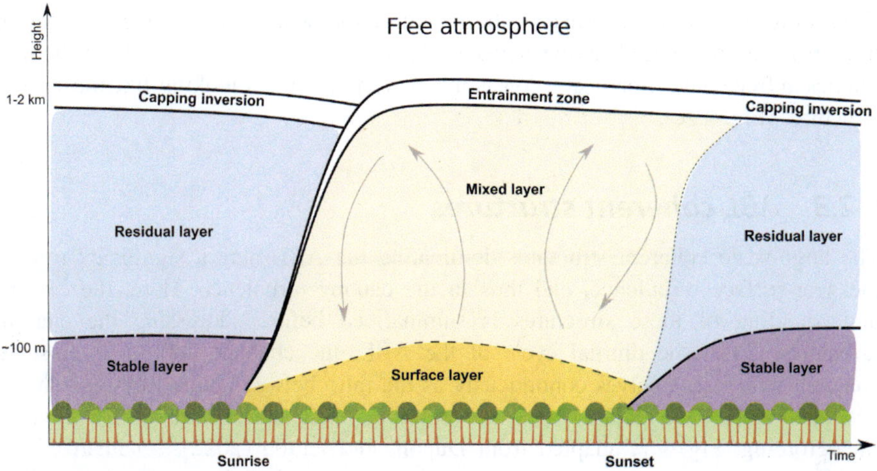

Figure 9.1 Idealized schematic of the ABL diurnal cycle. Adapted from Stull (1988) with permission from Kluwer Academic Publishers © 2023.

turbulence within the daytime remaining mixed layer, referred to as the residual layer, decays in the absence of additional production mechanisms.

9.2.2 ABL turbulence

The turbulence in the ABL manifests as high-frequency fluctuations (from milliseconds to several minutes) in all meteorological quantities (velocity, temperature, water vapor, and trace gases). These fluctuations are related to the three-dimensional eddies (disturbances) populating the flow, with sizes ranging from millimeters, corresponding to the dissipation scale, to several kilometers, approximately equal to the ABL depth (Smith, 1975; Katul, 2019). These eddies superpose on each other and interact with each other and with the mean flow, transferring and mixing meteorological quantities (Smith, 1975; Stull, 1988).

Despite the apparent chaotic motions of the atmospheric flow, some organized structures emerge as in laboratory turbulent flows, the so-called coherent structures, with scales that are significantly larger than those of the background eddies. These coherent structures are key elements for understanding the dynamics of turbulence in the ABL, while the background smaller eddies are more involved in processes of energy transfer to smaller eddies and in small-scale turbulence intermittency. These coherent structures carry most of the energy of the flow, defining the large scale, or integral scale, of the well-known turbulence eddy cascade from which the energy continually cascades down to smaller and smaller scales up to the viscous dissipation scale, where the viscosity damps out the small eddies. Coherent structures also play a significant role in the turbulent transport of momentum, heat, and mass within the ABL, as illustrated in the next section. Examples of such coherent structures are (1) hairpin vortices in near-wall turbulence that are composed of two legs with opposite signs of the streamwise vortices, a head with a spanwise vorticity, and an ejection motion in the low-speed region between the two legs (Fig. 9.2a), and (2) thermal plumes in free convection flows, consisting of warm rising air looking-like mushroom structures (Fig. 9.2c).

9.2.3 ABL coherent structures

The large-scale coherent structures dominating the ABL play a significant role in the near-surface turbulence, and thus in the canopy turbulence. Here, the current understanding of these structures is summarized before addressing the canopy turbulence. With the diurnal cycle of the ABL, the character of the large-scale coherent structures evolves continuously as the ratio between buoyancy- and shear-generated turbulence changes with the variations of the surface radiative and wind-shear forcing. Fig. 9.2, adapted from Dupont and Patton (2022), schematizes the evolution of these large-scale coherent structures and their interactions according to the thermal stratification: (a) the daytime near-neutral regime characterized by significant wind speed and small positive surface heat flux, (b) the forced convection regime, intermediate between the daytime near-neutral and the free convection regimes, occurring preferentially in the afternoon, (c) the free convection regime occurring preferentially in the morning, with the concomitant large heat flux from the surface and low wind speed, (d) the nocturnal near-neutral to stable regime corresponding to the weakly stable regime developing at night, and (e) the very stable regime occurring preferentially before sunrise under weak geostrophic wind forcing and clear sky.

- In *daytime near-neutral* regime (Fig. 9.2a), the ASL turbulence is dominated by shear-driven turbulence, the surface heat flux being small. This dominant wind shear is suspected to produce embedded coherent motions in the ASL: (1) packets of head-up hairpin vortices, known as large-scale motions (LSM), that scale with the ABL depth, and (2) very large-scale elongated meandering regions of low- and high-speed streaks, aligning the LSMs, known as very large-scale motions (VLSM), that scale with more than 10 times the ABL depth (Kim and Adrian, 1999; Hutchins and Marusic, 2007; Marusic and Monty, 2019). Above the ASL, the mixed-layer is dominated by longitudinal convective rolls aligned along the mean wind direction (Lemone, 1973; Moeng and Sullivan, 1994). These convective rolls are suspected to be somehow linked to the ASL shear-driven motions. Khanna and Brasseur (1998) suggested that buoyancy-driven upwelling motions, emanating from the warm surface along the ASL low-speed streaks, where the wind shear is weak, initiate the formation of these convective rolls. Other authors suggested that convective rolls may be triggered by some mechanisms, such as an inflection point in the wind profile of the Ekman layer, gravity waves in the free atmosphere, among others (Atkinson and Zhang, 1996; Young et al., 2002).
- In *free convection* regime (Fig. 9.2c), the ASL is dominated by rising plumes of warm air from the surface that merge in the mixed layer into narrow and

intense large-scale thermals. These thermals are organized into open cells with large but weaker downdrafts in the middle. Near the surface, the wind field exhibits underneath the large-scale updrafts and downdrafts, convergence and divergence regions, with low and high velocity, respectively.

- In *forced convection* regime (Fig. 9.2b), hairpin vortices with a larger upward tilt angle than in the near-neutral regime dominate the ASL. This upward tilt angle suggests a progressive transitioning with increasing instability of these shear-driven motions to buoyantly-driven motions (plumes), implying also a progressive transitioning of the convective rolls to thermals in the above mixed layer (Salesky and Anderson, 2018). In Fig. 9.2b, this regime is simply represented as a mix of leftover rolls and rising thermals.
- In *nocturnal near-neutral to stable* regime (Fig. 9.2d), the turbulent structures are similar to those during the daytime near-neutral regime, but the ASL turbulent

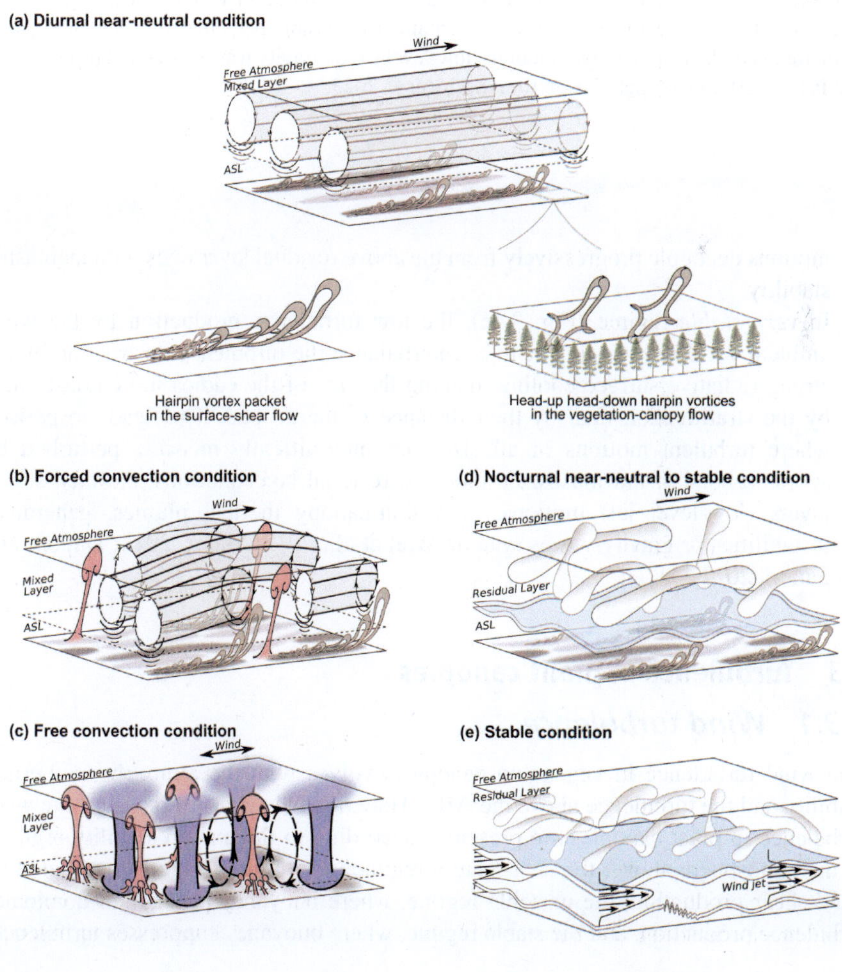

Figure 9.2 Idealized schematic of the main coherent motions present in the atmospheric boundary layer according to the thermal stratification. (a) Diurnal near-neutral conditions are characterized by longitudinal high- and low-speed streaks (VLSMs) in the ASL, combined with convective rolls within the mixed layer above. Near the surface (second row of figures), head-up hairpin vortex packets (LSMs) dominate above low-roughness surfaces (surface-shear flow) while head-up head-down hairpin vortices dominate the turbulence at canopy top (canopy flow). (b) Forced convection is intermediate between diurnal near-neutral (a) and free convection (c) conditions, with an increasing inclination of the ASL hairpin vortices toward the vertical before ultimately forming plumes that disturb the convective rolls. (c) Free convection conditions are characterized near the surface by rising plumes of warm air that merge into large-scale thermals (updrafts), compensated by large-scale downdrafts of cool air. Near the surface, patches of local wind shear form between the convergence/divergence regions induced by these large-scale updrafts and downdrafts. (d) Nocturnal near-neutral to stable conditions are characterized by a progressive decoupling with increasing stability between the near-surface turbulence and the above residual layer, reducing the influence of ABL-scale motions on near-surface turbulence, and thus reducing the size of the longitudinal streaks (VLSMs). (e) A stable condition is characterized by a decoupling between the near-surface layer and the above decaying residual layer. Near the surface, the turbulence is intermittently missing, perpetuated by sporadic bursts from, for example, residual-layer motions, low-level jet, or gravity waves. Adapted from Dupont and Patton (2022) with permission from Springer © 2023.

motions decouple progressively from the above residual layer ones with increasing stability.

- In *very stable* regime (Fig. 9.2e), the low turbulence production by the weak ambient wind shear is unable to counterbalance the turbulence destruction by the strong radiative surface cooling, making the size of the eddies more constrained by the stratification than by their distance to the ground. This leads to periods where turbulent motions of all sizes are intermittently missing, perturbed by sporadic bursts from mesoscale motions, residual-layer motions, elevated-shear-layers (low-level jet) motions, or within-canopy thermal plumes, generating instabilities or gravity waves (van de Wiel et al., 2012; Mahrt, 2014; Dupont and Patton, 2022).

9.3 Turbulence in plant canopies

9.3.1 Wind turbulence

The wind turbulence in vegetation canopies evolves with the atmospheric thermal stability and the turbulence above the ABL. Here the main characteristics of the wind turbulence in plant canopies are presented according to three main stability regimes of the atmosphere above: the near-neutral regime, where turbulence is dominated by wind-shear production, the unstable regime, where buoyancy production dominates turbulence production, and the stable regime, where buoyancy suppresses turbulence.

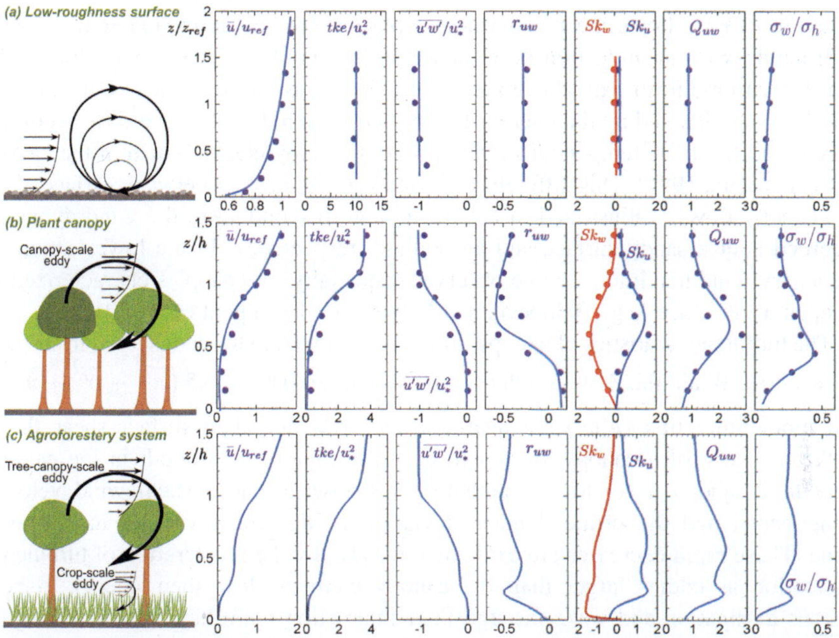

Figure 9.3 Main characteristics of the statistical wind turbulence vertical profiles over a low-roughness surface (a), a plant canopy (b), and an agroforestry system (c). From left to right: the mean horizontal velocity ($\overline{u}$), the turbulence kinetic energy (tke), the momentum flux ($\overline{u'w'}$), the correlation coefficient for momentum (r_{uw}), the skewnesses of the streamwise and vertical velocity components (Sk_u and Sk_w), the contribution fraction of sweep and ejection motions to the momentum flux (Q_{uw}), and the ratio between the vertical and horizontal velocity variances (σ_w/σ_h). The overbars are time averages in (a) and (b) and space-time averages in (c), and the primes are deviations from the averages. The left schematics summarize the main turbulent motions involved in each surface configuration. Figures in (a) result from the WIND-O-V's 2017 field experiment over a sandy bare surface (Dupont et al., 2018b), in (b) from the CHATS field experiment over an orchard (Patton et al., 2011; Dupont and Patton, 2012a), and in (c) from a numerical simulation (Dupont et al., 2022). Dots are measurement points.

9.3.1.1 Near-neutral regime

For conditions with significant wind in the presence of vegetation, flow momentum is absorbed not only at the ground surface, as in surface-shear flows, but also throughout the entire canopy depth by canopy elements such as foliage, branches, and trunk. This displaces the mean height at which momentum is absorbed from the ground surface in surface-shear flows to a location called the displacement height z_d in canopy flows (Kaimal and Finnigan, 1994). This momentum absorption is mainly explained by the form drag of canopy elements due to the pressure difference between the upwind and downwind faces of each canopy element, and to a lesser extent by the viscous drag induced by the thin viscous boundary layers developing on each canopy element. Hence, unlike the logarithmic mean velocity profile $\overline{u}(z)$ observed in surface-shear flows, in canopy flows the mean velocity profile exhibits an exponential decrease with depth, with an inflection point and a strong wind shear at canopy top, and a near logarithmic form only above the canopy. Here z is the height above the ground, and the mean denoted by the overbar is obtained from a double space-time averaging as required by the heterogeneous distribution of canopy elements within the canopy (Raupach et al., 1986). This difference of wind velocity profiles between surface-shear and canopy flows is illustrated in Fig. 9.3a–b, corresponding to the wind dynamics recorded over a sandy surface and an orchard, respectively. In the lower canopy, a secondary wind maximum can be observed, especially for canopies characterized by an open trunk space (e.g., Baldocchi and Meyers, 1988; Dupont et al., 2011).

The turbulence statistics of canopy flows also differ from those observed in surface-shear flows. While the flow turbulence kinetic energy ($tke = 0.5\left(\overline{u'^2} + \overline{v'^2} + \overline{w'^2}\right)$) and momentum flux ($\overline{u'w'}$) are near-constant with height in surface-shear flows, in the presence of a canopy, they both decrease rapidly with depth in the canopy. Here u, v, and w refer to the streamwise, spanwise, and vertical wind velocity components, and the prime denotes deviation of the quantity from its averaged value. These rapid decreases are explained, for tke, by the acceleration of turbulence dissipation as eddies larger than the canopy elements lose their energy to both viscous dissipation and wake through their interaction with the vegetation, and for $\overline{u'w'}$ by the momentum absorption by the vegetation (Fig. 9.3a–b). The efficiency of turbulence at transporting momentum is also maximum near the canopy top, and larger than in surface-shear flows. This is reflected by the correlation coefficient for momentum r_{uw} ($= \overline{u'w'}/\sqrt{\overline{u'^2}\,\overline{w'^2}}$), reaching -0.50 near canopy top against -0.25 in absence of canopy (Fig. 9.3a–b). Near the canopy top, the horizontal and vertical wind velocity fluctuations exhibit skewed distributions with the streamwise velocity skewness Sk_u around 1.3 and the vertical velocity skewness Sk_w around -1.2, while these distributions are symmetrical and near-Gaussian in surface-shear flows (Fig. 9.3a–b). Furthermore, the transfer of momentum at canopy top is dominated by sweep motions (downward fast motions) over ejection motions (upward slow motions), conversely to an equal contribution of both motions in surface-shear flows (Q_{uw} in Fig. 9.3a–b). Overall, the positive Sk_u and negative Sk_w, the domination of sweep motions in momentum transfer, and the higher value of $-r_{uw}$ at canopy top, suggest the presence of intermittent, energetic downward-moving gusts (sweep motions) that transport momentum efficiently.

These energetic gusts are related to the wind profile inflection point located near the canopy top, where the vertical gradient of u is maximum. This point often coincides with the canopy top, where the foliage density terminates abruptly. The wind profile inflection point plays a crucial role in canopy turbulence. It is at the origin of the development of Kelvin–Helmholtz instabilities, which produce canopy coherent eddy structures similar to coherent structures developing in mixing-layer flows resulting from the interaction between a slower and a faster flow (Raupach et al., 1996). Here, the slower flow is located within the canopy and the faster flow above the canopy. The inflection point is characterized by a shear length scale

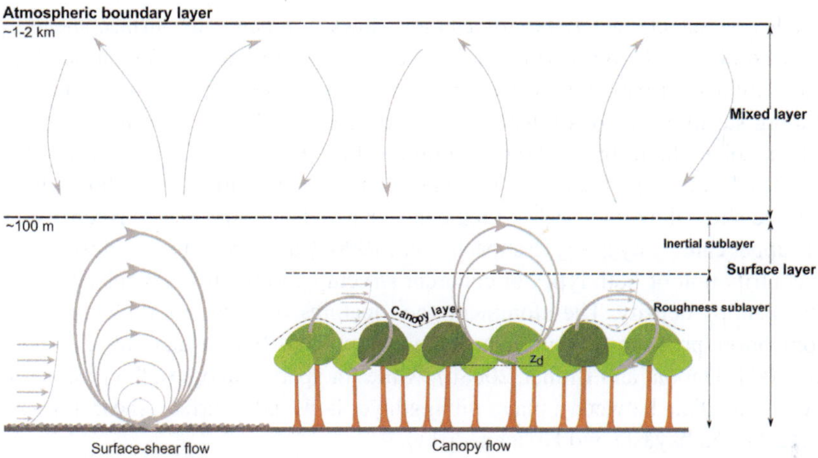

Figure 9.4 Schematic description of the atmospheric surface layer decomposition for surface-shear and canopy flows in near-neutral thermal stratification. In surface-shear flows, the turbulence is dominated by eddies scaling with the distance to the ground (on the left), while in canopy flows, the turbulence is dominated by mixing-layer type eddies (on the right), adding to the surface-shear flow eddies scaling with the distance to the displacement height z_d.

$L_s = \overline{u}(z = h)/(d\overline{u}(z = h)/dz)$, where h is the canopy height; L_s is approximately half the vorticity thickness. This latter quantity defines the scale of the canopy coherent structures that are responsible for the specific turbulence statistic properties of canopy flows mentioned previously.

These canopy mixing-layer type structures have been identified as downstream head-up ejection-generating hairpin vortices, paired with upstream head-down sweep-generating hairpin vortices (Finnigan et al., 2009, see also Fig. 9.2a). The head-up structure is stretched upward by the flow, ensuring a progressive transition toward the surface-shear flow above, while the downward progression of the head-down structures is constrained by both the canopy drag and the surface (Fig. 9.2a). The cycles of strong sweeps and weak ejections associated with this pair of head-up and head-down hairpin vortices are responsible for a substantial part of the turbulent transfer of momentum, heat, and mass between the canopy and the atmosphere. The mechanism by which canopy-scale coherent structures transition to surface-shear flow structures with increasing height above the canopy remains poorly constrained.

The region of the flow under the influence of these canopy coherent structures is defined as the roughness sublayer (RSL) (Fig. 9.4). In this sublayer, the flow is still under the influence of individual roughness elements of the canopy, inducing a horizontal variability of the mean flow. The RSL depth is about two to five times the roughness element height, including the canopy (Raupach et al., 1991). Above is the inertial sublayer (ISL) where the mean flow has equilibrated with the turbulence produced by the surface roughness elements. There, the flow is horizontally homogeneous and has similar characteristics to surface-shear flows.

Such distinct characteristics of canopy flows compared to surface-shear flows have been observed over a wide range of canopies with vastly different heights and vertical foliage distributions (Finnigan, 2000), and over different wind intensities, including storm conditions (Dupont et al., 2018a). This shows that in a majority of plant canopies, the turbulent flow is dominated by plane-mixing-layer type coherent structures developing at canopy top. These coherent structures add to those observed in surface-shear flows that scale with the distance to the ground, which is replaced here in the upper canopy by $z - z_d$ (Raupach et al., 1996; Katul and Chang, 1999) (Fig. 9.4). The contribution of both types of coherent structures to the flow turbulence depends on the canopy density. The aforementioned features of canopy turbulence typically become more pronounced with increasing canopy density (Novak et al., 2000; Poggi et al., 2004; Dupont and Brunet, 2008b), while for sparse canopies, the flow becomes more intermediate between a plane-mixing layer flow and a surface-shear flow (Poggi et al., 2004; Kobayashi and Hiyama, 2011).

The flow in the lower canopy shows less universal characteristics as it depends on the upper foliage vertical distribution (Dupont and Brunet, 2008b). The wake structures developing behind canopy elements accelerate the dissipation of turbulence, as observed from the exponential decrease of the turbulent kinetic energy profile within the canopy. This may also translate into a spectral short-cut of the large eddies into small ones without following the conventional picture of a continuous cascade, but with a steeper slope or with a secondary peak at high frequency (Finnigan, 2000). Specific smaller-scale turbulent structures may superpose on the canopy-scale coherent structures for particular morphologies of the subcanopy, such as in canopies with sparse trunk space, where vortices are shed behind each trunk of the canopy, similar to classical von Kármán vortex streets developing behind cylinders. This translates into a secondary peak in the inertial subrange region of the canopy wind spectra (Cava and Katul, 2008; Launiainen et al., 2007; Dupont et al., 2012). In agroforestry systems with a crop below a canopy of sparse trees, a second inflection point appears at crop top on the mean wind velocity profiles (Fig. 9.3c) (Dupont et al., 2022). As within the tree crown layer, this inflection point is associated with a decrease in the momentum and turbulence within the crop canopy, and an increase of $-r_{uw}$, Q_{uw}, and velocity skewnesses. By analogy with the single inflection point of a standalone canopy, this secondary inflection point at the crop top is responsible for the development of a second level of mixing-layer-type coherent motions, adding to the ones developing at the tree-canopy top. These double levels of coherent motions are expected to play a major role in momentum, energy, and mass transfers within agroforestry systems and with the above atmosphere.

9.3.1.2 Unstable conditions

With increasing instability, the wind shear at canopy top is attenuated by the enhanced vertical mixing, increasing the vertical extent of the canopy-top mixing-layer-type coherent structures. The main flow characteristics of the near-neutral regime persist in the forced convection regime up to about $(z - z_d)/L \approx -0.1$ (Fig. 9.5). In such a regime, the turbulence at the canopy top is still dominated by shear-driven motions,

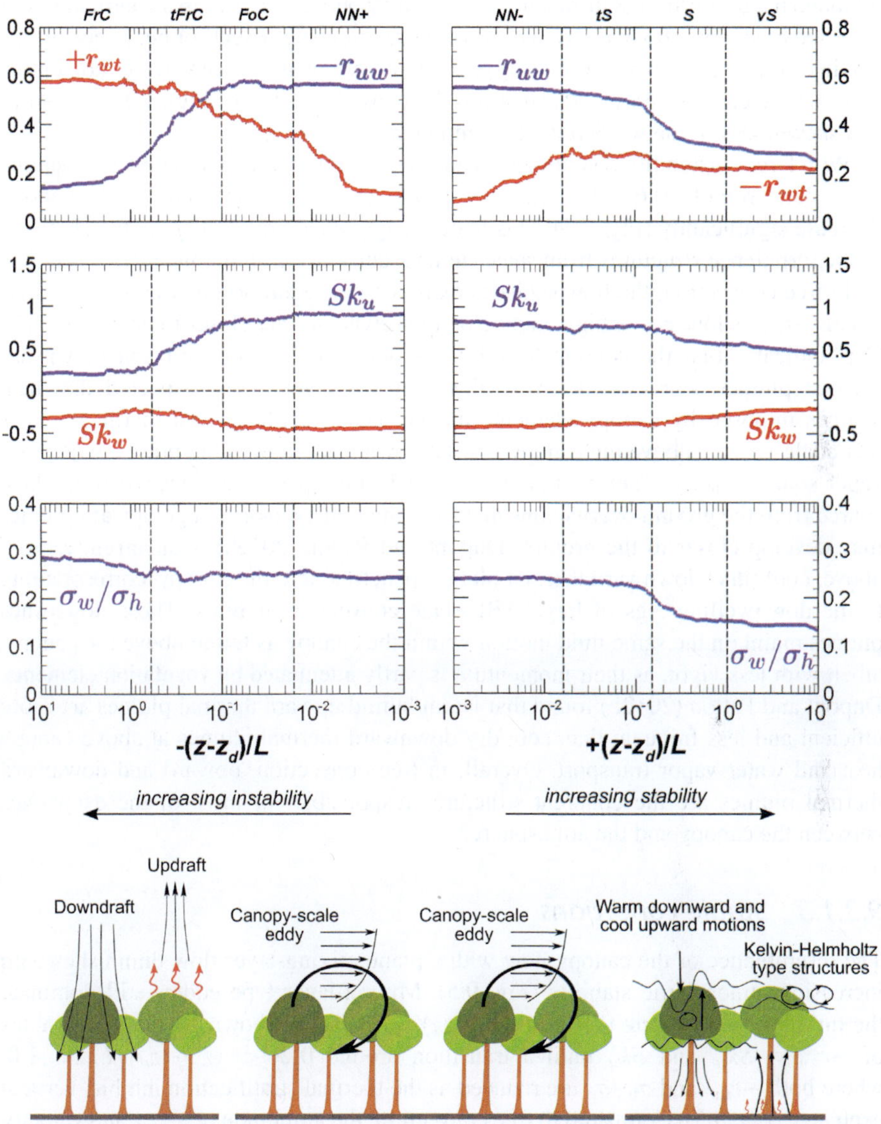

Figure 9.5 Variation of plant canopy turbulence characteristics with thermal stability at canopy top. From top to bottom: correlation coefficients for momentum and heat transfers (r_{uw} and r_{wt}), skewnesses of the streamwise and vertical velocity components (Sk_u and Sk_w), and ratio between the vertical and horizontal velocity variances (σ_w/σ_h). The bottom schematics summarize the main turbulent motions involved in each stability configuration. The dashed vertical lines delineate the stability regimes: free convection (FrC), transition to free convection (tFrC), forced convection (FC), diurnal near-neutral (NN+), nocturnal near-neutral (NN-), transition to stable (tS), stable (S), and very stable (vS). Figures result from the CHATS field experiment over an orchard (Dupont and Patton, 2022).

although the proportion of turbulence produced by buoyancy increases continuously as reflected by the progressive increase of the correlation coefficient for heat transfer r_{wt}. Mixing-layer type structures are still present and efficient at transporting momentum between the canopy and the above atmosphere, as shown by the high values of $-r_{uw}$ and skewnesses of the wind velocity components.

With larger instability conditions, an abrupt transition of the main flow characteristics occurs around $-0.6 < (z - z_d)/L < -0.1$, where $-r_{uw}$ and velocity skewnesses decrease significantly (Fig. 9.5). This feature suggests a rapid change in the character of the turbulence structures from shear- to buoyantly-driven motions.

In free convection, the flow statistics exhibit a new equilibrium with $-(z - z_d)/L$ where $-r_{uw}$ is low, r_{wt} is high, and velocity skewnesses are closer to zero (Fig. 9.5). This indicates that the canopy flow is now dominated by upward and downward thermal plumes. For canopies with dense crown layer, it is suspected that heat sources imposed by canopy elements induce local, small, upward thermal plumes that coalesce well above the canopy into large warm (humid in the presence of water vapor source) upward thermal plumes (Fig. 9.2). For sparse canopies where the heat source from the ground overwhelms that from the upper canopy, large upward plumes may develop closer to the ground (Dupont and Patton, 2012b). Concurrently, from above, cool (dry) downward thermal plumes penetrate into the canopy, corresponding to the downwelling legs of large ABL-scale convective motions. These downward plumes maintain the same fluid motion within the canopy as found above the canopy, albeit with less vigor, as their momentum is partly attenuated by vegetation elements. Dupont and Patton (2012b) found that warm/humid upward thermal plumes are more efficient and less frequent than cool/dry downward thermal plumes at above-canopy heat and water vapor transport. Overall, in free convection, upward and downward thermal plumes are the coherent structures responsible for most of the exchanges between the canopy and the atmosphere.

9.3.1.3 Stable conditions

The resemblance of the canopy flow with a plane mixing-layer flow diminishes with increasing atmospheric stability (Fig. 9.5). Mixing-layer type eddies still dominate the turbulence at canopy top up to $(z - z_d)/L \approx 0.2$ as shown by the large values of $-r_{uw}$, $-Sk_w$ and Sk_u until a transition around $0.2 < (z - z_d)/L < 1.0$, where both $-r_{uw}$ and σ_w/σ_h are reduced as the thermal stratification inhibits vertical exchange (Fig. 9.5). Compared to free convection, the amplitude of $-r_{uw}$ and velocity skewnesses remain larger. Hence, coherent eddy structures are still present in the stable regime, but they are weaker and more intermittent. They are suspected to remain confined within the upper canopy, not reaching the subcanopy, appearing more like Kelvin–Helmholtz type structures since canopy-top instabilities or gravity waves initiating these structures should be less perturbed by the background turbulence than in the other regimes.

For canopies with a dense crown layer, the subcanopy layer can be decoupled from the crown layer where the lower canopy is unstably stratified, dominated by thermal plume structures with narrower updrafts surrounded by wider downdrafts (Dupont and Patton, 2012b). Such decoupling is explained by the rapid radiative cooling of the leaves in the upper canopy, generating a stable layer above the crown, with an unstable layer in the trunk space, and by the limited penetration of canopy-top coherent structures.

9.3.1.4 Influence of plant motions

We may wonder how much impact plant motions have on the wind turbulence in plant canopies. Indeed, under the action of wind load, plants, branches, and foliage reconfigure and sway. The passage of the canopy coherent structures mentioned previously is responsible for the largest plant displacements (e.g., Gardiner, 1994; Schindler, 2008).

Plant reconfiguration, also known as plant foliage streamlining, represents the main impact of plant motions reported on the canopy flows. This process reduces the plant frontal area and the pressure component of drag and thus enhances the penetration of extreme gusts within canopies (e.g., Vollsinger et al., 2005; de Langre et al., 2012; Pan et al., 2014). This process has been described empirically by introducing a drag coefficient proportional to a power-law of the wind velocity, whose negative exponent is known as the Vogel number (Vogel, 1989).

A lock-in mechanism of canopy coherent structures onto the plant natural vibration frequency has also been reported with increasing wind speed in several studies based either on wind-tunnel experiments or linear stability analysis (Finnigan and Mulhearn, 1978a,b; Py et al., 2006; Gosselin and de Langre, 2009). However, this has never been confirmed from field measurements, nor from the modeling of waving crops and forest motion (Dupont et al., 2010, 2015, 2018a). The discrepancy between linear stability analysis and wind-plant interaction numerical simulations was attributed to the presence of a nonlinear saturation mechanism in the airflow simulations, independent of canopy motion, which was not considered in the linear stability analysis (Dupont et al., 2010). Only Dupont et al. (2018a) observed a peak in frequency on the wind velocity fluctuations related to the fundamental tree vibration mode, but it was near the edge of a forest. They suggested that in their edge flow, the velocity fluctuations induced by tree motions emerged due to the lower background turbulence coming from the upwind clearing as compared to the stand flow.

9.3.2 Scalar turbulence

Scalars released in a plant canopy are transported and dispersed by the turbulent flow within and outside the canopy. Here scalar refers to a quantity without specific direction, such as heat, water vapor, or a trace gas. Unlike water vapor and trace gases, which have a negligible impact on the flow, heat is an active scalar that produces convective motions in unstable conditions or attenuates turbulence in stable conditions. Like the flow velocity, scalar concentration exhibits a complex dynamics with fluctuations spanning over a broad range of space and time scales. However, scalar and momentum turbulence exhibit different statistical properties. These differences are exacerbated with increasing instability as (1) the mixing-layer type eddies transporting scalars at canopy top in near-neutral conditions are progressively replaced by thermal plumes, and as (2) the source/sink distributions between scalars and momentum diverge within the canopy. Overall, it is more difficult to draw some universal statistical behaviors of scalar turbulence within a canopy, compared to wind turbulence, as scalar statistics depend strongly on both the dynamics and the distribution of their sources/sinks, while wind statistics depend mostly on the canopy foliage distribution. Hereafter, we focus on three main differences between wind and scalar turbulence: (1) the presence of ramp-cliff patterns on scalar time series, (2) the dissimilarity in turbulent transport between scalar and momentum, and (3) the larger fine-scale intermittency of scalars.

9.3.2.1 Ramp-cliff patterns

Unlike longitudinal wind velocity time series, scalar time series at canopy top systematically show ramp-cliff patterns, irrespective of the thermal stability, and whether the canopy is a source or a sink of scalar (Gao et al., 1989; Paw et al., 1992; Fitzmaurice et al., 2004; Katul et al., 2006, 2013; Dupont et al., 2020). These ramp-cliff patterns refer to gradual rises (or decreases) of scalar concentration with weak fluctuations over scales commensurate with the canopy height, followed by relatively sharp drops (jumps) depending on whether the canopy is a source (sink) of the scalar under consideration (Katul et al., 2013). These structures have been the subject of intensive research due to their significant contribution to scalar turbulent fluxes (Gao et al., 1989; Katul et al., 2013). Such ramp-cliff patterns are observed in the most common turbulent flows (Warhaft, 2000).

An example of these ramp-cliff patterns is illustrated in Fig. 9.6, adapted from Dupont et al. (2020), showing temperature (θ) and longitudinal wind velocity (u)

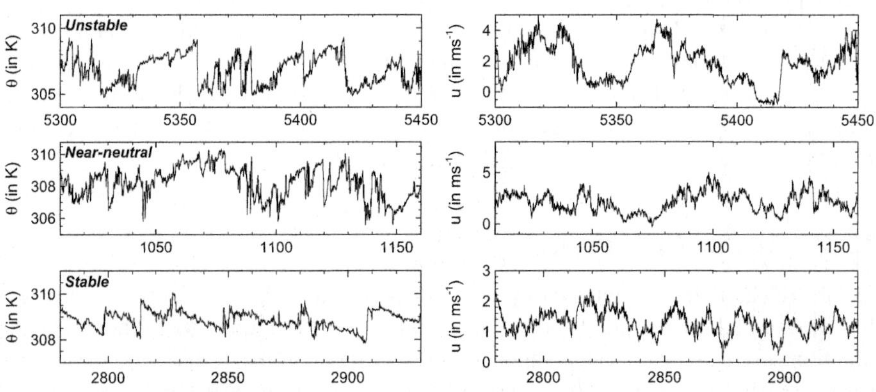

Figure 9.6 Visualization of ramp-cliff patterns in air temperature θ time series (left) at the top of a pine forest for different thermal stability conditions (unstable, near-neutral, and stable from top to bottom), by comparison with the streamwise velocity u time series (right). Time series are 150 s long. Adapted from Dupont et al. (2020) with permission from Cambridge University Press © 2023.

time series recorded at the top of a pine forest in unstable, near-neutral, and stable conditions. Unlike u, θ exhibits ramp-cliff patterns in the non-neutral regimes. In the unstable regime, these structures correspond to a gradual rise of air temperature, followed by a sharp drop, while in the stable regime, these θ ramp-cliff patterns are inverted due to the inversion of the temperature vertical gradient. In the near-neutral regime, the ramp-cliff patterns are not as well-defined for this transitional regime between uprising and decreasing ramps.

As mentioned in Dupont et al. (2020), these ramp-cliff patterns were initially interpreted as the signature of convective plumes (Taylor, 1958). They now appear related to the coherent structures developing at canopy top, corresponding to the anisotropic forcing of scalar fluctuations. By analogy with the surface renewal theory (Higbie, 1935), these patterns result from a parcel of fluid coming from above that remains in contact with the canopy top where scalar sources (sinks) are located, and whose amount of scalar increases (decreases) progressively (ramp), before being replaced (cliff) by another fluid parcel (Paw et al., 1995; Katul et al., 2006).

9.3.2.2 Turbulent transport dissimilarity

In the ABL, scalars and momentum are not necessarily transported similarly by turbulent motions. Dissimilarity between scalars themselves or between scalar and momentum is especially observed with increasing instability as the character of the main turbulent motions changes from shear- to buoyantly-driven motions (Li and Bou-Zeid, 2011). As an example, scalars co-located with heat sources are becoming preferentially released into local upward plumes with increasing instability, while momentum remains preferentially absorbed by the surface during downward sweeping gusts. Between scalars themselves, the dissimilarity in turbulent transport is explained by advection, unsteadiness, entrainment at the top of the ABL, and differences in sources/sinks (Li and Bou-Zeid, 2011). This is reflected by differences between scalars (1) in the turbulent exchange coefficient of the flux-gradient relationship well above the surface roughness elements that relates the scalar turbulent flux to the mean concentration vertical gradient, and (2) in the transport efficiency by turbulent motions.

In plant canopies, dissimilarity in turbulent transport between scalars themselves, or between scalars and momentum, is also observed. A first reason for this dissimilarity is the difference in spatial distribution between the scalar sources/sinks relative to the development areas of the main coherent structures, affecting the efficiency of turbulence in transporting scalars outside the canopy. Scalars emitted at the canopy crown level, where canopy-scale motion developed, are more efficiently transported and exchanged with the overlying atmosphere than scalars released at the ground. This has some implications regarding the residence times of trace gases within the canopy, with longer residence times for surface-emitted trace gases (Dupont and Patton, 2012b).

A second reason for explaining the dissimilarity in turbulent transport between scalars and momentum is the difference in the dynamics of their sources/sinks. The release of scalars by vegetation is not continuous, but depends on numerous environmental factors, leading to different turbulent transport between scalars. Scalar source should increase with the scalar gradient between the surface and the surrounding air, as well as with increasing wind velocity via the exchange coefficient or via a biological threshold process. This dynamics of scalar sources/sinks may explain the observed increase in scalar skewness, and to a lesser extent, scalar standard deviations, near the scalar sources/sinks in unstable conditions (Dupont and Patton, 2012a).

Even for apparently co-located sources/sinks of scalars at canopy scale, such as heat and water vapor in the canopy crown layer, heat was observed to be transported more efficiently by organized structures than water vapor (Dupont and Patton, 2012b). This feature was explained in Dupont and Patton (2012b) by (1) the dominant role played by heat in producing local upward thermal plumes, where water vapor may not be fully directly injected in, and thus transported similarly as heat, (2) the possible local dissimilarity between heat and water vapor source distribution, and (3) the time response for stomata to open/close that could generate a phase shift between water vapor release and thermal plume development.

9.3.2.3 Small-scale intermittency

As is typical in turbulent flows, scalars at the canopy top exhibit greater intermittency at fine scales than the longitudinal wind velocity (Schmitt et al., 1996; Dupont et al., 2020). This larger intermittency manifests itself in more intense and sporadic scalar fluctuations at small scales. Statistically, this leads to a larger spectrum of singularities (multifractals) and a non-Gaussian shape of the PDF of scalar fluctuation increments.

The possible cause of the larger fine-scale intermittency of scalar turbulence is often attributed to the ramp-cliff patterns observed in scalar fields (e.g., Warhaft, 2000; Zorzetto et al., 2018). In the traditional picture of the scale-by-scale cascade of energy from large to viscous-dissipation scales, the anisotropy of the large-scale motions is lost during the process. In the presence of ramp-cliff structures, it is believed that the cascading process is short-circuited with some coupling between large and small scales, the latter ones still carrying some signature from the forcing scale, explaining the high small-scale intermittency (Katul et al., 2003; Zorzetto et al., 2018; Dupont et al., 2020).

The persistence of the signature of large-scale ramp-cliff structures on small-scale fluctuations was observed well within the inertial convective subrange for temperature fluctuations recorded above surface roughness elements, in the surface-shear layer, by Zorzetto et al. (2018). This signature of the ramp-cliff patterns on small-scale temperature fluctuations explained their larger intermittency compared to velocity fluctuations. On the other hand, at canopy top, Dupont et al. (2020) found that the statistical contribution of the 'cliffs' is not significant enough to explain the stronger intermittency of temperature fluctuations. They suspected that the mixing-layer type flow prevailing at canopy top damps the memory of the large-scale motions as opposed to shear-layer flow located far above surface roughness elements. Their findings suggest that the scalar intermittency at fine scales has more of an internal origin, built from the fluctuating dissipation rate, than an external one coming from large-scale external forcing (cliffs).

9.3.3 Similarity theory to describe near-surface flow

The mean flow and turbulence characteristics in the ASL have been described using similarity theory for horizontally homogeneous surface roughness elements. The momentum and scalar turbulent fluxes are expressed in terms of vertical gradients of the transported quantities, known as the flux-gradient relationship, where the turbulent exchange coefficients relating the fluxes to the gradients are parameterized using similarity theory. The momentum and scalar profiles are then obtained by integrating the flux-gradient relationship.

The Monin-Obukhov Similarity Theory (MOST, Monin and Obukhov, 1954) assumes that the gradient and other turbulent statistics can be described using similarity universal functions dependent on a few scaling parameters, including the Obukhov length scale L, and whose constants are determined empirically from experimental datasets. The theory hypothesizes that the turbulent flow in the ASL depends mainly on local eddies, whose size scales with the distance z from the ground. In the presence of a plant canopy, an additional length scale has to be considered in order to characterize the mixing-layer type motions present in the RSL. The vorticity thickness of the flow at canopy top is commonly accepted as a scale suitable for these canopy motions (Harman and Finnigan, 2007, 2008; Finnigan et al., 2009).

Both MOST and corrected MOST in the RSL have been relatively well evaluated above uniform surfaces (Kaimal and Finnigan, 1994) and over several canopies (Harman and Finnigan, 2007, 2008; Weligepolage et al., 2012; Shapkalijevski et al., 2016), respectively. However, similarity universal functions exhibit often large scatter between experiments, and this scatter cannot solely be attributed to random errors. This suggests that all physical processes may not have been well accounted for in MOST, and additional nondimensional parameters may be required (e.g., Salesky and Chamecki, 2012). As reviewed in Dupont and Patton (2022), growing evidence suggests ABL-scale motion influence on near-surface flows, while MOST assumes that the near-surface turbulent flow depends only on local motions. This means that the similarity universal functions may have unknowingly integrated some of the ABL-scale motion influence, explaining some of their observed scatter between experiments, and this suggests a need to account for ABL-scale motions in near-surface turbulence similarity theories.

9.4 Influence of large-scale atmospheric motions on canopy turbulence

The influence of ABL-scale motions on both surface-shear and canopy flows has become evident over the last few decades, questioning the MOST hypothesis on the sole dependence of the ASL turbulent transport on local eddies. This influence of ABL-scale motions is visible in the spectra of the near-surface and canopy-top longitudinal wind velocity, describing the distribution of the velocity fluctuations (variance) as a function of the eddy size. Fig. 9.7 schematizes these spectra as deduced from Dupont and Patton (2022) spectral analysis above flat bare soils and vegetation canopies.

In both flows, the spectra are characterized at high frequencies by the famous inertial subrange with a $-5/3$ power law ($-2/3$ in premultiplied spectrum), regardless of the stability regime. In this frequency range, the flow kinetic energy produced at low frequencies is transferred from smaller to smaller frequencies without any production/dissipation of this energy. At low frequencies, a spectral gap separates three-dimensional (3D) boundary-layer turbulence from the two-dimensional (2D) mesoscale motions, these latter motions representing frontal passages. The influence of ABL-scale motions is visible at mid-frequencies, with overall a larger influence in surface-shear flows than in canopy flows.

In surface-shear flows, the influence of ABL-scale motions is characterized in windy conditions (near-neutral) by an extension of the spectrum, with a -1 power law

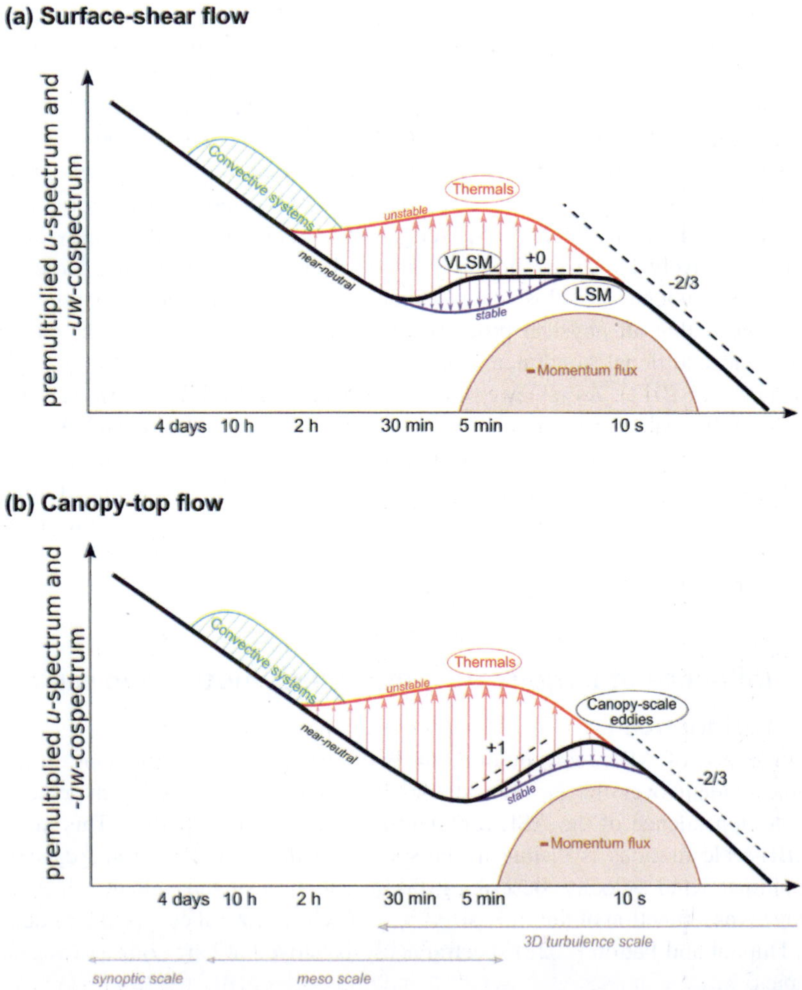

Figure 9.7 Schematic representation of the premultiplied spectrum shape of the near-surface longitudinal wind velocity u above a low-roughness surface (a) and at the top of a plant canopy (b), according to the thermal stratification (near-neutral in black, free convection in red, and stable in blue). At small time scales (< 10 s), both u-spectra exhibit an inertial subrange with the famous $-5/3$ power law ($-2/3$ in premultiplied spectrum) where the energy of u fluctuations is transferred down to smaller and smaller scales without production or dissipation. In near-neutral stratification, the spectrum is characterized in (a) by a plateau ($+0$ scaling) at the production scales (between 5 min and 10 s), delimited by very-large-scale motions (VLSM) and large-scale motions (LSM), and in (b) by a marked peak related to the mixing-layer type eddies developing at canopy top. With increasing instability, the spectral plateau in (a) and the spectral $+1$ slope in (b) of the near-neutral production subrange level off as the contribution of buoyantly-driven motions on u fluctuations increases over surface-shear and canopy-scale motions, respectively, extending the spectral inertial subrange to lower frequencies in the free convection regime, where thermals become the main producer of velocity fluctuations. With increasing stability, local-scale motions decouple progressively from the overhead ABL-scale motions (VLSM and thermals), forming a spectral gap between local- and large-scale motions in (a) and damping the contribution of canopy-scale motions to the u fluctuations in (b). The shape of the negative momentum-flux (uw) cospectrum represented by the brown parabola does not change much with thermal stratification, ABL-scale motions contributing on average weakly to the momentum turbulent transport. These figures were deduced from the spectral analysis of Dupont and Patton (2022). Figure (a) comes from Dupont (2022), reprinted with permission from American Geophysical Union © 2023.

(0 in premultiplied spectrum) resulting from the coexistence of VLSMs and LSMs at the low and high frequency limits of this scaling region, respectively. This influence represents more than 90% of the longitudinal velocity variance and is suspected to result from the impingement of large-scale eddies onto the ground (Dupont and Patton, 2022). The blockage by the surface of the downwelling portion of ABL-scale motions stretches them along the mean wind direction through pressure redistribution, converting downward vertical motions into horizontal motions (McNaughton and Laubach, 2000). This process leads to elongated streamwise structures at the surface, with a longitudinal scale larger than z. This explains the observed extra low-frequency energy of horizontal wind velocity spectrum as opposed to the vertical wind velocity spectrum. With increasing instability, shear-driven motions, i.e., LSMs and VLSMs, are progressively replaced by buoyantly-driven motions, i.e., rising plumes merging into thermals. This process induces a progressive leveling-off of the -1 power law (0 in Fig. 9.7a), extending the spectrum inertial subrange toward low frequencies in the free convection regime, where thermals become the main production source of longitudinal wind velocity fluctuations. This level-off reflects the increasing contribution of ABL-scale motions to the near-surface turbulence with instability. Conversely, with increasing stability, a gap forms on the 3D turbulence spectrum range of the wind velocity spectrum, reflecting a progressive decoupling of the near-surface turbulence with the overhead ABL-scale motions in the mixed layer (Fig. 9.7a).

In canopy flows, the influence of ABL-scale motion is lower than in surface-shear flows, but it becomes significant with increasing instability (Dupont and Patton, 2022). In near-neutral conditions, the canopy longitudinal wind velocity spectrum shows no -1 power law as in surface-shear flows but a marked peak, flanked by 0 and $-5/3$ power laws ($+1$ and $-2/3$ in premultiplied spectrum) on the low- and high-frequency sides, respectively (Fig. 9.7b). The absence of -1 power law in canopy flows suggests a negligible impact of ABL-scale motions on the near-canopy turbulence compared to the turbulence resulting from the canopy-scale eddies, representing less than 10% of longitudinal velocity variance according to Dupont and Patton (2022). With increasing instability, the influence of ABL-scale motions increases as the high-frequency portion of the 0 power law in near-neutral conditions transitions progressively to a $-5/3$ power law in free convection conditions ($+1$ and $-2/3$ in premultiplied spectrum), extending the inertial subrange and shifting the production subrange to lower frequencies. ABL-scale motions were found to contribute to more than 70% of longitudinal velocity variances in free convection conditions (Dupont and Patton, 2022). This transitioning is coherent with the increase of the vertical velocity skewness Sk_w from negative to near-zero values, and with the decrease of the momentum correlation coefficient $-r_{uw}$ (Fig. 9.5), highlighting the decreasing influence of canopy mixing-layer type eddies, replaced by intermittent rising buoyant plumes emanating from the underlying vegetation canopy and large-scale downdrafts penetrating within the canopy. Hence, in free convection, ABL-scale thermals become the main source of longitudinal wind velocity fluctuations at canopy top, with spatially-varying contributions, alternating between local convergence and divergence regions produced by thermals (Fig. 9.2c). With increasing stability, canopy-top turbulence becomes weaker and can decouple from the overhead motions as in surface-shear flows.

The turbulent transport of momentum and heat by ABL-scale motions (VLSMs or thermals) is found to generally be negligible for both surface-shear and canopy flows. The shape of the momentum- and heat-flux cospectra does not vary much with thermal stability (Fig. 9.7). As observed in Dupont and Patton (2022), ABL-scale motions may intermittently or locally transport momentum or heat vertically, but on average, this transport seems negligible. The near-surface vertical velocity component of these large-scale motions is too small to induce significant vertical transport. However, ABL-scale motions may spatially organize (or modulate) near-surface turbulent transport by local-scale motions as demonstrated from numerical simulations by Salesky and Anderson (2018) and Li et al. (2018) above rough surfaces and by Patton et al. (2016) and Perret and Patton (2021) at canopy top.

9.5 Investigations and applications

9.5.1 Investigations

Over the last decades, turbulence in plant canopies has been mainly investigated through field, laboratory, and numerical experiments. We give an overview of each approach.

Field measurements over plant canopies have benefited from both (1) the development of affordable fast-response sensors to measure turbulent fluctuations of the wind velocity components (three-dimensional sonic anemometers) and trace gases, and (2) research on quantifying energy and carbon exchanges between the biosphere and the atmosphere in the context of climate change. This research allowed the development of international networks of meteorological towers over a variety of vegetation ecosystems, such as Fluxnet (Baldocchi, 2020), with a continuous characterization of the canopy turbulent flow and an estimation of heat, water vapor, and carbon dioxide fluxes using the eddy-covariance technique. Sonic anemometers are usually fast enough to capture the main canopy-scale motions carrying the flow energy and involved in exchanges between the canopy and aloft. The measurement path length of sonic anemometers is, however, too large to capture eddies developing behind individual canopy elements. For such small-scale eddies, high-frequency techniques used in laboratories, such as hot wires, are still difficult to apply in outdoor conditions. The inconveniences of most field measurements are the prevalence of single-point measurements, which do not allow for obtaining a spatial view of the canopy turbulent fields, although campaigns with high-resolution single-point measurements have led to detailed vertical and horizontal meteorological profiles, e.g., the CHATS experiment (Patton et al., 2011). The recent development of Optical-fiber-based distributed sensing methods opens some perspectives to obtain high-spatial resolution horizontal or vertical profiles of meteorological variables based on temperature measurements along a fiber-optic cable (e.g., Thomas and Selker, 2021). Such an approach, however, is still under development and its sampling is limited to 1 Hz. Concurrently, the increase in high-resolution surface temperature images of vegetated canopies represents an indirect way to access a two-dimensional view of the canopy-top turbulence, although its relation to the wind field is not straightforward and limited to large-scale motions.

Laboratory experiments on turbulent flows in modeled canopies have been intensively developed in parallel to field experiments, using wind-tunnels or water channels (e.g., Raupach et al., 1986; Poggi et al., 2004). Such experiments have the advantages of acquiring turbulent flow information at high frequency in a controlled environment under stationary wind conditions. In such experiments, vegetation canopies are often represented as arrays of cylindrical stalks or "tombstones". The development of the particle image velocimetry (PIV) technique has further allowed access to two-dimensional views of instantaneous turbulence fields over canopies, revealing the coherent structures developing in canopy flows. The difficulties of laboratory measurements are to match some non-dimensional parameters in order to be comparable with reality. The Reynolds numbers and the ratio between the boundary-layer depth and the canopy height remain lower in laboratory experiments, with consequently a limited range of eddy scales as compared to the ABL. Furthermore, laboratory measurements are often limited to neutral conditions, as it is difficult to work with stratified flows in wind tunnels. Hence, large convective ABL-scale motions present above the ASL, and their influence on the ASL turbulence, are not represented in laboratory experiments.

Numerical experiments simulating the wind dynamics within and above vegetated canopies have been performed with plant canopies represented as porous media. Most of the airflow models resolve the momentum Navier–Stokes equation, using either a Reynolds-averaged approach (Li et al., 1990; Liu et al., 1996; Foudhil et al., 2005; Dupont and Brunet, 2006) or a large eddy simulation (LES) approach (e.g., Shaw and Schumann, 1992; Dupont and Brunet, 2008b,a), where the vegetation is accounted for through additional pressure and viscous drag terms. The Reynolds-averaged type models simulate only mean fields and model the turbulence, while LES-type models simulate instantaneous dynamical fields by modeling only the subgrid-scale turbulence. LES is thus capable of reproducing wind gusts in plant canopies and allows canopy turbulence to be analyzed in greater detail. However, it is expensive in computational time, and proper lateral boundary conditions are not easy to introduce, which makes LES techniques more difficult than Reynolds-averaged techniques to use in heterogeneous canopies. With increasing computational resources, LES became more and more popular and affordable over the last decade. It has been extensively validated over homogeneous canopies in neutral stability conditions. The main characteristics of the canopy turbulence were observed in good agreement with those derived from in situ and wind-tunnel datasets. LES has allowed characterizing the coherent eddy structures prevailing at canopy top from a composite average (Finnigan et al., 2009). Unstable conditions have been less investigated due to the need (1) to account in models for the thermo-hydric fluxes within the plant canopy, and (2) to consider a large spectrum of eddy scales ranging from canopy to ABL scales, which requires enormous computational time. Nonetheless, several recent studies bridged this gap. In particular, Patton et al. (2016) were able to simulate the complete interaction between ABL- and canopy-scale motions with increasing instability, revealing how the ABL-scale motions modulate canopy motions and the turbulent exchanges at the canopy-atmosphere interface. More recently, a Lattice Boltzmann method has been developed for large-eddy simulation of the canopy flow where the vegetation was considered as a momentum sink (Watanabe et al., 2020). Instead of solving conservation equations of a macroscopic wind velocity (Navier-Stokes equation), the Lattice Boltzmann method resolves the velocity distribution of microscopic fictive fluid particles. This new approach is expected to better deal with complex boundaries and to have better computational performance. However, this method is still limited to neutral conditions.

9.5.2 Applications: forest wind damage and particle dispersal

Among the various environmental issues, turbulence in plant canopies, wind damage in forests, and particle dispersal have probably received the most attention in research.

Wind storms are responsible for significant damage in forests, affecting forest ecosystems (e.g., Mitchell, 2013) and leading to severe loss in their economic value (Schelhaas et al., 2003; Gardiner et al., 2012; Hanewinkel et al., 2013; Fischer et al., 2013). Understanding and predicting the occurrence of tree damage in forests has been the subject of a significant amount of research, where turbulence in tree canopies plays a key role. One key issue to better predict the risk of forests is to improve the understanding of the wind-tree interaction under high-wind conditions.

To that purpose, several field experiments have been performed where both wind dynamics and tree motions were recorded (e.g., Gardiner, 1994; Rudnicki et al., 2008; Schindler et al., 2010; Dupont et al., 2018a). In parallel, simulations of the wind-tree interaction have been performed at individual-tree (Yang et al., 2020) and canopy scales (Dupont et al., 2010, 2015). At canopy scale, Dupont et al. (2010) simulated the full interaction between the turbulent wind and the motions of crop plants, reproducing the well-known phenomenon of "honami", typical of wave-like crop motions on windy days. Their model was then extended to tree motions in a forest under windstorm conditions, simulating large tree deflection and tree breakage (Dupont et al., 2015). Such simulations of wind-plant interaction open new perspectives for improving wind-risk models (Dupont, 2016).

The atmospheric dispersal of particles coming from plant canopies is also an important issue in terms of health, epidemic, and environmental impacts, where agricultural practices and landscape organization are key determinants (e.g., Aylor, 2017; Mahaffee et al., 2023). These particles are biotic and abiotic. They can be released by plants themselves, such as pollen, or by plant elements, such as pathogenic fungal spores from lesions under leaves, or by agricultural treatments, such as droplets of pesticide during spraying. Once released, particles either deposit rapidly on the ground or on neighborhood plants or escape from the canopy and are then transported over long distances from their emission point. Regardless of the type of particles, the wind dynamics and turbulence in plant canopies represent a major driver of particle dispersal. Unlike trace gases, particles can deviate from the flow due to their inertia and gravity (e.g., Chamecki and Meneveau, 2011). As for wind damage, advancements in numerical simulations that represent turbulent canopy flows offer new insights into understanding and predicting particle dispersal in complex landscapes (e.g., Pan et al., 2015; Dupont et al., 2022). This will enable a more thorough evaluation of agricultural practices that would be difficult to achieve through single-point measurements alone.

9.6 Future research challenges

Most of our understanding of turbulence in plant canopies has been gained from neutral thermal conditions over horizontally uniform plant canopies. This has enabled the identification of the crucial role played by the wind velocity inflection point at canopy top, generating canopy-scale coherent motions that condition turbulence statistics and drive all biosphere-atmosphere exchanges. These motions persist in forced-convection and transition-to-stable conditions until they transition to, or are replaced by, buoyantly driven motions with increasing instability, or they weaken with increasing stability.

An important research challenge in the field of canopy turbulence is to better understand the transition of canopy turbulence with increasing instability and stability. This transition has been an under-investigated area due to the challenges in reproducing non-neutral conditions in numerical and laboratory experiments. With increasing computational power, LES over vegetated canopies in unstable conditions becomes feasible with the perspective of simulating turbulent motions from canopy to

ABL scales (e.g., Patton et al., 2016). The frequent difference of stability between the lower and upper canopy represents another challenge to comprehend this transition, as it complicates the establishment of a universal description of canopy turbulence in non-neutral conditions because of its dependency on the upper-canopy foliage density. An additional challenge arises from the increasing influence of ABL-scale motions on canopy turbulence with instability, contradicting the hypotheses of the corrected MOST in the RSL used to describe the wind and scalar profiles and fluxes. More generally, a comprehensive understanding of the diurnal cycle interaction between ABL and canopy-scale motions is lacking, particularly during the morning and afternoon transitions when field point measurements are usually discarded due to the non-stationarity of the flow. A recent LES study has been conducted in this regard (Dupont et al., 2024). A future challenge will consist of accounting for ABL-scale motions in MOST.

Accounting for the impact of nearby heterogeneities on plant turbulence represents another challenge requiring further research. Landscapes are characterized by a significant fragmentation of the surface, with the presence of clearings, crops, roads, and forest plots of different heights. This heterogeneity of the surface, added to the topography, creates vertical and horizontal gradients of micrometeorological variables (wind, temperature, humidity), influencing energy and mass exchanges between the canopies and the atmosphere (Belcher et al., 2012). Canopy turbulence in the presence of heterogeneities has started to be investigated, such as around canopy edges (e.g., Yang et al., 2006; Dupont and Brunet, 2008a; Dupont et al., 2011; Boudreault et al., 2017), clearing-forest patterns (Poëtte et al., 2017), agroforestry systems (Dupont et al., 2022), or hills (e.g., Dupont et al., 2008; Finnigan et al., 2020), but mostly under simplified conditions and in neutral conditions. Understanding the role of landscape heterogeneities on plant turbulence is essential for future research to address environmental and ecological inquiries regarding agricultural and forestry practices, particularly in terms of the dispersion of contaminants, pollutants, genes, nutrients, forest wind risk, or wildfire propagation.

References

Atkinson, B.W., Zhang, J.W., 1996. Mesoscale shallow convection in the atmosphere. Reviews of Geophysics 34, 403–431. https://doi.org/10.1029/96rg02623.

Aylor, D.E., 2017. Aerial dispersal of pollen and spores. The American Phytopathological Society.

Baldocchi, D.D., 2020. How eddy covariance flux measurements have contributed to our understanding of global change biology. Global Change Biology 26, 242–260. https://doi.org/10.1111/gcb.14807.

Baldocchi, D.D., Meyers, T.P., 1988. Turbulence structure in a deciduous forest. Boundary-Layer Meteorology 43, 345–364. https://doi.org/10.1007/bf00121712.

Belcher, S.E., Harman, I.N., Finnigan, J.J., 2012. The wind in the willows: flows in forest canopies in complex terrain. Annual Review of Fluid Mechanics 44, 479–504. https://doi.org/10.1146/annurev-fluid-120710-101036.

Bonan, G.B., 2008. Forests and climate change: forcings, feedbacks, and the climate benefits of forests. Science 320, 1444–1449. https://doi.org/10.1126/science.1155121.

Boudreault, L.É., Dupont, S., Bechmann, A., Dellwik, E., 2017. How forest inhomogeneities affect the edge flow. Boundary-Layer Meteorology 162, 375–400. https://doi.org/10.1007/s10546-016-0202-5.

Brunet, Y., 2020. Turbulent flow in plant canopies: historical perspective and overview. Boundary-Layer Meteorology 177, 315–364. https://doi.org/10.1007/s10546-020-00560-7.

Cava, D., Katul, G.G., 2008. Spectral short-circuiting and wake production within the canopy trunk space of an Alpine hardwood forest. Boundary-Layer Meteorology 126, 415–431. https://doi.org/10.1007/s10546-007-9246-x.

Chamecki, M., Meneveau, C., 2011. Particle boundary layer above and downstream of an area source: scaling, simulations, and pollen transport. Journal of Fluid Mechanics 683, 1–26. https://doi.org/10.1017/jfm.2011.243.

Dupont, S., 2016. A simple wind–tree interaction model predicting the probability of wind damage at stand level. Agricultural and Forest Meteorology 224, 49–63. https://doi.org/10.1016/j.agrformet.2016.04.014.

Dupont, S., 2022. On the influence of thermal stratification on emitted dust flux. Journal of Geophysical Research. Atmospheres 127. https://doi.org/10.1029/2022jd037364.

Dupont, S., Argoul, F., Gerasimova-Chechkina, E., Irvine, M.R., Arneodo, A., 2020. Experimental evidence of a phase transition in the multifractal spectra of turbulent temperature fluctuations at a forest canopy top. Journal of Fluid Mechanics 896, A15. https://doi.org/10.1017/jfm.2020.348.

Dupont, S., Bonnefond, J.M., Irvine, M.R., Lamaud, E., Brunet, Y., 2011. Long-distance edge effects in a pine forest with a deep and sparse trunk space: in situ and numerical experiments. Agricultural and Forest Meteorology 151, 328–344. https://doi.org/10.1016/j.agrformet.2010.11.007.

Dupont, S., Brunet, Y., 2006. Simulation of turbulent flow in an urban forested park damaged by a windstorm. Boundary-Layer Meteorology 120, 133–161. https://doi.org/10.1007/s10546-006-9049-5.

Dupont, S., Brunet, Y., 2008a. Edge flow and canopy structure: a large-eddy simulation study. Boundary-Layer Meteorology 126, 51–71. https://doi.org/10.1007/s10546-007-9216-3.

Dupont, S., Brunet, Y., 2008b. Influence of foliar density profile on canopy flow: a large-eddy simulation study. Agricultural and Forest Meteorology 148, 976–990. https://doi.org/10.1016/j.agrformet.2008.01.014.

Dupont, S., Brunet, Y., Finnigan, J.J., 2008. Large-eddy simulation of turbulent flow over a forested hill: validation and coherent structure identification. Quarterly Journal of the Royal Meteorological Society 134, 1911–1929. https://doi.org/10.1002/qj.328.

Dupont, S., Défossez, P., Bonnefond, J.M., Irvine, M.R., Garrigou, D., 2018a. How stand tree motion impacts wind dynamics during windstorms. Agricultural and Forest Meteorology 262, 42–58. https://doi.org/10.1016/j.agrformet.2018.06.022.

Dupont, S., Gosselin, F., Py, C., de Langre, E., Hemon, P., Brunet, Y., 2010. Modelling waving crops using large-eddy simulation: comparison with experiments and a linear stability analysis. Journal of Fluid Mechanics 652, 5–44. https://doi.org/10.1017/s0022112010000686.

Dupont, S., Irvine, M.R., Bidot, C., 2024. Morning transition of the coupled vegetation canopy and atmospheric boundary layer turbulence according to the wind intensity. Journal of the Atmospheric Sciences 81, 1225–1249. https://doi.org/10.1175/jas-d-23-0201.1.

Dupont, S., Irvine, M.R., Bonnefond, J.M., Lamaud, E., Brunet, Y., 2012. Turbulent structures in a pine forest with a deep and sparse trunk space: stand and edge regions. Boundary-Layer Meteorology 143, 309–336. https://doi.org/10.1007/s10546-012-9695-8.

Dupont, S., Irvine, M.R., Motisi, N., Allinne, C., Avelino, J., Bagny Beilhe, L., 2022. Wind-flow dynamics and spore-like particle dispersal over agroforestry systems: impact of the tree density distribution. Agricultural and Forest Meteorology 327, 109214. https://doi.org/10.1016/j.agrformet.2022.109214.

Dupont, S., Patton, E.G., 2012a. Influence of stability and seasonal canopy changes on micrometeorology within and above an orchard canopy: the chats experiment. Agricultural and Forest Meteorology 157, 11–29. https://doi.org/10.1016/j.agrformet.2012.01.011.

Dupont, S., Patton, E.G., 2012b. Momentum and scalar transport within a vegetation canopy following atmospheric stability and seasonal canopy changes: the CHATS experiment. Atmospheric Chemistry and Physics 12, 5913–5935. https://doi.org/10.5194/acp-12-5913-2012.

Dupont, S., Patton, E.G., 2022. On the influence of large-scale atmospheric motions on near-surface turbulence: comparison between flows over low-roughness and tall vegetation canopies. Boundary-Layer Meteorology 184, 195–230. https://doi.org/10.1007/s10546-022-00710-z.

Dupont, S., Pivato, D., Brunet, Y., 2015. Wind damage propagation in forests. Agricultural and Forest Meteorology 214–215, 243–251. https://doi.org/10.1016/j.agrformet.2015.07.010.

Dupont, S., Rajot, J.L., Labiadh, M., Bergametti, G., Alfaro, S.C., Bouet, C., Fernandes, R., Khalfallah, B., Lamaud, E., Marticorena, B., Bonnefond, J.M., Chevaillier, S., 2018b. Aerodynamic parameters over an eroding bare surface: reconciliation of the law of the wall and eddy covariance determinations. Journal of Geophysical Research. Atmospheres 123, 4490–4508. https://doi.org/10.1029/2017jd027984.

Finnigan, J., 2000. Turbulence in plant canopies. Annual Review of Fluid Mechanics 32, 519–571. https://doi.org/10.1146/annurev.fluid.32.1.519.

Finnigan, J., Ayotte, K., Harman, I., Katul, G., Oldroyd, H., Patton, E., Poggi, D., Ross, A., Taylor, P., 2020. Boundary-layer flow over complex topography. Boundary-Layer Meteorology 177, 247–313. https://doi.org/10.1007/s10546-020-00564-3.

Finnigan, J.J., Mulhearn, P.J., 1978a. Modelling waving crops in a wind tunnel. Boundary-Layer Meteorology 14, 253–277. https://doi.org/10.1007/bf00122623.

Finnigan, J.J., Mulhearn, P.J., 1978b. A simple mathematical model of airflow in waving plant canopies. Boundary-Layer Meteorology 14, 415–431. https://doi.org/10.1007/bf00121050.

Finnigan, J.J., Shaw, R.H., Patton, E.G., 2009. Turbulence structure above a vegetation canopy. Journal of Fluid Mechanics 637, 387–424. https://doi.org/10.1017/s0022112009990589.

Fischer, A., Marshall, P., Camp, A., 2013. Disturbances in deciduous temperate forest ecosystems of the northern hemisphere: their effects on both recent and future forest development. Biodiversity and Conservation 22, 1863–1893. https://doi.org/10.1007/s10531-013-0525-1.

Fitzmaurice, L., Shaw, R.H., U, K.T.P., 2004. Three-dimensional scalar microfront systems in a large-eddy simulation of vegetation canopy flow. Boundary-Layer Meteorology 112, 107–127. https://doi.org/10.1023/b:boun.0000020159.98239.4a.

Foudhil, H., Brunet, Y., Caltagirone, J.P., 2005. A fine-scale k-ϵ model for atmospheric flow over heterogeneous landscapes. Environmental Fluid Mechanics 5, 247–265. https://doi.org/10.1007/s10652-004-2124-x.

Gao, W., Shaw, R.H., U, K.T.P., 1989. Observation of organized structure in turbulent flow within and above a forest canopy. Boundary-Layer Meteorology 47, 349–377. https://doi.org/10.1007/bf00122339.

Gardiner, B., Blennow, K., Carnus, J., Fleischer, P., Ingemarson, F., Landmann, G., Lindner, M., Marzano, M., Nicoll, B., Orazio, C., Peyron, J., Reviron, M., Schelhaas, M., Schuck,

A., Spielmann, M., Usbeck, T., 2012. Destructive Storms in European Forest: Past and forthcoming impacts. Technical Report.

Gardiner, B.A., 1994. Wind and wind forces in a plantation spruce forest. Boundary-Layer Meteorology 67, 161–186. https://doi.org/10.1007/bf00705512.

Gosselin, F., de Langre, E., 2009. Destabilising effects of plant flexibility in air and aquatic vegetation canopy flows. European Journal of Mechanics. B, Fluids 28, 271–282. https://doi.org/10.1016/j.euromechflu.2008.06.003.

Hanewinkel, M., Cullmann, D.A., Schelhaas, M.J., Nabuurs, G.J., Zimmermann, N.E., 2013. Climate change may cause severe loss in the economic value of European forest land. Nature Climate Change 3, 203–207. https://doi.org/10.1038/nclimate1687.

Harman, I.N., Finnigan, J.J., 2007. A simple unified theory for flow in the canopy and roughness sublayer. Boundary-Layer Meteorology 123, 339–363. https://doi.org/10.1007/s10546-006-9145-6.

Harman, I.N., Finnigan, J.J., 2008. Scalar concentration profiles in the canopy and roughness sublayer. Boundary-Layer Meteorology 129, 323–351. https://doi.org/10.1007/s10546-008-9328-4.

Higbie, R., 1935. The rate of absorption of a pure gas into a still liquid during short periods of exposure. Transactions of the American Institute of Chemical Engineers 31, 365–389.

Hutchins, N., Marusic, I., 2007. Evidence of very long meandering features in the logarithmic region of turbulent boundary layers. Journal of Fluid Mechanics 579, 1–28. https://doi.org/10.1017/s0022112006003946.

Kaimal, J.C., Finnigan, J.J., 1994. Atmospheric Boundary Layer Flows. Their Structure and Measurements. Oxford University Press, New-York. 289 pp.

Katul, G., Chang, W., 1999. Principal length scales in second-order closure models for canopy turbulence. Journal of Applied Meteorology 38, 1631–1643. https://doi.org/10.1175/1520-0450(1999)038<1631:plsiso>2.0.co;2.

Katul, G., Porporato, A., Cava, D., Siqueira, M., 2006. An analysis of intermittency, scaling, and surface renewal in atmospheric surface layer turbulence. Physica D. Nonlinear Phenomena 215, 117–126. https://doi.org/10.1016/j.physd.2006.02.004.

Katul, G.G., 2019. The anatomy of large-scale motion in atmospheric boundary layers. Journal of Fluid Mechanics 858, 1–4. https://doi.org/10.1017/jfm.2018.731.

Katul, G.G., Angelini, C., De Canditiis, D., Amato, U., Vidakovic, B., Albertson, J.D., 2003. Are the effects of large scale flow conditions really lost through the turbulent cascade? Geophysical Research Letters 30, 1164. https://doi.org/10.1029/2002GL015284.

Katul, G.G., Cava, D., Siqueira, M., Poggi, D., 2013. Scalar turbulence within the canopy sublayer. In: Coherent Flow Structures at Earth's Surface. John Wiley & Sons, Ltd, pp. 73–95. https://doi.org/10.1002/9781118527221.ch6.

Khanna, S., Brasseur, J.G., 1998. Three-dimensional buoyancy- and shear-induced local structure of the atmospheric boundary layer. Journal of the Atmospheric Sciences 55, 710–743. https://doi.org/10.1175/1520-0469(1998)055<0710:tdbasi>2.0.co;2.

Kim, K.C., Adrian, R.J., 1999. Very large-scale motion in the outer layer. Physics of Fluids 11, 417–422. https://doi.org/10.1063/1.869889.

Kobayashi, N., Hiyama, T., 2011. Stability dependence of canopy flows over a flat larch forest. Boundary-Layer Meteorology 139, 97–120. https://doi.org/10.1007/s10546-010-9572-2.

de Langre, E., Gutierrez, A., Cossé, J., 2012. On the scaling of drag reduction by reconfiguration in plants. Comptes Rendus. Mécanique 340, 35–40. https://doi.org/10.1016/j.crme.2011.11.005.

Launiainen, S., Vesala, T., MöLder, M., Mammarella, I., Smolander, S., Rannik, Ü., Kolari, P., Hari, P., Lindroth, A., Katul, G.G., 2007. Vertical variability and effect of stability on turbulence characteristics down to the floor of a pine forest. Tellus. Series B, Chemical and Physical Meteorology 59, 919–936. https://doi.org/10.1111/j.1600-0889.2007.00313.x.

LeMone, M.A., 1973. The structure and dynamics of horizontal roll vortices in the planetary boundary layer. Journal of the Atmospheric Sciences 30, 1077–1091. https://doi.org/10.1175/1520-0469(1973)030<1077:tsadoh>2.0.co;2.

Li, D., Bou-Zeid, E., 2011. Coherent structures and the dissimilarity of turbulent transport of momentum and scalars in the unstable atmospheric surface layer. Boundary-Layer Meteorology 140, 243–262. https://doi.org/10.1007/s10546-011-9613-5.

Li, Q., Gentine, P., Mellado, J.P., McColl, K.A., 2018. Implications of nonlocal transport and conditionally averaged statistics on Monin-Obukhov similarity theory and Townsend's attached eddy hypothesis. Journal of the Atmospheric Sciences 75, 3403–3431. https://doi.org/10.1175/jas-d-17-0301.1.

Li, Z., Lin, J.D., Miller, D.R., 1990. Air flow over and through a forest edge: a steady-state numerical simulation. Boundary-Layer Meteorology 51, 179–197. https://doi.org/10.1007/bf00120467.

Liu, J., Chen, J.M., Black, T.A., Novak, M.D., 1996. k-ϵ modelling of turbulent air flow downwind of a model forest edge. Boundary-Layer Meteorology 77, 21–44. https://doi.org/10.1007/bf00121857.

Mahaffee, W.F., Margairaz, F., Ulmer, L., Bailey, B.N., Stoll, R., 2023. Catching spores: linking epidemiology, pathogen biology, and physics to ground-based airborne inoculum monitoring. Plant Disease 107, 13–33. https://doi.org/10.1094/pdis-11-21-2570-fe.

Mahrt, L., 2014. Stably stratified atmospheric boundary layers. Annual Review of Fluid Mechanics 46, 23–45. https://doi.org/10.1146/annurev-fluid-010313-141354.

Marusic, I., Monty, J.P., 2019. Attached eddy model of wall turbulence. Annual Review of Fluid Mechanics 51, 49–74. https://doi.org/10.1146/annurev-fluid-010518-040427.

McNaughton, K., Laubach, J., 2000. Power spectra and cospectra for wind and scalars in a disturbed surface layer at the base of an advective inversion. Boundary-Layer Meteorology 96, 143–185. https://doi.org/10.1023/a:1002477120507.

Mitchell, S.J., 2013. Wind as a natural disturbance agent in forests: a synthesis. Forestry 86, 147–157. https://doi.org/10.1093/forestry/cps058.

Moeng, C.H., Sullivan, P.P., 1994. Comparison of shear- and buoyancy-driven planetary boundary layer flows. Journal of the Atmospheric Sciences 51, 999–1022. https://doi.org/10.1175/1520-0469(1994)051<0999:acosab>2.0.co;2.

Monin, A.S., Obukhov, A.M., 1954. Basic laws of turbulent mixing in the surface layer ofthe atmosphere. Transactions of the Academy of Sciences of the USSR, Geophysical Institute 24 (151), 163–187 (in Russian).

Novak, M.D., Warland, J.S., Orchansky, A.L., Ketler, R., Green, S., 2000. Wind tunnel and field measurements of turbulent flow in forests. Part I: uniformly thinned stands. Boundary-Layer Meteorology 95, 457–495. https://doi.org/10.1023/a:1002693625637.

Obukhov, A.M., 1946. Turbulence in an atmosphere with a non-uniform temperature. Tr. Inst. Teor. Geofiz. Akad. Nauk SSSR 1, 95–115. https://doi.org/10.1007/BF00718085.

Pan, Y., Chamecki, M., Isard, S.A., Nepf, H.M., 2015. Dispersion of particles released at the leading edge of a crop canopy. Agricultural and Forest Meteorology 211–212, 37–47. https://doi.org/10.1016/j.agrformet.2015.04.012.

Pan, Y., Follett, E., Chamecki, M., Nepf, H., 2014. Strong and weak, unsteady reconfiguration and its impact on turbulence structure within plant canopies. Physics of Fluids 26, 105102. https://doi.org/10.1063/1.4898395.

Patton, E.G., Finnigan, J.J., 2012. Canopy turbulence. In: Handbook of Environmental Fluid Dynamics, Volume One. CRC Press, pp. 311–327. https://doi.org/10.1201/b14241-28. chapter 24.

Patton, E.G., Horst, T.W., Sullivan, P.P., Lenschow, D.H., Oncley, S.P., Brown, O.J., Burns, S.P., Guenther, A.B., Held, A., Karl, T., Mayor, S.D., Rizzo, L.V., Spuler, S.M., Sun, J., Turnipseed, A., Allwine, E.J., Edburg, S.L., Lamb, B.K., Avissar, R., Calhoun, R.J., Kleissl, J., Massman, W.J., Paw U., K.T., Weil, J.C., 2011. The canopy horizontal array turbulence study. Bulletin of the American Meteorological Society 92, 593–611. https://doi.org/10.1175/2010bams2614.1.

Patton, E.G., Sullivan, P.P., Shaw, R.H., Finnigan, J.J., Weil, J.C., 2016. Atmospheric stability influences on coupled boundary layer and canopy turbulence. Journal of the Atmospheric Sciences 73, 1621–1647. https://doi.org/10.1175/jas-d-15-0068.1.

Paw, U., Qiu, J., Su, H.B., Watanabe, T., Brunet, Y., 1995. Surface renewal analysis: a new method to obtain scalar fluxes. Agricultural and Forest Meteorology 74, 119–137. https://doi.org/10.1016/0168-1923(94)02182-j.

Paw, U.K.T., Brunet, Y., Collineau, S., Shaw, R.H., Maitani, T., Qiu, J., Hipps, L., 1992. On coherent structures in turbulence above and within agricultural plant canopies. Agricultural and Forest Meteorology 61, 55–68. https://doi.org/10.1016/0168-1923(92)90025-y.

Perret, L., Patton, E.G., 2021. Stability influences on interscale transport of turbulent kinetic energy and Reynolds shear stress in atmospheric boundary layers interacting with a tall vegetation canopy. Journal of Fluid Mechanics 921, A14. https://doi.org/10.1017/jfm.2021.453.

Poggi, D., Porporato, A., Ridolfi, L., Albertson, J.D., Katul, G.G., 2004. The effect of vegetation density on canopy sub-layer turbulence. Boundary-Layer Meteorology 111, 565–587. https://doi.org/10.1023/b:boun.0000016576.05621.73.

Poëtte, C., Gardiner, B., Dupont, S., Harman, I., Böhm, M., Finnigan, J., Hughes, D., Brunet, Y., 2017. The impact of landscape fragmentation on atmospheric flow: a wind-tunnel study. Boundary-Layer Meteorology 163, 393–421. https://doi.org/10.1007/s10546-017-0238-1.

Py, C., de Langre, E., Moulia, B., 2006. A frequency lock-in mechanism in the interaction between wind and crop canopies. Journal of Fluid Mechanics 568, 425. https://doi.org/10.1017/s0022112006002667.

Raupach, M.R., 1981. Turbulence in and above plant canopies. Annual Review of Fluid Mechanics 13, 97–129. https://doi.org/10.1146/annurev.fl.13.010181.000525.

Raupach, M.R., Antonia, R.A., Rajagopalan, S., 1991. Rough wall turbulent boundary layers. Applied Mechanics Reviews 44, 1–25. https://doi.org/10.1115/1.3119492.

Raupach, M.R., Coppin, P.A., Legg, B.J., 1986. Experiments on scalar dispersion within a model plant canopy part I: the turbulence structure. Boundary-Layer Meteorology 35, 21–52. https://doi.org/10.1007/bf00117300.

Raupach, M.R., Finnigan, J.J., Brunet, Y., 1996. Coherent eddies and turbulence in vegetation canopies: the mixing-layer analogy. Boundary-Layer Meteorology 78, 351–382. https://doi.org/10.1007/BF00120941.

Rudnicki, M., Meyer, T.H., Lieffers, V.J., Silins, U., Webb, V.A., 2008. The periodic motion of lodgepole pine trees as affected by collisions with neighbors. Trees 22, 475–482. https://doi.org/10.1007/s00468-007-0207-2.

Salesky, S.T., Anderson, W., 2018. Buoyancy effects on large-scale motions in convective atmospheric boundary layers: implications for modulation of near-wall processes. Journal of Fluid Mechanics 856, 135–168. https://doi.org/10.1017/jfm.2018.711.

Salesky, S.T., Chamecki, M., 2012. Random errors in turbulence measurements in the atmospheric surface layer: implications for Monin–Obukhov similarity theory. Journal of the Atmospheric Sciences 69, 3700–3714. https://doi.org/10.1175/jas-d-12-096.1.

Schelhaas, M.J., Nabuurs, G.J., Schuck, A., 2003. Natural disturbances in the European forests in the 19th and 20th centuries. Global Change Biology 9, 1620–1633. https://doi.org/10.1046/j.1365-2486.2003.00684.x.

Schindler, D., 2008. Responses of Scots pine trees to dynamic wind loading. Agricultural and Forest Meteorology 148, 1733–1742. https://doi.org/10.1016/j.agrformet.2008.06.003.
Schindler, D., Vogt, R., Fugmann, H., Rodriguez, M., Schönborn, J., Mayer, H., 2010. Vibration behavior of plantation-grown Scots pine trees in response to wind excitation. Agricultural and Forest Meteorology 150, 984–993. https://doi.org/10.1016/j.agrformet.2010.03.003.
Schmitt, F., Schertzer, D., Lovejoy, S., Brunet, Y., 1996. Multifractal temperature and flux of temperature variance in fully developed turbulence. Europhysics Letters 34, 195–200. https://doi.org/10.1209/epl/i1996-00438-4.
Shapkalijevski, M., Moene, A.F., Ouwersloot, H.G., Patton, E.G., Vila-Guerau de Arellano, J., 2016. Influence of canopy seasonal changes on turbulence parameterization within the roughness sublayer over an orchard canopy. Journal of Applied Meteorology and Climatology 55, 1391–1407. https://doi.org/10.1175/jamc-d-15-0205.1.
Shaw, R.H., Schumann, U., 1992. Large-eddy simulation of turbulent flow above and within a forest. Boundary-Layer Meteorology 61, 47–64. https://doi.org/10.1007/bf02033994.
Smith, F., 1975. Turbulence in the atmospheric boundary layer. Science Progress 62, 127–151.
Stull, R.B., 1988. An Introduction to Boundary Layer Meteorology. Kluwer, Netherlands.
Taylor, R., 1958. Thermal structures in the lowest layers of the atmosphere. Australian Journal of Physics 11, 168. https://doi.org/10.1071/ph580168.
Thomas, C.K., Selker, J., 2021. Optical fiber-based distributed sensing methods. In: Springer Handbook of Atmospheric Measurements. Springer International Publishing, pp. 609–631.
Vogel, S., 1989. Drag and reconfiguration of broad leaves in high winds. Journal of Experimental Botany 40, 941–948. https://doi.org/10.1093/jxb/40.8.941.
Vollsinger, S., Mitchell, S.J., Byrne, K.E., Novak, M.D., Rudnicki, M., 2005. Wind tunnel measurements of crown streamlining and drag relationships for several hardwood species. Canadian Journal of Forest Research 35, 1238–1249. https://doi.org/10.1139/x05-051.
Warhaft, Z., 2000. Passive scalars in turbulent flows. Annual Review of Fluid Mechanics 32, 203–240. https://doi.org/10.1146/annurev.fluid.32.1.203.
Watanabe, T., Shimoyama, K., Kawashima, M., Mizoguchi, Y., Inagaki, A., 2020. Large-eddy simulation of neutrally-stratified turbulent flow within and above plant canopy using the central-moments-based lattice Boltzmann method. Boundary-Layer Meteorology 176, 35–60. https://doi.org/10.1007/s10546-020-00519-8.
Weligepolage, K., Gieske, A.S.M., van der Tol, C., Timmermans, J., Su, Z., 2012. Effect of sub-layer corrections on the roughness parameterization of a Douglas fir forest. Agricultural and Forest Meteorology 162–163, 115–126. https://doi.org/10.1016/j.agrformet.2012.04.017.
van de Wiel, B.J.H., Moene, A.F., Jonker, H.J.J., Baas, P., Basu, S., Donda, J.M.M., Sun, J., Holtslag, A.A.M., 2012. The minimum wind speed for sustainable turbulence in the nocturnal boundary layer. Journal of the Atmospheric Sciences 69, 3116–3127. https://doi.org/10.1175/jas-d-12-0107.1.
Wyngaard, J.C., 2010. Turbulence in the Atmosphere. Cambridge University Press.
Yang, B., Raupach, M.R., Shaw, R.H., U, K.T.P., Morse, A.P., 2006. Large-eddy simulation of turbulent flow across a forest edge. Part I: flow statistics. Boundary-Layer Meteorology 120, 377–412. https://doi.org/10.1007/s10546-006-9057-5.
Yang, M., Défossez, P., Dupont, S., 2020. A root-to-foliage tree dynamic model for gusty winds during windstorm conditions. Agricultural and Forest Meteorology 287, 107949. https://doi.org/10.1016/j.agrformet.2020.107949.
Young, G.S., Kristovich, D.A., Hjelmfelt, M.R., Foster, R.C., 2002. Rolls, streets, waves, and more: a review of quasi-two-dimensional structures in the atmospheric boundary layer. Bulletin of the American Meteorological Society 83, 997–1002. https://doi.org/10.1175/BAMS-83-7-Young.

Zorzetto, E., Bragg, A.D., Katul, G., 2018. Extremes, intermittency, and time directionality of atmospheric turbulence at the crossover from production to inertial scales. Physical Review Fluids 3, 094604. https://doi.org/10.1103/physrevfluids.3.094604.

Harnessing the atmospheric wind: analysis and modeling of wind farm flows

10

Marc Calaf[a] *and Giacomo Valerio Iungo*[b]
[a]Department of Mechanical Engineering, University of Utah, Salt Lake City, UT, United States, [b]Department of Mechanical Engineering, University of Texas at Dallas, Richardson, TX, United States

10.1 Introduction

Modern wind turbines generate electricity by converting the wind's axial momentum into angular momentum around the axis of the turbine, which is in turn connected to a power generator that will convert the mechanical energy into electricity (Burton et al., 2011). The conversion from axial to angular momentum is achieved by the lift forces generated by the aerodynamic efficiency of the blade airfoils. In this transformation, as the wind flow approaches the wind turbine rotor, it senses the blockage by a progressive increase in the pressure field and a slight reduction in the flow velocity. Once the air moves across the wind turbine blades, there is a sudden drop in the pressure field, a substantial reduction in mean momentum, and an increase in turbulence intensity. The latter is typically defined as the ratio of the standard deviation to the mean of the streamwise velocity component ($TI = \sigma_u/U$). The flow region downstream of the wind turbine rotor is denoted as the wake region, and its attributes depend mainly on the aerodynamic characteristics of the wind turbine blades (i.e., the rotor thrust coefficient), as well as the background ABL configuration and surface characteristics (Stevens and Meneveau, 2017). Further, the downstream evolution of wind turbine wakes and the gradual recovery to the freestream conditions via turbulent mixing are directly affected by the background turbulence characteristics of the incoming wind field (van der Laan et al., 2022a).

Traditionally, wind turbines have mainly operated within the atmospheric surface layer (ASL); however, with the continuous growth of wind turbine rotor diameters, some turbines now operate even above the ASL. Before the beginning of the current millennium, the average hub height and rotor diameter of wind turbines deployed in the USA were smaller than $\sim$ 50 m. Today, these statistics have doubled, with the largest wind turbine prototypes for offshore applications reaching rotor diameters of over 300 m and capacities closer to 20 MW (Ashuri et al., 2016; Koragappa and Verdin, 2024). Therefore the wind flows interacting with wind turbines are characterized by a wide range of time- and length-scales (e.g., height-dependent flow through a blade rotational cycle, flow variability through the ABL diurnal cycle, and the seasonal flow variations). Nonetheless, according to the International Electrotechnical Commission (IEC) 61400-1 standards, the probing and classification

Environmental Turbulence. https://doi.org/10.1016/B978-0-32-395873-8.00017-6
Copyright © 2026 Elsevier Inc. All rights are reserved, including those for text and data mining, AI training, and similar technologies.

of incoming wind conditions are standardized to averaging periods of 10 minutes (International Electrotechnical Commission, 2019).

To increase economic profitability and leverage advantageous economic scale factors, wind turbines are typically deployed in arrays, which are referred to as wind farms (Herbert-Acero et al., 2014; Hietanen et al., 2024). However, clustering of wind turbines leads to hazards related to wind turbine wake interactions. The wakes generated by upstream wind turbines may, under specific wind and atmospheric conditions, reach downstream machines, which are then affected by lower power capture and increased fatigue loads associated with the wake-generated turbulence. Therefore optimizing the layout of wind farms for specific wind resources and atmospheric conditions, together with logistic-economic constraints, is a critical task for the design of wind farms (Chowdhury et al., 2012; Chen et al., 2013; Shakoor et al., 2016; Santhanagopalan et al., 2018).

On larger spatial and temporal scales, the cumulative wake generated by a large array of wind turbines can travel downwind for very long distances, causing detrimental effects for neighboring wind farms, i.e., farm-to-farm interactions, especially for operations under stably stratified atmospheric conditions and offshore wind farms (Christiansen and Hasager, 2005; Nygaard, 2014; van der Laan et al., 2022b). Once again, the propagation and evolution of the wind farm wake will be highly dependent on local surface conditions (e.g., onshore vs. offshore) and local ABL conditions (Krishnamurthy et al., 2025).

When studying the interactions between wind turbines and the ABL, it is useful to consider the problem at different scales, namely, at a wind turbine scale and then at the wind farm scale. At the wind turbine scale, different flow characteristics are encountered, whether it is upstream of the wind turbine, in the near wake region, or in the far wake region downstream of the turbines. Similarly, the flow dynamics are significantly different in different regions of a wind farm. These regions were partially captured in the famous Horns Rev wind farm picture represented in Fig. 10.1, and their characterization in terms of ABL flow dynamics will be the partial focus of this chapter.

Specifically, this chapter aims to provide a characterization of the different flow regions and characteristics encountered around a single wind turbine, as well as within a large array of turbines, with a specific emphasis on their synergistic interaction with the background ABL configurations and surface conditions. As a result, this chapter is organized to cover the topics of the wind turbine wake, the wind turbine array boundary layer, and the perturbations induced by large wind farms on the canonical structure of the ABL, including some notes on offshore wind energy and simulation approaches. The chapter concludes with some perspectives and considerations about the remaining challenges in the study of turbulence in wind farms.

10.2 The wind turbine wake

The velocity field of a wind-turbine wake can generally be divided into (a) a near-wake region and (b) a far-wake region (Vermeer et al., 2003). The former typically has a length-scale of less than three rotor diameters (D) downwind from the turbine (Crespo et al., 1996), and the latter can expand beyond $7D$ downstream from the rotor disk, depending on atmospheric stability and local wind conditions.

Figure 10.1 Picture of the famous Horns Rev wind farm fog case, where the corresponding wind turbine wakes became visible to the naked eye by the water vapor tracer (Hasager et al., 2013).

Flow characteristics in the near wake region are strongly correlated with the specific blade aerodynamics, stalled flow, tip vortices, as well as the nacelle and tower geometries (Vermeer et al., 2003; Iungo et al., 2013; Viola et al., 2014; Ashton et al., 2016; Debnath et al., 2017). Further, the main vorticity structures generated from the rotation of the turbine blades are the helicoidal tip vortices detached from the blade tips, which in their turn induce an additional velocity field (Fukumoto and Okulov, 2005), and a hub vortex (Felli et al., 2011; Iungo et al., 2013; Viola et al., 2014). The hub vortex is a streamwise-oriented vorticity structure located approximately at the center of the wake, generated at the blade root and the turbine nacelle. Alternatively, the blade tip vortices are formed by the trailing vortices associated with the circulation distribution throughout the blade span (Sherry et al., 2010), which gradually rolls up, forming the tip vortices. These are then advected, forming a characteristic helicoidal structure as a result of the velocity induced by the blade rotation and the incoming velocity field (Medici and Alfredsson, 2006; Chamorro and Porté-Agel, 2009, 2010; Lu and Porté-Agel, 2011; Zhang et al., 2012). Eventually, through different instability mechanisms, these blade tip vortices break down (Okulov and Sorensen, 2004; Okulov and Sørensen, 2007), contributing to the characteristics of the far wake flow. See the sketch introduced in Fig. 10.2 for a graphical depiction of these flow regions.

The transition between the near and far wake cannot be precisely delineated (Sørensen, 2011), although it may be associated with the flow region where the vorticity structures remain concentrated and not significantly diffused, but can also

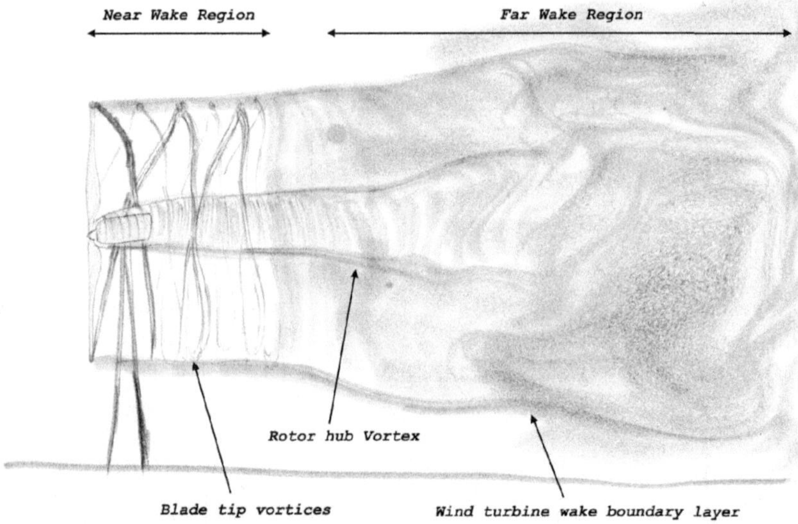

Figure 10.2 Sketch of the flow around a single wind turbine with the near-wake region, the far-wake region, the blade-tip, and rotor-hub vortices.

undergo different instability mechanisms (Medici and Alfredsson, 2008; Felli et al., 2011; Sarmast et al., 2014; Hong et al., 2014; Quaranta et al., 2015; Lignarolo et al., 2015). For the tip helicoidal vortices, the main instability processes are connected with the mutual induction among the adjacent vortices and the radial shear generated between the wake velocity field and the background velocity field (Widnall, 1972; Okulov and Sørensen, 2007; Ivanell et al., 2010). Downstream of the instability of the helicoidal tip vortices, the hub vortex also becomes unstable (Felli et al., 2011; Iungo et al., 2013; Kang et al., 2014), developing a single-helix counter-winding instability with a frequency roughly equal to 0.3 times the hub rotational frequency, which is independent of the number of blades (Viola et al., 2014; Ashton et al., 2016).

In the far-wake region, vortices are practically diffused, the pressure roughly recovers to the freestream condition, while the wake velocity field shows a smooth quasi-Gaussian velocity profile, which has spurred the development of analytical models based on the achieved self-similarity (Porté-Agel et al., 2020). Global wake instabilities also occur in the far-wake region, which are generally ascribed to the phenomenon of wake meandering (España et al., 2011; Larsen et al., 2007; Keck et al., 2014; Larsen et al., 2015; Foti et al., 2016). This long-wave instability is mainly affected by the characteristics of the incoming wind, atmospheric stability, and surface roughness. Further downstream, the wake gradually recovers the incoming wind velocity distribution through turbulent fluxes and mean momentum advection.

The momentum deficit (defined as the local momentum minus the upstream reference) that develops behind the wind turbine blades is mainly dependent on the efficiency of the wind turbine blades and their configuration to generate lift. However, the shape, intensity, and length of the wind turbine far-wake are strictly related to

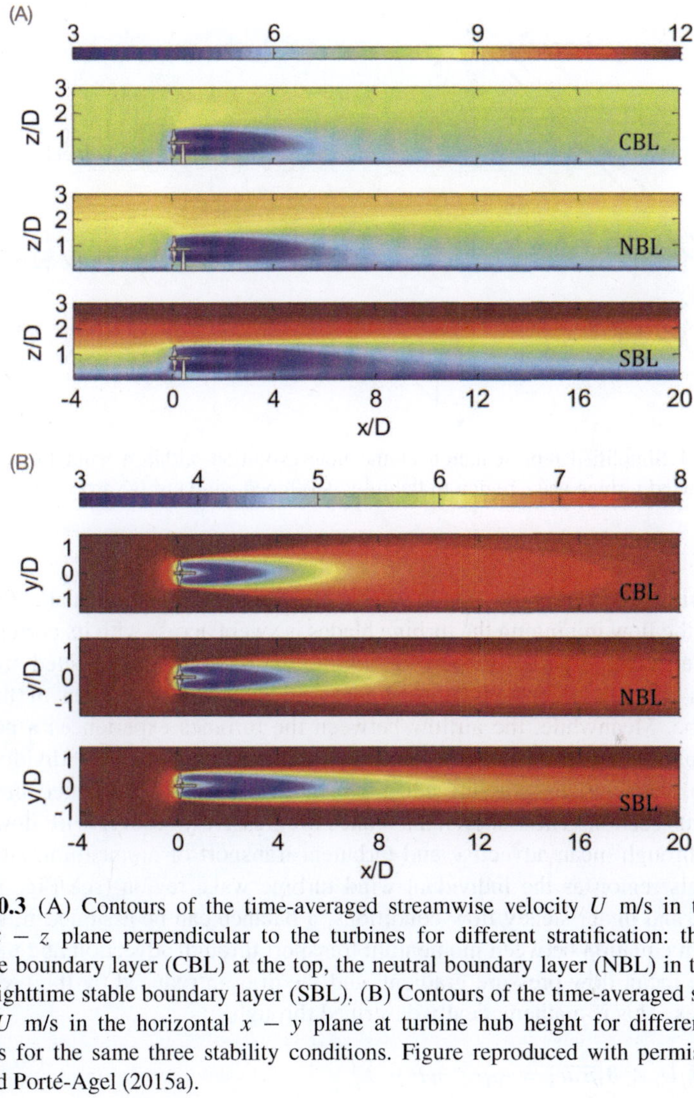

Figure 10.3 (A) Contours of the time-averaged streamwise velocity U m/s in the middle vertical $x - z$ plane perpendicular to the turbines for different stratification: the daytime convective boundary layer (CBL) at the top, the neutral boundary layer (NBL) in the middle, and the nighttime stable boundary layer (SBL). (B) Contours of the time-averaged streamwise velocity U m/s in the horizontal $x - y$ plane at turbine hub height for different stability conditions for the same three stability conditions. Figure reproduced with permission from Abkar and Porté-Agel (2015a).

the local atmospheric flow (i.e., intensities of the wind shear and veer) and the ABL stability. Fig. 10.3 illustrates the dependence on the wind turbine wake shape and intensity as a function of the local ABL stability.

10.3 The wind turbine array boundary layer

Similarly to what happens in forested canopies, as an air mass flows towards a wind farm, it senses a progressive increase in atmospheric pressure resulting from the flow obstruction produced by the wind turbines. As the air flow reaches the first wind

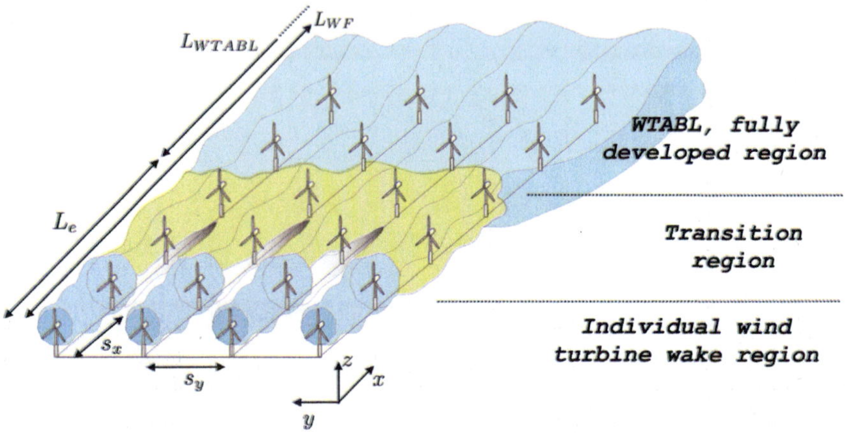

Figure 10.4 Simplified representation of the flow evolution within a wind farm, from the individual wind turbine wake region to the fully developed wind turbine array boundary layer, WTABL.

turbines within the wind farm, about 2/3 (i.e., the Betz limit, (Burton et al., 2011; Betz, 2014)) of the flow impinging the turbine blades is swept across with its corresponding kinetic energy being partially harvested, while the other 1/3 is deflected around the turbine as a result of conservation of mass and the increased area of the swept stream tube. Meanwhile, the airflow between the turbines experiences a noticeable acceleration. In the region behind each one of the upstream turbines, individual wakes form with characteristic pressure and momentum deficits, as explained in detail in the previous section. These individual wakes progressively recover with downstream distance through mean advective and turbulent transport of momentum. Herein, we refer to this region as the individual wind turbine wake region (see Fig. 10.4). At this stage, and under steady flow conditions, a balance can be assumed in the mean momentum equation between momentum transport through advection and turbulence fluxes, the mean flow pressure gradient, and the drag introduced by the presence of the turbines. This is mathematically described through

$$U_j \partial_j U_i + \partial_j \overline{u_i u_j} = -\rho^{-1} \partial_i P - F_i^t \tag{10.1}$$

where capital letters indicate mean flow, lower case represents zero-mean turbulent fluctuations, ρ is the mean air flow density, and i and j are indices representing the Cartesian coordinates $i = 1, 2, 3 = x, y, z$.

In the case of isolated wind turbines, the wake gradually recovers to the freestream wind conditions over a downstream extent (distance referred to as L_x) that is mainly governed by the atmospheric stability regime and background turbulence (see the changes in wake deficit and structure illustrated in Fig. 10.3). In contrast, when turbines are installed in large arrays with numerous turbines downstream of each other, the flow dynamics change, and one must consider the successive interactions of multiple wind turbine wakes, as well as the local atmospheric boundary-layer stability.

The evolution of multiple wind turbine wakes and associated wake interactions is crucial for the harvested power from the wind farm, as well as for the added fatigue loads on downwind wind turbines and the potential hazard to the reduced lifespan of the generators. See the schematic in Fig. 10.4 and the specific wind farm examples illustrated in Fig. 10.5.

Because turbines have traditionally been installed at distances (s_x) smaller than the distance needed for the wind turbine wake to fully recover ($s_x < L_x$) based on economic arguments, the upstream turbine wakes interact with the downstream wind turbines. As a result, the flow approaching the following downstream turbine is still partially depleted of momentum; furthermore, it has increased turbulence levels. With this weakened upstream flow, the downstream turbine extracts once again about 2/3 of the incoming perturbed momentum and produces a proportional wake, leading to

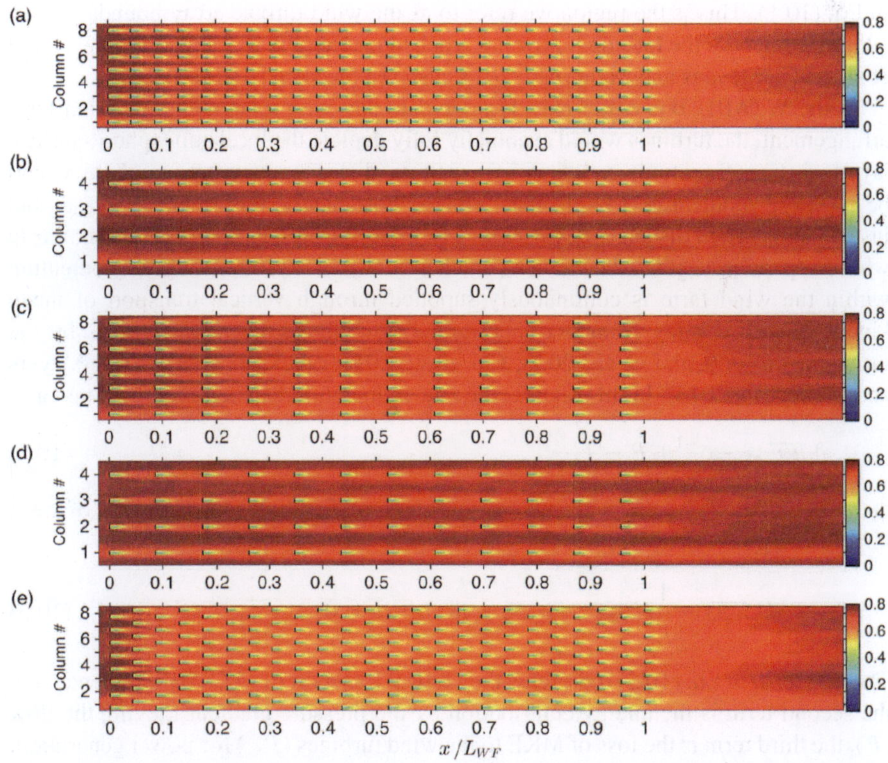

Figure 10.5 Mean wind velocity for different finite-size wind-farm configurations where the corresponding streamwise (s_x) and spanwise (s_y) spacing is changed. The velocity values are normalized by the far upstream velocity. Note that the figure represents only 25% of the full numerical domain. Results were obtained with large eddy simulations, with a domain of $\sim$ 100 km length and 3 km width, and a wind farm of length $\sim$ 20 km. The numerical resolution employed was $\sim$ 20 m in the horizontal directions, and 7 m in the vertical directions. Figure has been reproduced from Sharma et al. (2018).

a more severe downstream momentum deficit. This process is successively repeated with the perturbed flow interacting further downstream with the following turbines.

Additionally, as the upstream wind turbine wakes propagate downstream, they also expand radially, perpendicular to the mean wind direction (i.e., in the horizontal and vertical directions). As a result of the lateral expansion (L_y), when the wind turbine wake becomes larger than the lateral spacing between the turbines (s_y), the wakes of an upstream turbine not only interact with the immediate downstream turbine, but also with the wakes of the turbines located on the sides of the downstream turbine. Therefore the momentum deficit also propagates within the initial high-momentum channels formed between the turbines, thus progressively weakening the mean advective transport through which turbine momentum deficits could be initially recovered. Herein, we refer to this region as the wind turbine array transition region (see Fig. 10.4). As these successive wind turbine wake interactions continue, the flow progressively evolves towards a new equilibrium that differs from the one identified in Eq. (10.1). This is the region we refer to as the wind turbine array boundary layer (WTABL), where the turbulent flow has reached a new fully developed regime (Calaf et al., 2010; Stevens and Meneveau, 2017, also see the sketch in Fig. 10.4).

In the WTABL, one could naively anticipate that with a large enough wind farm arrangement, the turbines would eventually fully deplete the local atmospheric surface layer of its momentum, and thus at a certain distance downstream, there would be no more momentum to spin the turbines. Fortunately, this is not the case, and instead at a certain distance downstream a new equilibrium is reached (i.e., similar to what happens in vegetated canopies (Belcher et al., 2003, 2012)), where momentum within the wind farm is continuously supplied through vertical transport of mean kinetic energy coming from the unperturbed flow above the wind turbines by the turbulent fluxes (Calaf et al., 2010; Cal et al., 2010; Markfort et al., 2012; Meyers and Meneveau, 2013). In this regime, the new momentum balance can be written as,

$$\partial_z \overline{uw} = -\rho^{-1} \partial_x P - F_x^t, \tag{10.2}$$

and the corresponding mean kinetic energy equation, illustrating the exchange of mean kinetic energy is also simplified to

$$-\partial_z (U\, \overline{uw}) - \frac{1}{\rho} U \partial_x P - U\, F_x^t + (\overline{uw}) \partial_z U = 0. \tag{10.3}$$

The first term in Eq. (10.3) represents the vertical transport of MKE by turbulence (ϕ), the second term is the energy contribution by the pressure gradient moving the flow (P), the third term is the loss of MKE to the wind turbines (P_{wt}) for power generation, and the last term represents the rate at which kinetic energy is being lost by conversion to turbulent kinetic energy (ϵ).

Several studies (Calaf et al., 2010; Cal et al., 2010; Markfort et al., 2012) have shown that the main players in the MKE balance are the vertical transport of MKE by turbulence and the turbine-absorbed MKE, which, to a first-order approximation, compensate each other. This clearly illustrates that, well within a wind farm, the flow field reaches an equilibrium between its capacity of replenishment and extraction

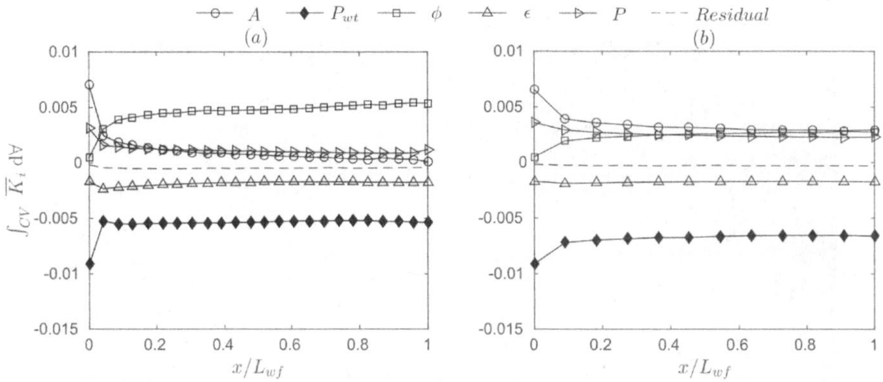

Figure 10.6 Terms of the mean kinetic energy budget (A: horizontal advection of MKE, P_{wt}: harvested power by the wind turbines, ϕ: turbulent flux of MKE, ϵ: dissipation of MKE, P: work due to the pressure gradient) for two different examples wind farm arrangement cases (left, AL-1 case with $s_x = s_y = 5.5D$; right, AL-4 case with $2s_x$, $2s_y$ spacing) as a function of the normalized wind farm length, $x = L_{wf}$. In addition to the MKE terms, the residual is illustrated with a dashed line. The y-axis represents the corresponding control volume integrated MKE budget terms at different downstream locations. Figure reproduced from Cortina et al. (2020).

by the turbines (Meyers and Meneveau, 2013) (also see the example presented in Fig. 10.6). The distance at which this new balance is obtained is here referred to as L_e, as a direct reference to the entrance length, but it has also been identified as L_c (Markfort et al., 2018) in similitude with ABL theory over vegetated canopies (Belcher et al., 2003, 2012).

Beyond the last row of wind turbines, the turbulent flow continues to evolve downstream of the wind farm with a significant momentum deficit that can propagate for tens of kilometers downstream (see Fig. 10.7). This phenomenon has sparked controversy when installing newer wind farms upstream or downstream of other pre-existing ones, coining the term of "*wind stealing*" or more commonly referred to as farm-to-farm interactions (Lundquist et al., 2019; Schneemann et al., 2020). Currently, little is known about the recovery process of the wind farm wake and its interactions at mesoscales, and there is a lack of simplified models that facilitate studying the process of farm-to-farm interactions (Nygaard, 2014; Moss et al., 2023; Nygaard et al., 2020; Stieren and Stevens, 2022).

In summary, the wind turbine array boundary layer presents many similarities to that of vegetated canopies. The details of the structure and variables used to describe its characteristics are presented in Fig. 10.8, with the corresponding induction zone (L_i), the entrance zone (L_e), the WTABL region (L_{WTABL}), and the wind farm wake region (L_w), together with the corresponding roughness-sublayer height (δ_r) and the corresponding internal boundary layer developed above (δ_{IBL}). The values of the corresponding length-scales depend on the wind farm arrangement, the incoming wind, the local atmospheric stratification, and the surface conditions. Fig. 10.9 provides an illustrative example of how the recovery of individual wind

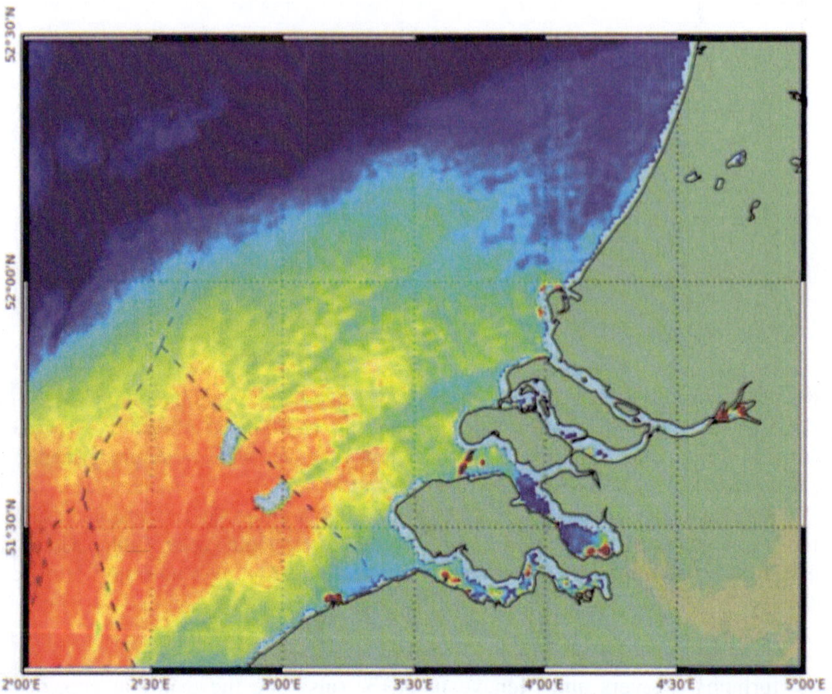

Figure 10.7 Image from the Synthetic Aperture Radar (SAR), depicting wind speed in the North Sea, with two wind farms in the border of the territorial waters of Belgium, with their corresponding wakes propagating downstream over large distances and into the territorial water of the Netherlands (borders are demarcated by dashed lines). Image obtained from Christiansen and Hasager (2005).

turbine wakes changes within a large wind farm based on ABL stability. As a result of this dependence, the overall energy harvested by the wind farm is also ABL stability dependent. In general, higher power outputs are often obtained during stably stratified flow conditions in which the wind farm is penetrated by a stark low-level jet.

Conceptual models have been proposed to describe the details of individual wind turbine wakes (i.e., near and far wake region) (Stevens and Meneveau, 2017), the entrance region (Markfort et al., 2018), as well as the structure and characteristics of the proper WTABL (Meneveau, 2012; Meyers and Meneveau, 2013; VerHulst and Meneveau, 2014). However, little is known about the wind farm impact region, the structure of the roughness sublayer developed above the wind turbines, the extent of the internal boundary layer developed above the wind farm, or the characteristics of the wind farm wake region. Therefore it is of paramount importance that we continue to expand our understanding of the flow around wind farms because this is not only critical in the design of new farms, and their corresponding installation with respect to other upstream/downstream existing installations, but also for the well-being of humans and wildlife, as well as the sustainability of activities that may develop in nearby areas.

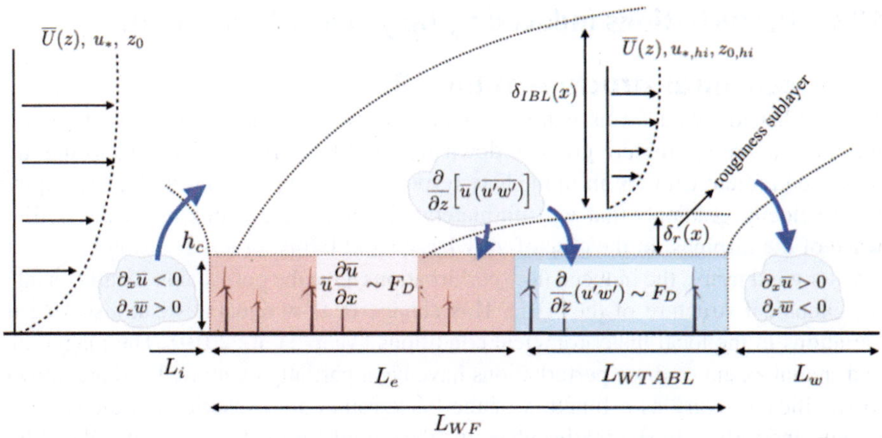

Figure 10.8 Schematic of the structure of the wind turbine array boundary layer, where L_i is the impact length, L_e is the entrance length, L_{WTABL} is the wind turbine array boundary layer length, L_w is the length of the wind farm's wake, and L_{WF} is the total length of the wind farm. Further, $\overline{U}(z)$ is the mean wind speed profile, u_* is the friction velocity, and z_0 is the roughness length. Variables with subindex *-hi* refer to the corresponding variables observed from above the WTABL.

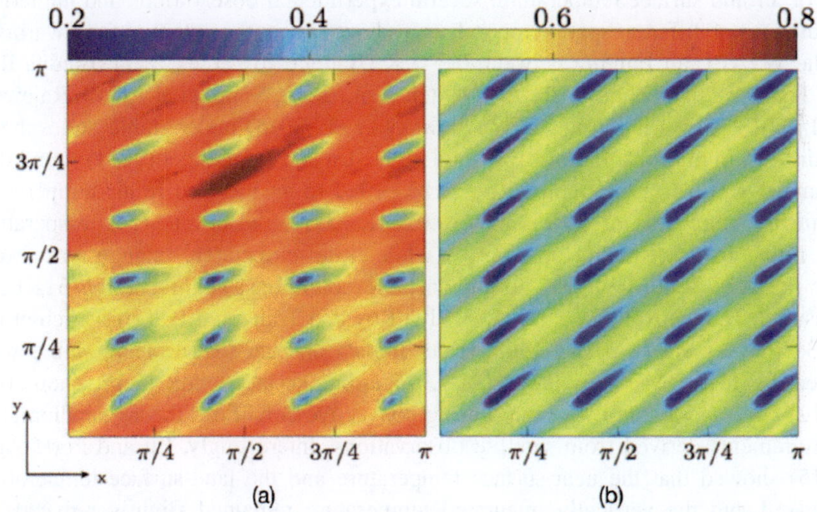

Figure 10.9 Horizontal slices of 10-min-averaged, normalized velocity magnitude fields in an idealized wind farm at hub-height for two characteristic periods during convective daytime (left), and stable nighttime (right) stratification. Data was obtained using large eddy simulations by Sharma et al. (2017).

10.4 Perturbations induced by large wind farms on the canonical structure of the ABL

As established, wind farms extract kinetic energy from the atmospheric flow and introduce a strong pressure gradient downstream of the turbines. This means that the pressure gradient triggers an immediate response in the mean wind field, developing large velocity gradients that will ultimately trigger enhanced mixing in the trailing wake of the turbines. In the case of very large wind farms, or a succession of several smaller wind farms, the induced flow perturbations have the potential to locally change the canonical structure of the ABL. This change in flow dynamics can also lead to variations in the local meteorological conditions (Veers et al., 2019). The magnitude and spatial extent of these perturbations have been partially quantified and are known to significantly vary as a function of the background atmospheric conditions, local terrain, and turbine type, greatly enhancing the complexity of the problem and making the development of universal relations a daunting task.

The perturbations investigated so far include changes in local air and surface temperature, near-surface vertical mixing, heat and moisture fluxes, formation of the low-level jet, height of the boundary layer, downwind ABL perturbations, as well as changes in cloud cover and precipitation. Further details on each of these are provided below, with the list not being exhaustive, but rather illustrative of the potential changes induced by the interaction of different wind farm and ABL configurations.

For air and surface temperature, several experimental observations and numerical simulations at different spatiotemporal scales have shown the potential for a warming of the WTABL air ranging between 0.2–2 K (Baidya Roy et al., 2004; Baidya Roy and Traiteur, 2010; Zhou et al., 2012; Fitch et al., 2013; Smith et al., 2013; Rajewski et al., 2013; Cervarich et al., 2013; Slawsky et al., 2015) under stable atmospheric conditions. In high-resolution LES simulations, Lu and Porté-Agel (2011) found that enhanced vertical mixing caused by turbines led to an increase in air temperature within the rotor area of approximately 0.5 K. Above the turbine, the temperature was found to decrease by up to 1 K, while little temperature change was found near the surface. In contrast, during unstable conditions, slight cooling has been measured (Baidya Roy et al., 2004; Baidya Roy and Traiteur, 2010; Cervarich et al., 2013; Slawsky et al., 2015), with greater impact on near-surface temperature with increasing TKE induced by the turbines. Regarding surface temperature, Zhou et al. (2012) observed a small warming during convective conditions through radiometric measurements derived from satellite observations. Interestingly, Lu and Porté-Agel (2015) showed that the near-surface temperature and the land-surface temperature increased, but the vertically integrated temperature remained slightly reduced. In their work, they also noted that the warming effect on the land surface is due to the redistribution of available energy between the soil and the air.

These variations in near-surface and surface temperature are explained by the enhanced vertical mixing promoted by the turbines within and above the rotor disk height, where the vertical fluxes of mean kinetic energy are of the same order of magnitude as the power extracted by the turbines (Calaf et al., 2010; Cal et al., 2010), as well as due to a large reduction in momentum flux (30%) near the surface (Calaf et al., 2010; Lu and Porté-Agel, 2011). For example, during convective conditions, it has been observed (Fitch et al., 2013) that wind farms can cause an increase of TKE throughout the depth of the BL with a maximum increase of 33% in the upper half of the rotor area. During stable conditions, the relative increase in TKE is much greater, given that ambient turbulence is low, and the power production by the turbines is greater. At the surface, changes in mixing can also lead to perturbations of the local heat, moisture, and CO_2 fluxes.

For example, during daytime, a five-fold increase in CO_2 and water vapor flux was observed at a distance 3 to 5 diameters downstream of a given turbine (Rajewski et al., 2014). At night, the signature of these changes was much more attenuated; however, there remained a strong statistical signal of changes in CO_2 and water vapor. These changes in scalar mixing respond to the balance between the momentum transport above the wind turbines and the reduced momentum transport near the surface, as partially explained in (Calaf et al., 2011).

Above the WTABL, significant perturbations have also been identified at the top of the ABL, with changes in the height of the ABL and the entrainment flux near the top, both of which have been related to the wind farm loading and the background stratification (Abkar and Porté-Agel, 2013; Sharma et al., 2017; Allaerts and Meyers, 2015, 2017). In stable conditions, the ABL height was found to increase by almost a factor of four above the wind farm, while during convective conditions the change was otherwise negligible (Fitch et al., 2013). Fig. 10.11 illustrates the change in ABL height, measured in numerical simulations of an infinite wind farm through a realistic diurnal cycle (Sharma et al., 2017). Results illustrate an increase of more than 250 m during stable conditions and of about 100 m during the convective regime. At the same time, the ratio between entrainment-flux and surface-flux has been shown to almost double (e.g., from 29% to 48%) due to the presence of a large wind farm (Lu and Porté-Agel, 2015). The enhanced entrainment at the top of the boundary layer is indicative of an increased downward flux of relatively warmer air in the entrainment layer. In the presence of large wind farms, this change in vertical mixing was also shown to modify the overall thickness of the entrainment zone and to alter the growth of the convective boundary layer (see Figs. 10.10 and 10.11).

Out of all these features induced by the presence of large wind farms, probably the one of most relevance, at least for the wind energy industry, relates to the perturbations induced on the LLJ. Several studies (Lu and Porté-Agel, 2011; Fitch et al., 2013; Sharma et al., 2017) have shown that, depending on the wind farm loading (i.e., number of wind turbines, spacing, and thrust coefficient), the LLJ formed during stably stratified periods can either flow through the wind turbines, thus leading to a high energy output, or be vertically displaced to a region above the turbines. In the later case, energy output can be severely affected, with a much lower energy production than the one developers could have anticipated. At this time, we do not have an analytical model capable of a priori determining under what configurations this might or might not occur, and thus one would require detailed numerical simulations to assess the issue beforehand.

Interestingly, the ABL perturbations induced by large wind farms are not necessarily limited to the vertical space immediately above the wind farm, but can

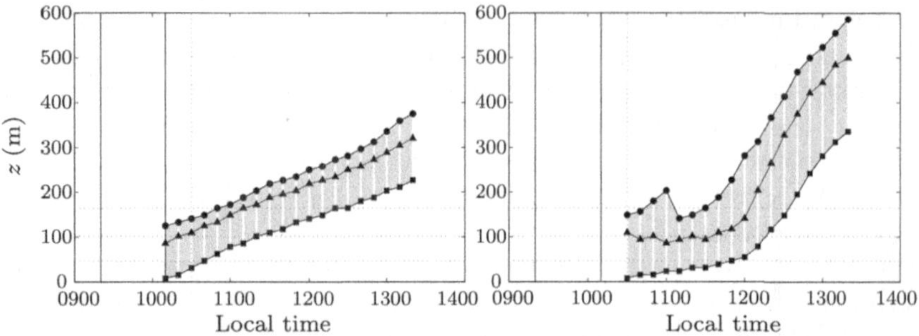

Figure 10.10 Left: Entrainment region for an ABL study case with no wind turbines. Right: same diurnal cycle study case, including a wind farm. The lower edge of the shaded region (marked by square symbols) represents the height where the heat flux becomes zero, and up to which the ABL is subjected to a positive sensible heat flux. The triangle symbols mark the height where the maximum value of absolute (negative/downward) heat flux is found in the entrainment region. The upper edge of the entrainment region, marked by circular, symbols is the height where the heat flux becomes zero once again. Each subplot contains 3 vertical lines, which, beginning from the left, represent the "sunrise" (solid), morning transition for the no wind turbine case (solid), and for the wind farm case (dotted), respectively. The three horizontal lines represent the rotor disk region of the wind farm. Figure has been reproduced from Sharma et al. (2017).

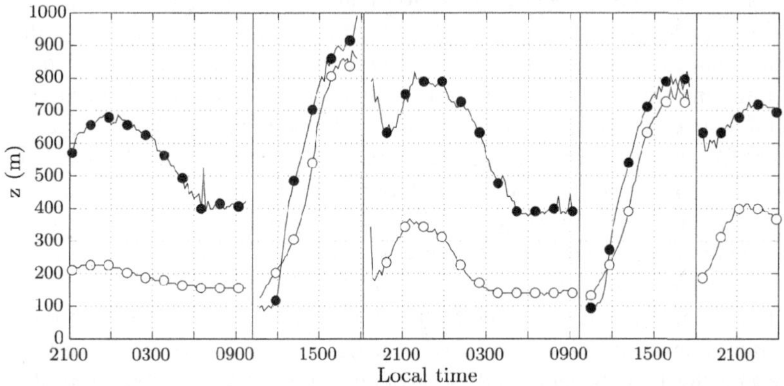

Figure 10.11 Temporal evolution of the height of the ABL during a diurnal cycle. The open markers represent the no-wind turbine case, and the closed markers represent the wind farm case. Note that the ABL height is computed every 10 min based on bulk statistics. The four vertical lines mark the transition points at 1010 h and 1810 h on Day 2 and 0940 h and 1800 h on Day 3. Figure has been reproduced from Sharma et al. (2017).

also propagate considerable distances downstream. Different studies have identified wind farm wakes propagating considerable distances ($\sim$ 20 km) during convective conditions, and even further during stable conditions (up to 4-fold increase) (Fitch et al., 2012, 2013; Baidya Roy, 2011).

Last but not least, the list of detected perturbations induced by wind farms also includes changes to the lifting condensation level (LCL), and cloud formation and precipitation. However, this remains the least studied phenomenon, with results from only a couple of very large (unrealistic) wind farm study cases (Kirk-Davidoff and Keith, 2008; Fiedler and Bukovsky, 2011), and a preliminary analysis of Sharma (2018), where large-eddy simulations were used to investigate the impact that wind farms of different loading characteristics might have on stratocumulus-topped boundary layers (STBL). Stratocumulus is one of the most common cloud types found in the ABL (Wood, 2012), having a significant impact on the global radiation and energy balance (Hartmann et al., 1992). In spite of their ubiquitous presence, they are quite challenging to account for in weather prediction and climate models, mostly because of their shallowness. These clouds are usually thin with a thickness of $\mathcal{O}(100m)$ and are found near the top of the ABL if the lifting condensation level is below the capping inversion. In the analysis of Sharma (2018), different configurations of the wind turbine array were implemented using the actuator disk with rotation model in the open source LES code PyCLES, (Pressel et al., 2015) to solve the cloud dynamics. Results seemed to indicate a potential decrease in cloud fraction (i.e., metric representing the percentage of surface area covered by the cloud layer) on the order of $\sim 5\%$, as well as in the liquid water path ($\sim 20\%$), which is a crucial quantity that defines the optical properties of the cloud and its interaction with radiation. Finally, it was also found that the cloud base and the cloud top can be shifted vertically upward, with an overall reduction in the cloud thickness. Although these results remain illustrative of a simple exploratory study, and hence they should be taken with a grain of salt, they nonetheless suggest that further and more rigorous analyses should be developed in the near future, even if it is only to clarify the conceptual understanding of this interaction.

10.5 Notes on offshore wind energy

At first sight, operations of wind turbines in the offshore environment could be considered less complex to characterize and predict than their onshore counterpart due to the lack of heterogeneity in the terrain coverage and different levels of complexity of the topography. On the other hand, the marine atmospheric boundary layer (MABL) is affected by wave motion (Jiménez and Dudhia, 2018) and the heterogeneity of sea surface temperature (SST), leading to strong thermal stratification (Shaw et al., 2022). These features lead to complex wind fields that govern the performance of offshore wind turbines.

One of the substantial differences between the onshore and marine ABL is associated with the high thermal inertia of the water relative to other surfaces, e.g., the onshore terrain. Therefore the MABL is less affected by the diurnal cycle of the solar radiation. However, the slow variability of the SST has an influence on the overall atmospheric stability. Namely, unstable convective conditions are observed when cold air moves over the warmer water (Archer et al., 2016), or alternatively, stably stratified conditions occur when warmer air moves over colder water. In addition, offshore wind turbines can be exposed to climatological conditions very different from the onshore environment, with extreme weather events such as hurricanes, internal boundary layers (Smedman et al., 1997; Batchvarova et al., 1999; Barthelmie and Palutikof, 1996; Floors et al., 2011; Krishnamurthy et al., 2023; Gadde and Stevens, 2021), and stark low-level jets (Mahrt, 1999, 2014; Colle et al., 2016; Pichugina et al., 2017; Debnath et al., 2020). The later often occur with a velocity peak at heights of the order of $\sim$ 100 m, plunging turbines in simultaneous positive and negative wind shear, which might be a hazard for the performance and the reliability of the turbines.

For offshore wind energy, atmospheric stability also plays an important role in the overall energy yield, with stability-dependent wake interactions leading to power losses and wind farm wakes affecting wind resource for neighboring wind farms. For offshore regions where continental winds move towards wind farms, such as for the U.S. East coast or East Asia, the role of atmospheric stability on wind farm performance will be more significant than for European offshore wind farms, where westerly winds traveling across the Atlantic Ocean are generally in very close equilibrium with the SST, thus leading to near-neutral stratification (Sathe et al., 2011; Dörenkämper et al., 2015). Along the North American Atlantic coast, it has been observed that western currents can lead to stably stratified conditions, thus exacerbating wake effects, wind plant blockage (Mahrt et al., 2001; Angevine et al., 2006), and the downstream extent of wind farm wakes up to tens of kilometers (Christiansen and Hasager, 2005).

The technology used to deploy offshore wind turbines also plays an important role in the MABL/wind-turbine interaction. Fixed-bottom wind turbines are typically installed for water depths up to $\sim$ 30 m, while for deeper waters, where wind resources are more profitable, wind turbines are currently installed on floating platforms installed with different technologies, such as spar, semi-submersible, and tension-leg (Otter et al., 2022). Therefore for floating offshore wind turbines (FOWTs), the nonlinear interactions between the hydrodynamic forcing exerted on the floating platform and the aerodynamic forcing generated through the interaction of the incoming wind field and the turbine rotor are crucial for the overall performance of FOWT and wake generation (Johlas et al., 2019; Bayati et al., 2018; Li et al., 2022b).

Simulating offshore wind turbine operations remains a great challenge. This is because in offshore conditions, it is necessary to simultaneously encompass the effects associated with MABL, the ocean, land-sea breezes (Bou-Zeid et al., 2020), and the interaction with the turbine structure, together with the generation of wakes and their interactions within the farm. The limitations to overcome involve many aspects of the simulation framework, such as the variability in the aerodynamic roughness length associated with the wave motion, the wide range of scales to be simulated, and thus the associated computational costs needed to solve a multi-physics and multiscale problem (Haupt et al., 2019; Allaerts et al., 2020; Draxl et al., 2021).

Observations of MABL for offshore wind energy applications are also very limited. A good source of field data is the datasets collected in the European North Sea with the three FINO towers (Finger, 2007), while the IJmuiden (Kalverla et al., 2017) and the ASIT sites have been operating off the U.S. northeastern coast (Kirincich, 2020), mainly probing the MABL surface layer. More recently, floating wind LiDARs have become available, facilitating the collection of velocity profiles up to a height of about 200 m (Gottschall et al., 2014; Viselli et al., 2019; Shaw et al., 2020).

10.6 Simulation approaches of wind turbine and farm flows

In an ideal world, where concerns about computational resources are not an issue (i.e., potentially the landscape in the era of quantum computing), one would solve the atmospheric flow dynamics and thermodynamics, and its interactions with wind turbines at very fine resolutions in time and space ($\sim$ associated Kolmogorov scales) by solving the set of defining state equations (e.g., Navier–Stokes equations, scalar transport equations, equation of state, etc.). Unfortunately, this direct approach remains out of reach at the present time, and thus alternatives are employed both in academia and industry. For instance, a common approach is Large-Eddy Simulation (LES), where the range of turbulent scales directly simulated is reduced, with the smaller scales being modeled and not resolved. To further reduce the computational costs of wind turbine flow simulations, the Reynolds-averaged Navier–Stokes (RANS) approach is also commonly used, where the effects of the full turbulence spectrum on the averaged flow field are accounted for through a closure model. For wind energy problems requiring hundreds or even thousands of full wind-farm-flow simulations, such as for layout optimization and turbine control applications, only models with even lower computational cost than RANS (i.e., runtime lower than 1 s per wind farm simulation) can be considered. For these later cases, only engineering wake models are currently a satisfactory modeling approach, where the wake velocity fields and their overlap can be predicted with limited accuracy through simple analytical or empirical equations. Finally, a novel alternative for predicting wind turbine wakes and wind farm operations is represented by machine-learning (ML) models. Once trained on available datasets, ranging from field data, numerical simulations, and laboratory experiments, these models enable more accurate predictions than traditional engineering wake models, yet with similar computational costs. More details on the above-mentioned modeling approaches are provided next.

10.6.1 Large eddy simulations of wind farm flows

In LES (Lesieur et al., 2005; Stoll et al., 2020), the energy-containing turbulent eddies are resolved (spatially and temporally), while the smaller turbulent eddies responsible for the transfer of the turbulent kinetic energy (TKE) down the inertial range and into dissipation are modeled with turbulence closure models. In this computational approach, the turbulent flow is resolved instantaneously, at different points in space, and thus statistical approaches (i.e., ensemble, time, and/or space averaging) are needed to extract relevant information. Due to the common high spatiotemporal resolution employed in LES frameworks, the interaction of wind turbines with the ABL flow can be represented through different approaches based on momentum theory and blade element theory (Burton et al., 2011; Stevens and Meneveau, 2017).

The simplest approach based on momentum theory is the so-called actuator disk approach, where a wind turbine is represented as a 'pseudo' drag disk that extracts momentum from the instantaneous turbulent flow through

$$F_t = -\frac{1}{2}\rho C_T\, A_t\, U_\infty^2 \tag{10.4}$$

where C_T is the rotor thrust coefficient, A_t is the frontal area of the wind turbine rotor disk, and U_∞ is the incoming unperturbed wind speed at hub height (Burton et al., 2011; Jimenez et al., 2007). The main advantage of this model is the low requirement in terms of numerical grid resolution; one only needs enough numerical points so that the turbine rotor disk appears and behaves like a disk (and not another geometrical shape), exerting a drag on the flow. Early on, Wu and Porté-Agel (2011) found that the rotor disk needed to be covered by at least eight grid points per direction to produce acceptable results. This approach has been shown to generate good results in the far wake region of a wind turbine (Wu and Porté-Agel, 2011), and it is therefore appropriate to study multiple wind turbine wake interactions in large wind farms (Calaf et al., 2010).

Alternatively, the next approach in terms of complexity is the so-called actuator disk with rotation approach, where a wind turbine is equally represented as a 'pseudo' drag disk that extracts momentum from the instantaneous turbulent flow as a function of a thrust coefficient (as in the conventional actuator disk), and it also enforces wake rotation as the flow crosses the rotor disk region through a tangential force. In this case, the forces at the rotor disk region are computed as (Martínez-Tossas et al., 2015),

$$f_t = -\frac{1}{2}\rho C'_T \langle\overline{u}\rangle_d^2 \tag{10.5}$$

$$f_\theta = -\frac{1}{2}\rho C'_p \langle\overline{u}\rangle_d \frac{\langle\overline{u}\rangle_d}{\Omega r} \tag{10.6}$$

Here f_t is the standard drag force per unit area and f_θ is the tangential force per unit area at a distance r from the center. The rotational speed of the blades is given by Ω, the overbar denotes averaging in time, and the angled brackets denote averaging over the rotor-disk region. Note that the thrust and power coefficients used here have been modified from their tabulated values for a fixed airfoil (denoted without the primes), $C'_T = C_T/(1-a)^2$ and $C'_p = C_p/(1-a)^3$, where a is the axial induction factor ($a = 1/3$ in the theoretical Betz limit (Burton et al., 2011)) such that the velocity used in the calculation is that at the rotor disk, and not the unperturbed upstream one. This approach can also be used in Eq. (10.4) if one wishes to represent a turbine within a wind farm, where there is no unperturbed upstream flow. Embedded in the formulations of the thrust and tangential forces are the assumptions that the blades are optimally adapted to the time averaged incoming wind velocity such that the tip-speed ratio ($\lambda = \Omega R/\overline{u}$, where R is the radius of the turbine), and the ratio of lift and drag coefficients of the blades are kept constant using active pitch control. This model does not capture blade-scale fluid motions such as tip vortices. However, the numerical study of Wu and Porté-Agel (2011) showed that the actuator disk with rotation model can partially capture the near-wake processes without numerically resolving the flow over the wind turbine blades.

Finally, the most complete and realistic way to model wind turbines in an LES framework is through the use of the so-called actuator line approach based on blade element theory. This is the most computationally expensive approach since each wind turbine blade is modeled as a line of radially varying force that rotates with time. This implies that the numerical resolution needed must be sufficient to track the rotation of the blades, both in time and space. To compute the force, the model requires the radially varying velocity along the line of action to compute the corresponding velocity magnitude and angle of attack. These are then used as inputs to appropriate airfoil tables to obtain lift and drag coefficients along the line. Unlike the previously described actuator disk models, due to its discrete blade-like forces, the actuator line approach is able to reproduce flow features such as root and tip vortices and blade-local axial induction zones. Nonetheless, despite its enhanced capacity to resolve flow features, the actuator line model does not replicate finer flow features like the blade boundary layer and separation at high angles of attack since the surface of each blade is not modeled. This approach is highly appropriate for studying the near wake region and the interaction of the vortex structures generated by the wind turbine blades and the background ABL flow. Specifically, the forces generated by the actuator line approach are locally computed as

$$f_\theta = \frac{\rho c U_{rel}^2}{2rd\theta da}(C_L \sin\Phi - C_D \cos\Phi) \tag{10.7}$$

$$f_t = \frac{\rho c U_{rel}^2}{2rd\theta da}(C_L \cos\Phi + C_D \sin\Phi) \tag{10.8}$$

where f_t and f_θ are the thrust and tangential forces, respectively, applied at each blade element. Here $U_{rel} = (U_\theta - \Omega R, U_a)$ is the local velocity with respect to the rotating blade, with the corresponding tangential and axial components (no radial component). Further, $\Phi = tan^{-1}(U_a/\Omega r - U_\theta)$ is the angle between U_{rel} and the rotor plane, and C_L and C_D are the lift and drag coefficients, which are a function of α (the angle of attack, $\alpha = \Phi - \gamma$, where γ is the local pitch angle) and the Reynolds number (Re). The blade chord length is indicated by c, and $d\theta$ and da are differential units in the corresponding reference frame such that the volume of a blade element is defined as $dv = rdrd\theta da$. More details on the actuator line model for horizontal axis wind turbines and its implementation in LES frameworks can be found in the foundational work of (Sørensen and Shen, 2002; Jimenez et al., 2007), as well as in more recent works (Lu and Porté-Agel, 2011; Martínez-Tossas et al., 2015). Interestingly, the actuator line model has also been used to represent smaller, vertical-axis wind turbines in atmospheric LES frameworks (Hezaveh et al., 2017, 2018; Kadum et al., 2020). In this case, however, the forces are first computed as a normal and tangential component, which are then projected in the Cartesian (x, y) grid depending on whether the blades are in the upwind or downwind zone. Specific details of the algorithm can be found in Hezaveh et al. (2017), with the main constraint being the needed increase in numerical resolution since vertical axis wind turbines are, in general, much smaller than horizontal axis wind turbines.

10.6.2 Reynolds-averaged Navier–Stokes models for wind energy

Although the LES technique provides great insights into complex high-Reynolds-number fluid-dynamic problems, current computational constraints limit the broad applicability of this technique to study flows in very large domains ($>$ 10 km in horizontal directions) (Voller and Porté-Agel, 2002; Stoll et al., 2020). Naively, the natural progression in the development of computational approaches would consist of additional spatial filtering of the smaller-scale turbulence to larger and larger scales, thus allowing for a reduction in computational demand in terms of resolution at the smaller scales and facilitating the use of the resources to represent larger domains. This approach can indeed be used as long as the spatial filtering operation resolves part of the inertial regime in the turbulence energy spectral distribution for which the universal behavior of turbulence is well known, and hence the filtered-out turbulence can be properly modeled. However, once filtering occurs at length scales beyond those of the inertial regime and into the neighborhood of the energy-containing turbulent eddies, the LES approach breaks down due to a lack of appropriate turbulence closure models. This regime of uncertainty was coined by Wyngaard as the '*Terra Incognita*' (Wyngaard, 2004) and later re-branded as the Gray Zone (Honnert et al., 2020; Calaf et al., 2022). Another challenge with this approach for modeling wind farms is that the grid scale will rapidly become larger than the turbine diameter so that the rotor disc can no longer be resolved. To overcome this theoretical conundrum, the Reynolds-averaged Navier–Stokes (RANS) approach is commonly used, where one solves the equations for the mean flow, and the effect of turbulence is fully modeled with more or less complex turbulence closure models (Stoll et al., 2020). The main difference between the turbulence closure models in RANS and LES is that in the former, the models must include the effect of turbulence at all scales, while in the latter, turbulence models only need to account for the transfer of energy occurring at the inertial scales and/or the dissipation of turbulent kinetic energy, depending on the scale of filtering. Therefore in RANS, regardless of the numerical resolution, one only obtains a solution of the mean flow, and the numerical resolution is traditionally coarser than 1 km, and thus representation of wind turbines requires alternative parameterization approaches. Another difference between RANS and LES is that a steady-state RANS solver only needs to iterate until the equations are satisfied, while an LES solver must always integrate in time over multiple large-scale eddy turnover times to converge the statistics, further increasing the cost of LES relative to RANS.

At the coarsest range of numerical simulations, turbines cannot be resolved, and wind farms are represented by the so-called *enhanced surface roughness* approach. This approach has often been used in climate models where the numerical resolution is comparable to the size of a full wind farm with many wind turbines ($\sim$ 50 km) (Keith et al., 2004; Baidya Roy et al., 2004; Wang and Prinn, 2010). In this case, wind farms are represented through a change in the characteristic surface roughness, similar to what is done to represent the different types of land cover characteristics. At present, several parameterizations have been developed (Lettau, 1969; Frandsen, 1992; Calaf et al., 2010; Meneveau, 2012; Peña and Rathmann, 2014) with reasonable success in capturing variations in the wind farm configurations and the forcing effects of the ABL (e.g., stratification, Coriolis, etc.)

At an increased level of accuracy, there exists another parameterization traditionally used in weather forecast models (e.g., the weather research and forecasting (WRF) model) where the effect of wind turbines is included through an elevated momentum sink of the mean flow, converting kinetic energy into electricity and TKE (Baidya Roy et al., 2004; Fitch et al., 2012; Cervarich et al., 2013). One of the most used turbine models is that of Fitch et al. (2012), which improved on previous models (Baidya Roy et al., 2004; Blahak et al., 2010) by computing the atmospheric drag of turbines based on the thrust coefficient of modern commercial turbines. In addition, the source of TKE was imposed to vary with wind speed, reflecting the amount of energy extracted from the atmosphere by the turbines that do not produce electrical energy. A latter refinement by Volker et al. (2015) also included a parameterization for the wake expansion within the numerical grid cell that contains turbines. In addition, the turbines' TKE source was defined from random fluctuations around the ensemble-averaged velocity, instead of around the grid-cell-averaged velocity as done in the previous parametrizations. Although both approaches have been shown to work similarly well, both can overestimate the momentum deficit by more than 20% when several turbines fall within the same individual grid cell. This overestimation particularly occurs when coarser grid resolutions ($\sim$ 3 km or larger) are used, a resolution often used in traditional numerical weather prediction models. A latter iteration of this approach was introduced by Abkar and Porté-Agel (2015b) to correct for this limitation. In this case, the momentum deficit at a grid cell is proportionally weighted for the different turbines within the grid cell, according to their modified local velocity. Within their analysis, the weight values were extracted by comparing with LES data.

At a slightly improved level of accuracy, we find the recently implemented *distributed drag force* approach, also referred to as the actuator wind farm (AWF) approach (van der Laan et al., 2022b). The AWF represents the effect of complete wind farms using distributed drag forces, where the drag force distribution is computed as the superposition of two-dimensional Gaussian functions centered at the wind turbine locations to smooth out the spurious effects of imposing punctual wind turbines in coarse numerical grids. In this case, the effect of the wind farm on the flow is represented by

$$F_{WF}(x, y, z) = \frac{1}{2}\rho C_{T,WF} A(x, y, z)|U|U_i \tag{10.9}$$

where ρ is the air density, $C_{T,WF}$ is the wind farm thrust coefficient, $A(x, y, z)$ is the force distribution representing the density of wind turbines, and U_i is the local velocity vector. It is $A(x, y, z)$ that gets smoothed out with a two-dimensional

Gaussian function to homogenize the effect of individual wind turbines within a large wind farm.

Lastly, at the highest degree of resolution, it is also possible to implement the actuator disk model described above for LES configurations. In this case, each wind turbine is individually represented and thus provides the highest accuracy results. The challenge of this method is the needed numerical resolution, given that each wind turbine requires a horizontal cell size in the order of $D/8$, where D is the rotor diameter (Wu and Porté-Agel, 2011; van der Laan et al., 2022b). This approach is used in numerical weather prediction models in combination with multiple nesting strategies that progressively facilitate the needed numerical resolution. Often, the higher-resolution inner domain is then run in LES mode.

10.6.3 Engineering models for wind turbine wakes

The RANS-based approaches of the previous section are broadly used by industry and academia to investigate the contemporary effects of wind farm wake-wake interactions where large domains are needed, or to investigate the objective interaction between wind farms and realistic atmospheric flow conditions that include thermal stratification, low-level jets, land-sea breeze, topography, etc. Therefore this modeling approach arguably provides the most realistic representation of real wind farm flow configurations. However, despite its numerous benefits, RANS simulations remain too computationally expensive for daily engineering applications that require optimization of wind farm layouts for decision making. Therefore simpler alternatives have been sought out and developed.

Engineering wake models are used to predict the downstream evolution of wind turbine wakes as a function of the incoming freestream wind speed, turbulence intensity, and turbine operating conditions (e.g., the rotor thrust coefficient), while avoiding the computational costs associated with high-fidelity models (e.g., LES and RANS). The accuracy of the predictions obtained with engineering wake models relies on the definition and tuning of the various model parameters, such as the wake growth rate in the downstream direction (Peña et al., 2016; Carbajo Fuertes et al., 2018; Zhan et al., 2020; Vahidi and Porté-Agel, 2022) and the length of the near wake (Zhan et al., 2020; Vahidi and Porté-Agel, 2022; Vermeulen, 1980; Bastankhah and Porté-Agel, 2016), namely the minimum downwind distance from the turbine rotor where engineering wake models are generally applicable.

Engineering wake models can be coupled with power curve models to generate estimates of turbine power capture and predict the annual energy production of a wind farm (Lydia et al., 2014). These simplified models can also be used to run thousands of simulations required for the optimization of a wind farm layout, which is a difficult task for high-fidelity models (Santhanagopalan et al., 2018; Mosetti et al., 1994; Lackner and Elkinton, 2007; Blanco, 2009). In addition, wake interactions can also be estimated from engineering models using superposition techniques (Bastankhah et al., 2021). However, it has recently been found that engineering wake models may experience reduced accuracy in the very far-wake region (i.e., distances larger than

$10D$), which is relevant for farm-to-farm interaction problems (Pettas et al., 2021; Stieren and Stevens, 2022; Puccioni et al., 2023; Moss et al., 2023). To overcome these limitations, engineering models have been adjusted by retuning their parameters Nygaard et al. (2020); Pedersen et al. (2022).

In general, engineering wake models can be divided into two main categories: (i) empirical models, and (ii) analytical models. Empirical models have been used to estimate the variation of the wake-center velocity deficit with the streamwise distance from the turbine rotor (Emeis, 2010; Platis et al., 2021; Baker and Walker, 1984; Högström et al., 1988; Magnusson and Smedman, 1999; Barthelmie et al., 2003; Zhang et al., 2013; Iungo and Porté-Agel, 2014). Based on these models, the velocity deficit is generally assumed to have a power-law relationship with the downstream distance x, which is written as

$$\frac{\Delta U}{U_\infty} = A\left(\frac{x}{D}\right)^N \tag{10.10}$$

where A and N are coefficients obtained from experimental and numerical data, and U_∞ is the reference freestream velocity.

Unlike empirical models, whose equations are obtained solely by fitting experimental or numerical data, analytical wake models are derived from flow governing equations and thus have, in principle, a superior ability to capture the wake flow physics. For the analytical wake models, the streamwise velocity deficit induced by the rotor, which is considered to be positioned perpendicularly to the mean wind direction, is assumed to be axisymmetric with respect to the axis of the rotor and have a self-similar distribution in the wake cross section. Therefore the mean velocity $U(x, y, z)$ in the wake can be determined by the following equation

$$U(x, y, z) = U_0(y, z) - \Delta U \tag{10.11}$$

where U_0 is the incoming wind velocity field, and ΔU is the velocity deficit induced by the turbine rotor. The latter is typically a function of the radial distance from the center of the wake, r, the downstream distance, x, and the rotor thrust coefficient, C_T. Since it is assumed that in the far-wake the velocity deficit presents a self-similar behavior, it can be expressed as the product of a streamwise function, C, and a self-similar shape function, Γ, namely

$$\Delta U/U_h = C\,(C_T, I_\infty, x/D)\;\Gamma(r/D) \tag{10.12}$$

Here I_∞ is the freestream turbulence intensity, C is the function representing the maximum velocity deficit, ΔU_{max}, for each downwind location normalized by the reference wind speed at the hub height, U_h, which can be assumed equal to U_∞ or calculated as an equivalent rotor velocity (Wagner et al., 2009; Choukulkar et al., 2016). The function Γ is the self-similar distribution of the velocity deficit over a cross-plane, which is defined as the velocity deficit normalized by the maximum value at the center of the wake section.

One of the classical wake models for wind turbine wake velocity deficit was proposed by Jensen (1983), then further developed by Katic et al. (1986). This model assumes a linearly expanding wake with a uniform profile for the velocity deficit, termed 'top-hat shape'. The only constraint imposed by the Jensen model is mass conservation of the resulting flow. In the Jensen model, the expression of the velocity deficit is given by

$$\frac{\Delta U}{U_h} = \frac{1 - \sqrt{1 - C_T}}{(1 + 2kx/D)^2} \tag{10.13}$$

where the wake diameter, D_w, is assumed to grow linearly with downstream distance,

$$D_w = D + 2kx \tag{10.14}$$

The wake expansion coefficient, k, is defined in analogy with a jet spreading within shear flows (Pope, 2000). According to the wind atlas analysis and application program (WAsP, (Mortensen et al., 1993)), the value of the wake expansion coefficient, k, is suggested to be equal to 0.075 and 0.05 for onshore and offshore wind farms, respectively. However, a lower k value of 0.03 was found to achieve a better agreement with the SCADA data collected for the Nysted offshore wind farm (Barthelmie and Jensen, 2010). An empirical formula to predict the wake expansion factor based on the incoming wind turbulence intensity is (Peña et al., 2016)

$$k = 0.4\, I_\infty. \tag{10.15}$$

This expression has subsequently been revised based on LiDAR measurements (Zhan et al., 2020),

$$k = 0.48\, I_\infty - 0.01. \tag{10.16}$$

An improvement to the Jensen model was later proposed by considering the balance of momentum in addition to the mass conservation applied to a control volume, including the turbine rotor (Frandsen et al., 2006). Still assuming a top-hat wake velocity profile, the Frandsen model is expressed as

$$\frac{\Delta U}{U_h} = \frac{1}{2}\left(1 - \sqrt{1 - \frac{2C_T}{\beta + \alpha x/D}}\right) \tag{10.17}$$

In this case, α is the respective wake expansion factor, which is about one order of magnitude larger than the factor k used in the Jensen model (Eq. (10.13)), and β is the near-wake cross-sectional expansion calculated as (Frandsen et al., 2006)

$$\beta = \frac{1 + \sqrt{1 - C_T}}{2\sqrt{1 - C_T}} \tag{10.18}$$

The Jensen and Frandsen models generally underestimate the maximum wake velocity deficit and overestimate the wake edges as a consequence of the top-hat velocity profile used. This feature is shown in Fig. 10.12, where the mean wake velocity field retrieved from measurements collected with a scanning Doppler LiDAR is reproduced with the Jensen model (Zhan et al., 2020). In Fig. 10.12b, it is shown that while the average downstream evolution of the velocity deficit is captured well, the transverse variability of the velocity field is completely missed. Therefore to better

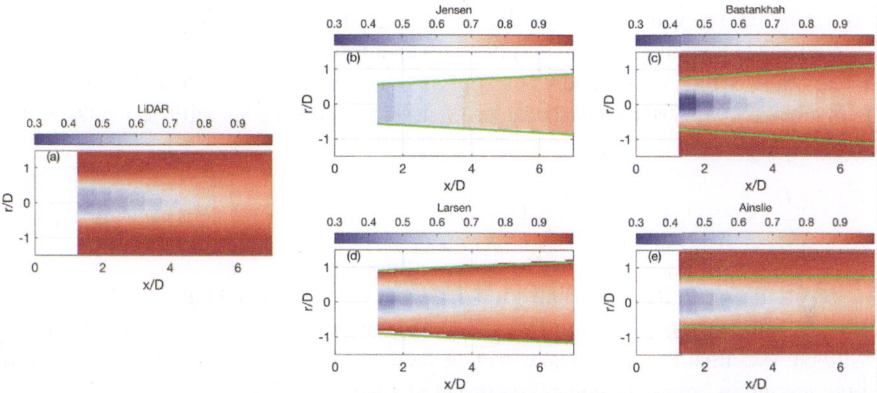

Figure 10.12 Modeling LiDAR wake measurements collected for a hub-height wind speed normalized with the turbine rated wind speed between 0.76 and 0.81, and $13.5\% \leq TI \leq 19.4\%$: a) LiDAR data Zhan et al. (2020); b) Jensen model Katic et al. (1986); c) Gaussian model Bastankhah and Porté-Agel (2014); d) Larsen model Larsen and Réthoré (2009); e) Ainslie model Ainslie (1988). The green lines represent the wake width, D_w.

approximate the spatial variability in the cross-plane of the mean velocity deficit as observed experimentally (Iungo et al., 2013; Chamorro and Porté-Agel, 2009; Zhan et al., 2020) and through numerical simulations (Xie and Archer, 2015), a Gaussian distribution of the wake velocity deficit was proposed (Bastankhah and Porté-Agel, 2014) (see Fig. 10.12c),

$$\frac{\Delta U}{U_h} = \left(1 - \sqrt{1 - \frac{C_T}{8(D_W/D)^2}}\right) exp\left(-\frac{r^2}{2D_W^2}\right) \tag{10.19}$$

Similarly to the Jensen model, the wake width is assumed to grow linearly starting from the location x_{NW} where the near-wake region is considered to transition to the far-wake region, and the mean wake velocity field achieves a self-similar regime (Vahidi and Porté-Agel, 2022; Vermeulen, 1980; Sørensen et al., 2015),

$$D_W = k^*(x - x_{NW}) + D/\sqrt{8}. \tag{10.20}$$

Here the wake growth rate is generally a function of the incoming turbulence intensity (Carbajo Fuertes et al., 2018). For instance, using LES results for neutral stability conditions, $k^* = 0.38I_\infty + 0.004$ was obtained (Niayifar and Porté-Agel, 2016). A physics-based model to predict k^* was also proposed by leveraging Taylor's diffusion theory and the turbulent mixing layer theory (Vahidi and Porté-Agel, 2022). Several Gaussian wake models have been proposed to attempt to achieve increased accuracy in wake predictions (Gao et al., 2016; Ishihara and Qian, 2018), also including wake effects in the presence of yaw steering of the turbine rotor (Bastankhah and Porté-Agel, 2016).

Several other models have also been proposed to reproduce the transition from a nearly top-hat shape of the velocity profile in the near-wake to a self-similar Gaussian profile in the far wake, such as the super-Gaussian model (Shapiro et al., 2019; Blondel and Cathelain, 2020). This approach enables recovery of mass and momentum conservation, which is violated when applying near-wake corrections in the near-wake (Ishihara and Qian, 2018). For the super-Gaussian model, the velocity deficit is defined as

$$\frac{\Delta U}{U_0} = C(x)\, exp\left(-\frac{r^n}{2D_w^n}\right) \tag{10.21}$$

where for $n = 2$, the Gaussian model is recovered (Bastankhah and Porté-Agel, 2014). As for other engineering wake models, a linear expansion of the wake width is assumed, specifically with a wake growth factor proportional to the friction velocity of the incoming boundary layer (Shapiro et al., 2019).

Another possible velocity deficit distribution for the near-wake region is proposed with the double-Gaussian wake model (Keane et al., 2016; Schreiber et al., 2020), which is derived similarly to the Frandsen and Gaussian models (Frandsen et al., 2006; Bastankhah and Porté-Agel, 2014) by applying conservation of mass and momentum in the context of the actuator disk theory and assuming a double-Gaussian distribution for the velocity deficit in the wake. Indeed, the presence of two minima in the streamwise velocity profiles close to the rotor disk had previously been observed in experimental measurements and high-fidelity CFD simulations (Wang et al., 2017).

Modeling the wake turbulence intensity is also important since the wake-generated turbulence increases the fatigue loading of downwind turbines. For engineering wake models, the added turbulence intensity is typically assumed axisymmetric, and the effects of the incoming shear layer are later considered to account for the non-symmetry in the vertical direction. The velocity standard deviation in any position, $\sigma_U(x, y, z)$, downwind the turbine is determined by the ambient turbulence standard deviation, $\sigma_{U0}(y, z)$, and the wake-added turbulence standard deviation, $\Delta u'$, as shown in the following equation (Frandsen, 2007):

$$\sigma_U(x, y, z) = \sqrt{\sigma_{U0}^2(y, z) + \Delta u'^2} \tag{10.22}$$

The value of $\Delta u'$ typically reaches its maximum in the range of two to four rotor diameters downstream at the top-tip level, coinciding with the transition between the near-wake and the far-wake regions. A correction term, $\delta(z)$, can be added to consider the weakening of turbulence intensity below hub height, z_h,

$$\delta(z) = \begin{cases} 0 & z \geq z_h \\ I \sin^2\left(\pi \frac{z_h - z}{z_h}\right) & z < z_h \end{cases} \tag{10.23}$$

It should be noted that in conventional wake models, the added turbulence intensity is defined using the mean velocity at hub height, U_h, as a reference. Therefore turbulence intensity can be defined in reference to the local mean velocity, U, and is expressed as follows

$$I_{local} = I\frac{U_h}{U} \tag{10.24}$$

For several models, e.g., Crespo (Crespo et al., 1996) and Frandsen (Frandsen, 2007) (the latter is the turbulence wake model referred to in the IEC61400-1 standard (International Electrotechnical Commission, 2019)), the wake turbulence intensity is only a function of the rotor thrust coefficient. For the Crespo model (Crespo et al., 1996), the wake turbulence intensity is assumed to have two different constant values in the near- and far-wake, where the transitions are assumed to occur at $x/D = 3$. In contrast, the Frandsen model (Frandsen, 2007) assumes wake turbulence intensity as a continuous function of the streamwise position. For the Ishihara model (Ishihara and Qian, 2018), the wake turbulence intensity is assumed to have a double-Gaussian distribution in the cross-plane, and to vary with the rotor thrust coefficient and the streamwise position.

10.6.4 Machine-learning modeling of wind-farm flows

With the growing availability of observations of the atmospheric boundary layer (ABL), atmospheric turbulence, wind turbine operations, and wind turbine wakes, machine learning (ML), or more generally data-driven modeling, has become a great opportunity to better understand interactions between the ABL and wind turbine arrays, with the overarching goal of improving accuracy of forecasting capabilities, optimize design of wind power plants, and maximize their profitability.

Turbine operations are typically monitored through supervisory control and data acquisition (SCADA) by probing power capture, wind parameters, and parameters of various engineering components. SCADA data are typically recorded as statistics calculated over 10-minute periods. The wind conditions are directly monitored through a nacelle-mounted anemometer, but also from meteorological (met) towers installed in the proximity of the wind power plant (Lundquist et al., 2017). Remote sensing instruments, such as LiDARs (Puccioni et al., 2023; Iungo and Porté-Agel, 2014) and radars (Hirth et al., 2012), have been used with increased frequency to monitor the incoming wind conditions, but also to measure wind turbine wakes. Other less common measuring systems for wind energy applications, typically used only for research purposes, are uncrewed aerial vehicles (Al-Sabban et al., 2013), aircraft (Platis et al., 2018), and satellites (Christiansen and Hasager, 2005).

At present, the majority of the existing data-driven models have been developed for canonical idealized fluid dynamics problems (Brunton et al., 2020; Vinuesa and Brunton, 2022; Vinuesa et al., 2023); their implementation for a multi-dimensional, multi-scale complex flow, such as that associated with wind farm operations, is not trivial, especially considering the variability in accuracy of the data collected from real wind farm operations.

In wind energy, ML has been widely applied for data filtering (Manobel et al., 2018; Yesilbudak, 2018), modeling power curves (Wang et al., 2019; Lydia et al., 2013; Zhou et al., 2014; Pei and Li, 2019), wakes (Ti et al., 2020, 2021; Zhang and Zhao, 2022; Ashwin Renganathan et al., 2022), wind-farm flows (Zehtabiyan-Rezaie et al., 2022; Howland and Dabiri, 2019; Park and Park, 2019; Bentsen et al., 2022; Yan et al., 2019), as well as forecasting power and wind speed (Liu et al., 2022; Duan et al., 2021; Ezzat, 2020; Zhang and Zhao, 2023). Data-driven models of wind farm flows can be classified as physics-based or purely data-driven models (Zehtabiyan-Rezaie et al., 2022), depending on whether a set of governing equations of the flow physics is tuned on the available datasets or the data are leveraged to develop empirical functions based on a set of input parameters, respectively.

An ML model enables mapping an input dataset to selected outputs with the aim of producing predictions of the output parameters or classifying the set of provided inputs with a specific label. Therefore ML algorithms can be classified as supervised, unsupervised, and reinforcement learning. For unsupervised ML, the outputs are unknown, and the data are processed to generate predictions of the most probable output, classifications, e.g., for cluster analysis, or to reduce the dimensionality of the input data, such as through autoencoders. For supervised ML models, unknown features embedded within the training dataset are identified by leveraging the associated output parameters. These identified features are then leveraged to perform predictions of new sets of input parameters. Finally, reinforcement learning merges characteristics of both unsupervised and supervised ML algorithms without the need for labeling outputs as a function of input sets, and the modeling goal is to find a balance between exploration of unknown data subsets and leveraging the current training (exploitation) (Sutton and Barto, 2018; Iosifidis and Tefas, 2022).

ML can be used to develop surrogate models of wind farm flows trained on simulation data, such as from RANS (Ti et al., 2020; Wilson et al., 2017) or LES (Shao et al., 2019; Zhang et al., 2021). Several models used convolutional neural networks (CNN) to predict both background flow and wakes (see, e.g., Zhang and Zhao, 2022; Li et al., 2022a). Other works used graph neural networks (GNN), which take advantage of the graph theory to learn from data estimated on unstructured grids, thus avoiding the need for a uniform grid during the training phase of the ML model (Li et al., 2023).

Developing ML models from real-world experimental data still represents a challenge due to the noisy nature of the data typically available, the unspecified or variable accuracy of the instruments, and the incomplete statistical convergence of the data collected (Moss et al., 2024). Nonetheless, surrogate wind farm models have also been developed from experimental data. For instance, LiDAR measurements of individual wakes collected for different conditions of the incoming wind and thrust coefficient of the turbine rotor were leveraged to develop an ML wake model (Ashwin Renganathan et al., 2022). First, a low-dimensional latent space was generated through a deep autoencoder from the LiDAR measurements to generate a computationally tractable problem. Then, a Gaussian process model was used to map the input parameters, namely incoming wind speed and turbulence intensity, to the latent space parameter. Finally, the predicted wake velocity field is generated through the decoding component of the autoencoder.

ML wind farm models can be developed for diagnostic purposes, namely to reproduce and predict wind turbine operations for a specific wind turbine array, which is leveraged for the training of the model itself. Albeit these models are not generalizable to other wind projects (i.e., with different wind farm layout, wind turbine characteristics, or site wind resources), they can predict with high accuracy and low computational costs the operations of wind turbines, in terms of mean wind speed, power capture, as well as turbulence statistics if turbulence data are available in the training set (Iungo et al., 2022; Moss et al., 2022). These predictions can be very valuable for early detection of under-performing wind turbines and for carrying out timely maintenance.

An effective ML wind farm model should provide predictive capabilities for arbitrary inflow conditions, wind farm layouts, turbine characteristics, and site wind resources. An ML wind farm model should predict wind parameters (Zhang and Zhao, 2021b,a), power capture (Optis and Perr-Sauer, 2019; Japar et al., 2014), and possibly, turbulence parameters (e.g., turbulence intensity), not only at the wind farm or turbine level, but also over the spatial field of interest (Iungo et al., 2022), thus providing detailed information about wakes and wake interactions (Ti et al., 2021; Zhang and Zhao, 2022).

Among recent research efforts to develop a generalizable ML wind farm model, the use of GNNs to learn spatial dependencies among neighboring wind turbines while keeping flexibility, i.e., generalizability, in the wind farm layout, is one of the most promising research paths (Park and Park, 2019; Bentsen et al., 2022). Another approach consists of encapsulating information on the wind farm layout, and thus on the interactions among neighboring wind turbines, by including several geometrical parameters, for instance, the shaded area due to upstream turbines and their weighted distance (Ghaisas and Archer, 2016). Albeit a very pragmatic approach, this method can produce reasonable results with accuracy comparable to that of engineering wake models, yet providing the advantage of being generalizable to different wind farms.

10.7 Perspectives and remaining challenges in the study of turbulence in wind farms

Over the past twenty years or more, the wind energy community has made giant strides in the advancement of the science and technology needed for producing electric power from wind. While at first it was mostly interpreted as a pure engineering challenge, it later became clear that for the robust development and generalized implementation of wind energy, there was a need for a more holistic understanding of wind energy science. This would necessarily have to include the synergistic interaction with the atmospheric flow and the local land/sea-atmosphere conditions. This has led to a solid understanding of the physics of the wind turbine wake, with the different vorticity generating processes, instabilities, mixing, and dependence on atmospheric stability, and vertical wind veer. With time, we have also come to develop a robust understanding of the flow dynamics in large wind farms, whether these are installed over flat terrain (homogeneous or heterogeneous), in complex terrain, or even in certain offshore conditions. At present, however, there remain many open research foci that should continue to be investigated to enable a wider spread of wind energy around the world. Some of these include:

- The study of wind farm wake-wake interactions. With a broader spread and implementation of wind energy, it is inevitable that wind farms are installed in the vicinity of each other, which could lead to unwanted outcomes unless the proper research is developed. For example, what is the impact of wind turbine arrangements within a wind farm on the overall wind farm wake characteristics? How does the local ABL flow and surface conditions affect wind farm wakes? Specific numerical and scale experiments should be developed to investigate this problem.
- As suggested in the corresponding section above, there is an imperative need to further investigate wind energy in offshore conditions, both using onsite experimental data and idealized simulations. It is, for example, critical to better understand offshore wind conditions in regions strongly affected by the nearby onshore flows, like the north-eastern coast of the United States. It is also urgent to better understand the role of ocean currents and changes in ocean surface temperature in the near-surface flows.
- While our computing resources do not cease to increase, there is an imperative need to continue developing faster, more efficient, and reliable computing approaches for industry purposes. The goal is to facilitate optimization of wind turbine arrays for both real-time wind operations as well as initial installation. In view of the current and most recent developments, it is clear that artificial intelligence will play a major role in this arena.
- To date, there is a considerable body of work that has looked at the problem from the atmospheric side of things, that is, how does wind energy perturb the local ABL. However, there remains plenty of work to be done, for example, on the understanding of how wind farms perturb cloud formation and the precipitation process. Also, based on the large number of wind farms already installed around the world, it would be useful to understand whether considering their presence or not in weather prediction models appreciably changes the local weather predictions. This would, for example, address the question of whether renewable energy canopies should be included in weather forecast models in the same way that we include vegetated and urban canopies.

References

Abkar, M., Porté-Agel, F., 2013. The effect of free-atmosphere stratification on boundary-layer flow and power output from very large wind farms. Energies (ISSN 1996-1073) 6 (5), 2338–2361. https://doi.org/10.3390/en6052338.

Abkar, M., Porté-Agel, F., 2015a. Influence of atmospheric stability on wind-turbine wakes: a large-eddy simulation study. Physics of Fluids 27 (3).

Abkar, M., Porté-Agel, F., 2015b. A new wind-farm parameterization for large-scale atmospheric models. Journal of Renewable and Sustainable Energy (ISSN 1941-7012) 7 (1), 013121. https://doi.org/10.1063/1.4907600.

Ainslie, J.F., 1988. Calculating the flowfield in the wake of wind turbines. Journal of Wind Engineering and Industrial Aerodynamics 27 (1–3), 213–224.

Al-Sabban, W.H., Gonzalez, L.F., Smith, R.N., 2013. Wind-energy based path planning for unmanned aerial vehicles using Markov decision processes. In: 2013 IEEE International Conference on Robotics and Automation, pp. 784–789.

Allaerts, D., Meyers, J., 2015. Large eddy simulation of a large wind-turbine array in a conventionally neutral atmospheric boundary layer. Physics of Fluids (ISSN 1070-6631) 27 (6). https://doi.org/10.1063/1.4922339, 065108.

Allaerts, D., Meyers, J., 2017. Boundary-layer development and gravity waves in conventionally neutral wind farms. Journal of Fluid Mechanics (ISSN 0022-1120) 814, 95–130. https://doi.org/10.1017/jfm.2017.11.

Allaerts, D., Quon, E., Draxl, C., Churchfield, M., 2020. Development of a time–height profile assimilation technique for large-eddy simulation. Boundary-Layer Meteorology 176 (3), 329–348. https://doi.org/10.1007/s10546-020-00538-5.

Angevine, W.M., Hare, J.E., Fairall, C.W., Wolfe, D.E., Hill, R.J., Brewer, W.A., White, A.B., 2006. Structure and formation of the highly stable marine boundary layer over the Gulf of Maine. Journal of Geophysical Research. Atmospheres 111 (D23).

Archer, C.L., Colle, B.A., Veron, D.L., Veron, F., Sienkiewicz, M.J., 2016. On the predominance of unstable atmospheric conditions in the marine boundary layer offshore of the us northeastern coast. Journal of Geophysical Research. Atmospheres 121 (15), 8869–8885.

Ashton, R., Viola, F., Camarri, S., Gallaire, F., Iungo, G.V., 2016. Hub vortex instability within wind turbine wakes: effects of wind turbulence, loading conditions, and blade aerodynamics. Physical Review Fluids 1 (7), 073603.

Ashuri, T., Martins, J.R.R.A., Zaaijer, M.B., van Kuik, G.A.M., van Bussel, G.J.W., 2016. Aeroservoelastic design definition of a 20 mw common research wind turbine model. Wind Energy 19 (11), 2071–2087.

Ashwin Renganathan, S., Maulik, R., Letizia, S., Iungo, G.V., 2022. Data-driven wind turbine wake modeling via probabilistic machine learning. Neural Computing & Applications, 1–16.

Baidya Roy, S., 2011. Simulating impacts of wind farms on local hydrometeorology. Journal of Wind Engineering and Industrial Aerodynamics (ISSN 0167-6105) 99 (4), 491–498. https://doi.org/10.1016/j.jweia.2010.12.013.

Baidya Roy, S., Traiteur, J.J., 2010. Impacts of wind farms on surface air temperatures. Proceedings of the National Academy of Sciences of the United States of America (ISSN 1091-6490) 107 (42), 17899–17904. https://doi.org/10.1073/pnas.1000493107.

Baidya Roy, S., Pacala, S.W., Walko, R.L., 2004. Can large wind farms affect local meteorology? Journal of Geophysical Research D: Atmospheres (ISSN 0148-0227) 109, 1–6.

Baker, R.W., Walker, S.N., 1984. Wake measurements behind a large horizontal axis wind turbine generator. Solar Energy 33 (1), 5–12.

Barthelmie, R.J., Palutikof, J.P., 1996. Coastal wind speed modelling for wind energy applications. Journal of Wind Engineering and Industrial Aerodynamics 62 (2–3), 213–236.

Barthelmie, R.J., Folkerts, L., Ormel, F.T., Sanderhoff, P., Eecen, P.J., Stobbe, O., Nielsen, N.M., 2003. Offshore wind turbine wakes measured by sodar. Journal of Atmospheric and Oceanic Technology 20 (4), 466–477.

Barthelmie, R.J., Jensen, L.E., 2010. Evaluation of wind farm efficiency and wind turbine wakes at the nysted offshore wind farm. Wind Energy 13 (6), 573–586.

Bastankhah, M., Porté-Agel, F., 2014. A new analytical model for wind-turbine wakes. Renewable Energy 70, 116–123.

Bastankhah, M., Porté-Agel, F., 2016. Experimental and theoretical study of wind turbine wakes in yawed conditions. Journal of Fluid Mechanics 806, 506–541.

Bastankhah, M., Welch, B.L., Martínez-Tossas, L.A., King, J., Fleming, P., 2021. Analytical solution for the cumulative wake of wind turbines in wind farms. Journal of Fluid Mechanics 911, A53.

Batchvarova, E., Cai, X., Gryning, S.-E., Steyn, D., 1999. Modelling internal boundary-layer development in a region with a complex coastline. Boundary-Layer Meteorology 90, 1–20.

Bayati, I., Bernini, L., Zanotti, A., Belloli, M., Zasso, A., 2018. Experimental investigation of the unsteady aerodynamics of fowt through piv and hot-wire wake measurements. Journal of Physics: Conference Series 1037, 052024.

Belcher, S.E., Jerram, N., Hunt, J.C.R., 2003. Adjustment of a turbulent boundary layer to a canopy of roughness elements. Journal of Fluid Mechanics (ISSN 1469-7645) 488, 369–398. https://doi.org/10.1017/s0022112003005019.

Belcher, S.E., Harman, I.N., Finnigan, J.J., 2012. The wind in the willows: flows in forest canopies in complex terrain. Annual Reviews Fluid Mechanics 44, 479–504.

Bentsen, L.Ø., Dilp Warakagoda, N., Stenbro, R., Engelstad, P., 2022. Wind park power prediction: attention-based graph networks and deep learning to capture wake losses. Journal of Physics. Conference Series 2265 (2), 022035.

Betz, A., 2014. Introduction to the Theory of Flow Machines. Elsevier.

Blahak, U., Goretzki, B., Meis, J., 2010. A simple parameterization of drag forces induced by large wind farms for numerical weather prediction models. In: Proc. European Wind Energy Conf. and Exhibition. EWEC, pp. 186–189.

Blanco, M.I., 2009. The economics of wind energy. Renewable and Sustainable Energy Reviews 13 (6–7), 1372–1382.

Blondel, F., Cathelain, M., 2020. An alternative form of the super-Gaussian wind turbine wake model. Wind Energy Science Discussions 2020, 1–16.

Bou-Zeid, E., Anderson, W., Katul, G.G., Mahrt, L., 2020. The persistent challenge of surface heterogeneity in boundary-layer meteorology: a review. Boundary-Layer Meteorology 177 (2–3), 227–245. https://doi.org/10.1007/s10546-020-00551-8.

Brunton, S.L., Noack, B.R., Koumoutsakos, P., 2020. Machine learning for fluid mechanics. Annual Review of Fluid Mechanics 52, 477–508.

Burton, T., Jenkins, N., Sharpe, D., Bossanyi, E., 2011. Wind Energy Handbook. John Wiley & Sons.

Cal, R.B., Lebrón, J., Castillo, L., Kang, H.S., Meneveau, C., 2010. Experimental study of the horizontally averaged flow structure in a model wind-turbine array boundary layer. Journal of Renewable and Sustainable Energy (ISSN 1941-7012) 2, 1–25. https://doi.org/10.1063/1.3289735.

Calaf, M., Meneveau, C., Parlange, M., 2011. Large Eddy Simulation study of a fully developed thermal wind-turbine array boundary layer. In: Kuerten, H., Geurts, B., Armenio, V., Fröhlich, J. (Eds.), Direct and Large-Eddy Simulation VIII. In: ERCOFTAC Series, vol. 15. Springer, Dordrecht. https://doi.org/10.1007/978-94-007-2482-2_38.

Calaf, M., Meneveau, C., Meyers, J., 2010. Large eddy simulation study of fully developed wind-turbine array boundary layers. Physics of Fluids 22 (1), 015110.

Calaf, M., Vercauteren, N., Katul, G.G., Giometto, M.G., Morrison, T.J., Margairaz, F., Boyko, V., Pardyjak, E.R., 2022. Boundary-layer processes hindering contemporary numerical weather prediction models. Boundary-Layer Meteorology (ISSN 0006-8314), 1–26. https://doi.org/10.1007/s10546-022-00742-5.

Carbajo Fuertes, F., Markfort, C.D., Porté-Agel, F., 2018. Wind turbine wake characterization with nacelle-mounted wind lidars for analytical wake model validation. Remote Sensing 10 (5), 668.

Cervarich, M.C., Baidya Roy, S., Zhou, L., 2013. Spatiotemporal structure of wind farm-atmospheric boundary layer interactions. Energy Procedia 40, 530–536. https://doi.org/10.1016/j.egypro.2013.08.061.

Chamorro, L.P., Porté-Agel, F., 2009. A wind-tunnel investigation of wind-turbine wakes: boundary-layer turbulence effects. Boundary-Layer Meteorology 132, 129–149.

Chamorro, L.P., Porté-Agel, F., 2010. Effects of thermal stability and incoming boundary-layer flow characteristics on wind-turbine wakes: a wind-tunnel study. Boundary-Layer Meteorology 136, 515–533.

Chen, Y., Li, H., Jin, K., Song, Q., 2013. Wind farm layout optimization using genetic algorithm with different hub height wind turbines. Energy Conversion and Management 70, 56–65.

Choukulkar, A., Pichugina, Y., Clack, C.T.M., Calhoun, R., Banta, R., Brewer, A., Hardesty, M., 2016. A new formulation for rotor equivalent wind speed for wind resource assessment and wind power forecasting. Wind Energy 19 (8), 1439–1452.

Chowdhury, S., Zhang, J., Messac, A., Castillo, L., 2012. Unrestricted wind farm layout optimization (uwflo): investigating key factors influencing the maximum power generation. Renewable Energy 38 (1), 16–30.

Christiansen, M.B., Hasager, C.B., 2005. Wake effects of large offshore wind farms identified from satellite sar. Remote Sensing of Environment 98 (2–3), 251–268.

Colle, B.A., Sienkiewicz, M.J., Archer, C., Veron, D., Veron, F., Kempton, W., Mak, J.E., 2016. Improving the mapping and prediction of offshore wind resources (impowr): experimental overview and first results. Bulletin of the American Meteorological Society 97 (8), 1377–1390.

Cortina, G., Sharma, V., Torres, R., Calaf, M., 2020. Mean kinetic energy distribution in finite-size wind farms: a function of turbines' arrangement. Renewable Energy (ISSN 0960-1481) 148, 585–599. https://doi.org/10.1016/j.renene.2019.10.148.

Crespo, A., Hernández, J., et al., 1996. Turbulence characteristics in wind-turbine wakes. Journal of Wind Engineering and Industrial Aerodynamics 61 (1), 71–85.

Debnath, M., Santoni, C., Leonardi, S., Iungo, G.V., 2017. Towards reduced order modelling for predicting the dynamics of coherent vorticity structures within wind turbine wakes. Philosophical Transactions of the Royal Society A: Mathematical, Physical and Engineering Sciences 375 (2091), 20160108.

Debnath, M., Doubrawa, P., Optis, M., Hawbecker, P., Bodini, N., 2020. Extreme wind shear events in us offshore wind energy areas and the role of induced stratification. Wind Energy Science Discussions 2020, 1–22.

Dörenkämper, M., Optis, M., Monahan, A., Steinfeld, G., 2015. On the offshore advection of boundary-layer structures and the influence on offshore wind conditions. Boundary-Layer Meteorology 155, 459–482.

Draxl, C., Allaerts, D., Quon, E., Churchfield, M., 2021. Coupling mesoscale budget components to large-eddy simulations for wind-energy applications. Boundary-Layer Meteorology 179 (1), 73–98. https://doi.org/10.1007/s10546-020-00584-z.

Duan, J., Zuo, H., Bai, Y., Duan, J., Chang, M., Chen, B., 2021. Short-term wind speed forecasting using recurrent neural networks with error correction. Energy 217, 119397.

Emeis, S., 2010. A simple analytical wind park model considering atmospheric stability. Wind Energy 13 (5), 459–469.

España, G., Aubrun, S., Loyer, S., Devinant, P., 2011. Spatial study of the wake meandering using modelled wind turbines in a wind tunnel. Wind Energy 14 (7), 923–937. https://doi.org/10.1002/we.515.

Ezzat, A.A., 2020. Turbine-specific short-term wind speed forecasting considering within-farm wind field dependencies and fluctuations. Applied Energy 269, 115034.

Felli, M., Camussi, R., Di Felice, F., 2011. Mechanisms of evolution of the propeller wake in the transition and far fields. Journal of Fluid Mechanics 682, 5–53.

Fiedler, B.H., Bukovsky, M.S., 2011. The effect of a giant wind farm on precipitation in a regional climate model. Environmental Research Letters 6, 045101.

Finger, A., 2007. The Role of the Research Platforms FINO 1-3 in the Technical and Ecological Support Research on Offshore Wind Energy Use.

Fitch, A.C., Olson, J.B., Lundquist, J.K., Dudhia, J., Gupta, A.K., Michalakes, J., Barstad, I., 2012. Local and mesoscale impacts of wind farms as parameterized in a mesoscale NWP model. Monthly Weather Review (ISSN 0027-0644) 140 (2005), 3017–3038. https://doi.org/10.1175/MWR-D-11-00352.1.

Fitch, A.C., Lundquist, J.K., Olson, J.B., 2013. Mesoscale influences of wind farms throughout a diurnal cycle. Monthly Weather Review (ISSN 0027-0644) 141 (2005), 2173–2198. https://doi.org/10.1175/MWR-D-12-00185.1.

Floors, R., Gryning, S.-E., Pena, A., Batchvarova, E., 2011. Analysis of diabatic flow modification in the internal boundary layer. Meteorologische Zeitschrift 20 (6), 649.

Foti, D., Yang, X., Guala, M., Sotiropoulos, F., 2016. Wake meandering statistics of a model wind turbine: insights gained by large eddy simulations. Physical Review Fluids 1, 044407. https://doi.org/10.1103/PhysRevFluids.1.044407.

Frandsen, S., 1992. On the wind speed reduction in the center of large clusters of wind turbines. Journal of Wind Engineering and Industrial Aerodynamics (ISSN 0167-6105) 39, 251–265. https://doi.org/10.1016/0167-6105(92)90551-K.

Frandsen, S., 2007. Turbulence and Turbulence-Generated Structural Loading in Wind Turbine Clusters.

Frandsen, S., Barthelmie, R., Pryor, S., Rathmann, O., Larsen, S., Højstrup, J., Thøgersen, M., 2006. Analytical modelling of wind speed deficit in large offshore wind farms. Wind Energy: An International Journal for Progress and Applications in Wind Power Conversion Technology 9 (1–2), 39–53.

Fukumoto, Y., Okulov, V.L., 2005. The velocity field induced by a helical vortex tube. Physics of Fluids (ISSN 1070-6631) 17, 1–19. https://doi.org/10.1063/1.2061427.

Gadde, S.N., Stevens, R.J.A.M., 2021. Interaction between low-level jets and wind farms in a stable atmospheric boundary layer. Physical Review Fluids 6 (1), 014603.

Gao, X., Yang, H., Lu, L., 2016. Optimization of wind turbine layout position in a wind farm using a newly-developed two-dimensional wake model. Applied Energy 174, 192–200.

Ghaisas, N.S., Archer, C.L., 2016. Geometry-based models for studying the effects of wind farm layout. Journal of Atmospheric and Oceanic Technology 33 (3), 481–501.

Gottschall, J., Wolken-Möhlmann, G., Viergutz, T., Lange, B., 2014. Results and conclusions of a floating-lidar offshore test. Energy Procedia 53, 156–161.

Hartmann, D.L., Ockert-Bell, M.E., Michelsen, M.L., 1992. The effect of cloud type on earth's energy balance: Global analysis. Journal of Climate 5 (11), 1281–1304.

Hasager, C.B., Rasmussen, L., Peña, A., Jensen, L.E., Réthoré, P.-E., 2013. Wind farm wake: the horns rev photo case. Energies (ISSN 1996-1073) 6 (2), 696–716. https://doi.org/10.3390/en6020696.

Haupt, S.E., Kosovic, B., Shaw, W., Berg, L.K., Churchfield, M., Cline, J., Draxl, C., Ennis, B., Koo, E., Kotamarthi, R., Mazzaro, L., Mirocha, J., Moriarty, P., Muñoz-Esparza, D., Quon, E., Rai, R.K., Robinson, M., Sever, G., 2019. On bridging a modeling scale gap: mesoscale to microscale coupling for wind energy. Bulletin of the American Meteorological Society 100 (12), 2533–2550. https://doi.org/10.1175/BAMS-D-18-0033.1.

Herbert-Acero, J.F., Probst, O., Réthoré, P.-E., Larsen, G.C., Castillo-Villar, K.K., 2014. A review of methodological approaches for the design and optimization of wind farms. Energies 7 (11), 6930–7016.

Hezaveh, S.H., Bou-Zeid, E., Lohry, M.W., Martinelli, L., 2017. Simulation and wake analysis of a single vertical axis wind turbine. Wind Energy (ISSN 1095-4244) 20 (4), 713–730. https://doi.org/10.1002/we.2056.

Hezaveh, S.H., Bou-Zeid, E., Dabiri, J., Kinzel, M., Cortina, G., Martinelli, L., 2018. Increasing the power production of vertical-axis wind-turbine farms using synergistic clustering. Boundary-Layer Meteorology (ISSN 0006-8314) 169 (2), 275–296. https://doi.org/10.1007/s10546-018-0368-0.

Hietanen, A.I., Snedker, T.H., Dykes, K., Bayati, I., 2024. A novel techno-economical layout optimization tool for floating wind farm design. Wind Energy Science 9 (2), 417–438.

Hirth, B.D., Schroeder, J.L., Gunter, W.S., Guynes, J.G., 2012. Measuring a utility-scale turbine wake using the TTUKa mobile research radars. Journal of Atmospheric and Oceanic Technology 29 (6), 765–771. https://doi.org/10.1175/JTECH-D-12-00039.1.

Högström, U., Asimakopoulos, D.N., Kambezidis, H., Helmis, C.G., Smedman, A., 1988. A field study of the wake behind a 2 mw wind turbine. Atmospheric Environment (1967) 22 (4), 803–820.

Hong, J., Toloui, M., Chamorro, L.P., Guala, M., Howard, K., Riley, S., Tucker, J., Sotiropoulos, F., 2014. Natural snowfall reveals large-scale flow structures in the wake of a 2.5-mw wind turbine. Nature Communications 5 (1), 4216.

Honnert, R., Efstathiou, G.A., Beare, R.J., Ito, J., Lock, A., Neggers, R., Plant, R.S., Shin, H.H., Tomassini, L., Zhou, B., 2020. The atmospheric boundary layer and the "gray zone" of turbulence: a critical review. Journal of Geophysical Research. Atmospheres (ISSN 2169-897X) 125 (13). https://doi.org/10.1029/2019jd030317.

Howland, M.F., Dabiri, J.O., 2019. Wind farm modeling with interpretable physics-informed machine learning. Energies 12 (14), 2716.

International Electrotechnical Commission, 2019. IEC standard 61400-1, wind turbines, Part 1: design requirements, 61400-1. http://fs.fish.govt.nz/Page.aspx?pk=7&sc=SUR. Ed. 4, 2018.

Iosifidis, A., Tefas, A., 2022. Deep Learning for Robot Perception and Cognition. Academic Press.

Ishihara, T., Qian, G.-W., 2018. A new Gaussian-based analytical wake model for wind turbines considering ambient turbulence intensities and thrust coefficient effects. Journal of Wind Engineering and Industrial Aerodynamics 177, 275–292.

Iungo, G.V., Porté-Agel, F., 2014. Volumetric lidar scanning of wind turbine wakes under convective and neutral atmospheric stability regimes. Journal of Atmospheric and Oceanic Technology 31 (10), 2035–2048. https://doi.org/10.1175/JTECH-D-13-00252.1.

Iungo, G.V., Viola, F., Camarri, S., Porté-Agel, F., Gallaire, F., 2013. Linear stability analysis of wind turbine wakes performed on wind tunnel measurements. Journal of Fluid Mechanics 737, 499–526.

Iungo, G.V., Maulik, R., Renganathan, S.A., Letizia, S., 2022. Machine-learning identification of the variability of mean velocity and turbulence intensity for wakes generated by onshore wind turbines: cluster analysis of wind lidar measurements. Journal of Renewable and Sustainable Energy 14 (2).

Ivanell, S., Mikkelsen, R., Sørensen, J.N., Henningson, D., 2010. Stability analysis of the tip vortices of a wind turbine. Wind Energy 13 (8), 705–715.

Japar, F., Mathew, S., Narayanaswamy, B., Lim, C.M., Hazra, J., 2014. Estimating the wake losses in large wind farms: a machine learning approach. In: ISGT 2014. IEEE, pp. 1–5.

Jensen, N.O., 1983. A Note on Wind Generator Interaction. Risø National Laboratory.

Jimenez, A., Crespo, A., Migoya, E., Garcia, J., 2007. Advances in large-eddy simulation of a wind turbine wake. Journal of Physics. Conference Series (ISSN 1742-6596) 75, 012041. https://doi.org/10.1088/1742-6596/75/1/012041.

Jiménez, P.A., Dudhia, J., 2018. On the need to modify the sea surface roughness formulation over shallow waters. Journal of Applied Meteorology and Climatology 57 (5), 1101–1110.

Johlas, H.M., Martínez-Tossas, L.A., Schmidt, D.P., Lackner, M.A., Churchfield, M.J., 2019. Large Eddy simulations of floating offshore wind turbine wakes with coupled platform motion. Journal of Physics: Conference Series 1256, 012018.

Kadum, H., Cal, R.B., Quigley, M., Cortina, G., Calaf, M., 2020. Compounded energy gains in collocated wind plants: energy balance quantification and wake morphology description. Renewable Energy (ISSN 0960-1481) 150, 868–877. https://doi.org/10.1016/j.renene.2019.12.077.

Kalverla, P.C., Steeneveld, G.-J., Ronda, R.J., Holtslag, A.A.M., 2017. An observational climatology of anomalous wind events at offshore meteomast ijmuiden (North Sea). Journal of Wind Engineering and Industrial Aerodynamics 165, 86–99.

Kang, S., Yang, X., Sotiropoulos, F., 2014. On the onset of wake meandering for an axial flow turbine in a turbulent open channel flow. Journal of Fluid Mechanics 744, 376–403.

Katic, I., Højstrup, J., Jensen, N.O., et al., 1986. A Simple Model for Cluster Efficiency. European Wind Energy Association Conference and Exhibition, vol. 1. A. Raguzzi, Rome, Italy, pp. 407–410.

Keane, A., Aguirre, P.E.O., Ferchland, H., Clive, P., Gallacher, D., 2016. An analytical model for a full wind turbine wake. Journal of Physics: Conference Series 753, 032039.

Keck, R.-E., de Maré, M., Churchfield, M.J., Lee, S., Larsen, G., Aagaard Madsen, H., 2014. On atmospheric stability in the dynamic wake meandering model. Wind Energy 17 (11), 1689–1710. https://doi.org/10.1002/we.1662.

Keith, D.W., Decarolis, J.F., Denkenberger, D.C., Lenschow, D.H., Malyshev, S.L., Pacala, S., Rasch, P.J., 2004. The influence of large-scale wind power on global climate. Proceedings of the National Academy of Sciences of the United States of America (ISSN 0027-8424) 101 (46), 16115–16120. https://doi.org/10.1073/pnas.0406930101.

Kirincich, A., 2020. A metocean reference station for offshore wind energy research in the us. Journal of Physics. Conference Series 1452, 012028.

Kirk-Davidoff, D.B., Keith, D.W., 2008. On the climate impact of surface roughness anomalies. Journal of the Atmospheric Sciences (ISSN 0022-4928) 65, 2215–2234. https://doi.org/10.1175/2007JAS2509.1.

Koragappa, P., Verdin, P.G., 2024. Design and optimisation of a 20 mw offshore wind turbine blade. Ocean Engineering 305, 117975.

Krishnamurthy, R., Fernando, H.J.S., Alappattu, D., Creegan, E., Wang, Q., 2023. Observations of offshore internal boundary layers. Journal of Geophysical Research. Atmospheres 128 (8), e2022JD037425.

Krishnamurthy, R., Newsom, R.K., Kaul, C.M., Letizia, S., Pekour, M., Hamilton, N., Chand, D., Flynn, D., Bodini, N., Moriarty, P., 2025. Observations of wind farm wake recovery at an operating wind farm. Wind Energy Science (ISSN 2366-7451) 10 (2), 361–380. https://doi.org/10.5194/wes-10-361-2025.

Lackner, M.A., Elkinton, C.N., 2007. An analytical framework for offshore wind farm layout optimization. Wind Engineering 31 (1), 17–31.

Larsen, G.C., Madsen, H.A., Bingöl, F., Mann, J., Ott, S., Sørensen, J.N., Okulov, V., Troldborg, N., Nielsen, M., Thomsen, K., Larsen, T.J., Mikkelsen, R., 2007. Dynamic wake meandering modeling. Technical report. Risø National Laboratory.

Larsen, G.C., Machefaux, E., Chougule, A., 2015. Wake meandering under non-neutral atmospheric stability conditions – theory and facts. Journal of Physics. Conference Series 625 (1), 012036. https://doi.org/10.1088/1742-6596/625/1/012036.

Larsen, G.C., Réthoré, P.E., 2009. A Simple Stationary Semi-Analytical Wake Model. Risø National Laboratory for Sustainable Energy, vol. 510. Technical University of Denmark, Roskilde, Denmark.

Lesieur, M., Metais, O., Comte, P., 2005. Large-Eddy Simulations of Turbulence. Cambridge University Press.

Lettau, H., 1969. Note on aerodynamic roughness-parameter estimation on the basis of roughness element description. Journal of Applied Meteorology 8, 828–832.

Li, R., Zhang, J., Zhao, X., 2022a. Dynamic wind farm wake modeling based on a bilateral convolutional neural network and high-fidelity LES data. Energy 258, 124845.

Li, S., Zhang, M., Piggot, M.D., 2023. End-to-end wind turbine wake modelling with deep graph representation learning. Applied Energy 339, 120928.

Li, Z., Dong, G., Yang, X., 2022b. Onset of wake meandering for a floating offshore wind turbine under side-to-side motion. Journal of Fluid Mechanics 934, A29.

Lignarolo, L.E.M., Ragni, D., Scarano, F., Ferreira, C.S., Van Bussel, G.J.W., 2015. Tip-vortex instability and turbulent mixing in wind-turbine wakes. Journal of Fluid Mechanics 781, 467–493.

Liu, G., Zhang, S., Zhao, H., Liu, J., Liang, G., Zhao, J., Sun, G., 2022. Super-resolution perception for wind power forecasting by enhancing historical data. Frontiers in Energy Research 10.

Lu, H., Porté-Agel, F., 2015. On the impact of wind farms on a convective atmospheric boundary layer. Boundary-Layer Meteorology (ISSN 0006-8314) 157 (1), 81–96. https://doi.org/10.1007/s10546-015-0049-1.

Lu, H., Porté-Agel, F., 2011. Large-eddy simulation of a very large wind farm in a stable atmospheric boundary layer. Physics of Fluids 23 (6), 065101.

Lundquist, J.K., DuVivier, K.K., Kaffine, D., Tomaszewski, J.M., 2019. Costs and consequences of wind turbine wake effects arising from uncoordinated wind energy development. Nature Energy 4 (1), 26–34. https://doi.org/10.1038/s41560-018-0281-2.

Lundquist, J.K., Wilczak, J.M., Ashton, R., Bianco, L., Brewer, W.A., Choukulkar, A., Clifton, A., Debnath, M., Delgado, R., Friedrich, K., Gunter, S., Hamidi, A., Iungo, G.V., Kaushik, A., Kosović, B., Langan, P., Lass, A., Lavin, E., Lee, J.C.-Y., McCaffrey, K.L., Newsom, R.K., Noone, D.C., Oncley, S.P., Quelet, P.T., Sandberg, S.P., Schroeder, J.L., Shaw, W.J., Sparling, L., Martin, C.S., Pe, A.S., Strobach, E., Tay, K., Vanderwende, B.J., Weickmann,

A., Wolfe, D., Worsnop, R., 2017. Assessing state-of-the-art capabilities for probing the atmospheric boundary layer: the XPIA field campaign. Bulletin of the American Meteorological Society 98 (2), 289–314. https://doi.org/10.1175/BAMS-D-15-00151.1.

Lydia, M., Selvakumar, A.I., Kumar, S.S., Kumar, G.E.P., 2013. Advanced algorithms for wind turbine power curve modeling. IEEE Transactions on Sustainable Energy 4 (3), 827–835.

Lydia, M., Kumar, S.S., Selvakumar, A.I., Kumar, G.E.P., 2014. A comprehensive review on wind turbine power curve modeling techniques. Renewable and Sustainable Energy Reviews 30, 452–460.

Magnusson, M., Smedman, A.-S., 1999. Air flow behind wind turbines. Journal of Wind Engineering and Industrial Aerodynamics 80 (1–2), 169–189.

Mahrt, L., Vickers, D., Edson, J., Wilczak, J.M., Hare, J., Højstrup, J., 2001. Vertical structure of turbulence in offshore flow during rasex. Boundary-Layer Meteorology 100, 47–61.

Mahrt, L., 1999. Stratified atmospheric boundary layers. Boundary-Layer Meteorology 90, 375–396.

Mahrt, L., 2014. Stably stratified atmospheric boundary layers. Annual Review of Fluid Mechanics 46, 23–45.

Manobel, B., Sehnke, F., Lazzús, J.A., Salfate, I., Felder, M., Montecinos, S., 2018. Wind turbine power curve modeling based on Gaussian processes and artificial neural networks. Renewable Energy 125, 1015–1020.

Markfort, C.D., Zhang, W., Porté-Agel, F., 2012. Turbulent flow and scalar transport through and over aligned and staggered wind farms. Journal of Turbulence 13 (1), N33. https://doi.org/10.1080/14685248.2012.709635.

Markfort, C.D., Zhang, W., Porté-Agel, F., 2018. Analytical model for mean flow and fluxes of momentum and energy in very large wind farms. Boundary-Layer Meteorology (ISSN 0006-8314) 166 (1), 31–49. https://doi.org/10.1007/s10546-017-0294-6.

Martínez-Tossas, L.A., Churchfield, M.J., Leonardi, S., 2015. Large eddy simulations of the flow past wind turbines: actuator line and disk modeling. Wind Energy 18 (6), 1047–1060. https://doi.org/10.1002/we.1747.

Medici, D., Alfredsson, P.H., 2008. Measurements behind model wind turbines: further evidence of wake meandering. Wind Energy: An International Journal for Progress and Applications in Wind Power Conversion Technology 11 (2), 211–217.

Medici, D., Alfredsson, P.H., 2006. Measurements on a wind turbine wake: 3d effects and bluff body vortex shedding. Wind Energy 9 (3), 219–236.

Meneveau, C., 2012. The top-down model of wind farm boundary layers and its applications. Journal of Turbulence 13 (13), N7. https://doi.org/10.1080/14685248.2012.663092.

Meyers, J., Meneveau, C., 2013. Flow visualization using momentum and energy transport tubes and applications to turbulent flow in wind farms. Journal of Fluid Mechanics 715, 335–358. https://doi.org/10.1017/jfm.2012.523.

Mortensen, N.G., Landberg, L., Troen, I., Petersen, E.L., 1993. Wind atlas analysis and application program (wasp).

Mosetti, G., Poloni, C., Diviacco, B., Poloni, C., Diviacco, B., 1994. Optimization of wind turbine positioning in large windfarms by means of a genetic algorithm. Journal of Wind Engineering and Industrial Aerodynamics 51 (1), 105–116.

Moss, C., Maulik, R., Iungo, G.V., 2022. Modeling Wind Turbine Performance and Wake Interactions with Machine Learning.

Moss, C., Maulik, R., Moriarty, P., Iungo, G.V., 2023. Predicting wind farm operations with machine learning and the p2d-rans model: a case study for an awaken site. Wind Energy.

Moss, C., Maulik, R., Iungo, G.V., 2024. A call for enhanced data-driven insights into wind energy flow physics. Theoretical and Applied Mechanics Letters 14 (1), 100488.

Niayifar, A., Porté-Agel, F., 2016. Analytical modeling of wind farms: a new approach for power prediction. Energies 9 (9), 741.

Nygaard, N.G., 2014. Wakes in very large wind farms and the effect of neighbouring wind farms. Journal of Physics: Conference Series 524, 012162.

Nygaard, N.G., Steen, S.T., Poulsen, L., Pedersen, J.G., 2020. Modelling Cluster Wakes and Wind Farm Blockage. Journal of Physics: Conference Series 1618, 062072.

Okulov, V.L., Sorensen, J.N., 2004. Instability of a vortex wake behind wind turbines. Dokl. Phys. 49, 772–777.

Okulov, V.L., Sørensen, J.N., 2007. Stability of helical tip vortices in a rotor far wake. Journal of Fluid Mechanics (ISSN 0022-1120) 576, 1–25. https://doi.org/10.1017/s0022112006004228.

Optis, M., Perr-Sauer, J., 2019. The importance of atmospheric turbulence and stability in machine-learning models of wind farm power production. Renewable and Sustainable Energy Reviews 112, 27–41.

Otter, A., Murphy, J., Pakrashi, V., Robertson, A., Desmond, C., 2022. A review of modelling techniques for floating offshore wind turbines. Wind Energy 25 (5), 831–857.

Park, J., Park, J., 2019. Physics-induced graph neural network: an application to wind-farm power estimation. Energy 187, 115883.

Pedersen, J.G., Svensson, E., Poulsen, L., Nygaard, N.G., 2022. Turbulence optimized park model with gaussian wake profile. Journal of Physics: Conference Series 2265, 022063.

Pei, S., Li, Y., 2019. Wind turbine power curve modeling with a hybrid machine learning technique. Applied Sciences 9 (22), 4930.

Peña, A., Rathmann, O., 2014. Atmospheric stability-dependent infinite wind-farm models and the wake-decay coefficient. Wind Energy 17 (8), 1269–1285. https://doi.org/10.1002/we.1632.

Peña, A., Réthoré, P.-E., van der Laan, M.P., 2016. On the application of the Jensen wake model using a turbulence-dependent wake decay coefficient: the sexbierum case. Wind Energy 19 (4), 763–776.

Pettas, V., Kretschmer, M., Clifton, A., Cheng, P.W., 2021. On the effects of inter-farm interactions at the offshore wind farm alpha ventus. Wind Energy Science Discussions 2021, 1–25.

Pichugina, Y.L., Brewer, W.A., Banta, R.M., Choukulkar, A., Clack, C.T.M., Marquis, M.C., McCarty, B.J., Weickmann, A.M., Sandberg, S.P., Marchbanks, R.D., et al., 2017. Properties of the offshore low level jet and rotor layer wind shear as measured by scanning Doppler lidar. Wind Energy 20 (6), 987–1002.

Platis, A., Siedersleben, S.K., Bange, J., Lampert, A., Bärfuss, K., Hankers, R., Cañadillas, B., Foreman, R., Schulz-Stellenfleth, J., Djath, B., Neumann, T., Emeis, S., 2018. First in situ evidence of wakes in the far field behind offshore wind farms. Scientific Reports 8 (1), 2163.

Platis, A., Hundhausen, M., Mauz, M., Siedersleben, S., Lampert, A., Bärfuss, K., Djath, B., Schulz-Stellenfleth, J., Canadillas, B., Neumann, T., et al., 2021. Evaluation of a simple analytical model for offshore wind farm wake recovery by in situ data and weather research and forecasting simulations. Wind Energy 24 (3), 212–228.

Pope, S.B., 2000. Turbulent Flows. Cambridge University Press.

Porté-Agel, F., Bastankhah, M., Shamsoddin, S., 2020. Wind-turbine and wind-farm flows: a review. Boundary-Layer Meteorology 174 (1), 1–59. https://doi.org/10.1007/s10546-019-00473-0.

Pressel, K.G., Kaul, C.M., Schneider, T., Tan, Z., Mishra, S., 2015. Large-Eddy simulation in an anelastic framework with closed water and entropy balances. Journal of Advances in Modeling Earth Systems 7 (3), 1425–1456.

Puccioni, M., Iungo, G.V., Moss, C., Solari, M.S., Letizia, S., Bodini, N., Moriarty, P., 2023. LiDAR measurements to investigate farm-to-farm interactions at the AWAKEN experiment. Journal of Physics. Conference Series (ISSN 1742-6596) 2505 (1), 012045. https://doi.org/10.1088/1742-6596/2505/1/012045.

Quaranta, H.U., Bolnot, H., Leweke, T., 2015. Long-wave instability of a helical vortex. Journal of Fluid Mechanics 780, 687–716.

Rajewski, D.A., Takle, E.S., Lundquist, J.K., Oncley, S., Prueger, J.H., Horst, T.W., Rhodes, M.E., Pfeiffer, R., Hatfield, J.L., Spoth, K.K., Doorenbos, R.K., 2013. Crop wind energy experiment (CWEX): observations of surface-layer, boundary layer, and mesoscale interactions with a wind farm. Bulletin of the American Meteorological Society (ISSN 0003-0007) 94 (5), 655–672. https://doi.org/10.1175/BAMS-D-11-00240.1.

Rajewski, D.A., Takle, E.S., Lundquist, J.K., Prueger, J.H., Pfeiffer, R.L., Hatfield, J.L., Spoth, K.K., Doorenbos, R.K., 2014. Changes in fluxes of heat, H2O, and CO2 caused by a large wind farm. Agricultural and Forest Meteorology (ISSN 0168-1923) 194, 175–187. https://doi.org/10.1016/j.agrformet.2014.03.023.

Santhanagopalan, V., Rotea, M.A., Iungo, G.V., 2018. Performance optimization of a wind turbine column for different incoming wind turbulence. Renewable Energy 116, 232–243.

Sarmast, S., Dadfar, R., Mikkelsen, R.F., Schlatter, P., Ivanell, S., Sørensen, J.N., Henningson, D.S., 2014. Mutual inductance instability of the tip vortices behind a wind turbine. Journal of Fluid Mechanics 755, 705–731.

Sathe, A., Gryning, S.-E., Peña, A., 2011. Comparison of the atmospheric stability and wind profiles at two wind farm sites over a long marine fetch in the North Sea. Wind Energy 14 (6), 767–780.

Schneemann, J., Rott, A., Dörenkämper, M., Steinfeld, G., Kühn, M., 2020. Cluster wakes impact on a far-distant offshore wind farm's power. Wind Energy Science 5 (1), 29–49. https://doi.org/10.5194/wes-5-29-2020. https://wes.copernicus.org/articles/5/29/2020/.

Schreiber, J., Balbaa, A., Bottasso, C.L., 2020. Brief communication: a double-Gaussian wake model. Wind Energy Science 5 (1), 237–244.

Shakoor, R., Hassan, M.Y., Raheem, A., Wu, Y.-K., 2016. Wake effect modeling: a review of wind farm layout optimization using Jensen's model. Renewable and Sustainable Energy Reviews 58, 1048–1059.

Shao, Z., Wu, Y., Li, L., Han, S., Liu, Y., 2019. Multiple wind turbine wakes modeling considering the faster wake recovery in overlapped wakes. Energies 12 (4), 680.

Shapiro, C.R., Starke, G.M., Meneveau, C., Gayme, D.F., 2019. A wake modeling paradigm for wind farm design and control. Energies 12 (15), 2956.

Sharma, V., Parlange, M.B., Calaf, M., 2017. Perturbations to the spatial and temporal characteristics of the diurnally-varying atmospheric boundary layer due to an extensive wind farm. Boundary-Layer Meteorology (ISSN 0006-8314) 162 (2), 255–282. https://doi.org/10.1007/s10546-016-0195-0.

Sharma, V., Cortina, G., Margairaz, F., Parlange, M.B., Calaf, M., 2018. Evolution of flow characteristics through finite-sized wind farms and influence of turbine arrangement. Renewable Energy (ISSN 0960-1481) 115, 1196–1208. https://doi.org/10.1016/j.renene.2017.08.075.

Sharma, V., 2018. Numerical Experiments of Atmospheric Boundary Layer flows: Interplay between distributed drag elements and buoyancy effects. PhD thesis. École Polytechnique Fédérale de Lausanne, EPFL.

Shaw, W., Berg, L., Debnath, M., Deskos, G., Draxl, C., Ghate, V., Hasager, C., Kotamarthi, R., Mirocha, J., Muradyan, P., et al., 2022. Scientific challenges to characterizing the wind resource in the marine atmospheric boundary layer. Wind Energy Science Discussions 2022, 1–47.

Shaw, W.J., Draher, J., Garcia Medina, G., Gorton, A.M., Krishnamurthy, R., Newsom, R.K., Pekour, M.S., Sheridan, L.M., Yang, Z., 2020. General analysis of data collected from doe lidar buoy deployments off Virginia and New Jersey. Technical report. Pacific Northwest National Lab. (PNNL), Richland, WA (United States).

Sherry, M., Sheridan, J., Lo Jacono, D., 2010. Horizontal axis wind turbine tip and root vortex measurements. In: 15th International Symposium on Applications of Laser Techniques to Fluid Mechanics.

Slawsky, L.M., Zhou, L., Roy, S.B., Xia, G., Vuille, M., Harris, R.A., 2015. Observed thermal impacts of wind farms over Northern Illinois. Sensors (ISSN 1424-8220) 15 (7), 14981–15005. https://doi.org/10.3390/s150714981.

Smedman, A.-S., Bergström, H., Grisogono, B., 1997. Evolution of stable internal boundary layers over a cold sea. Journal of Geophysical Research: Oceans 102 (C1), 1091–1099.

Smith, C.M., Barthelmie, R.J., Pryor, S.C., 2013. In situ observations of the influence of a large onshore wind farm on near-surface temperature, turbulence intensity and wind speed profiles. Environmental Research Letters (ISSN 1748-9326) 8, 034006. https://doi.org/10.1088/1748-9326/8/3/034006.

Sørensen, J.N., Mikkelsen, R.F., Henningson, D.S., Ivanell, S., Sarmast, S., Andersen, S.J., 2015. Simulation of wind turbine wakes using the actuator line technique. Philosophical Transactions of the Royal Society A: Mathematical, Physical and Engineering Sciences 373 (2035), 20140071.

Sørensen, J.N., 2011. Instability of helical tip vortices in rotor wakes. Journal of Fluid Mechanics 682, 1–4.

Sørensen, J.N., Shen, W.Z., 2002. Numerical modelling of wind turbine wakes. Journal of Fluids Engineering (ISSN 0098-2202) 124 (2), 393–399.

Stevens, R.J.A.M., Meneveau, C., 2017. Flow structure and turbulence in wind farms. Annual Review of Fluid Mechanics (ISSN 0066-4189) 49 (1), 1–29. https://doi.org/10.1146/annurev-fluid-010816-060206.

Stieren, A., Stevens, R.J.A.M., 2022. Impact of wind farm wakes on flow structures in and around downstream wind farms. Flow 2, E21.

Stoll, R., Gibbs, J.A., Salesky, S.T., Anderson, W., Calaf, M., 2020. Large-Eddy simulation of the atmospheric boundary layer. Boundary-Layer Meteorology (ISSN 0006-8314), 1–41. https://doi.org/10.1007/s10546-020-00556-3.

Sutton, R.S., Barto, A.G., 2018. Reinforcement Learning: An Introduction. MIT Press.

Ti, Z., Deng, X.W., Yang, H., 2020. Wake modeling of wind turbines using machine learning. Applied Energy 257, 114025.

Ti, Z., Deng, X.W., Zhang, M., 2021. Artificial neural networks based wake model for power prediction of wind farm. Renewable Energy 172, 618–631.

Vahidi, D., Porté-Agel, F., 2022. A physics-based model for wind turbine wake expansion in the atmospheric boundary layer. Journal of Fluid Mechanics 943, A49.

van der Laan, M.P., Baungaard, M., Kelly, M., 2022a. Brief communication: a clarification of wake recovery mechanisms. Wind Energy Science Discussions 2022, 1–8.

van der Laan, M.P., García-Santiago, O., Kelly, M., Meyer Forsting, A., Dubreuil-Boisclair, C., Seim, K.S., Imberger, M., Peña, A., Sørensen, N.N., Réthoré, P.-E., 2022b. A new RANS-based wind farm parametrization and inflow model for wind farm cluster modeling. Wind Energy Science Discussions 2022, 1–38. https://doi.org/10.5194/wes-2022-112.

Veers, P., Dykes, K., Lantz, E., Barth, S., Bottasso, C.L., Carlson, O., Clifton, A., Green, J., Green, P., Holttinen, H., Laird, D., Lehtomäki, V., Lundquist, J.K., Manwell, J., Marquis, M., Meneveau, C., Moriarty, P., Munduate, X., Muskulus, M., Naughton, J., Pao, L., Paquette, J., Peinke, J., Robertson, A., Rodrigo, J.S., Sempreviva, A.M., Smith, J.C., Tuohy, A., Wiser, R., 2019. Grand challenges in the science of wind energy. Science (ISSN 0036-8075) 366 (6464), eaau2027. https://doi.org/10.1126/science.aau2027.

VerHulst, C., Meneveau, C., 2014. Large eddy simulation study of the kinetic energy entrainment by energetic turbulent flow structures in large wind farms. Physics of Fluids (ISSN 1089-7666) 26, 1–20. https://doi.org/10.1063/1.4865755.

Vermeer, L.J., Sørensen, J.N., Crespo, A., 2003. Wind turbine wake aerodynamics. Progress in Aerospace Sciences 39, 467–510. https://doi.org/10.1016/S0376-0421(03)00078-2.

Vermeulen, P.E.J., 1980. An experimental analysis of wind turbine wakes. In: 3rd International Symposium on Wind Energy Systems, pp. 431–450.

Vinuesa, R., Brunton, S.L., 2022. Emerging trends in machine learning for computational fluid dynamics. Computing in Science & Engineering 24 (5), 33–41.

Vinuesa, R., Brunton, S.L., McKeon, B.J., 2023. The transformative potential of machine learning for experiments in fluid mechanics. Nature Reviews Physics 5 (9), 536–545.

Viola, F., Iungo, G.V., Camarri, S., Porté-Agel, F., Gallaire, F., 2014. Prediction of the hub vortex instability in a wind turbine wake: stability analysis with eddy-viscosity models calibrated on wind tunnel data. Journal of Fluid Mechanics 750, R1.

Viselli, A., Filippelli, M., Pettigrew, N., Dagher, H., Faessler, N., 2019. Validation of the first lidar wind resource assessment buoy system offshore the northeast United States. Wind Energy 22 (11), 1548–1562.

Volker, P.J.H., Badger, J., Hahmann, A.N., Ott, S., 2015. The explicit wake parametrisation v1.0: a wind farm parametrisation in the mesoscale model wrf. Geoscientific Model Development (ISSN 1991-9603) 8, 3715–3731. https://doi.org/10.5194/gmd-8-3715-2015.

Voller, V.R., Porté-Agel, F., 2002. Moore's law and numerical modeling. Journal of Computational Physics (ISSN 0021-9991) 179 (2), 698–703. https://doi.org/10.1006/jcph.2002.7083.

Wagner, R., Antoniou, I., Pedersen, S.M., Courtney, M.S., Jørgensen, H.E., 2009. The influence of the wind speed profile on wind turbine performance measurements. Wind Energy: An International Journal for Progress and Applications in Wind Power Conversion Technology 12 (4), 348–362.

Wang, C., Prinn, R.G., 2010. Potential climatic impacts and reliability of very large-scale wind farms. Atmospheric Chemistry and Physics Discussions (ISSN 1680-7324) 9, 19081–19102. https://doi.org/10.5194/acpd-9-19081-2009.

Wang, J., Foley, S., Nanos, E.M., Yu, T., Campagnolo, F., Bottasso, C.L., Zanotti, A., Croce, A., 2017. Numerical and experimental study of wake redirection techniques in a boundary layer wind tunnel. Journal of Physics: Conference Series 854, 012048.

Wang, Y., Hu, Q., Li, L., Foley, A.M., Srinivasan, D., 2019. Approaches to wind power curve modeling: a review and discussion. Renewable and Sustainable Energy Reviews 116, 109422.

Widnall, S.E., 1972. The stability of a helical vortex filament. Journal of Fluid Mechanics 54 (4), 641–663.

Wilson, B., Wakes, S., Mayo, M., 2017. Surrogate modeling a computational fluid dynamics-based wind turbine wake simulation using machine learning. In: 2017 IEEE Symposium Series on Computational Intelligence (SSCI). IEEE, pp. 1–8.

Wood, R., 2012. Stratocumulus clouds. Monthly Weather Review 140 (8), 2373–2423.

Wu, Y.T., Porté-Agel, F., 2011. Large-Eddy simulation of wind-turbine wakes: evaluation of turbine parametrisations. Boundary-Layer Meteorology 138, 345–366. https://doi.org/10.1007/s10546-010-9569-x.

Wyngaard, J.C., 2004. Toward numerical modeling in the "Terra Incognita". Journal of the Atmospheric Sciences (ISSN 0022-4928) 61 (14), 1816–1826. https://doi.org/10.1175/1520-0469(2004)061<1816:tnmitt>2.0.co;2.

Xie, S., Archer, C., 2015. Self-similarity and turbulence characteristics of wind turbine wakes via large-eddy simulation. Wind Energy 18 (10), 1815–1838.

Yan, C., Pan, Y., Archer, C.L., 2019. A general method to estimate wind farm power using artificial neural networks. Wind Energy 22 (11), 1421–1432.

Yesilbudak, M., 2018. Implementation of novel hybrid approaches for power curve modeling of wind turbines. Energy Conversion and Management 171, 156–169.

Zehtabiyan-Rezaie, N., Iosifidis, A., Abkar, M., 2022. Data-driven fluid mechanics of wind farms: a review. Journal of Renewable and Sustainable Energy 14 (3), 032703.

Zhan, L., Letizia, S., Iungo, G.V., 2020. Optimal tuning of engineering wake models through lidar measurements. Wind Energy Science 5 (4), 1601–1622.

Zhang, J., Zhao, X., 2021a. Spatiotemporal wind field prediction based on physics-informed deep learning and lidar measurements. Applied Energy 288, 116641.

Zhang, J., Zhao, X., 2021b. Three-dimensional spatiotemporal wind field reconstruction based on physics-informed deep learning. Applied Energy 300, 117390.

Zhang, J., Zhao, X., 2022. Wind farm wake modeling based on deep convolutional conditional generative adversarial network. Energy 238, 121747.

Zhang, J., Zhao, X., 2023. Digital twin of wind farms via physics-informed deep learning. Energy Conversion and Management 293, 117507.

Zhang, W., Markfort, C.D., Porté-Agel, F., 2012. Near-wake flow structure downwind of a wind turbine in a turbulent boundary layer. Experiments in Fluids 52, 1219–1235.

Zhang, W., Markfort, C.D., Porté-Agel, F., 2013. Wind-turbine wakes in a convective boundary layer: a wind-tunnel study. Boundary-Layer Meteorology 146, 161–179.

Zhang, Z., Santoni, C., Herges, T., Sotiropoulos, F., Khosronejad, A., 2021. Time-averaged wind turbine wake flow field prediction using autoencoder convolutional neural networks. Energies 15 (1), 41.

Zhou, J., Guo, P., Wang, X.-R., 2014. Modeling of wind turbine power curve based on Gaussian process. In: 2014 International Conference on Machine Learning and Cybernetics, vol. 1, pp. 71–76.

Zhou, L., Tian, Y., Roy, S.B., Thorncroft, C., Bosart, L.F., Hu, Y., 2012. Impacts of wind farms on land surface temperature. Nature Climate Change (ISSN 1758-678X) 2 (7), 539–543. https://doi.org/10.1038/nclimate1505.

Turbulence in the ocean mixed layer

11

Qing Li[a] *and Bicheng Chen*[b]
[a]Earth, Ocean and Atmospheric Sciences Thrust, The Hong Kong University of Science and Technology (Guangzhou), Guangzhou, Guangdong, China, [b]College of Ocean and Earth Sciences, Xiamen University, Xiamen, Fujian, China

11.1 Introduction

The ocean mixed layer is a layer of water in the upper ocean immediately beneath the air-sea interface, where vigorous turbulence occurs and, as a result, water properties such as temperature and salinity are well mixed. The thickness of this layer, or the mixed-layer depth, has a global mean of around 50 m with large spatial variability (e.g., de Boyer Montégut et al., 2004), ranging anywhere from a few meters (e.g., near a river mouth) to as deep as a kilometer (e.g., in deep convection regions in Labrador Sea). Meanwhile, the mixed-layer depth at a particular location may vary at different temporal scales. Throughout most parts of the ocean, it varies most on a seasonal time scale: shallower in summer, when the surface water is stabilized by intense solar radiation, and deeper in winter, when solar radiation is weak and winds are strong. On the diurnal time scale, a shallow mixed layer can also develop on top of the deeper seasonal mixed layer during the daytime, followed by a breakdown of this daytime mixed layer due to nighttime convection.

The ocean mixed layer serves as a buffer between the fast-varying atmosphere and the slow-varying ocean interior, modulating the exchange of heat, energy, and trace gases (e.g., CO_2) between the two reservoirs. By affecting the air-sea heat and carbon fluxes, processes in the ocean mixed layer can influence the regional and global weather and climate. In addition, the ocean mixed layer overlaps with the euphotic zone where enough sunlight is received for photosynthesis. Thus variations of the mixed-layer depth also have a significant impact on biogeochemical processes in the ocean.

It is often helpful to distinguish the *mixed layer* where water properties have been well mixed by either ongoing or previous turbulent mixing events and the *mixing layer* (sometimes referred to as the *boundary layer*) where active turbulent mixing is occurring (e.g., Brainerd and Gregg, 1995; Franks, 2015). They often overlap with each other when a developing ocean mixed layer deepens as a result of active turbulent mixing. However, they can be quite different under the condition of diurnal heating when a daytime mixing layer develops on top of the seasonal mixed layer (e.g., Sutherland et al., 2014), or in spring restratification when a shallower mixing layer develops on top of the deep winter mixed layer. In both cases, turbulent mixing only occurs in the shallower mixing layer but not necessarily in the deeper mixed layer, although the water properties there may still be well mixed due to previous

Environmental Turbulence. https://doi.org/10.1016/B978-0-32-395873-8.00018-8
Copyright © 2026 Elsevier Inc. All rights are reserved, including those for text and data mining, AI training, and similar technologies.

turbulent mixing events. Mixed layers from previous mixing events are important in preconditioning water properties for future mixing events, but are not the focus of this chapter. Here we are focusing on a developing ocean mixed layer, so we are not making the distinction between mixed layer and mixing layer for simplicity. Throughout this chapter, we are using the terms *ocean mixed layer* and *ocean surface boundary layer* interchangeably unless otherwise noted.

Turbulence in the ocean mixed layer is characterized by its various driving forces arising from complex processes at the air-sea interface, as well as the constraints of its development by the strong stratification underneath. The primary driving forces include wind, surface gravity waves, and surface buoyancy flux, all varying at different temporal and spatial scales. Each of the three primary driving forces generates turbulence with distinctive structures and features, as discussed later in this chapter. Combined, turbulence in the ocean mixed layer under realistic forcing exhibits complex structures and significant unsteadiness. From below, the development of the turbulence is constrained by the strong stratification at the bottom of the mixed layer. Turbulent mixing at this lower interface brings up dense water and converts turbulent kinetic energy (TKE) into potential energy, which suppresses the turbulence.

Perhaps the most distinctive feature of turbulence in the ocean mixed layer, differentiating it from boundary layer turbulence in other environments, is the unique forcing from ocean surface gravity waves. Given the large density difference between air and water, the air-sea interface is traditionally approximated as a solid wall, so that theories for wall-bounded turbulence can be applied. However, this air-sea interface is not a wall but a free surface where ocean surface gravity waves develop, which significantly modify the turbulence structures. The interaction between ocean surface gravity waves and the background turbulence in the ocean mixed layer extracts energy from the waves and generates a specific form of turbulence known as the Langmuir turbulence (Langmuir, 1938; Leibovich, 1983; Thorpe, 2004). Langmuir turbulence has unique features that differentiate it from the wind-driven shear turbulence and convective turbulence in a traditional wall boundary layer. As will be discussed in Section 11.2.2, it has a significant impact on the turbulent mixing in the ocean mixed layer.

Note that when referring to the effects of ocean surface gravity waves in this chapter, we are focusing on non-breaking waves. Breaking of surface gravity waves provides another TKE source that energizes turbulent mixing near the ocean surface. As a result, observed TKE dissipation below breaking waves is one or two orders of magnitude higher than that close to a wall (Agrawal et al., 1992; Terray et al., 1996). More details on breaking waves and their effects are to be discussed in Chapter 12. Another related topic that is excluded from the discussion here is the ocean mixed layer in shallow coastal waters, where the surface boundary layer and bottom boundary layer strongly interact with each other, or even merge to form a mixed layer throughout the whole water column.

The rest of this chapter focuses on the three primary types of turbulence in the ocean mixed layer, including wind-driven shear turbulence, convective turbulence, and Langmuir turbulence. We start from theories describing the dynamics and statistics of turbulence in the ocean mixed layer in Section 11.2, including a traditional view of a wall boundary layer in Section 11.2.1 and an extended view of a wavy boundary layer in Section 11.2.2. We then briefly discuss the observational tools and modeling techniques to study such turbulence in Section 11.3, as well as the application in understanding the role of ocean mixed layer turbulence in broader environmental and climate problems. We close the chapter in Section 11.4 by providing some discussions on remaining issues and future research directions on turbulence in the ocean mixed layer.

11.2 Theoretical foundations

Away from sharp frontal regions in the ocean, large-scale flows and water properties in the ocean mixed layer often have a much smaller horizontal gradient than vertical gradient. This is a direct result of their vast horizontal dimensions (tens to hundreds of kilometers or larger) versus their vertical extent (tens to hundreds of meters). Thus the impact of small-scale turbulence in the ocean mixed layer on the mean flow and stratification is mostly through vertical mixing. In this case, horizontal homogeneity can be assumed to simplify the problem, and the focus is placed on the vertical mixing. Classic boundary layer theories are often useful in describing the turbulence in the ocean mixed layer.

We start by describing the turbulence in the ocean mixed layer from the perspective of a wall boundary layer in Section 11.2.1. Theories of wall-bounded turbulence yield useful results that can explain many aspects of the turbulence in the ocean mixed layer. The effects of a free surface on the turbulence in the ocean mixed layer are then discussed in Section 11.2.2.

11.2.1 A wall boundary layer – air-sea interface as a rigid-lid

Traditionally, the ocean mixed layer is conceptually modeled as a wall boundary layer, bounded by the air-sea interface. Since the seawater is about a thousand times denser than the air, this air-sea interface is nearly flat. Fig. 11.1 shows a schematic of the ocean mixed layer as a wall boundary layer.

The wind imposes a wind stress at the air-sea interface, similar to the stress of a moving wall. This wind stress drives shear turbulence in the ocean mixed layer, which exhibits self-similarity across scales in a region near the wall boundary. The mean velocity is influenced by the rotation of the Earth and follows an Ekman spiral. Meanwhile, the entire column of water in the ocean mixed layer may also rotate in an inertial oscillation, driving a shear near the base of the mixed layer and contributing to the entrainment of dense water from below into the mixed layer.

Surface buoyancy flux results from air-sea exchange of heat due to surface cooling or warming, and freshwater due to precipitation or evaporation. This resembles the effect of a heated or cooled plate in an idealized setup. The surface buoyancy loss drives convective plumes that penetrate the strong stratification below the mixed layer.

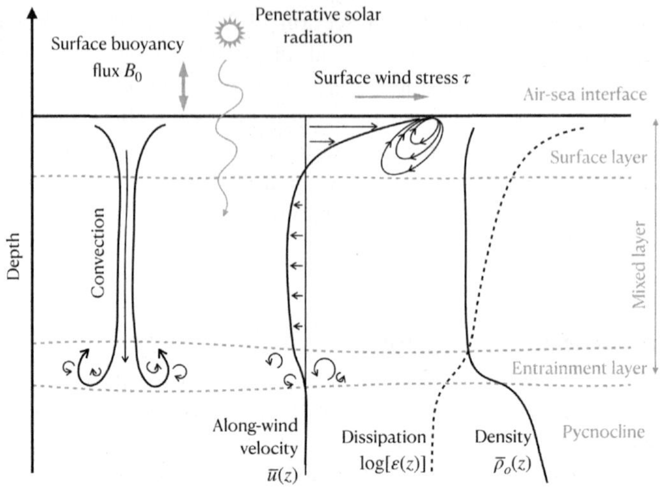

Figure 11.1 Schematic of the ocean mixed layer as a wall boundary layer The ocean mixed layer is bounded by the air-sea interface from above and the pycnocline from below. Adjacent to the air-sea interface is a surface layer where similarity of turbulence across scales is expected. Near the base of the mixed layer, there is an entrainment layer where dense water from below is mixed up. The driving forces include surface wind stress τ and surface buoyancy flux B_0, and penetrative solar radiation whose light transmittance depends on the turbidity of the water. Along-wind velocity has the strongest shear near the surface and a weaker shear in the entrainment layer. Its shear structure is shaped by the Ekman spiral due to the rotation of the Earth.

These plumes are a primary driving force of the entrainment at the base of the ocean mixed layer. In addition, penetrative solar radiation heats the water throughout a finite depth and affects the stability of the water column differently than a surface buoyancy flux.

With this wall boundary layer assumption, ideas and theories for wall-bounded turbulence can be applied to describe turbulence in the ocean mixed layer. Note that while the theories briefly presented below are in the context of the ocean mixed layer, they originate from the applications to describe turbulence in the atmospheric boundary layer. Indeed, many ocean mixed layer turbulence models are based on their atmospheric counterpart (e.g., Large et al., 1994), for which a wall boundary layer assumption is perhaps more appropriate. See, e.g., Wyngaard (2010), for a more detailed discussion on these topics in the context of atmospheric boundary layer turbulence.

11.2.1.1 Monin–Obukhov similarity theory

The Monin–Obukhov similarity theory (MOST) is among the most widely applied theories describing the planetary boundary layer turbulence bounded by a solid, flat, and horizontally homogeneous surface, in a layer adjacent to the boundary referred to as the *surface layer* (see Fig. 11.1). The theory is built on the idea of a universal length scale in the atmospheric surface layer by Obukhov (1946) and formally derived by Monin and Obukhov (1954). A historical survey of MOST is given by Foken (2006).

Basically, the MOST describes the relation between vertical gradients of mean quantities (e.g., current speed, temperature and other scalars) within the surface layer and the surface wind and buoyancy forcings. The presence of the wall introduces a length scale that describes the turbulence in the wall boundary layer—the distance from the wall $|z|$ ($z \leq 0$). A scaling for the mean shear and, thus the universal log-law for the mean velocity (see Chapter 1) follows if $|z|$ is the only relevant length scale. An obvious question to ask is whether the surface forcing introduces another relevant length scale whose effects can be described using a somewhat universal relation?

To define such a length scale in the surface layer, let us assume that the turbulence in the surface layer is only affected by the surface wind via the water-side friction velocity $u_* = \sqrt{\tau/\rho_o}$, with τ the wind stress and ρ_o the seawater density, and the surface buoyancy flux B_0. Then this turbulent length scale L can be written as a function of u_* and B_0,

$$L = L(u_*, B_0) \tag{11.1}$$

Since u_* has a unit of ms^{-1} and B_0 has a unit of $\mathrm{m}^2\mathrm{s}^{-3}$, and using dimensional analysis, a length scale, the Obukhov length (Obukhov, 1946), is formed,

$$L_O = -\frac{u_*^3}{\kappa B_0} \tag{11.2}$$

where $\kappa \approx 0.4$ is a dimensionless constant, the von Kármán constant.

Immediately, it is seen that the magnitude of L_O increases with increasing u_*, but decreases with increasing $|B_0|$. To some extent, $|L_O|$ (or more precisely $|L_O|/|z|$) measures the relative importance of the wind and surface buoyancy forcings. This will become obvious after we introduce the TKE budget below. The sign of L_O is determined by the sign of B_0. Thus L_O is negative in unstable conditions (e.g., with surface cooling or evaporation), positive in stable conditions (e.g., with surface warming or precipitation), and infinite at neutral conditions ($B_0 = 0$).

Now we can define a dimensionless depth $\zeta = |z|/L_O$ that measures the distance from the surface. The next step is to nondimensionalize the vertical gradients of the mean quantities. It is natural to nondimensionalize the vertical shear of the mean along-wind velocity $\partial_z \overline{u}$ by u_* and $|z|$, i.e., $\frac{|z|}{u_*}\partial_z \overline{u}$, where the overline represents a horizontal average. To nondimensionalize the buoyancy gradient $\partial_z \overline{b}$, a buoyancy scale can be defined from B_0 and u_*, i.e., $b_* = B_0/u_*$, so that the dimensionless buoyancy gradient is $\frac{|z|u_*}{B_0}\partial_z \overline{b}$. Similarly, one can define a tracer scale $\lambda_* = \Lambda_0/u_*$ for any tracer λ, where Λ_0 is the surface flux of tracer λ, such that the dimensionless tracer gradient is $\frac{|z|u_*}{\Lambda_0}\partial_z \overline{\lambda}$.

The MOST implies that the dimensionless gradients of the mean quantities can be written as universal functions of ζ, such that,

$$\frac{\kappa |z|}{u_*}\partial_z \overline{u} = \phi_m(\zeta) \tag{11.3}$$

and

$$\frac{\kappa|z|}{\lambda_*}\partial_z\overline{\lambda} = \phi_s(\zeta) \tag{11.4}$$

The functions $\phi_m(\zeta)$ and $\phi_s(\zeta)$ for momentum and scalar tracers are assumed to be universal for all homogeneous, quasi-steady surface layers, determined by the empirical relations fitted from measurements in boundary layer experiments.

The most widely used universal functions of Businger et al. (1971) are based on the Kansas experiments for atmospheric surface layers,

$$\phi_m(\zeta) = \begin{cases} 1+4.7\zeta, & 0 \le \zeta \\ (1-15\zeta)^{-1/4}, & \zeta < 0 \end{cases} \tag{11.5}$$

$$\phi_s(\zeta) = \begin{cases} 0.74+4.7\zeta, & 0 \le \zeta \\ 0.74(1-9\zeta)^{-1/2}, & \zeta < 0 \end{cases} \tag{11.6}$$

A similar form is adopted by Large et al. (1994) to best fit the available measurements in the oceanic surface layer,

$$\phi_m(\zeta) = \begin{cases} 1+5\zeta, & 0 \le \zeta \\ (1-16\zeta)^{-1/4}, & -0.2 \le \zeta < 0 \\ (1.26-8.38\zeta)^{-1/3}, & \zeta < -0.2 \end{cases} \tag{11.7}$$

$$\phi_s(\zeta) = \begin{cases} 1+5\zeta, & 0 \le \zeta \\ (1-16\zeta)^{-1/2}, & -1 \le \zeta < 0 \\ (-28.86-98.96\zeta)^{-1/3}, & \zeta < -1 \end{cases} \tag{11.8}$$

Fig. 11.2 compares these two sets of universal functions.

In neutral condition ($\zeta = 0$), $\phi_m = \phi_s = 1$, MOST reduces to the classic *law of the wall*. Integration of (11.3) suggests that the mean along-wind velocity varies with the distance $|z|$ from the wall boundary according to,

$$\overline{u}_0 - \overline{u}(z) = \frac{u_*}{\kappa}\ln\left(\frac{|z|}{z_0}\right) \tag{11.9}$$

where $\overline{u}_0 = \overline{u}(0)$ is the mean velocity at the surface and $z_0 > 0$ is a constant of integration known as the *roughness length*, describing the roughness effect.[1] In stable conditions ($\zeta > 0$), both ϕ_m and ϕ_s are linear functions of ζ. Integration of (11.3) with (11.7) for $\zeta > 0$ yields,

$$\overline{u}_0 - \overline{u}(z) = \frac{u_*}{\kappa}\left[\ln\left(\frac{|z|}{z_0}\right) + 5\zeta\right] \tag{11.10}$$

[1] The relation of z_0 to the actual roughness geometry is still a topic of fundamental turbulence research.

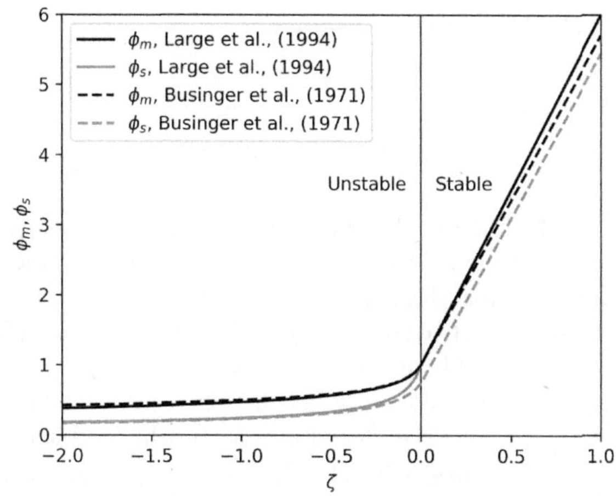

Figure 11.2 A comparison of universal functions for MOST Visualization of the universal functions in Large et al. (1994) and Businger et al. (1971). Solid lines in black and gray show ϕ_m and ϕ_s as in (11.7) and (11.8), and dashed lines show ϕ_m and ϕ_s as in (11.5) and (11.6).

This suggests that the mean velocity decreases faster than a logarithmic profile in stable conditions, as stable stratification suppresses the vertical transport of momentum. In unstable conditions ($\zeta < 0$), the universal functions of both (11.5) and (11.7) are harder to integrate. But in weakly unstable conditions or close enough to the wall ($\zeta < 0$ and $|\zeta| \ll 1$), for example, (11.7) can be approximated by a linear function of ζ, so that the mean velocity profile is in a similar form as (11.10),

$$\overline{u}_0 - \overline{u}(z) \approx \frac{u_*}{\kappa}\left[\ln\left(\frac{|z|}{z_0}\right) + 4\zeta\right] \tag{11.11}$$

so that the mean velocity decreases more slowly than a logarithmic profile. This is due to the enhanced vertical momentum transport by convective motions. In strongly unstable conditions or far enough away from the wall ($\zeta \ll -1$), the flow tends to be dominated by convection that is independent of surface wind stress.

Note that while the universal functions (11.7) and (11.8) have been applied to the oceanic surface layer with success in many scenarios, recent observational and modeling evidence highlights the need for modification of MOST for oceanic surface layers in the presence of ocean surface waves (e.g., Large et al., 2019; Zheng et al., 2021). This is still a topic of ongoing research, which we will briefly discuss in Section 11.4.1.

11.2.1.2 Turbulent kinetic energy budget

It is often helpful to distinguish different energy sources that drive turbulence in the boundary layer. The governing equation for the TKE, $e = \frac{1}{2}\left(\overline{u'^2} + \overline{v'^2} + \overline{w'^2}\right)$, is derived from the Reynolds averaged equations,

$$\partial_t e = \underbrace{-\overline{w'\boldsymbol{u}'_h} \cdot \partial_z \overline{\boldsymbol{u}}_h}_{P} \underbrace{+\overline{w'b'}}_{B} \underbrace{-\frac{1}{\rho_o}\partial_z \overline{w'p'} - \partial_z \overline{w'e}}_{T} \underbrace{-\varepsilon}_{D} \tag{11.12}$$

where $\boldsymbol{u}_h = (u, v)$ is the horizontal component of the velocity, and p is the pressure. A prime represents the fluctuation from the mean. The terms on the right-hand side represent the shear production (P), buoyancy production (B), pressure correlation and TKE transport (the sum of the two are denoted as T), and dissipation (D), respectively. In deriving (11.12), horizontal homogeneity is assumed.

As an example, we use (11.12) to discuss the physical meaning of the Obukhov length. Consider the TKE budget in the surface layer. The turbulent momentum flux $\overline{w'\boldsymbol{u}'_h}$ scales with u_*^2. Combined with (11.3), the shear production term in (11.12) scales with $u_*^3/(\kappa|z|)$. The buoyancy production term in (11.12) scales with the surface buoyancy flux B_0. If we define a depth z_e at which the shear production equals the buoyancy production,

$$\frac{u_*^3}{\kappa|z_e|} = B_0 \tag{11.13}$$

we get

$$|z_e| = \frac{u_*^3}{\kappa B_0} = |L_O| \tag{11.14}$$

Therefore $|L_O|$ is a rough estimate of the depth at which the shear production and buoyancy production are comparable. For depth much shallower than $|L_O|$, shear production dominates the TKE budget due to the strong velocity shear near the surface. The velocity shear decays quickly with depth. Eventually, for depth much deeper than $|L_O|$, buoyancy starts to dominate. The effects of buoyancy take the form of buoyancy production due to coherent convective plumes in unstable conditions (Chapter 2) and buoyancy destruction in stable conditions (Chapter 3).

11.2.1.3 Wind-driven shear turbulence and convective turbulence

In the ocean mixed layer, two fundamental types of turbulence are wind-driven shear turbulence and convective turbulence. This corresponds to the two production terms in the TKE Eq. (11.12) for the wall boundary layer, the shear production and buoyancy production.

Wind-driven shear turbulence can be understood as shear turbulence driven by a moving wall from above, with surface wind stress $\tau = \rho_o u_*^2$ at the wall as the upper boundary condition. A natural velocity scale for the wind-driven shear turbulence is u_*. In neutral stratification under steady surface wind stress, the turbulent boundary layer keeps deepening. The presence of planetary rotation puts a limit on this wind-driven boundary layer growth. Under steady surface wind stress and constant rotation rate f, the canonical solution of an *Ekman spiral* of the velocity profile can be derived, prohibiting the wind-driven shear from growing indefinitely. The depth over which the Ekman spiral decays therefore gives an upper limit of the boundary layer depth under wind-driven shear turbulence in neutral stratification (e.g., Large et al., 1994). This depth depends on the turbulent viscosity, but in general it scales with u_*/f.

In reality, stratification from below the ocean mixed layer may constrain the development of the Ekman layer. Other factors, like convection, discussed below may also strongly affect the ocean mixed layer. So u_*/f does not usually give a good estimate of the ocean mixed layer depth. However, the Ekman spiral of the velocity affects the shear available in driving ocean mixed layer turbulence through the shear production in (11.12). The inertial oscillation due to the planetary rotation may also strongly affect the development of wind-driven shear turbulence in the ocean mixed layer, and contribute to the velocity shear near the base of the mixed layer that drives the entrainment. For example, rotating wind due to the passage of a storm can sometimes resonate with inertial currents, generating intense shear turbulence and rapid mixed-layer deepening (e.g., D'Asaro, 1985; Crawford and Large, 1996; Skyllingstad et al., 2000).

Fig. 11.3 gives an example of wind-driven shear turbulence in an idealized large eddy simulation (LES) under rotation. The turbulent structures illustrated by the snapshot of the simulated vertical velocity field show typical features of shear turbulence near a solid boundary, showing irregular structures elongated in the wind direction. The horizontally averaged velocity variance in panel (b) also shows maximum variability in the wind direction and a minimum variability in the vertical. The TKE budget of wind-driven shear turbulence, as shown in panel (c), is dominated by a balance between the shear production P and dissipation D, with a relatively small contribution from TKE transport T that redistributes TKE in the vertical. This suggests that the shear production is largely dissipated locally in wind-driven shear turbulence.

While the wind-driven shear turbulence in the ocean mixed layer under the horizontally homogeneous assumption is essentially local in the sense that it is the local shear that drives the turbulence, convective turbulence often forms coherent structures (convective plumes) and exhibits nonlocal features in which the driving force of the turbulence is at a different location (destabilizing buoyancy flux due to cooling or evaporation at the ocean surface). Both the vertical and horizontal extent of these coherent structures grow as the mixed layer deepens. Thus a reasonable length scale for the convective turbulence is the mixed layer depth h. Combined with the surface buoyancy flux B_0, we can define a velocity scale for the convective turbulence $w_* = (B_0 h)^{1/3}$ (Deardorff, 1970).

Fig. 11.4 gives an example of the convective turbulence in an idealized LES under rotation forced by constant surface cooling. The vertical velocity snapshot clearly shows the convective cells consisting of intense and narrow downwelling plumes and weak and wide upwelling regions. These convective plumes can extend down to the bottom of the mixed layer. As a result, horizontally averaged velocity variance in panel (b) shows maximum variation in the vertical direction, especially in the middle of the

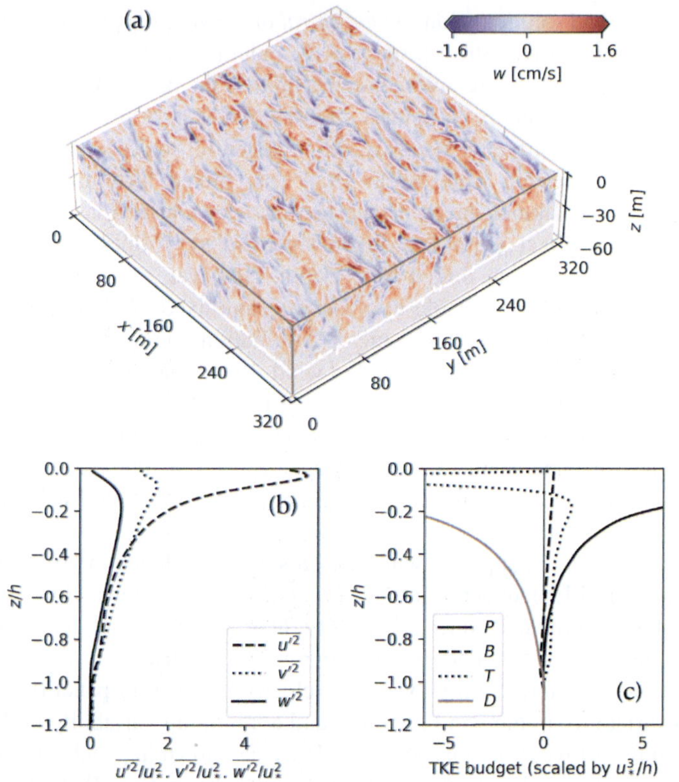

Figure 11.3 An LES solution of wind-driven shear turbulence (a): Snapshot of the vertical velocity in an idealized LES of wind-driven shear turbulence forced by 5 m s^{-1} surface wind in the x-direction and a weak 5 W m^{-2} surface cooling. A Coriolis parameter $f = 10^{-4}$ s^{-1} was used. (b) Vertical profiles of horizontally averaged $\overline{u'^2}$ (dashed), $\overline{v'^2}$ (dotted), and $\overline{w'^2}$ (solid), normalized by u_*^2.(c) normalized TKE budget scaled by u_*^3/h, including shear production (P; solid in black), buoyancy production (B; dashed in black), transport (T; dotted in black), and dissipation (D; solid in gray). The vertical axis is normalized by the mean mixed layer depth defined by the depth of maximum stratification in (b) and (c).

mixed layer, whereas the horizontal variation is symmetric in x- and y-directions and is relatively weak except near the surface and the base of the mixed layer where the vertical variation is bounded by the surface boundary and bottom stratification. This is in contrast with the wind-driven shear turbulence for which the vertical velocity variance is the weakest. The TKE budget of convective turbulence shown in panel (c) is dominated by a balance among the buoyancy production B, dissipation D, and TKE transport T. This also contrasts with wind-driven shear turbulence for which the TKE transport is relatively small. The TKE transport is the primary energy source for entrainment near the base of the mixed layer in convective turbulence, where the mixing of dense water from below results in buoyancy destruction rather than production.

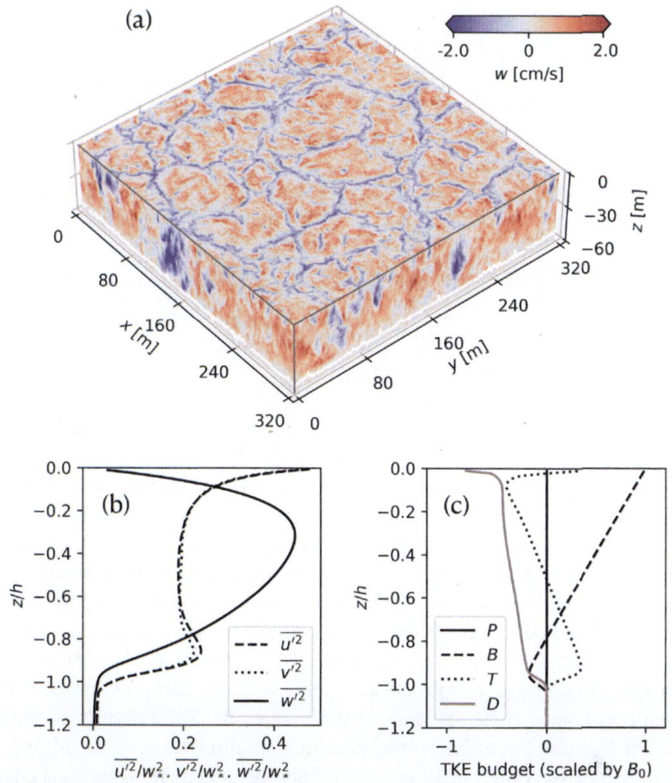

Figure 11.4 An LES solution of convective turbulence Same as Fig. 11.3, but for an idealized LES of convective turbulence forced by a 100 W m^{-2} surface cooling without wind. Vertical profiles in (b) are normalized by w_*^2, and the TKE budget is scaled by B_0.

11.2.2 A wavy boundary layer – effects of a free-surface

While a wall boundary-layer assumption is simple and theories based on it explain many features of turbulence in the ocean mixed layer, in reality, the air-sea interface is not a wall but a free surface. An important difference is the presence of surface gravity waves. Wave orbital motions and the resulting Stokes drift (to be introduced below) tilt and stretch existing vortices in the ocean mixed layer, thereby modifying the Reynolds stresses (Teixeira and Belcher, 2002). Indeed, increasing observational evidence suggests that ocean surface waves at the air-sea interface significantly influence turbulence in the ocean mixed layer (e.g., D'Asaro, 2001; D'Asaro et al., 2014). Interactions between ocean surface waves and vortices in the ocean mixed layer manifest as turbulence with a unique structure known as *Langmuir turbulence* (Langmuir, 1938; Leibovich, 1983; McWilliams et al., 1997; Thorpe, 2004), distinct from the wall-bounded wind-driven turbulence and convective turbulence (e.g., Li et al., 2005; Li and Fox-Kemper, 2020). Fig. 11.5 shows an updated schematic of the ocean mixed layer with an emphasis on the effects of ocean surface waves.

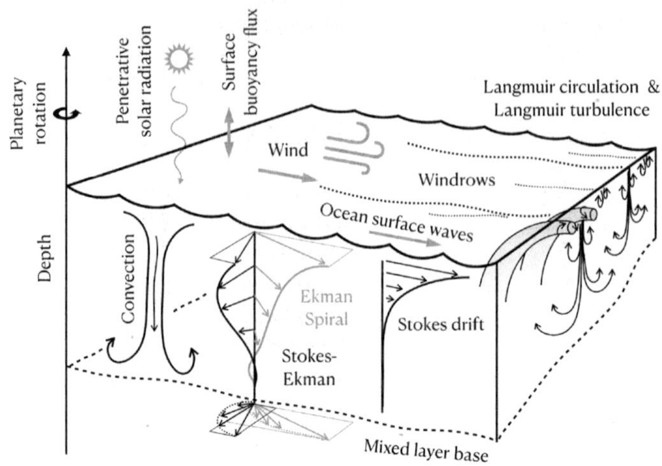

Figure 11.5 Schematic of the ocean mixed layer as a wavy boundary layer This schematic is an extension of Fig. 11.1 to include the effects of ocean surface waves. In particular, the presence of ocean surface waves results in a mean drift velocity, known as Stokes drift, which decays rapidly with depth. The vertical shear of Stokes drift is a key factor in the formation of Langmuir circulation and Langmuir turbulence, which are characterized by parallel counter-rotating vortices tilted and stretched by Stokes drift shear towards the direction of the wind and waves, and by the resulting deeply penetrating downward jets. At the surface, Langmuir circulation and Langmuir turbulence can be seen from the associated bands of surface convergence regions, or windrows, that trap foams, algae, or other floating materials. The presence of the Stokes drift induces an extra Coriolis force due to the planetary rotation, thereby modifying the Ekman spiral, as illustrated by the Eulerian mean velocity profiles and their projections on a horizontal plane in black (Stokes–Ekman) and grey (Ekman).

11.2.2.1 Stokes drift

The *Stokes drift* is the mean velocity when following a specific water parcel as it travels with the wave orbital motions – the net drift of a water parcel when averaged over several wave periods. This net drift arises from the fact that the trajectory of a water parcel following the wave orbital motions is not closed. For a monochromatic wave with wavenumber k, angular frequency ω, and amplitude a, propagating in water of depth H, the Stokes drift is,

$$u^S = \frac{\omega k a^2 \cosh[2k(z+H)]}{2\sinh^2(kH)} \tag{11.15}$$

or, in the limit of deep-water waves ($kH \gg 1$),

$$u^S = \omega k a^2 \exp(2kz) \tag{11.16}$$

It is readily seen that the Stokes drift decays rapidly with depth, with an e-folding depth of $\delta^S \sim 1/2k$. Thus Stokes drift for long waves (small k) decays slower than short waves (large k).

Ocean surface waves are composed of a spectrum of monochromatic waves of different wavenumbers and frequencies in different directions (e.g., Holthuijsen, 2007). The total Stokes drift is estimated by integrating (11.15) or (11.16) over the wave spectrum. Given that Stokes drift for short waves is more confined to near the surface than long waves, the total Stokes drift usually decays much faster than that of a monochromatic wave.

A recent review of the fundamentals and progress in laboratory and field observations of Stokes drift is given by van den Bremer and Breivik (2018). More discussions about Stokes drift and the derivation of (11.15) and (11.16) can be found in classic textbooks such as Phillips (1977).

11.2.2.2 The Craik–Leibovich equations

While Stokes drift decays rapidly with depth, its influence on flows can extend much deeper into the ocean mixed layer. Such influence of Stokes drift on the dynamics of turbulent flows in the upper ocean is described by the wave-phase-averaged, or Craik–Leibovich (CL), equations (Craik and Leibovich, 1976; Leibovich, 1983),

$$\begin{aligned} \partial_t \boldsymbol{u} + (\boldsymbol{u} \cdot \nabla)\boldsymbol{u} &= -\boldsymbol{f} \times \boldsymbol{u}^L + \boldsymbol{b} + \boldsymbol{D}^u - \nabla\left(p + \frac{1}{2}\left|\boldsymbol{u}^L\right|^2 - \frac{1}{2}|\boldsymbol{u}|^2\right) \\ &\quad - (\nabla \times \boldsymbol{u}) \times \boldsymbol{u}^S \\ \nabla \cdot \boldsymbol{u} &= 0 \end{aligned} \tag{11.17}$$

for the momentum, and,

$$\partial_t \lambda + (\boldsymbol{u}^L \cdot \nabla)\lambda = D^\lambda \tag{11.18}$$

for the tracer λ. Here the Lagrangian velocity $\boldsymbol{u}^L = \boldsymbol{u} + \boldsymbol{u}^S$ is the sum of the Eulerian (wave-phase-averaged) velocity $\boldsymbol{u}$ and Stokes drift $\boldsymbol{u}^S$. Diffusion of the momentum and tracer are represented by $\boldsymbol{D}^u$ and D^λ, respectively. The assumption here is that turbulent flows evolve on a timescale much longer than the wave orbital motions, so that their dynamics can be described in a wave-phase-averaged sense; but much shorter than the evolution of the mean wave statistics, so that Stokes drift is not affected by the turbulent flows and can be specified.

Stokes drift enters the dynamics of the wave-phase-averaged flows in (11.17) through three terms: (1) a Coriolis–Stokes force (included in the first term on the right hand side), (2) a pressure head (the fourth term), and (3) a *vortex force* (the last term). It also contributes to the advection of tracers in (11.18). The effect of the vortex force is to modulate the velocity components that are perpendicular to the Stokes drift, or equivalently, to modulate the vorticity ($\boldsymbol{\omega} = \nabla \times \boldsymbol{u}$) in the direction of Stokes drift. This can be seen more clearly in the vorticity equation by taking the curl of (11.17),

$$\partial_t \boldsymbol{\omega} + (\boldsymbol{u} \cdot \nabla)\boldsymbol{\omega} = \cdots - (\boldsymbol{u}^S \cdot \nabla)\boldsymbol{\omega} + (\boldsymbol{\omega} \cdot \nabla)\boldsymbol{u}^S \tag{11.19}$$

in which we are only explicitly writing out the terms corresponding to the vortex force for simplicity. The contribution of the vortex force to the vorticity budget is through advection (the first term after the ellipsis) and vortex stretching (the second term).

There are multiple mathematically equivalent forms of the CL-equations, each emphasizing different aspects of the dynamics (e.g., Holm, 1996). Among these different forms, a particularly useful one is derived and discussed in detail in Suzuki and Fox-Kemper (2016):

$$\partial_t \boldsymbol{u} + (\boldsymbol{u}^L \cdot \nabla)\boldsymbol{u} = -\boldsymbol{f} \times \boldsymbol{u}^L + \boldsymbol{b} + \boldsymbol{D}^u - \nabla p - u_j^L \nabla u_j^S \tag{11.20}$$

where the last term represents the force related to the Stokes drift shear, thereby the name *Stokes shear force*. In index notation, the last term is $-u_j^L \partial_i u_j^S$ and acts in the direction of the gradient of Stokes drift. The physical meaning of the Stokes shear force is clear: water parcels moving in (against) the direction of Stokes drift experience a force in the down-gradient (up-gradient) direction of the Stokes drift shear; whereas water parcels moving perpendicular to the Stokes drift do not experience a force. Given that the primary variation of Stokes drift is in the vertical, the Stokes shear force is largest in the vertical. As will be discussed below in Section 11.2.2.4, this Stokes shear force strongly affects the turbulence structure in the ocean mixed layer.

11.2.2.3 Stokes production of TKE

Another way to look at the influence of Stokes drift is to examine its effect on the energetics of the ocean mixed layer. The TKE budget equation with the influence of Stokes drift is derived from the CL equations,

$$\partial_t e = \underbrace{-\overline{w'\boldsymbol{u}_h'} \cdot \partial_z \boldsymbol{u}^S}_{P^S} \underbrace{-\overline{w'\boldsymbol{u}_h'} \cdot \partial_z \overline{\boldsymbol{u}}_h}_{P} \underbrace{+\overline{w'b'}}_{B} \underbrace{-\frac{1}{\rho_o}\partial_z \overline{w'p'} - \partial_z \overline{w'e}}_{T} \underbrace{-\varepsilon}_{D} \tag{11.21}$$

Compared to (11.12), there is an additional term P^S that is in a similar form as the shear production P, but with the vertical shear of the Stokes drift rather than the vertical shear of the Eulerian velocity. This term represents a distinct source of TKE, known as the *Stokes production*.

While in a similar form as the shear production, the Stokes production affects the TKE in a very different way. By writing out the budget equations for the three components of the TKE ($\overline{u'^2}$, $\overline{v'^2}$, and $\overline{w'^2}$) separately, one can easily find that, unlike the shear production which contributes to the budget of $\overline{u'^2}$ and $\overline{v'^2}$, the Stokes production only appears in the $\overline{w'^2}$ budget equation. In other words, Stokes production produces vertical velocity variance. This is a major difference between the Stokes drift shear and the Eulerian shear and affects the anisotropy of the turbulence significantly (Li et al., 2005; Li and Fox-Kemper, 2020).

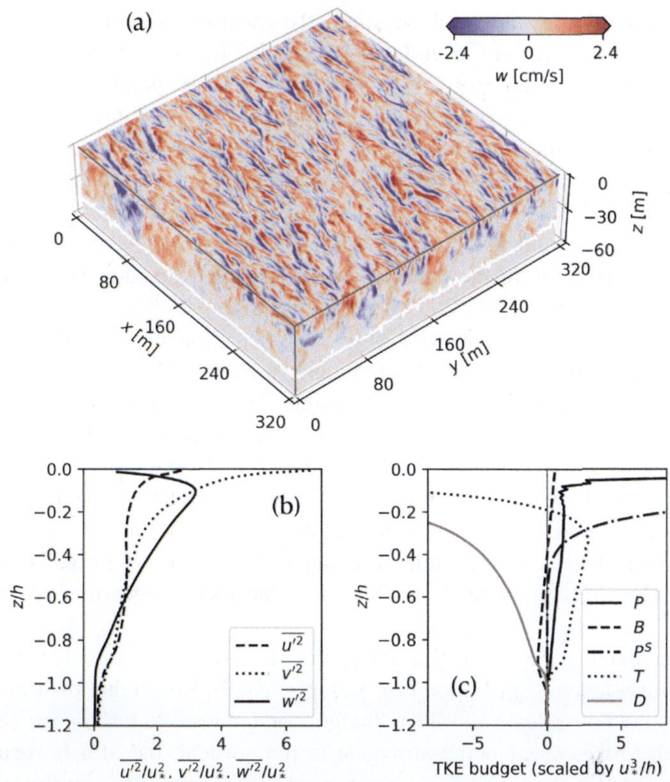

Figure 11.6 An LES solution of Langmuir turbulence Same as Fig. 11.3, but for an idealized LES of Langmuir turbulence forced by 5 m s^{-1} surface wind and a monochromatic deep-water wave with an amplitude of $a = 0.8$ m and a wavenumber of $k = 2\pi/60$ m^{-1}, both in x-direction, as well as a weak 5 W m^{-2} surface cooling. Note the additional dash-dotted line in (c) showing the Stokes production (P^S) in the TKE budget.

11.2.2.4 Langmuir turbulence

Bands of floating materials (e.g., foams due to wave breaking) aligned parallel to the wind and waves have long been recognized as evidence of Langmuir circulation (Langmuir, 1938; Thorpe, 2004). This conceptualized flow structure, characterized by pairs of counter-rotating vortices, which trap floating materials in parallel bands of surface convergence zones, is also supported by direct measurements (e.g., Weller and Price, 1988; Plueddemann et al., 1996; Smith, 1998). Theories of wave-phase averaged flows, in particular the CL equations in their various forms (Craik and Leibovich, 1976; Leibovich, 1983; Holm, 1996; Suzuki and Fox-Kemper, 2016), were developed to explain the formation of Langmuir circulation. Numerical solutions of the CL equations in LES show the characteristic structures of Langmuir circulation, with rich irregular patterns and unsteady structures, representing a distinct turbulent regime known as Langmuir turbulence (McWilliams et al., 1997).

Fig. 11.6 shows an example of the simulated Langmuir turbulence in an idealized LES forced by constant wind and waves. As its laminar counterpart, Langmuir turbulence is characterized by a flow structure that has parallel bands of narrow surface convergence zones (blue shading in panel (a)) where intense downward jets penetrate deep into the mixed layer, with wider and weaker upwelling returning flow in between (red shading). These deeply penetrating jets also have a horizontal component in the direction of wind and waves. Therefore, as illustrated in panel (b), the Reynolds stress associated with Langmuir turbulence has an anisotropic structure with larger components in the vertical and cross-wind direction but smaller in the along-wind direction (Li et al., 2005; Li and Fox-Kemper, 2020). This is also a direct result of the fact that, as discussed above, Stokes production in the TKE budget only directly contributes to the vertical velocity variance. Thus, Langmuir turbulence is very efficient in generating vertical mixing. The TKE budget in panel (c) also demonstrates this point. In the upper part of the mixed layer, the TKE budget is dominated by Stokes production P^S, dissipation D, and the transport term T acting as a TKE sink (negative). This transport term becomes the TKE source (positive) in the lower part of the mixed layer, where Stokes production goes to zero due to the rapid decay of Stokes drift. It is interesting to compare Fig. 11.6c with Fig. 11.3c in which the transport term T plays a much smaller role in the TKE budget of wind-driven shear turbulence.

To some extent, the vertical mixing due to Langmuir turbulence resembles convective turbulence (e.g., Li and Fox-Kemper, 2020). Recall that the transport term T also plays a key role in the TKE budget for convective turbulence (Fig. 11.4). Indeed, with Stokes shear being strongest in the vertical, the Stokes shear force in (11.20) is largest in the vertical momentum equation, acting in a similar way as the buoyancy. The difference is that, unlike buoyancy, which exerts a force throughout the mixed layer well into the stratification until the buoyancy of a water parcel matches the environment, the Stokes shear force is nonzero only in the near-surface regions where Stokes drift shear is non-zero. A useful analogy is to think of the Stokes shear force as a kick near the surface, versus a continuous push by the buoyancy force throughout the mixed layer. In this sense, a significant portion of the effects of Langmuir turbulence on vertical mixing is *nonlocal*, meaning that the turbulent fluxes are nonproportional to the local gradients. Thus, it is sometimes convenient to represent the mixing effect of Langmuir turbulence in a similar way as the convective turbulence (e.g., Li and Fox-Kemper, 2017).

11.3 Investigations and applications

11.3.1 Observations

Systematic observation of turbulence in the ocean mixed layer involves simultaneous measurements of meteorological fields (e.g., wind, precipitation, solar radiation, and air-sea fluxes), water property profiles (e.g., temperature and salinity), ocean surface waves, and various turbulence statistics. Such observations are complex and difficult to conduct, so a variety of observational techniques have been developed, each with assumptions at different levels of applicability. Observational techniques commonly used in measuring ocean turbulence are introduced in textbooks such as Thorpe (2005). Frajka-Williams et al. (2022) provide a recent review of advances and new frontiers in observational techniques for measuring ocean mixing, focusing on new platforms enabled by the wide use of autonomous underwater vehicles such as floats and gliders. These new platforms greatly expand the temporal and spatial coverage of turbulence measurements in the ocean. The application of these new observational platforms in observing turbulence in the ocean mixed layer is discussed in, e.g., Fox-Kemper et al. (2022). Readers are referred to these books and references therein for a comprehensive review of observational techniques used in measuring ocean turbulence. Here, we only briefly discuss a few direct measurements of turbulence in the ocean mixed layer. Many indirect observational techniques are omitted here for brevity.

A useful direct measurement of turbulence in the ocean, including the ocean mixed layer, is the TKE dissipation rate ε. Estimating the dissipation rate requires simultaneous measurements of velocity shear in three orthogonal directions, which is extremely difficult in the oceanic environment. In practice, isotropy is often assumed so that

$$\varepsilon = \frac{15}{2}\nu\overline{(\partial_z u')^2} \tag{11.22}$$

where ν is the kinematic viscosity of seawater. This reduces the measurement requirement to only the velocity shear in one direction. This can be done using microstructure profilers with a resolution that is fine enough to capture the small-scale velocity shear.

When the influences of surface waves and buoyancy flux are negligible, one can also estimate the vertical turbulent viscosity K_u from ε by assuming a balance between the shear production and dissipation in the TKE budget (11.12) in steady state, so that

$$-\overline{w'u'}\partial_z\overline{u} = \varepsilon \tag{11.23}$$

and

$$K_u = \frac{-\overline{w'u'}}{\partial_z\overline{u}} = \frac{\varepsilon}{(\partial_z\overline{u})^2} \tag{11.24}$$

where we are assuming $\overline{v} = v' = 0$ for demonstration purposes. This is not a bad assumption for pure wind-driven shear turbulence as shown in Fig. 11.3c, but fails for convective turbulence and Langmuir turbulence when significant TKE transport occurs (Figs. 11.4c and 11.6c).

Similarly, the decay rate of temperature variance χ can be estimated from the microstructure of the temperature gradient,

$$\chi = 6\kappa_\theta\overline{(\partial_z\theta')^2} \tag{11.25}$$

where θ is the temperature and κ_θ is the molecular diffusivity of heat. Assuming a steady state for the temperature variance, one can estimate the vertical turbulent diffusivity of heat K_θ by a balance between the mean gradient production and decay of the temperature variance,

$$-2\overline{w'\theta'}\partial_z\overline{\theta} = \chi \tag{11.26}$$

and

$$K_\theta = \frac{-\overline{w'\theta'}}{\partial_z\overline{\theta}} = \frac{\chi}{2\left(\partial_z\overline{\theta}\right)^2} \tag{11.27}$$

11.3.2 Modeling techniques

In addition to observations, numerical models have played a significant role in improving our understanding of turbulence in the ocean mixed layer. The typical velocity scale for the turbulent flows in the ocean mixed layer is $U \sim 0.1$ m s^{-1}. Assuming that it varies at a length scale of $L \sim 10$ m, the Reynolds number is $Re = UL/\nu \sim 10^6$. For such high-Reynolds-number problems, direct numerical simulation (DNS) is prohibitively expensive. To make matters worse, processes that contribute to turbulence in the ocean mixed layer often have length scales of hundreds of meters or more and time scales of hours to days. Since it is often the large, energy-containing turbulent structures that we are most interested in, large eddy simulation (LES) is a natural choice for modeling turbulence in the ocean mixed layer.

The development of LES was pioneered by applications in studying the atmospheric boundary layer turbulence (e.g., Smagorinsky, 1963; Lilly, 1967; Deardorff, 1974). Its application to the oceanic boundary layer for buoyancy- and shear-driven turbulence follows the same principles. A major advance in the application of LES to ocean mixed-layer turbulence is the inclusion of the effects of ocean surface waves by solving the CL equations (Skyllingstad and Denbo, 1995; McWilliams et al., 1997). It has greatly improved our understanding of turbulence in the wavy ocean mixed layer, especially Langmuir turbulence, including its energetics (e.g., Grant and Belcher, 2009), turbulence structures such as anisotropy (e.g., Li et al., 2005; Li and Fox-Kemper, 2020), influence in the Ekman layer (McWilliams et al., 2012), evolution under misaligned wind and waves (e.g., Van Roekel et al., 2012; McWilliams et al., 2014), its role in a boundary layer forced at multiple time scales (Pham et al., 2023), and dynamics in the presence of submesoscale eddies and fronts (e.g., Hamlington et al., 2014; Sullivan and McWilliams, 2019). Such LES studies of Langmuir turbulence provide the foundation for developing parameterizations of turbulent mixing in the ocean mixed layer (e.g., Li et al., 2019) and testing them in complex settings (e.g., Johnson et al., 2023).

11.3.3 Parameterizations in ocean general circulation models

Turbulence in the ocean mixed layer is too small to be directly resolved in regional and global ocean general circulation models (GCMs) and climate models. However, its effects on modulating the exchange of heat, momentum, and trace gases between the atmosphere and the deep ocean are so important that they need to be parameterized in these large-scale GCMs. Given the hydrostatic assumption and the large aspect ratio of the grid cells (the horizontal dimension over the vertical) commonly used in GCMs, the most important effect of ocean mixed-layer turbulence to be parameterized is the vertical turbulent fluxes of momentum and tracers. In general, such a parameterization can be written in the form,

$$\overline{w'\lambda'} = \mathcal{F}\left(\partial_z\overline{\lambda}, \partial_z\overline{\boldsymbol{u}}, \partial_z\overline{b}, \boldsymbol{p}\right) \tag{11.28}$$

where $\overline{()}$ is an ensemble mean, commonly the average over a grid cell and a time step as implemented in GCMs, and $()'$ is the sub-grid scale fluctuation. Here λ can be either a tracer (e.g., temperature, salinity, dissolved oxygen, etc.) or a component of the velocity. The vertical flux of λ is written as a function of vertical gradient of the mean field $\overline{\lambda}$, the mean velocity $\overline{\boldsymbol{u}}$, and the mean buoyancy $\overline{b}$, which are all resolved fields in GCMs, as well as a set of other parameters $\boldsymbol{p} = (p_1, p_2, \ldots)$. The functional form $\mathcal{F}$ can either be diagnostic, representing a simple relation between the vertical fluxes and parameters, or prognostic, representing a set of governing equations to be solved.

The simplest parameterization adopts the concept of *turbulent diffusion*,

$$\overline{w'\lambda'} = -K_\lambda\partial_z\overline{\lambda} + \Gamma_\lambda \tag{11.29}$$

representing the vertical turbulent flux as the sum of a *down-gradient* component and a *nonlocal* component Γ_λ. The nonlocal component represents any flux that is not proportional to the local gradient, resulting from unresolved coherent structures such as convective plumes. From dimensional analysis, the turbulent diffusivity K_λ can be written as the product of a velocity scale q, a length scale l and a non-dimensional coefficient C_λ,

$$K_\lambda = C_\lambda q l \tag{11.30}$$

Depending on how K_λ is determined, a parameterization written in this simple form of (11.29) can represent first-moment closure schemes (e.g., Large et al., 1994), which are based on *similarity theory* and empirical relationships under various conditions, and algebraic second-moment closure schemes (e.g., Mellor and Yamada, 1982; Rodi, 1987; Umlauf and Burchard, 2003), which evolve equations for q and l (usually functions of q and l such as the TKE $e \propto q^2$ and dissipation rate of TKE $\varepsilon \propto q^3/l$) but algebraically determines C_λ. Some second-moment closures and higher-order-moment closures evolve a system of equations for second moments (or covariances) and higher-order moments and close the system at higher order (Mellor and Yamada, 1982). With higher-order moments in the system, the whole equation set becomes progressively more complex and harder to solve. See, e.g., Burchard et al. (2008) for a more detailed review of different vertical mixing parameterizations commonly used in regional and global ocean GCMs.

One of the major recent advances in parameterizing vertical mixing by turbulence in the ocean mixed layer is the inclusion of the effects of Langmuir turbulence. As discussed in Section 11.2.2.4, the presence of Langmuir turbulence significantly alters the turbulent structures and TKE budget within the mixed layer, thereby affecting the vertical mixing by modifying each of q, l and C_λ in (11.30). Ideas incorporating the effects of Langmuir turbulence range from enhancing the vertical velocity scale q by a Langmuir number dependent enhancement factor in KPP (e.g., McWilliams and Sullivan, 2000), to accounting for the Langmuir turbulence modified TKE budget in an energetics based planetary boundary layer (ePBL) framework (Reichl and Li, 2019), to including Stokes production of TKE and Stokes modification of length scale (e.g., Kantha and Clayson, 2004) and further accounting for the Stokes drift modification of C_λ and a down-Stokes drift momentum flux (Harcourt, 2015) in algebraic second-moment closures. See Li et al. (2019) for a review and comparison of various such ideas and some representative Langmuir turbulence parameterizations in a consistent numerical framework under a broad range of forcing scenarios. Implementations of Langmuir turbulence parameterizations in global ocean GCMs are promising, showing a significant bias reduction in simulated mixed-layer depth and other climate variables (e.g., Li et al., 2016). However, as demonstrated by Li et al. (2019), large uncertainties exist in different existing Langmuir turbulence parameterizations. Future observational and modeling studies to further improve the parameterization framework and constrain the free parameters are warranted.

11.3.4 Applications in environmental sciences

The turbulent mixing in the ocean mixed layer has gained broad interest in environmental and ecosystem sciences over recent decades. It is one of the key physical processes that affect substance transport, which is critical to the life cycle of oceanic organisms and the dispersion of marine pollutants. Here, we briefly discuss two applications.

One application is the dispersion of spilled oil. Thick oil slicks floating at the ocean surface are highly toxic to oceanic organisms and can damage the ocean ecosystem. For example, the Deepwater Horizon oil spill in the Gulf of Mexico in 2010, which is the largest marine oil spill in history, has polluted more than 112,000 km^2 of ocean surface (Beyer et al., 2016). When such a disaster happens, an effective method of removing the spilled oil from the sea surface is essential for the remediation strategies. One of the most effective ways, applicable for a wide range of ocean conditions, is to use oil dispersants (National Research Council, 2005). The application of oil dispersants reduces the surface tension of oil so that the thick oil slicks or large oil droplets can be easily broken into smaller droplets by turbulence at the sea surface, which can then be mixed into deeper layers. LES has been used to understand the effects of oil dispersants and turbulent mixing on the dispersion and transport of spilled oil (e.g., Chen et al., 2018).

Another application is to estimate the nutrient supply for the living organisms in the ocean and to predict the total biomass production in oceanic farms. For example,

Frieder et al. (2022) combine a macroalgal growth model and a turbulent flow field from LES (Yan et al., 2021) to predict the within-farm kelp biomass. The drag provided by the large suspended kelp can also alter the micro-scale flow structure within and near the kelp farm. On one hand, the existence of the kelp farm reduces the speed of the near-surface current within the farm, creating a mean divergent flow around it, pushing the nutrients away from the kelp. On the other hand, the rows of the kelp farm can intensify Langmuir turbulence within the farm, mixing the ambient nutrients into the kelp farm field from the bottom. The interaction between these two effects can greatly modulate nutrient distribution, thereby modifying the distribution of predicted kelp biomass within the farm field.

11.4 Future perspectives and open research challenges

Our understanding of turbulence in the ocean mixed layer has greatly improved over the past few decades. This is a result of both the advances in observational techniques, in particular new platforms that allow longer and wider temporal and spatial coverage, and the tremendous increase in computational power, which allows numerical simulations of ocean mixed-layer turbulence under various complex conditions. The interaction with surface gravity waves at the air-sea interface marks a key feature that differentiates turbulence in the ocean mixed layer from that in the atmospheric boundary layer or other environments. While it has increasingly been appreciated that the presence of ocean surface waves significantly modifies the turbulence in the ocean mixed layer, theories for turbulence in such a wavy boundary layer are still in development, and challenges remain. Here we highlight a few of such challenges.

11.4.1 A similarity theory for the wavy boundary layer?

The presence of ocean surface waves at the air-sea interface significantly alters the turbulent structure in the ocean mixed layer, including the surface layer. The classic Monin–Obukhov similarity theory used to describe the similarity of velocity and tracer gradients near a wall may fail for a wavy boundary layer (e.g., Zheng et al., 2021). Stokes drift associated with ocean surface waves introduces an additional velocity scale in the physical problem, but no new dimensions. According to the Buckingham Pi Theorem, there is one more independent dimensionless quantity that this physical problem should depend on. A natural choice of this dimensionless quantity is the turbulent Langmuir number $\mathrm{La}_t = (u_*/u_0^S)^{1/2}$ (McWilliams et al., 1997). Modifications to MOST, which account for the effects of ocean surface waves by multiplying the universal function $\phi_\lambda(\zeta)$ in (11.3) by a Stokes similarity function that depends on La_t, are explored using data from an LES in Large et al. (2019). While providing promising evidence that this extended MOST can reasonably describe the surface layer under the influence of ocean surface waves in their LES forced by realistic wind, waves, and buoyancy fluxes in the Southern Ocean, its applicability to broader scenarios requires careful validations against observations and model simulations.

One may also argue that, in addition to the new velocity scale represented by the surface value of Stokes drift, a length scale can also be introduced to describe the decay depth scale of the Stokes drift profile (e.g., Kukulka and Harcourt, 2017). Again, this introduces a new dependent variable in the physical problem, but no additional dimensions. A new independent dimensionless quantity can be formed, for example, by taking the ratio of the distance $|z|$ from the air-sea interface and this new length scale. However, how to incorporate this new length scale in a similarity theory remains unknown.

11.4.2 Interactions with other processes

We have been discussing turbulence in the ocean mixed layer in a horizontally homogeneous sense, focusing on its role in vertical mixing. The assumption is that vertical mixing and horizontal mixing in the ocean are dominated by different processes with clear scale separation. Recent discovery of ocean submesoscale processes suggests that the ocean is filled with flows at intermediate scale, which largely occur in the ocean mixed layer and overlap with ocean boundary layer turbulence in spatial and temporal scales (e.g., McWilliams, 2016). Interactions between boundary layer turbulence and submesoscale processes have been shown to strongly influence the turbulent mixing in the ocean mixed layer (e.g., Hamlington et al., 2014; Sullivan and McWilliams, 2019). Such interactions challenge the conventional ocean-mixing parameterizations in ocean GCMs as horizontal resolution increases, and the assumption of scale separation between vertical and horizontal mixing that is often made by these parameterizations starts to break down. Future improvement in understanding such cross-scale interactions requires concurrent observation of turbulent flows at different scales, involving multiple observational platforms working together (e.g., Buckingham et al., 2019), and multiscale numerical simulations with turbulence-resolving high resolution and a large enough domain that allows the development of submesoscale processes (e.g., Hamlington et al., 2014; Sullivan and McWilliams, 2019).

11.4.3 A unified ocean mixed layer turbulence parameterization?

Over the past two decades, there have been significant advances in ocean mixed-layer turbulence parameterizations by incorporating the effects of ocean surface waves, in particular, Langmuir turbulence. Various parameterizations have been proposed and tested in different cases, and some have already been applied to global climate simulations and projections. See a review and comparison of a short list of such parameterizations under various conditions in Li et al. (2019). Even within this non-exhaustive list of Langmuir turbulence parameterizations, large discrepancies are found. This highlights the limitations and uncertainties in our understanding of Langmuir turbulence and its role in the ocean mixed layer.

In addition, such parameterization should account for interactions between boundary-layer turbulence and the other processes in the ocean discussed above. Horizontal homogeneity is often assumed in the development of ocean mixed-layer turbulence parameterizations, with attention focused on the effects of vertical mixing. This is a good assumption when a clear scale separation exists between vertical and horizontal mixing. That is often the case for global ocean GCMs and some regional ocean GCMs as well, in which the vertical spacing of the grid cells is $1 \sim 10$ m, and the horizontal spacing is $1 \sim 100$ km. However, with the increase of computational power, modern regional ocean simulations employ grid cells with horizontal spacing of a few hundred meters. As the horizontal resolution increases in these simulations, it is necessary to reexamine the scale-separation assumption between vertical and horizontal mixing, especially in frontal regions where strong coupling between the two can occur (e.g., Sullivan and McWilliams, 2019).

Another trend in ocean GCM development is the use of unstructured grids (such as MPAS-Ocean, FESOM). In those models, unstructured grids with regionally refined horizontal resolution can be used, and the resolved flow structure is at different spatial scales. In addition to affecting the representation of horizontal mixing, it may affect the representation of vertical mixing, especially in regions with fine flow structures in coastal oceans. A unified ocean mixed-layer turbulent mixing parameterization that automatically accounts for such effects is needed.

References

Agrawal, Y.C., Terray, E.A., Donelan, M.A., Hwang, P.A., Williams, A.J., Drennan, W.M., Kahma, K.K., Krtaigorodskii, S.A., 1992. Enhanced dissipation of kinetic energy beneath surface waves. Nature 359, 219–220. https://doi.org/10.1038/359219a0.

Beyer, J., Trannum, H.C., Bakke, T., Hodson, P.V., Collier, T.K., 2016. Environmental effects of the deepwater horizon oil spill: a review. Marine Pollution Bulletin 110, 28–51. https://doi.org/10.1016/j.marpolbul.2016.06.027.

National Research Council, 2005. Oil Spill Dispersants: Efficacy and Effects. National Academies Press.

Brainerd, K.E., Gregg, M.C., 1995. Surface mixed and mixing layer depths. Deep-Sea Research. Part 1. Oceanographic Research Papers 42, 1521–1543. https://doi.org/10.1016/0967-0637(95)00068-H.

Buckingham, C.E., Lucas, N., Belcher, S., Rippeth, T., Grant, A., Sommer, J.L., Ajayi, A.O., Naveira Garabato, A.C., 2019. The contribution of surface and submesoscale processes to turbulence in the open ocean surface boundary layer. Journal of Advances in Modeling Earth Systems 11, 4066–4094. https://doi.org/10.1029/2019MS001801.

Burchard, H., Craig, P.D., Gemmrich, J.R., van Haren, H., Mathieu, P.P., Meier, H.E.M., Smith, W.A.M.N., Prandke, H., Rippeth, T.P., Skyllingstad, E.D., Smyth, W.D., Welsh, D.J.S., Wijesekera, H.W., 2008. Observational and numerical modeling methods for quantifying coastal ocean turbulence and mixing. Progress in Oceanography 76, 399–442. https://doi.org/10.1016/j.pocean.2007.09.005.

Businger, J.A., Wyngaard, J.C., Izumi, Y., Bradley, E.F., 1971. Flux-profile relationships in the atmospheric surface layer. Journal of the Atmospheric Sciences 28, 181–189. https://doi.org/10.1175/1520-0469(1971)028<0181:FPRITA>2.0.CO;2.

Chen, B., Yang, D., Meneveau, C., Chamecki, M., 2018. Numerical study of the effects of chemical dispersant on oil transport from an idealized underwater blowout. Physical Review Fluids 3, 083801. https://doi.org/10.1103/PhysRevFluids.3.083801.

Craik, A.D.D., Leibovich, S., 1976. A rational model for Langmuir circulations. Journal of Fluid Mechanics 73, 401–426. https://doi.org/10.1017/S0022112076001420.

Crawford, G.B., Large, W.G., 1996. A numerical investigation of resonant inertial response of the ocean to wind forcing. Journal of Physical Oceanography 26, 873–891. https://doi.org/10.1175/1520-0485(1996)026<0873:ANIORI>2.0.CO;2.

D'Asaro, E.A., 1985. The energy flux from the wind to near-inertial motions in the surface mixed layer. Journal of Physical Oceanography 15, 1043–1059. https://doi.org/10.1175/1520-0485(1985)015<1043:TEFFTW>2.0.CO;2.

D'Asaro, E.A., 2001. Turbulent vertical kinetic energy in the ocean mixed layer. Journal of Physical Oceanography 31, 3530–3537. https://doi.org/10.1175/1520-0485(2002)031<3530:TVKEIT>2.0.CO;2.

D'Asaro, E.A., Thomson, J., Shcherbina, A.Y., Harcourt, R.R., Cronin, M.F., Hemer, M.A., Fox-Kemper, B., 2014. Quantifying upper ocean turbulence driven by surface waves. Geophysical Research Letters 41, 102–107. https://doi.org/10.1002/2013GL058193.

de Boyer Montégut, C., Madec, G.C., Fischer, A.S., Lazar, A., Iudicone, D., 2004. Mixed layer depth over the global ocean: an examination of profile data and a profile-based climatology. Journal of Geophysical Research 109, C12003. https://doi.org/10.1029/2004JC002378.

Deardorff, J.W., 1970. Convective velocity and temperature scales for the unstable planetary boundary layer and for Rayleigh convection. Journal of the Atmospheric Sciences 27, 1211–1213.

Deardorff, J.W., 1974. Three-dimensional numerical study of the height and mean structure of a heated planetary boundary layer. Boundary-Layer Meteorology 7, 81–106. https://doi.org/10.1007/BF00224974.

Foken, T., 2006. 50 years of the Monin–Obukhov similarity theory. Boundary-Layer Meteorology 119, 431–447. https://doi.org/10.1007/s10546-006-9048-6.

Fox-Kemper, B., Johnson, L., Qiao, F., 2022. Ocean near-surface layers. In: Meredith, M., Naveira Garabato, A. (Eds.), Ocean Mixing. Elsevier, pp. 65–94. https://doi.org/10.1016/B978-0-12-821512-8.00011-6.

Frajka-Williams, E., Brearley, J.A., Nash, J.D., Whalen, C.B., 2022. New technological frontiers in ocean mixing. In: Meredith, M., Naveira Garabato, A. (Eds.), Ocean Mixing. Elsevier, pp. 345–361. https://doi.org/10.1016/B978-0-12-821512-8.00021-9.

Franks, P.J.S., 2015. Has Sverdrup's critical depth hypothesis been tested? Mixed layers vs. turbulent layers. ICES Journal of Marine Science 72, 1897–1907. https://doi.org/10.1093/icesjms/fsu175.

Frieder, C.A., Yan, C., Chamecki, M., Dauhajre, D., McWilliams, J.C., Infante, J., McPherson, M.L., Kudela, R.M., Kessouri, F., Sutula, M., Arzeno-Soltero, I.B., Davis, K.A., 2022. A macroalgal cultivation modeling system (MACMODS): evaluating the role of physical-biological coupling on nutrients and farm yield. Frontiers in Marine Science 9. https://doi.org/10.3389/fmars.2022.752951.

Grant, A.L.M., Belcher, S.E., 2009. Characteristics of Langmuir turbulence in the ocean mixed layer. Journal of Physical Oceanography 39, 1871–1887. https://doi.org/10.1175/2009JPO4119.1.

Hamlington, P.E., Van Roekel, L.P., Fox-Kemper, B., Julien, K., Chini, G.P., 2014. Langmuir–submesoscale interactions: descriptive analysis of multiscale frontal spindown simulations. Journal of Physical Oceanography 44, 2249–2272. https://doi.org/10.1175/JPO-D-13-0139.1.

Harcourt, R.R., 2015. An improved second-moment closure model of Langmuir turbulence. Journal of Physical Oceanography 45, 84–103. https://doi.org/10.1175/JPO-D-14-0046.1.

Holm, D.D., 1996. The ideal Craik-Leibovich equations. Physica D. Nonlinear Phenomena 98, 415–441. https://doi.org/10.1016/0167-2789(96)00105-4.

Holthuijsen, L.H., 2007. Waves in Oceanic and Coastal Waters. Cambridge University Press, New York.

Johnson, L., Fox-Kemper, B., Li, Q., Pham, H.T., Sarkar, S., 2023. A finite-time ensemble method for mixed layer model comparison. Journal of Physical Oceanography 53, 2211–2230. https://doi.org/10.1175/JPO-D-22-0107.1.

Kantha, L.H., Clayson, C.A., 2004. On the effect of surface gravity waves on mixing in the Oceanic mixed layer. Ocean Modelling 6, 101–124. https://doi.org/10.1016/S1463-5003(02)00062-8.

Kukulka, T., Harcourt, R.R., 2017. Influence of Stokes drift decay scale on Langmuir turbulence. Journal of Physical Oceanography 47, 1637–1656. https://doi.org/10.1175/JPO-D-16-0244.1.

Langmuir, I., 1938. Surface motion of water induced by wind. Science 87, 119–123.

Large, W.G., Mcwilliams, J.C., Doney, S.C., 1994. Oceanic vertical mixing: a review and a model with a nonlocal boundary layer parameterization. Reviews of Geophysics 32, 363–403. https://doi.org/10.1029/94RG01872.

Large, W.G., Patton, E.G., DuVivier, A.K., Sullivan, P.P., Romero, L., 2019. Similarity theory in the surface layer of large-eddy simulations of the wind, wave and buoyancy forced Southern Ocean. Journal of Physical Oceanography 49, 2165–2187. https://doi.org/10.1175/JPO-D-18-0066.1.

Leibovich, S., 1983. The form and dynamics of Langmuir circulations. Annual Review of Fluid Mechanics 15, 391–427.

Li, M., Garrett, C., Skyllingstad, E.D., 2005. A regime diagram for classifying turbulent large eddies in the upper ocean. Deep-Sea Research. Part 1. Oceanographic Research Papers 52, 259–278. https://doi.org/10.1016/j.dsr.2004.09.004.

Li, Q., Fox-Kemper, B., 2017. Assessing the effects of Langmuir turbulence on the entrainment buoyancy flux in the ocean surface boundary layer. Journal of Physical Oceanography 47, 2863–2886. https://doi.org/10.1175/JPO-D-17-0085.1.

Li, Q., Fox-Kemper, B., 2020. Anisotropy of Langmuir turbulence and the Langmuir-enhanced mixed layer entrainment. Physical Review Fluids 5, 013803. https://doi.org/10.1103/PhysRevFluids.5.013803.

Li, Q., Reichl, B.G., Fox-Kemper, B., Adcroft, A., Belcher, S., Danabasoglu, G., Grant, A., Griffies, S.M., Hallberg, R.W., Hara, T., Harcourt, R., Kukulka, T., Large, W.G., McWilliams, J.C., Pearson, B., Sullivan, P., Van Roekel, L., Wang, P., Zheng, Z., 2019. Comparing ocean surface boundary vertical mixing schemes including Langmuir turbulence. Journal of Advances in Modeling Earth Systems 11, 3545–3592. https://doi.org/10.1029/2019MS001810.

Li, Q., Webb, A., Fox-Kemper, B., Craig, A., Danabasoglu, G., Large, W.G., Vertenstein, M., 2016. Langmuir mixing effects on global climate: WAVEWATCH III in CESM. Ocean Modelling 103, 145–160. https://doi.org/10.1016/j.ocemod.2015.07.020.

Lilly, D.K., 1967. The representation of small-scale turbulence in numerical simulation experiments. In: Proceedings of the IBM Scientific Computing Symposium on Environmental Sciences, pp. 195–210.

McWilliams, J.C., 2016. Submesoscale currents in the ocean. Proceedings of the Royal Society A: Mathematical, Physical and Engineering Science 472, 20160117. https://doi.org/10.1098/rspa.2016.0117.

McWilliams, J.C., Huckle, E., Liang, J.H., Sullivan, P.P., 2012. The wavy Ekman layer: Langmuir circulations, breaking waves, and Reynolds stress. Journal of Physical Oceanography 42, 1793–1816. https://doi.org/10.1175/JPO-D-12-07.1.

McWilliams, J.C., Huckle, E., Liang, J.H., Sullivan, P.P., 2014. Langmuir turbulence in swell. Journal of Physical Oceanography 44, 870–890. https://doi.org/10.1175/JPO-D-13-0122.1.

McWilliams, J.C., Sullivan, P.P., 2000. Vertical mixing by Langmuir circulations. Spill Science and Technology Bulletin 6, 225–237. https://doi.org/10.1016/S1353-2561(01)00041-X.

McWilliams, J.C., Sullivan, P.P., Moeng, C.H., 1997. Langmuir turbulence in the ocean. Journal of Fluid Mechanics 334, 1–30. https://doi.org/10.1017/S0022112096004375.

Mellor, G.L., Yamada, T., 1982. Development of a turbulence closure model for geophysical fluid problems. Reviews of Geophysics and Space Physics 20, 851–875.

Monin, A.S., Obukhov, A.M., 1954. Basic laws of turbulent mixing in the surface layer of the atmosphere. Contrib. Geophys. Inst. Acad. Sci. USSR 24, 163–187.

Obukhov, A., 1946. Turbulentnost' v temperaturnoj-neodnorodnoj atmosfere (turbulence in an atmosphere with a non-uniform temperature). Trudy Inst. Theor. Geofiz. AN SSSR 1, 95–115.

Pham, H.T., Sarkar, S., Johnson, L., Fox-Kemper, B., Sullivan, P.P., Li, Q., 2023. Multi-scale temporal variability of turbulent mixing during a monsoon intraseasonal oscillation in the Bay of Bengal: AN LES STUDY. Journal of Geophysical Research: Oceans 128, e2022JC018959. https://doi.org/10.1029/2022JC018959.

Phillips, O.M., 1977. The Dynamics of the Upper Ocean, second ed. Cambridge University Press, Cambridge-London-New York-Melbourne.

Plueddemann, A.J., Smith, J.A., Farmer, D.M., Weller, R.A., Crawford, W.R., Pinkel, R., Vagle, S., Gnanadesikan, A., 1996. Structure and variability of Langmuir circulation during the surface waves processes program. Journal of Geophysical Research: Oceans 101, 3525–3543. https://doi.org/10.1029/95JC03282.

Reichl, B.G., Li, Q., 2019. A parameterization with a constrained potential energy conversion rate of vertical mixing due to Langmuir turbulence. Journal of Physical Oceanography 49, 2935–2959. https://doi.org/10.1175/JPO-D-18-0258.1.

Rodi, W., 1987. Examples of calculation methods for flow and mixing in stratified fluids. Journal of Geophysical Research: Oceans 92, 5305–5328. https://doi.org/10.1029/JC092iC05p05305.

Skyllingstad, E.D., Denbo, D.W., 1995. An ocean large-eddy simulation of Langmuir circulations and convection in the surface mixed layer. Journal of Geophysical Research 100, 8501. https://doi.org/10.1029/94JC03202.

Skyllingstad, E.D., Smyth, W.D., Crawford, G.B., 2000. Resonant wind-driven mixing in the ocean boundary layer. Journal of Physical Oceanography 30, 1866–1890. https://doi.org/10.1175/1520-0485(2000)030<1866:RWDMIT>2.0.CO;2.

Smagorinsky, J., 1963. General circulation experiments with the primitive equations I. The basic experiment. Monthly Weather Review 91, 99–164. https://doi.org/10.1175/1520-0493(1963)091<0099:GCEWTP>2.3.CO;2.

Smith, J.A., 1998. Evolution of Langmuir circulation during a storm. Journal of Geophysical Research: Oceans 103, 12649–12668. https://doi.org/10.1029/97JC03611.

Sullivan, P.P., McWilliams, J.C., 2019. Langmuir turbulence and filament frontogenesis in the Oceanic surface boundary layer. Journal of Fluid Mechanics 879, 512–553. https://doi.org/10.1017/jfm.2019.655.

Sutherland, G., Reverdin, G., Marié, L., Ward, B., 2014. Mixed and mixing layer depths in the ocean surface boundary layer under conditions of diurnal stratification. Geophysical Research Letters 41, 8469–8476. https://doi.org/10.1002/2014GL061939.

Suzuki, N., Fox-Kemper, B., 2016. Understanding Stokes forces in the wave-averaged equations. Journal of Geophysical Research: Oceans 121, 3579–3596. https://doi.org/10.1002/2015JC011566.

Teixeira, M.A.C., Belcher, S.E., 2002. On the distortion of turbulence by a progressive surface wave. Journal of Fluid Mechanics 458, 229–267. https://doi.org/10.1017/S0022112002007838.

Terray, E., Donelan, M., Agrawal, Y., Drennan, W., Kahma, K., Williams, A., Hwang, P., Kitaigorodskii, S., 1996. Estimates of kinetic energy dissipation under breaking waves. Journal of Physical Oceanography 26, 792–807. https://doi.org/10.1175/1520-0485(1996)026<0792:EOKEDU>2.0.CO;2.

Thorpe, S.A., 2004. Langmuir circulation. Annual Review of Fluid Mechanics 36, 55–79. https://doi.org/10.1146/annurev.fluid.36.052203.071431.

Thorpe, S.A., 2005. The Turbulent Ocean. Cambridge University Press, Cambridge.

Umlauf, L., Burchard, H., 2003. A generic length-scale equation for geophysical turbulence models. Journal of Marine Research 61, 235–265. https://doi.org/10.1357/002224003322005087.

van den Bremer, T.S., Breivik, Ø., 2018. Stokes drift. Phil. Trans. R. Soc. A 376, 20170104. https://doi.org/10.1098/rsta.2017.0104.

Van Roekel, L., Fox-Kemper, B., Sullivan, P.P., Hamlington, P.E., Haney, S.R., 2012. The form and orientation of Langmuir cells for misaligned winds and waves. Journal of Geophysical Research 117, C05001. https://doi.org/10.1029/2011JC007516.

Weller, R.A., Price, J.F., 1988. Langmuir circulation within the Oceanic mixed layer. Deep-Sea Research. Part 1. Oceanographic Research Papers 35, 711–747. https://doi.org/10.1016/0198-0149(88)90027-1.

Wyngaard, J.C., 2010. Turbulence in the Atmosphere. Cambridge University Press, New York.

Yan, C., McWilliams, J.C., Chamecki, M., 2021. Generation of attached Langmuir circulations by a suspended macroalgal farm. Journal of Fluid Mechanics 915. https://doi.org/10.1017/jfm.2021.111.

Zheng, Z., Harcourt, R.R., D'Asaro, E.A., 2021. Evaluating Monin-Obukhov scaling in the unstable Oceanic surface layer. Journal of Physical Oceanography 51, 911–930. https://doi.org/10.1175/JPO-D-20-0201.1.

Turbulence at the ocean-atmosphere interface

12

On waves, drops, bubbles and turbulence

Luc Deike
Princeton University, Princeton, NJ, United States

12.1 Introduction

The ocean and atmosphere are turbulent environments that are coupled through surface waves. Surface waves at the ocean-atmosphere interface have a large effect on climate and weather by controlling or modulating fluxes of momentum, energy, heat, and mass. Turbulent wind blowing at the ocean surface generates waves, which can travel across the ocean, leading to a multi-scale system of their own. Surface waves act as a dynamical rough surface, modulating the atmospheric boundary layer by enhancing the drag and controlling momentum fluxes, with properties evolving depending on the wind speed and its alignment with the local sea state. With strong wind forcing, wave amplitude can reach values of several meters with steep waves that will break, limiting the wave height, enhancing turbulence generation in the upper ocean, together with entraining bubbles in the water and ejecting sea spray droplets in the air. Droplet and bubble dynamics will be affected by the turbulent flow surrounding them, modifying their dynamics compared to quiescent flows.

Hence, understanding ocean-atmosphere coupling at small scales requires understanding the atmospheric and oceanic turbulent boundary layers and their coupling through waves, while fluxes also require understanding multi-phase turbulent flows with drops moving in the air and bubbles moving in the water. Waves by themselves are multiscale, and their statistical representations present similarities to turbulent systems. The problem is therefore very rich, and a full accounting would require its own book. In the present chapter, we discuss various aspects of turbulent flows at the ocean atmosphere interface, insisting on the role of waves, including wave breaking, entrainment and dynamics of air bubbles, and briefly discuss spray ejection. We present scaling analysis, sometimes inherited from turbulence, and discuss when classic results from single-phase turbulent flow can be applied and when they fail due to the specificity of air-sea coupling.

The chapter is aimed as an introduction to some aspects of the ocean-atmosphere coupling, with some focus on small scale process, and presents some open topics. The main goal is to provide the necessary tools to describe fluxes between the ocean and the atmosphere. For that, we will need to understand the input of momentum and energy from the wind to the waves, which can be characterized as a turbulent boundary-layer problem over a moving surface and involve understanding the wind profile over the waves. The mirror problem will involve the injection of energy into

Environmental Turbulence. https://doi.org/10.1016/B978-0-32-395873-8.00019-X
Copyright © 2026 Elsevier Inc. All rights are reserved, including those for text and data mining, AI training, and similar technologies.

the water column due to wave breaking, and significant differences in time and length scale exist due to the differences in density. In all of that, drops and bubbles will come into play, and while tiny, they will provide new pathways for the exchange of heat and mass.

Understanding of the wind profile in the atmospheric surface layer has advanced significantly in the past decades, and several chapters of the present book are dedicated to the flow over rough surfaces (cities, vegetation, etc.), and the applicability of Monin–Obukhov similarity theory for scaling the vertical profiles of wind, buoyancy, and turbulent dissipation. However, the wind profile and turbulent flux profiles over the ocean have revealed significant differences with static boundaries, as ocean waves carry a large amount of energy and momentum, and can create important variations in air-sea exchange and wind and current profiles. The occurrence of airflow separation effects modulates the wind turbulent boundary layer, affecting the overall momentum flux into the ocean. At moderate to high wind speeds, breaking events start to dominate air-sea fluxes, with wave breaking being a transitional process from laminar to turbulent flow on the water side. The surface experiences dramatic changes, including sea spray ejection in the atmosphere and air entrainment into the ocean water. Once ejected, sea spray droplets and particles are transported into the marine boundary layer, where they can evaporate, leaving sea salt and biological particles that can serve as cloud condensation nuclei. At very high wind speeds, the amount of drops in the air is expected to contribute to the overall drag.

On the ocean side, air bubbles are entrained, can break in the turbulent flow, and contribute to gas exchange, as well as sea spray production as they burst at the surface. Describing drops and bubbles at the ocean atmosphere interface requires understanding the movement of inertial finite-size particles of heavier (drops) or lighter (bubbles) density than the carrier turbulent flow.

Exchange of momentum, heat, mass, and energy at the ocean-atmosphere interface is therefore a complex multiscale problem. The current formulations for ocean atmosphere fluxes rely on empirical correlations mostly based on wind speed, and open questions remain on the detailed local mechanism controlling mass, heat, and momentum exchanges, with the respective role and feedback between the turbulent wind, waves, drops, and outer flow; and their impact on regional and global dynamics.

In the present chapter, we discuss various components of the turbulent air flow at the ocean-atmosphere interface, focusing on the multi-phase coupling and components between waves, wave breaking, turbulent air flow, turbulent water flow, and dynamics of drops and bubbles. Scaling arguments are discussed for the separate sub-problems, and a discussion on how to bring the various processes together into a unified multi-scale framework are presented. The applicability of theoretical arguments against observations in the field (or the lack of them) is discussed. Exchange of gas between turbulent flow is discussed with and without bubbles.

We start by discussing the physics of ocean surface waves, their forcing and growth by the turbulent wind, nonlinear wave interaction, and the resulting wave spectrum (Section 12.2). In Section 12.3, we describe the physics of a single breaking wave, including energy dissipation, air entrainment, and bubble dynamics under water.

Section 12.4 presents the classic Phillips framework for representing wave-breaking dynamics and statistics in the field, and for constraining air-sea fluxes of momentum, energy, and mass under breaking-dominated conditions. In the remaining section, we discuss how to apply the physics of breaking waves, drops, and bubbles to the fluxes of momentum (Section 12.5.1), gas such as carbon dioxide and oxygen (Section 12.5.2), and spray (Section 12.5.3).

12.2 Ocean surface waves

Ocean waves propagate at the sea surface and modulate the atmospheric and oceanic turbulent boundary layers. We introduce briefly some of the necessary results of gravity and capillary wave dynamics, in particular the dispersion relation, time and length scales, before diving into the statistical description of ocean waves, and the main source terms. Empirical and theoretical wind-wave growth relationships are discussed, together with nonlinear interaction. The topics covered in this section are described in more detail in various books, including Phillips (1977); Komen et al. (1996); Janssen (2004).

12.2.1 Linear wave theory and dispersion relation

The classic water-wave theory considers a free-surface flow. The density of water is much larger than that of air, and neglecting viscous dissipation, the equations for a free-surface flow are obtained from the Navier–Stokes equations, assuming irrotational wave motion, which leads to the definition of a velocity potential. The velocity potential follows the Laplace equation, while kinematic and dynamic (pressure) boundary conditions are provided at the air-sea interface (together with bottom boundary conditions, here we consider infinite water depth), and nonlinear terms are considered through the boundary conditions.

The water surface elevation is written as $\eta(x, y, t)$, and the propagation of ocean gravity waves can be described by linear wave theory as long as the wave slope ak is small (nominally $ak \ll 1$), with k the wave number and a the wave amplitude. The linearized equations admit stable wave solutions as Fourier modes, $\eta \sim ae^{i(kx-\omega t)}$, with ω the angular frequency, and $c = \omega/k$ the phase speed, and performing linear stability analysis, we obtain the gravity wave dispersion relation

$$\omega = \sqrt{gk} \tag{12.1}$$

or, equivalently, $c = \sqrt{g/k}$ in deep water. Longer wavelengths travel faster than shorter ones. When waves approach the shore, the water depth modifies the dispersion relation and the propagation. We will stay in deep water open ocean conditions in the present chapter, with shallow water conditions described elsewhere, e.g., Bosboom and Stive (2021).

Ocean gravity waves are therefore dispersive, i.e., waves of different wavelengths travel at different speeds. Contrary to ocean waves, acoustic waves, or electromagnetic waves have a constant speed for all wavelengths in a given medium, and the speed depends on the medium's properties.

At small scales, capillary effects will become important and modify the wave propagation, and the full dispersion relation becomes

$$\omega = \sqrt{gk + \frac{\gamma}{\rho}k^3} \tag{12.2}$$

with γ the surface tension. Gravity and surface tension are balanced for $\lambda_c = 2\pi l_c = 2\pi\sqrt{\frac{\gamma}{g\rho}}$, which is about 1.7 cm in water.

When considering the coupling with the surrounding airflow characterized by a speed U, it is important to note that the relative velocity between the waves and the wind profile will differ for each wave frequency component.

12.2.2 Wave spectrum. Observations and scalings

At the ocean surface, under the action of wind and nonlinear effects, the sea presents complex patterns that can be described in terms of the superposition of many waves interacting with each other. Statistical description of the waves becomes necessary, and the directional wave spectrum at the ocean surface $F(\mathbf{k})$ describes the wave energy density per scale, with $\langle\eta^2\rangle = \int F(\mathbf{k})d\mathbf{k}$, the variance of the wave surface elevation. From a practical standpoint, describing the wave spectrum is the main task for understanding and predicting the time and space evolution of the wave field.

The wave spectrum can be observed at sea by wave buoys or wave gauges, which provide a temporal signal of the wave elevation $\eta(t)$ at a given location. The frequency spectrum $\Phi(\omega)$ can then be obtained by the Fourier transform of the autocorrelation signal. Similarly, spatial measurements of the wave elevation $\eta(x, y)$ will provide the spatial wave spectrum $F(\mathbf{k})$ and/or the angle integrated wave spectra $\phi(k)$. The temporal and spatial spectra are related by the wave variance as

$$\langle\eta^2\rangle = \int F(\mathbf{k})d\mathbf{k} = \int \phi(k)dk = \int \Phi(\omega)d\omega \tag{12.3}$$

A typical spatial wave spectrum is sketched in Fig. 12.1 in the wave-number space. The typical wave spectrum for wind sea exhibits a peak, with associated wave number k_p (or angular frequency ω_p, both linked by the wave dispersion relation). The significant wave height is defined as $H_s = 4\sqrt{\langle\eta^2\rangle}$ and characterizes the mean wave energy. Above the peak frequency or wavenumber, an energy decay is observed, which can be described by a power law, with scaling in the open ocean being from $\Phi(\omega) \propto \omega^{-4}$ to $\Phi(\omega) \propto \omega^{-5}$, or in an equivalent way in wave-number $\phi(k) \propto k^{-5/2}$ to $\phi(k) \propto k^{-3}$; with a transition to steeper spectrum at higher frequency/wavenumbers. Proposed scaling laws are based on the dominant processes governing the ocean surface wave dynamics and statistics, i.e., wind input, nonlinear wave-wave interaction, and wave breaking. We will discuss some of the physical arguments and analogies for the derivation of the classic hydrodynamical turbulence spectra.

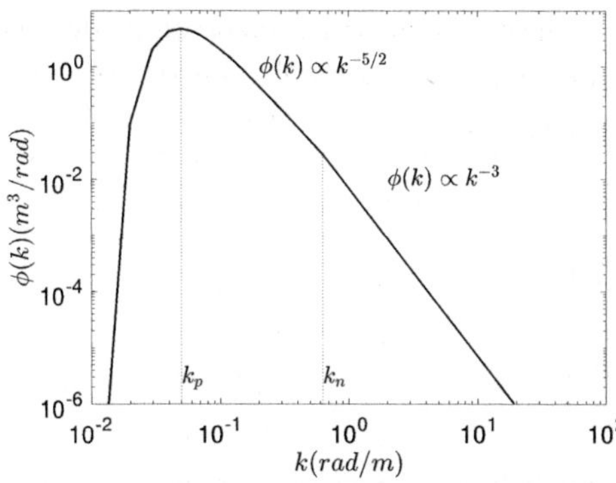

Figure 12.1 Sketch of a typical wave spectrum in the k-space. The peak of the spectrum k_p corresponds to the longest wave at the ocean surface, with scales of O(100 m), followed by an equilibrium range, $\phi(k) \propto k^{-5/2}$ until higher frequencies (wavelength of O(10 m)) and the saturation range $\phi(k) \propto k^{-3}$ where wave breaking dominates. The saturation range connects to the capillary scales (O(1 cm to 1 mm)).

Separately, operational wave forecast will focus on understanding and predicting the peak frequency $\omega_p = 2\pi f_p$ and significant wave height H_s as a function of the wind speed and distance (or time) since the wind has been blowing; assuming a similarity shape for the wave spectrum. Under such a similarity hypothesis, two non-dimensional parameters will characterize the wind-wave field: the overall non-dimensional wave amplitude, or slope $k_p H_s$, and the wave age, or ratio of wave over wind speed c_p/u_*.

12.2.3 Fetch limited relationships

Before diving into various theories to quantify the evolution of the wave spectra in space and time, as a function of the various source terms influencing wave dynamics, it is useful to discuss the classic fetch-limited relationships. Fetch-limited relationships have a long history and have aimed at empirically describing the evolution of the wind wave spectra's two main characteristics, i.e., the peak period $T_p = 1/f_p$, and a metric for the wave amplitude, the significant wave height H_s, which provides an estimate of the total energy in the wave field. The modern version of these relationships was initially derived by Sverdrup and Munk as part of World War II planning and relies on dimensional analysis (Sverdrup and Munk, 1947; Snodgrass et al., 1966). They are still used today and can be rationalized through the wave action equation (Janssen, 2004; Zakharov et al., 2015).

The assumption is that the wave spectrum will keep a self-similar shape, so that knowledge of the significant wave height H_s and wave peak frequency f_p is enough to determine the wave statistics. Such an assumption will be reasonable as long as the waves are formed by the direct action of the wind, and we are not in the presence of mixed or crossing seas. Then, the wave properties are assumed to depend only on a characteristic wind speed U (typically the wind speed at 10 m, U_{10} or the wind friction velocity u_*) and a fetch parameter, which is either the distance over which the wind has been blowing F at a given time (fetch limited conditions), or the duration at a given distance (duration limited conditions). We take here the approach of Toba (1972, 1973, 1978) and present these relationships in terms of the wind friction velocity u_*. Dimensional analysis includes the wave restoring force, i.e., gravity, and then suggests

$$\frac{gH_s}{u_*^2} = G\left(\frac{gF}{u_*^2}\right) \tag{12.4}$$

$$\frac{f_p u_*}{g} = H\left(\frac{gF}{u_*^2}\right) \tag{12.5}$$

with the functions G and H be determined empirically. Measurements in laboratory, as well as over lakes or specific ocean conditions where the fetch limited conditions are well observed have strongly supported these relationships and led to empirical form of the type $\frac{gH_s}{u_*^2} \propto \left(\frac{gF}{u_*^2}\right)^{\zeta_1}$ with $\zeta_1 \approx 1/2$ and $\frac{f_p u_*}{g} \propto \left(\frac{gF}{u_*^2}\right)^{\zeta_2}$ with $\zeta_2 \approx -1/3$, with some variability in the exponent depending on the data sets being considered. The appropriate conditions correspond to the wave field developing in space (or time) with the wave amplitude (or energy) growing as the wind keeps forcing them, and as the wave field grows, the peak frequency is shifted to lower values, i.e., longer wavelength, which is attributed to both the wind forcing effect and the nonlinear interactions between the wave modes.

Combining both relationships to eliminate the fetch leads to a link between the wave spectrum peak frequency, the significant wave height, and the wind speed. These relationships were proposed by Toba, who derived them from the similarity hypothesis on the wave spectrum, and read

$$\frac{gH_s}{u_*^2} = c_T\left(\frac{f_p u_*}{g}\right)^{-3/2} \tag{12.6}$$

with $c_T \approx 0.062$ known as the Toba constant. This final relationship is very robust when considering various data sets and can serve as a justification to keep only one of the non-dimensional wind-wave spectra non-dimensional parameters, i.e., an effective slope $k_p H_s$, and a wave age c_p/u_* since both quantities are linked by Eq. (12.6).

12.2.4 Wave dynamics and source terms to the wave action equation

We refer to classic textbooks on wave dynamics for a full description of the wave dynamics and statistics, and the wave action equation (Komen et al., 1996; Janssen, 2004). Locally, the wave field is related to the action of wind, leading to wave growth as it inputs energy and momentum into the waves; this will consist of an interaction between the turbulent atmospheric boundary layer and the wave field. Once waves are formed, nonlinear interaction through the scales (both forward and inverse cascade being possible for surface gravity waves) will occur; this will be the realm of weak-wave turbulence (Zakharov et al., 1992, 2012; Nazarenko, 2011). Finally, when wave

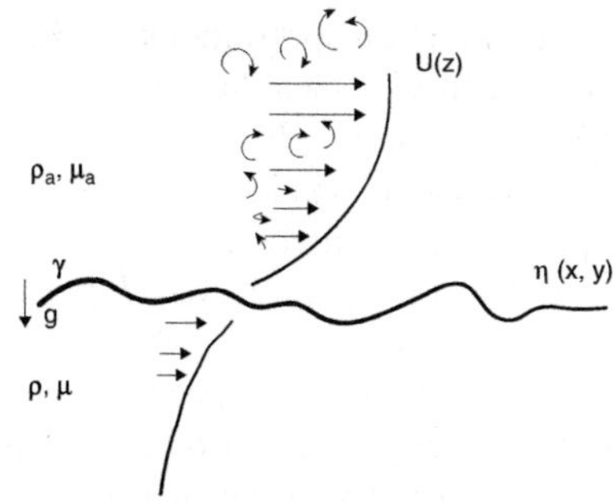

Figure 12.2 Sketch of the wind wave current coupling.

amplitude (or more precisely wave slope) becomes high enough, waves will break, dissipate energy, and generate turbulence in the oceanic boundary layer. Hence, the ocean surface is a dynamical coupler between the oceanic and atmospheric turbulent boundary layers while being itself a turbulent system.

Each of these processes (input, weak nonlinear energy transfer among waves, and dissipation) is, in principle, dynamically coupled. However, within a statistical description, the spectral evolution of the wave spectrum can be obtained by solving the wave action equation, relating the Lagrangian derivative of the spectral density following a wave packet at its group speed in both physical and spectral spaces to source terms able to input, output, and transfer energy through the scales (Komen et al., 1996; Janssen, 2004):

$$\frac{dN(\mathbf{k})}{dt} = (S_{wind} + S_{diss} + S_{nl})/\omega \tag{12.7}$$

where $N(\mathbf{k}) = F(\mathbf{k})/\omega$ is the wave action.

Each source term can then be studied independently and is described in the following sections. The wind forcing is closely related to understanding momentum fluxes, wind wave growth, and the wind velocity profile above the waves (Miles, 1957; Plant, 1982; Komen et al., 1996). The nonlinear interaction term, S_{nl}, transfers energy through the scales through a four-wave resonant process for gravity waves (Hasselmann, 1962) and is analogous to energy transfer across the scales in turbulence. Energy dissipation mainly occurs through wave breaking, limiting wave height, and injecting turbulent kinetic energy into the water column (Phillips, 1985; Romero et al., 2012; Sutherland and Melville, 2013).

12.2.5 Generation of waves by wind

The generation of waves by wind and the onset of this process have been longstanding research topics. The dynamics of the wind wave interaction is a coupled two-phase flow and sketched in Fig. 12.2. The wind (of much smaller density) blows across a moving, wavy water surface $\eta(x, y, t)$ (of much higher density), and the structure of the atmospheric turbulent boundary layer is altered. Because of the large density ratio, the coupling is weak. The fundamental question of how wind inputs energy and momentum into the waves has two main applications: finding the correct growth rate of the waves at various scales, and understanding the feedback of the waves on the wind profile and turbulent boundary layer.

We present a brief discussion on wind wave growth theory, dimensional constraints, as well as remaining open questions, summarizing the discussion in Pizzo et al. (2021); Wu et al. (2022). Initial theoretical attempts were developed by Helmholtz and Kelvin in terms of a shear instability. Such shear-flow instability will happen at the interface between two layers of fluid moving at different speeds. Keeping the assumptions used to derive the gravity-capillary waves dispersion relation above (inviscid irrotational flow in the water, high density ratio), but now considering a constant velocity profile in the air U, the generalized dispersion relation is obtained, and the least stable wave on the margin of stability is obtained as $\lambda_c = 2\pi l_c = 2\pi\sqrt{\frac{\gamma}{g\rho}}$. For air-water, conditions above the onset of the growth of the wave mode are obtained if the wind exceeds 6.5 m/s (and the first wave mode excited is at the minimum phase speed). However, various observations at sea or at the surface of lakes have shown that wave patterns start to occur at much smaller wind speeds (with speeds as low as 1 or 2 m/s being discussed in the literature).

This prompted Phillips (1957) to propose another theory for the onset of waves, due to the pressure fluctuations acting on the water surface. Considering the effect of random pressure fluctuations due to the turbulence on the (inviscid) water surface elevation, the initial onset and growth can then be described, with the mean square surface elevation $\bar{\eta}^2$ growing linearly

$$\bar{\eta}^2 \propto \frac{\bar{p}'^2 t}{U_c g \rho_w^2} \tag{12.8}$$

with the mean square pressure fluctuations $\bar{p}'^2$ and U_c, the typical convection speed of the turbulent structures. The proportionality coefficient can be determined by assumptions on the turbulence pressure field. Phillips' theory provides an onset criterion but does not determine the fastest growing mode. Once waves start to form and propagate, another mechanism needs to be invoked to explain the observation of various growth rates for various wavelengths at a given wind speed (Phillips, 1977; Perrard et al., 2019; Li and Shen, 2022).

Wave amplification and growth were first described by Jeffreys (1925), which argued that the air flowing over the water waves is deflected by the moving surface wave geometry. Such a process led to sheltering, resulting in a pressure difference across the moving wave so that work can be done by the wind. When waves become steep enough, air flow separation can occur. The sheltering idea postulates the dominant role of the wind pressure in amplifying the waves, and is not limited to small amplitudes. Miles (1957) proposed an instability mechanism following the original shear instability idea, but accounting for a realistic wind profile and introducing the concept of critical height, where wind speed equals phase speed at which a vortex force is acting. Miles critical layer theory is a linear stability analysis, so that the wave amplitude is assumed small, the airflow is assumed to be inviscid and laminar, and as a result of that assumption, the forcing comes solely from the pressure.

In both Miles critical layer instability theory and Jeffreys' sheltering hypothesis, the dominant wind stress terms come from the pressure stress ($p_s \boldsymbol{n}$) in the normal direction, while the viscous (tangential) stress is assumed to be negligible in the wave growth. The energy input rate is then written as

$$S_{wind} \approx c\langle p_s \frac{\partial \eta}{\partial x} \rangle \tag{12.9}$$

where S_{wind} denotes the wave-averaged rate of energy input flux. The angular brackets denote averaging over one wavelength, and to be nonzero, requires a phase shift between the pressure profile and the wave profile. Understanding this phase shift for various conditions remains a critical question to get a quantitative prediction of the wave amplification. The derivative in x assumes that the waves are predominantly 2D and traveling in the wind speed (x direction). The sheltering mechanism or shear layer instability theories need information on the structure of the pressure field near the surface. Jeffreys and Miles made different assumptions to relate the pressure field to the boundary layer mean flow, leading to estimates of the energy input rate (see Wu et al. (2022) for a detailed discussion).

The applicability of the Miles mechanism has been questioned in the context of steep waves, particularly regarding the role of turbulence in regimes of relatively high u_*/c, where the critical height might be very close to the surface. On the contrary, the sheltering idea is not limited to small amplitude waves, but the pre-factor is not provided by the theory.

The growth rate is often normalized by the total wave energy E and angular frequency ω, i.e., the definition of growth rate per radian $\gamma = S_{wind}/(\omega E)$, and is based on the assumption that the growth is exponential. Considering the definitions of wave energy and the gravity wave dispersion relation, $E = \frac{1}{2}\rho_w g a^2$, $\omega = kc = \sqrt{gk}$, and using the assumption that the form drag is related to the wave slope as $F_p \sim (ak)^2$ (which is not really observed at high ak), leads to

$$F_p = \frac{1}{2}\beta(ak)^2 \rho_a u_*^2 \tag{12.10}$$

where β is a non-dimensional constant related to the wave growth model (Jeffreys or Miles) and the assumed (or measured) wind profile. Considering the link between wind input and form drag $S_{wind} \approx cF_p$, we obtain the non-dimensional growth rate

$$\gamma = \frac{S_{wind}}{\omega E} = \frac{cF_p}{\omega E} = \beta \frac{\rho_a}{\rho_w}\left(\frac{u_*}{c}\right)^2 \tag{12.11}$$

It is worth noticing that this relationship, widely used in the literature, presents some strong self-correlation between the normalization of S_{wind} by ω in the left-hand side and the phase speed $c = \omega/k$ on the right-hand side. The resulting $(u_*/c)^2$ scaling is often taken as an indirect proof of Miles' theory. We caution that while the $(u_*/c)^2$ scaling seems to hold, there is a wide scatter in the β value at a given value of u_*/c, with sometimes over an order of magnitude variation, which can be partially explained by considering a slope dependent coefficient β as in Belcher's theory (Belcher and Hunt, 1993; Belcher, 1999).

12.2.6 Nonlinear interaction and weak-wave turbulence

When the wave amplitude is finite, nonlinear effects will start to occur, sharpening the wave profile and leading to wave-wave nonlinear interactions. Nonlinear wave dynamics has been the topic of many mathematical developments in trying to understand the route to wave breaking and the exchange of energy between various modes (Zakharov et al., 1992, 2012; Nazarenko, 2011; Newell and Rumpf, 2011). The occurrence of bound waves, i.e., modes locked by nonlinearities to travel at the same speed, is directly observable at sea through radar, and is a necessary component of understanding remote sensing of the ocean. One example of bound waves can be understood in terms of nonlinear expansion, such as Stokes waves, which are solutions of the full nonlinear potential flow associated with irrotational wave theory; while other examples include parasitic capillary waves riding on top of longer gravity waves (Fedorov and Melville, 1998). Weakly nonlinear interactions will be characterized by wave-wave interactions, depending on the dispersion relation, and lead to wave components exchanging energy with each other. It has been shown and measured that wave modes close in spectral space are more efficient at exchanging wave energy/wave action, so the cascade process is local.

For surface pure gravity waves, 3-wave interactions are theoretically prohibited because of the curvature of the dispersion relation (i.e., $\omega = \sqrt{gk}$), so that 4-wave interactions have been shown to be dominant; while capillary waves allow for 3-wave interactions (as we have $\omega = \sqrt{\gamma/\rho k^3}$).

As a consequence of 4-wave interactions being the leading term for weakly nonlinear surface gravity waves, both the wave action and wave energy are conserved, leading to the theoretical prediction of an inverse and a direct cascade, similar to two-dimensional hydrodynamical turbulence (Zakharov et al., 1992, 2012). The inverse cascade can be related to the frequency downshift of the wave energy spectrum as the wave field develops from a calm sea to a fully developed spectrum, and yields formulations of the fetch-limited relationships compatible with the historical description.

The direct cascade picture can be used to describe the wave spectrum from the peak of the wave spectrum k_p to a higher frequency, say a transition wave number k_n. The direct cascade from weak-wave turbulence is a steady state solution of the kinetic equation (the kinetic equation can be seen as similar to the wave action with the more rigorous definition of the wave action) (Zakharov et al., 1992, 2012), with the nonlinear term acting to redistribute energy from the largest scale to the smallest scale where dissipation occurs. It assumes no forcing and no dissipation within the inertial range, in a similar fashion to hydrodynamic turbulence. Introducing a wave energy flux P constant through the scales, and with dimension $[L^3T^{-3}]$, and using the fact that the spectrum shall scale with the energy flux $\phi(k) \propto P^{1/(N-1)}$ where N is the leading nonlinear interaction order, we have by dimensional analysis (Zakharov et al., 2012)

$$\phi(k) = C_{KZ}P^{1/3}g^{-1/2}k^{-5/2} \tag{12.12}$$

This steady state solution is referred to as the Kolmogorov–Zakharov solution, and the prefactor C_{KZ} can be determined theoretically (Zakharov et al., 1992, 2012). In the context of wind waves, it is important to note that the energy flux can be related dimensionally to the wind velocity by $u_* \propto P^{1/3}$.

Scaling theories have been proposed independently to describe wind-wave spectra and have yielded results compatible with those from weak-wave turbulence in the equilibrium range. Considering that the spectrum is constrained by a balance between the three main source terms of the wave action equation (wind input, nonlinear interactions, and energy dissipation by breaking), led to the concept of the wave *equilibrium range* (Toba, 1972; Phillips, 1985),

$$\phi(k) = \beta_T u_* g^{-1/2} k^{-5/2}, \text{ for } k_p < k < k_n \tag{12.13}$$

where the Toba constant, β_T, is determined empirically. Empirical fit has been proposed as a function of the wave age $\beta_T \approx 0.016(c_p/u_*)^{0.53}$ (Romero and Melville, 2010), which effectively modifies the scaling in wind friction velocity. The idea of equilibrium between the wind forcing, energy dissipation from breaking, as well as nonlinear interaction, is a physical picture of the fact that all processes are broad banded, and happen at the same time.

Overall, the equilibrium range $\phi(k) \propto k^{-5/2}$ is well observed in the field, together with its corresponding frequency spectrum, $\Phi(\omega) = \beta_T u_* g \omega^{-4}$ both related by $\phi(k)dk = \phi(\omega)d\omega$. At high frequency, the spectrum is observed to steepen to a $\phi(k) \propto k^{-3}$ or $\Phi(\omega) \propto \omega^{-5}$ spectrum, which is called the saturation range.

12.2.7 Saturation spectrum by breaking waves

At the ocean surface, energy is mostly dissipated through wave breaking. Phillips (1958, 1985) proposed a simple scaling law to describe the saturation of the wave spectrum due to breaking. Considering that the wave crest gets chopped off due to breaking, and that the limit in wave height is then constrained by gravity, Phillips proposed that the saturation spectrum be determined solely by gravity. This yields searching for the frequency spectrum as $\Phi(\omega) \propto g^{\alpha}\omega^{\zeta}$ with α and ζ exponents to be determined by dimensional analysis. Since $\Phi(\omega)$ has dimension of $[L^2T]$, we obtain,

$$\Phi(\omega) \propto g^2\omega^{-5} \tag{12.14}$$

The wave number spectrum can be obtained using the identity $\Phi(\omega)d\omega = \phi(k)dk$ and the dispersion relation for gravity waves, leading to

$$\phi(k) \propto k^{-3} \tag{12.15}$$

As discussed by Phillips, this analysis considers that the limiting spectra due to breaking are independent of the wind velocity and that this limiting configuration is characterized by the occurrence of sharp wave crests. These sharp wave crests could be seen as simple discontinuities in gradient of the wave elevation at one or more points, and it can be shown that the Fourier transform of such function is asymptotically proportional to k^{-2} so that $F(\mathbf{k}) \propto f(\theta)k^{-4}$ as stated above. A similar result is also discussed independently as a spectrum formed by singularities by Kuznetsov (2004) and Nazarenko et al. (2010). The transition from equilibrium to saturation, k_n, is itself a function of wind and sea state.

12.2.8 Capillary wave spectrum

Most of the above discussion does not consider capillary waves, which will start to be important at centimetric scales and will effectively serve as a physical cut-off for wave breaking, the dynamics transitioning from large-scale breakers that entrain air to micro-breakers and parasitic capillary waves. The wave spectrum in the capillary range will follow different scaling laws, which are important for remote sensing. The dynamics of capillary waves riding on a longer wave is modulated by the carrier wave, and is important in understanding the roughness of the sea surface.

Measurements of capillary wave spectrum at the ocean surface or in wind wave tanks are relatively limited, and some studies have reported a maximum in the saturation spectrum, with an increase of the saturation spectrum at the capillary scale and a maximum at high wave number (typically 1500 rad/m). Other studies have reported a spectrum similar to the saturation range, or even steeper.

Scalings obtained by weak wave turbulence (which assumes the dominance of nonlinear wave interaction), because of three wave interaction, lead to $\phi(k) \propto P^{1/2}\left(\frac{\gamma}{\rho}\right)^{-3/4} k^{-15/4}$, and have been well documented in wave tank experiments in the absence of wind forcing (Falcon and Mordant, 2022). Separately, an assumption of saturation of the wave spectrum similar to the Phillips limiting crest spectrum would lead to a saturation in $\phi(k) \propto k^{-3}$ in continuity with the gravity spectra. Eventually, viscous dissipation will become dominant and lead to a sharp drop off.

12.3 Wave breaking dynamics: energy dissipation, air entrainment and bubble dynamics

We now discuss the physics of breaking waves, focusing on the post-breaking flow dynamics. As discussed above, the higher frequency part of the wave spectrum is dominated by wave breaking at moderate to high wind speeds. We first discuss theoretical scalings for single wave breaking before coming back to the statistical description. We discuss energy dissipation and air entrainment, which will control fluxes of energy and mass in the upper ocean. The focus here is on the properties of the breaking waves, once they are formed. The route to breaking is in itself a fascinating topic, and will occur in deep water when there is a local wave energy excess, either through linear focusing (i.e., several modes interfere in a constructive way to generate this local wave energy excess), or through nonlinear wave focusing (i.e., different wave modes interact to create a wave modulation which will cause local wave energy excess). In any case, the process of wave breaking is characterized by a threshold, and finding such a universal threshold remains an open topic. Three types of criteria are proposed in the literature: geometric (i.e., based on the wave slope), kinematic (based on the ratio of the local velocity to the wave speed), or dynamic (Perlin et al., 2013).

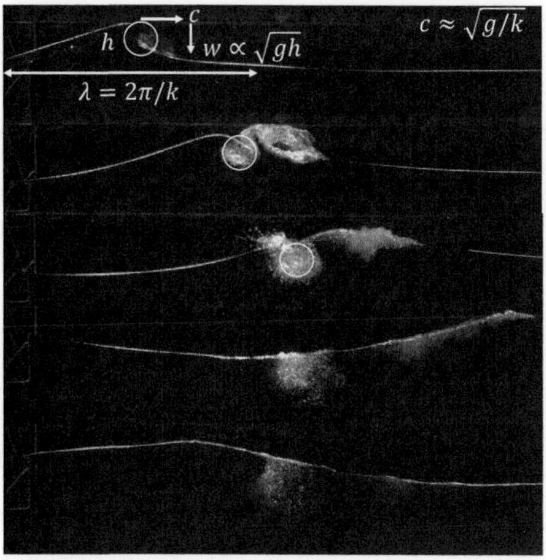

Figure 12.3 Laboratory experiments on a single breaking wave obtained from a focusing wave packet, illustrating the characteristic scales of a breaker: the speed c and height at breaking h. A falling jet of speed $w \propto \sqrt{gh}$ impacts the water surface, entraining air bubbles and generating turbulence in a cross-section area $\sim h^2$. Reproduced with permission from Deike (2022).

Here we focus on the dynamics once the wave has started to break, and the discussion on the route to breaking leads to the identification of characteristic scales of breaking, i.e., the speed of the breaker and the slope of the breaker, defined as the ratio of the breaker's amplitude to its wavelength. We note that the bandwidth of the breaking packet is a key parameter in understanding breaking occurrence.

12.3.1 Energy dissipation

We consider a breaking wave of height at breaking h, characteristic wavenumber k, and speed c, as sketched in Fig. 12.3. As the water at the crest has a faster velocity than the trough, a jet will form and collapse back onto the free surface. Breaking waves closely matching this vision can be generated in laboratory experiments by linear focusing packets, while their local geometry is very similar to shallow water breaking waves. These various observations support the idea of a universal breaking process, regardless of the way the breaker has been generated.

We characterize the scales of the turbulence generated by such a model-breaking process. The local turbulent dissipation rate generated in the water column after wave breaking impact can be described by a turbulent inertial scaling, adapted from the classic Taylor argument. The classic Taylor argument links the local kinematic turbulence dissipation rate ε to the integral length scale of the turbulence L, and the fluctuations at the injection scale U,

$$\varepsilon \propto \frac{U^3}{L} \tag{12.16}$$

In the case of breaking waves in deep water, the integral length scale is constrained by the breaker height $\sim h$, while the fluctuations at the injection scale are given by the jet falling velocity w. Assuming that the jet is falling due to gravity from its height h, the velocity at impact is estimated as $w \propto \sqrt{gh}$. The local turbulence dissipation rate under a breaking wave can then be written as

$$\varepsilon \propto \sqrt{gh}^3/h \tag{12.17}$$

Practically, in the ocean atmosphere, we are interested in the dissipation rate per unit length of breaking crest, ϵ_l, assuming that the breaking zone extends down the forward face of the wave over a fixed fraction of its amplitude and that its shape is geometrically similar for waves of different scales. The total dynamic dissipation rate per unit breaking crest length, ϵ_l is then related to the local turbulent dissipation rate ε considering that the turbulence is confined to a volume $\mathcal{V} = AL_c$, of cross section $A \propto \pi h^2/4$ (Drazen et al., 2008), and L_c the length of breaking crest, so that $\epsilon_l = \rho A \varepsilon$. This yields

$$\epsilon_l = b\rho c^5/g \tag{12.18}$$

where $b = \chi S^{5/2}$, with χ an $O(1)$ constant, where we have used the linear dispersion of gravity waves $c = \sqrt{g/k}$. We have assumed that the speed of the breaker follows the linear dispersion relation, which is reasonable, although the breaker tends to experience a slowdown.

The non-dimensional breaking parameter b can be seen as a non-dimensional energy dissipation (scaled by the breaker speed) and is a function of the breaking slope. Laboratory measurements of the energy lost by breaking have shown good agreement with the proposed scaling, and a semi-empirical scaling accounting for a breaking threshold defined by a critical slope S_0 (Romero et al., 2012) can then be written,

$$b = \chi_T (S - S_0)^{5/2} \tag{12.19}$$

where $\chi_T \approx 0.4$ and $S_0 \approx 0.08$ describe most of the existing experimental and numerical data. This latter equation can be seen as a simple model for energy dissipation by breaking waves, with an empirical breaking threshold. Note that historically, Duncan (1981) and Phillips (1985) described the dissipation rate due to breaking, per unit length of breaking crest, ϵ_l, assuming that the breaking zone extends down the forward face of the wave over a fixed fraction of its amplitude and that its shape is geometrically similar for waves of different scales, which introduced the non-dimensional breaking parameter, b, first assumed to be a non-dimensional constant, but shown to vary over several orders of magnitude by extensive experimental scrutiny (Duncan, 1981; Rapp and Melville, 1990; Drazen et al., 2008; Tian et al., 2010), when varying the breaking wave slope $S = hk$.

Following the above discussion, a turbulent patch characterized by $\varepsilon \propto \sqrt{gh}^3/h$ is generated over a depth $\sim h$ by every breaking wave. Since the slope of breaking waves $S = hk$ is constrained within a relatively narrow interval (typically between 0.2 to 0.5

depending on the way breaking is triggered), the magnitude of the turbulence and the strength of the breaking wave is implicitly controlled by the speed of the breaker c which sets the overall scale of the breaker, as initially postulated by Duncan and Phillips and summarized by Eq. (12.18). The turbulence kinetic energy will then decay in time, following classic turbulence scaling $\sim t^{-1}$. The scaling for the turbulence being generated by a single breaking wave will be used in a subsequent section to quantify the vertical profile of turbulence under an ensemble of breaking waves.

12.3.2 Air entrainment and bubble breakup in the turbulent flow beneath the breaking wave

As the wave breaks, air bubbles are being entrained into the water column as illustrated in Fig. 12.3. Constraints on the total amount of air entrained, as well as knowledge of the bubble size distribution, are required to compute mass fluxes associated with bubbles, such as gas and sea spray emissions by bubble bursting. Smaller bubbles are more effective at transferring mass than larger bubbles when filling the same total volume (most mass exchange processes scale with the surface area available for transfer).

The geometry of the breaking wave provides a constraint on the total amount of air that can be entrained, with the area of the air cavity cross-section scaling as $A \propto h^2$, with entrainment of a relatively homogeneous bubble cloud up to a depth $\sim h$. The bubble size distribution can then be determined by dimensional analysis. We consider that the flow beneath the breaking wave can be modeled as a three-dimensional homogeneous and isotropic turbulent flow, characterized by an integral length scale $\sim h$, and a turbulence dissipation rate $\varepsilon \propto \sqrt{gh}^3/h$. The problem to be considered now is the breakup of bubbles inside such a turbulent flow.

Before proceeding to the turbulent cascade argument, let us recall the classic critical size, or Kolmogorov–Hinze scale (Hinze, 1955). While turbulence will tend to deform and break the bubbles, surface tension will act as a restoring/stabilizing force, defining a critical length scale, the Kolmogorov–Hinze scale R_H below which the bubbles could in principle not be broken up by the turbulent flow anymore. The critical scale is obtained by dimensional analysis, balancing the stresses imposed by the turbulence at the bubble scale and surface tension, (Hinze, 1955; Rivière et al., 2021, 2022)

$$R_H = C(\gamma/\rho)^{3/5}\varepsilon^{-2/5} \tag{12.20}$$

where C is a dimensionless constant O(1). The definition of the Kolmogorov-Hinze scale is equivalent to the definition of a critical Weber number, comparing turbulence and surface tension forces. The breakup time of a bubble of size R_b (assumed to be in the turbulence inertial range) is given by the eddy turnover time at that size, (given by the ratio of the size of the bubble R_b to the turbulent velocity fluctuations at this scale $\Delta v \sim (\varepsilon R_b)^{1/3}$)

$$t_c(R_b) \propto \epsilon^{-1/3} R_b^{2/3} \tag{12.21}$$

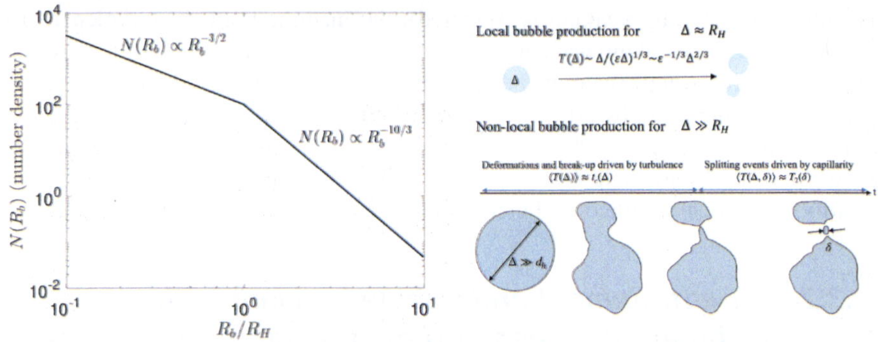

Figure 12.4 Sketch of a typical bubble size distribution under a breaking wave. The Hinze scale R_H (O(1–2mm) under breaking waves) separates a steep $N(R_b) \propto R_b^{-10/3}$ distribution while sub-Hinze bubbles are described by a $N(R_b) \propto R_b^{-3/2}$ scaling. Super-Hinze size distribution can be described by a turbulent breakup cascade with breakup time controlled by the eddy turnover time at the parent size, while the sub-Hinze distribution is controlled by the capillary breakup of elongated filaments formed when large bubbles are breaking. Adapted with permission from Rivière et al. (2022).

The turbulent flow presents large fluctuations leading to a broad range of breakup time for the same turbulent conditions, which makes the definitions of critical conditions challenging, so that the Hinze scale can be considered a soft limit. The Hinze scale is defined in practice in a statistical sense, for a given time (or spatial) domain of observation, with different large-scale fluctuation conditions considered in various experiments and simulations (Rivière et al., 2021, 2022; Ni, 2024).

12.3.3 Bubble size distribution under breaking waves

We now describe the size distribution under a breaking wave by considering the bubbles larger or smaller than the Kolmogorov–Hinze scale. Starting with $R_b > R_H$, we consider a local turbulent breakup cascade within the turbulence inertial range and a *direct cascade* process: air is injected at large scales (large bubbles) by the entrainment process and turbulent fluctuations break them into smaller bubbles, until the Kolmogorov–Hinze scale is reached.

Such a local cascade picture, akin to turbulence formalism, was formalized by Garrett et al. (2000). A steady-state solution of the population balance equation is considered by assuming that the breakup rates are controlled by the eddy turnover time at the scale of the parent bubble, with the breakup cascade local in scale and similarity through the scales. In this self similar model, a parent of size R_b breaks, in a time given by $t_c(R_b) = \epsilon^{-1/3} R_b^{2/3}$, into m fragments of equal diameter $m^{-1/3} R_b$. Each of these child bubbles then breaks in a time $\epsilon^{-1/3}(m^{-1/3} R_b)^{2/3}$. We obtain an increasing number of $m \times m^{-2/9} = m^{7/9}$ bubbles per unit of time, which yields the bubble density $N(m^{-1/3} R_b) = m^{1/3} m^{7/9} N(R_b)$. Assuming $N(R_b) \propto R_b^{\alpha}$, we have $-\alpha/3 = 1/3 + 7/9$ which gives $\alpha = -10/3$. This argument holds as long as

$R_b > R_H$. For $R_b < R_H$, the self-similar argument cannot be applied anymore, and surface tension needs to be considered.

Garrett et al. (2000) also proposed a dimensional analysis perspective of the bubble breakup cascade. Consider a *steady* turbulent breakup cascade for the bubble size distribution (per unit volume) $N(R_b)$, R_b the bubble radius, assuming a *constant* air flow rate per unit volume of water Q, $N(r) \propto Q$, and the process to be controlled only by the local (time-averaged) turbulent dissipation rate ε, so that dimensional analysis yields

$$N(R_b) \propto Q\varepsilon^{-1/3} R_b^{-10/3} \tag{12.22}$$

While the above cascade picture presents R_H as a cut-off below which surface tension opposes to bubble deformation, smaller bubbles are always observed below breaking waves, and in turbulent flow in general. Indeed, capillarity is the driving force for the destabilization of elongated, locally columnar-like bubbles, or ligament-like pieces of bubbles (Villermaux, 2020; Villermaux et al., 2022). Small bubbles can therefore be stripped from large deformable bubbles with $R_b > R_H$. Such a splitting process will be much faster as the ligament is thinner. The capillary time scale associated with a ligament of size δ is $T(\delta) \sim \sqrt{\rho\delta^3/\gamma}$, and is much faster than the associated turbulence time at the same scale $T(\delta) \ll t_c(\delta)$. As such, the turbulence will appear frozen and will only set the initial conditions, letting the surface-driven breakup drive the dynamics and control the small bubble production (Rivière et al., 2022). The small bubbles with $R_b < R_H$ have bypassed the cascade, which can be seen as a nonlocal production process.

Rivière et al. (2022) describes the rate of change of the bubble size distribution by integration over a capillary breakup production kernel $F(u)$, providing the number of children, with a production rate limited by the capillary time scale,

$$\partial_t N(R_b) \propto \frac{1}{T(R_b)} \int_{R_H}^{\infty} du \frac{N(u)F(u)}{u} \propto R_b^{-3/2} \tag{12.23}$$

In the above equation, only the capillary time scale acts as the rate of production and is the only term that depends on the bubble size R_b (which will be valid as long as the parent bubbles are large enough). It follows that the size distribution can be obtained by time integration and $N(R_b) \propto R_b^{-3/2}$ at any given time, with the prefactor/total number of bubbles increasing with time.

We therefore have the scalings for the two regimes of bubble size,

$$N(R_b < R_H) \sim R_b^{-3/2}, \quad \text{and} \quad N(R_b > R_H) \sim R_b^{-10/3} \tag{12.24}$$

one controlled by an inertial turbulence breakup time scale, and the other by a capillary time scale.

The two processes will happen simultaneously, and the above power laws can be seen as a mean description of the distribution as long as sufficiently large bubbles are fed into the system. Once there is no more supply of large bubbles, the super-Hinze size distribution will steepen until all (or most) bubbles are stable. We note that the above discussion neglects the effect of gravity, which for large bubbles could influence the breakup processes by providing another source of deformation, while also leading the bubbles to rise. The scaling described by Eq. (12.24) has been observed in multiple laboratory experiments and numerical simulations for large-scale breaking waves and air cavity breaking up in turbulent flows (Deane and Stokes, 2002; Deike et al., 2016; Mostert et al., 2022; Rivière et al., 2022; Ruth et al., 2022).

12.3.4 Volume of air entrained

The geometric constraints of breaking waves can be used to provide an estimate of the total amount of air entrained. When a wave breaks, it entrains a tube of air of cross-sectional area $A \approx \pi h^2/4$, creating a turbulent bubble cloud over a similar area for short times after the breaking onset, with h the breaking height, i.e., the vertical distance between the crest of the breaker and its point of impact on the surface below. The maximum volume of air entrained is then simply given by $V_0 = AL_c = \pi h^2 L_c/4$ with L_c the length of the breaking crest. While observations immediately after breaking confirm this simple idea, part of the bubble cloud quickly rises to the surface so that the effective bubble volume entrained is smaller. This observation can be explained by considering that globally (integrating over space and time of the event) the work done against buoyancy forces in entraining the bubbles is proportional to the mechanical energy dissipated, which was originally suggested by experimental laboratory measurements (Lamarre and Melville, 1991) and confirmed through high-fidelity direct numerical simulation (Deike et al., 2016; Mostert et al., 2022). The balance between energy dissipation and buoyancy forces can be translated into an integral equation matching energy dissipation and buoyancy forces. Assuming low void fraction, a homogeneous bubble cloud in space (over the area $\propto h^2$) rising with a mean velocity weighted by the surrounding turbulent flow W, the ratio of buoyancy and turbulence can be expressed as a simple non-dimensional ratio $\varepsilon/(Wg)$, with W a characteristic rise velocity of the bubble cloud.

The number of bubbles and their size distribution can then be described assuming a homogeneous bubble cloud. We consider the bubble size distribution averaged in time over the active breaking time, over which the assumption of similarity cascade developed for the bubble size distribution holds. The total bubble size distribution can then be written as

$$\mathcal{N}(R_b) = V_0 N(R_b) = AL_c N(R_b) \tag{12.25}$$

The shape of the size distribution has been described by (Eq. (12.24)). The prefactor is given by the geometric constraints and an energy balance argument (Deike et al., 2016; Mostert et al., 2022). Introducing the largest bubble in the system R_M, the size distribution for super-Hinze scale bubbles can be described by:

$$\mathcal{N}(R_b/R_H) = \frac{B}{2\pi}\frac{\varepsilon}{Wg}\left(\frac{R_b}{R_H}\right)^{-10/3}\frac{L_c R_M^{4/3}}{R_H^{7/3}} \tag{12.26}$$

This relation can be rigorously related to the scaling analysis (Eq. (12.22)) by considering the time scale for the large air pocket to break (Deike et al., 2016; Mostert et al., 2022).

Integrating over all bubble sizes (with the above Hinze scale bubbles contained more than 95% of the volume (Deane and Stokes, 2002)), we get the total volume of air entrained (considering $R_M \gg R_H$)

$$\mathcal{V} = \int_{R_H}^{R_M} \frac{4\pi}{3} R_b^3 \mathcal{N}(R_b) dR_b \tag{12.27}$$

Recalling the turbulence dissipation rate scaling ($\varepsilon \propto (\sqrt{(gh)})^3/h$), assuming $R_M \propto h$, and using the linear dispersion relation $c = \sqrt{g/k}$, the volume of air entrained can be written as,

$$\mathcal{V} = B L_c \frac{(hk)^{5/2} c^5}{W g^2} \tag{12.28}$$

Defining the volume of air entrained per unit length of breaking crest L_c, per unit time, for breakers traveling at a speed c, $v_l(c) = \mathcal{V}/(L_c \tau_b)$, with $\tau_b \approx h/W$ an approximate bubble plume life time, which leads to

$$v_l(c) \propto B(hk)^{3/2} c^3/g \tag{12.29}$$

which we will use later when considering sea spray emission and gas transfer from ocean wave breaking.

12.4 Dynamics and statistics of breaking waves at the ocean surface

Wave breaking occurs at the ocean surface at moderate to high wind speeds, with significant impacts on the transfer of momentum, energy, and mass between the ocean and the atmosphere (Melville, 1996; Deike, 2022). Visually, it manifests as white-capping, widely observable at sea above a certain wind speed. Breaking acts as an energy sink for the waves: it limits the wave height by transferring the excessive wave energy into underwater turbulence and currents, therefore influencing the upper-ocean dynamics as well. Air-sea fluxes in breaking-dominated seas can be computed from knowledge of the breaking-wave distribution combined with the individual breaking-wave dynamics. We introduce in this section the classic representation of breaking waves in the field through the geometric whitecap coverage and length of crest of breaking waves moving at a certain speed from Phillips (1985) and discuss how to compute the associated fluxes of energy, momentum, and mass towards the upper ocean (see also Romero (2019)).

12.4.1 Whitecap coverage and breaking distribution

The whitecap coverage, $\mathcal{W}$, is defined as the white patches of bubbles formed after breaking at the ocean surface and has been used as a simple proxy for the amount of breaking waves at the sea surface. The presence of whitecaps is also used as a visual metric for the state of the sea in the naval Beaufort scale. Whitecaps start to occur for wind speed at 10 m of about 7 m/s and their presence increases as wind speed increases.

Despite its apparent simplicity, accurately measuring and predicting the whitecap coverage as a function of wind and wave parameters has remained an active field of research. Measurements remain challenging and lead to very large scatter when presented as a function of wind speed only. The scatter is related in parts to the dependency of breaking waves on the sea-state, to the local breaking intensity, to the details of the image processing (with the amount of sunlight and inclination being a particular challenge, while making measurements on a moving platform is never easy), as well as the possible role of water contamination on the bubble plume life time. More elaborate classifications of the whitecap coverage accounting for dynamical features, in terms of active and passive whitecaps, have been proposed but suffer similar challenges in objective definitions and consistency in field image processing.

As a consequence, multiple parameterizations of the whitecap coverage have been proposed, typically of the functional form, $\mathcal{W} \propto (U_{10} - U_{th})^{\zeta}$ with ζ an exponent between 2 and 3, and U_{th} an empirical threshold. Since whitecap coverage provides only geometric information, relationships with the wind or the local wave field are determined empirically. Moreover, any connection between the whitecap coverage and air-sea fluxes will also require fitting to data, as it has been attempted for gas exchange, sea spray aerosols, or energy fluxes.

A description of the breaking wave field, connected to the wave dynamics, is therefore desired. Phillips (1985) introduced the breaking waves distribution of length of crest $\Lambda(\boldsymbol{c_b})$, where $\Lambda(\boldsymbol{c_b})d\boldsymbol{c_b}$ is the expected length per unit sea surface area of breaking fronts moving with speeds in the range of $(\boldsymbol{c_b}, \boldsymbol{c_b} + d\boldsymbol{c_b})$. Intuitively, this metric corresponds to the breaking fronts at the ocean surface when looking from above, considering that their speed is going to characterize the amount of breakers at that particular scale, and represents the fact that, as the wind blows, at any instant, the fronts of the breakers define a distribution of isolated lines. As initially discussed by Phillips, the scales of the breakers vary widely, from large scale breakers with length of breaking crests of O(100m) with extended whitecaps and patches of foam that will inject large amount of energy down in the water column; down to small scale breakers at scales close to parasitic capillary waves with crest of O(10cm), which barely entrain air and will only affect a thin layer below the surface.

The independent variable, breaking front propagating speed $\boldsymbol{c_b}$, can be directly observed and, as discussed before, will provide the scale of the breaker and therefore a direct link to energy dissipation. The wavenumber and breaker speed will be linked through the assumption that c_b is proportional to the wave phase speed c, which in turn relates to k by the linear dispersion relation $c = \sqrt{g/k}$, a hypothesis that is still being debated. Several observational studies have been conducted, which provide measurements of the $\Lambda(c)$ distribution, and its moments, made possible by technical advancement, including ship-borne and air-borne visible and infrared imagery, while modeling the breaking distribution. The azimuth-integrated distribution is then $\Lambda(c) = \int \Lambda(\mathbf{c}) c d\theta$.

12.4.2 The moments of the breaking distribution

The distribution of the length of the crest of breaking waves provides a scale-by-scale representation of the breaking field. Physical properties of breaking waves can then be inferred from the moments of $\Lambda(c)$. By dimensional analysis, since $\Lambda(c)dc$ is the length of breaker at a certain speed (per unit ocean surface area), then the zeroth moment gives the total length of breaking fronts per unit surface ocean area (dimension L/L^2),

$$\mathcal{L} = \int \Lambda(\mathbf{c}) d\mathbf{c} \tag{12.30}$$

The first moment gives the fraction of total surface area turned over per unit time per unit surface ocean area (dimension $1/T$),

$$\mathcal{R} = \int c \Lambda(\mathbf{c}) d\mathbf{c} \tag{12.31}$$

and quantifies the total number of breaking waves of all scales passing a certain point per unit time.

The whitecap coverage represents the area covered by white foam due to breaking waves, is dimensionless, and reads,

$$\mathcal{W} = \int \tau_c c \Lambda(\mathbf{c}) d\mathbf{c} \propto \int c^2 \Lambda(\mathbf{c}) d\mathbf{c} \tag{12.32}$$

where τ_c is the characteristic whitecap lifetime that can be related to the wave breaking properties following $1/\tau_c \propto ck$ so that the whitecap coverage is related to the second moment.

The rate of entrainment of air per unit area of ocean surface, V_A (units of a volume per area per time), can be written as,

$$V_A = \int v_l(c) \Lambda(c) dc, \text{ with } v_l(c) = V/(L_c \tau_b) \tag{12.33}$$

where $v_l(c)$ is the volume of air entrained by breakers moving at a speed between c and $c + dc$, per unit time, per unit length of breaking crest, which we introduced when discussing air entrainment by a single breaking wave V. Assuming the plume life-time, $\tau_b \propto h/W$, we have

$$V_A = \int B \frac{b}{(hk)} \frac{c^3}{g} \Lambda(c) dc \tag{12.34}$$

The volume of air entrained by breaking waves (a volumetric analog of the whitecap coverage) is thus given by the third moment of the breaking distribution, modulated by a factor that depends on wave slope, i.e., the breaking strength.

Leveraging our understanding of the energy dissipation by a single breaker with speed c, we can infer that the fifth moment gives the total amount of energy dissipation due to breaking waves per unit surface area,

$$S_{diss} = \int \frac{b\rho}{g} c^5 \Lambda(\mathbf{c}) d\mathbf{c} \tag{12.35}$$

where b is the breaking parameter, which depends on the wave slope. Finally, the fourth moment relates to the momentum flux,

$$\mathcal{M} = \int \frac{b\rho}{g} c^4 \Lambda(\mathbf{c}) d\mathbf{c} \tag{12.36}$$

so that under breaking dominated conditions, $\mathcal{M} = \int \frac{b\rho}{g} c^4 \Lambda(\mathbf{c}) d\mathbf{c} \approx \tau = \rho_a u_*^2$.

Since the breaking distribution is physically closely related to the wave dynamics, and to the wave spectrum (Romero, 2019; Wu et al., 2023), the above formulations for air-sea fluxes naturally contains sea-state variability through the sea state variability of the $\Lambda(c)$ distribution.

The breaking parameter has been shown to depend on the breaking slope for individual breakers. Two approaches have been used to evaluate Eq. (12.35) from field data. The first assumption is to consider a constant or averaged breaking parameter (or given as a function of some wave parameter such as waveage) while the second assumption is to estimate the local slope from the saturation spectrum, $hk \equiv \sqrt{B(k)} = \sqrt{\phi(k)k^3}$, to obtain a spectrally defined breaking parameter $b(k) = A_T(\sqrt{B(k)} - \sqrt{B_T})^{5/2}$, where A_T is a non-dimensional coefficient and B_T a breaking threshold. The coefficients A_T and B_T have orders of magnitude constrained by laboratory experiments and values adjusted to balance the wind input term used in spectral models (Romero et al., 2012; Romero, 2019). The link between wavenumber and breaking speed is then established through the dispersion relation. Both approaches have led to energy budgets in the upper ocean that are reasonably well closed.

12.4.3 Scalings of the breaking distribution

Phillips (1985) originally introduced the breaking distribution together with an energy balance for the functional form of $\Lambda(c)$, as summarized in Wu et al. (2023). The wave action balance equation involves the divergence of the nonlinear energy flux S_{nl}, the wind input S_{wind}, and dissipation due to breaking S_{diss}, written as (Phillips, 1985)

$$S_{nl} \propto gk^{-3}B^3(k), \quad S_{wind} \propto gk^{-3}(\frac{u_*}{c})^2 B(k), \quad \text{and } S_{diss} \propto gk^{-3} f(B(k)) \tag{12.37}$$

with the saturation $B(k) = k^3\phi(k)$, and $f(B(k))$ a functional dependence solely on $B(k)$ (assuming that breaking and consequent dissipation '*are the result of local excesses, however these excesses are produced*'). The balance between S_{nl} and S_{diss}

leads to $f(B) \propto B^3$, and therefore $S_{diss} \propto gk^{-3}B^3$. The breaking front distribution $\Lambda(c)$ is then obtained by writing the equality between dissipation in the k-space and the c-space: $\epsilon(k)dk = \epsilon(c)dc$. With $\epsilon(k)dk = (S_{diss}\omega)dk$; and $\epsilon(c)dc = bg^{-1}c^5\Lambda(c)dc$, where b is the non-dimensional breaking parameter (Duncan, 1981; Phillips, 1985). Substituting a spectral shape of $\phi(k) \propto k^{-5/2}$ into the S_{diss} then leads to $\Lambda(c) \propto c^{-6}$. Considering the equilibrium range, $S_{nl} \propto S_{wind}$ (Phillips, 1985), gives $\phi(k) \propto u_* g^{-1/2} k^{-2.5}$, which leads to

$$\Lambda(c) \propto u_*^3 g c^{-6} \tag{12.38}$$

While the $\Lambda(c) \propto c^{-6}$ has been observed in various data sets, the wind-only scaling was never confirmed by observations, leading to various empirical wave age type scalings such as,

$$\Lambda(c) c_p^3 g^{-1} (c_p/u_*)^{1/2} \propto (c/\sqrt{gH_s})^{-6} \tag{12.39}$$

which improved the collapse between data sets. More recently, a scaling only based on the mean square slope has been proposed by Wu et al. (2023). Finally, we comment that the scaling $\Lambda(c) \propto c^{-6}$ is based on a balance between the various source terms in the wave action equation, which corresponds to the equilibrium range for the wave spectrum, $\phi(k) \propto k^{-5/2}$; while the breaking dominated saturation range follows $\phi(k) \propto k^{-3}$ which would yield a shallower breaking distribution, $\Lambda(c) \propto c^{-5}$ which could be close to some observations for strong breaking wave fields.

12.4.4 Turbulent flow beneath an ensemble of breaking waves

It has long been recognized that under an active wave field including breaking waves, the classic law of the wall for the upper ocean turbulent flow, which leads to a turbulence dissipation rate profile with depth $\varepsilon(z) \propto \frac{u_*^3}{\kappa z}$ does not represent the flow, and presents an energy deficit (Terray et al., 1996; Gemmrich and Farmer, 2004; Sutherland and Melville, 2015).

Under strongly forced wind conditions, and prevalence of breaking waves, one can invoke the assumption of equilibrium between the mean energy input from the wind S_{wind} and the total dissipation due to breaking S_{diss}. Sutherland and Melville (2013, 2015); Thomson et al. (2016) performed field measurements, jointly measuring the sea surface breaking statistics and the near-surface upper ocean using underwater turbulence acoustic sensors to demonstrate such budget-closing, with a relative balance between wind input S_{wind} and the energy due to breaking S_{diss}, estimated either from the fifth moment of the breaking distribution, or the integration of the underwater turbulence over some depth $\sim \int dz \varepsilon(z)$. The observed profile was close to $\varepsilon(z) \propto z^{-1}$ within one significant wave height below the water surface.

Recently, Scapin et al. (2026) performed direct numerical simulation of wind-forced breaking waves and directly reported the balance between wind input (evaluated during the growth stage) and energy dissipation (evaluated during the breaking stage). Observations from the field and from high fidelity direct numerical simulations therefore suggest

$$\varepsilon(z) \propto \frac{S_{wind}}{\rho z} \propto \frac{S_{diss}}{\rho z} \tag{12.40}$$

The underwater dissipation under active breaking waves can therefore be fully described by the knowledge of the breaking distribution and breaking parameter.

We comment that based on the wind input, a simplified scaling can also be discussed: $\varepsilon(z) \sim u_*^2 c/z$ since $S_{wind} \sim u_*^2 c$, which explains why the wall-layer scaling (which leads to $\varepsilon(z) \sim u_*^3/z$) fails in breaking dominated configuration.

The scaling of dissipation profiles with wind input and with breaking-induced dissipation are consistent with the understanding of wind-wave growth and dissipation by breaking combined with the balance between wind input and breaking. Wave breaking occurs when fluid inertia overcomes restoring forces such as gravity. After the onset of breaking, wind input no longer affects the dissipation rate; notably, the total wind energy transferred to the waves remains central in determining the breaking conditions.

12.5 Application to ocean-atmosphere fluxes and open challenges

The following section discusses how our understanding of fundamental turbulent processes at the ocean-atmosphere interface can be used to improve the representation of air-sea fluxes through parameterization of momentum, gas exchange, and sea spray emissions. Such parameterizations are widely used in ocean, atmosphere, or coupled models at various scales as well as observational-based products. The accuracy of the air-sea fluxes formulation is critical in the representation of various weather in operational weather modeling, tropical cyclone intensification and trajectories, as well as for climate questions, including ocean circulation, carbon storage, and oxygen cycle. Historically, these parameterizations have been based on bulk formulas, expressed as a function of wind speed. As the spatial resolution of models and observations increases, uncertainties in the bulk formula and high-frequency variability become more and more important. Improved physics leads to improved representation in parameterization, which leads to improved predictability. Great progress has been made, thanks to technological advances, including satellite and field measurements, and computational capabilities at various scales. Rigorous integration of the novel observations at multiple scales within air-sea fluxes formulations remains an ongoing challenge. The following section highlights three types of fluxes central to weather and climate modeling and understanding. Momentum flux, affecting the atmospheric turbulent boundary layer over the ocean, is central to offshore wind energy, as well as critical to understand upper ocean turbulence. Gas flux and sea spray emissions, central to the chemical coupling between the ocean and the atmosphere, modulate atmosphere chemistry, cloud processes as well as the ocean biogeochemical cycle and ecology.

12.5.1 Momentum fluxes

Investigations of atmospheric turbulence over the oceans have shown that the interaction of wind with the ocean surface waves results in an air-side turbulent boundary layer that differs significantly from the turbulent boundary layer over surface terrestrial land. This is exemplified by considering the surface roughness. Over land, the surface roughness can often be treated as a constant determined by the vegetation or city height. Over the ocean, the surface roughness, or equivalently the surface drag, is determined by the wave field, itself largely determined by the wind. The stronger the winds, the rougher the seas, with strong modulation at a given wind speed depending on the specificity of the sea state (presence of swell, cross-wind waves will alter the drag). Therefore the exchange of momentum and energy is largely governed by the wave field near the ocean surface.

The wave's influence is the strongest close to the ocean surface, in a layer called the wave boundary layer. Physical oceanographers and marine meteorologists often divide the ocean boundary layer into the surface layer, where wind shear and buoyancy—stratification govern the turbulent flow (i.e., like a Monin–Obukhov layer; see other chapters of this book), and a wave boundary layer where additional scaling parameters are required for similarity. These additional scaling parameters will be wave-related, have been the subject of extensive studies, and are still an active area of research, and relate to the wave slope or the wave age.

The exchange of momentum between the atmosphere and ocean is difficult to measure directly over the ocean. Instead, as for other ocean-atmosphere fluxes, bulk formulas have been introduced and relate the fluxes to more easily measured averaged quantities, primarily wind speed (typically taken at 10 m, $U_{10} = U(z = 10m)$), temperature, and humidity. The flux is then obtained through a transfer coefficient, and the exchange of momentum at the ocean surface, considering dimensional argument is expected to scale as the wind speed squared (Edson et al., 2013; Ayet and Chapron, 2022),

$$\tau = \rho_a \langle u'w' \rangle = \rho_a u_*^2 = \rho_a C_D U_{10}^2 \tag{12.41}$$

where τ is the momentum flux (going from the air side to the water side), $\rho_a \langle u'w' \rangle$ is the wind stress (obtained in the field from eddy covariance measurement averaged over time windows of typically 30 min), u_* the wind friction velocity, and C_D the drag coefficient. Knowledge of the momentum flux requires the measurement of the wind friction velocity u_*, while in many practical purposes, only the wind speed U_{10} is provided (this is the case in large-scale modeling frameworks, or in some observational data sets). As a consequence, a parameterization for the drag coefficient is required.

For simplicity, the present discussion will consider neutral conditions; we refer the reader to other chapters of this book for discussion of the Monin–Obukhov similarity theory that has to be considered in the case of non-neutral conditions.

The turbulent boundary layer over a rough surface (wavy or not) has been extensively described to follow the log-law,

$$U(z) = \frac{u_*}{\kappa} \ln\left(\frac{z}{z_0}\right) \tag{12.42}$$

with $\kappa \approx 0.41$ the Von Karman constant, and z_0 the roughness length. Finding the roughness length is then equivalent to finding the drag coefficient

$$C_D = \frac{u_*^2}{U_{10}^2} = \frac{\kappa^2}{\ln(10/z_0)^2} \tag{12.43}$$

Various processes at the ocean surface can then control the roughness length. Considering a relatively smooth surface (at low wind speed), and assuming that the viscous shear dominates, leads to proposing by dimensional analysis $z_0 \propto \nu_a/u_*$ with a prefactor experimentally determined to be ≈ 0.11. When the wind speed increases, and the surface becomes rough, the Charnock coefficient has been proposed, leveraging again dimensional analysis and considering gravity as the restoring force controlling the roughness elements, and

$$z_0 = \alpha_{CH} u_*^2/g \tag{12.44}$$

with α_{CH} the Charnock parameter, which has to be determined empirically. While the initial assumption was for this parameter to be a constant, laboratory and field data have suggested a dependency of the Charnock parameter on various wave parameters, mainly the wave age c/u_*, misalignment between wind and waves, and wave steepness.

Determining the mean wind profile, and hence finding the roughness length or drag coefficient for a wide range of sea states and wave spectra, is tightly linked to understanding wind wave growth, as both questions deal with the coupling of the waves (potentially growing) with the turbulent boundary layer. Separately, finding the momentum flux τ is critical to parameterization of upper ocean processes as wind drives current and turbulence. While bulk formula and initial empirical parameterizations have proven extremely reliable to understand large-scale ocean circulations, the quantification of submesoscale processes, including fronts, filaments, or Langmuir turbulence at scales from 100 m to 10 km, requires a finer understanding of the momentum flux and the modulation by the wave field. As discussed above, in the case of strong breaking field, the momentum flux can be related to the breaking wave field through the $\Lambda(c)$ distribution, while other regimes of active research deal with interaction between the waves and the sub-mesoscale currents and temperature structure, and their effect on the wind profile, while mixed seas with swell and wind wave superposed also impact the drag coefficient, requiring to go beyond the equilibrium hypothesis.

12.5.2 Gas exchange

Gas exchange at an air-water interface depends not only on the concentration differences between the species but also on the efficiency of the transfer process. The efficiency is controlled by a variety of processes in the air and water near the interface, with a critical role played by the fluid underlying turbulence. The flux is usually written as (Wanninkhof, 1992; Takahashi et al., 1997; Wanninkhof et al., 2009; Ho et al., 2011)

$$F = k_w \Delta C = k_w S \Delta p \tag{12.45}$$

with ΔC representing the concentration difference between the air and water, which can be related to the gas solubility S and the gas partial pressure difference Δp. Understanding the gas transfer velocity efficiency, k_w in terms of the atmospheric and oceanic controlling parameters is then key to understanding exchanges of gases critical to the climate system, such as carbon dioxide, oxygen, and dimethylsulfide. As mentioned in the introduction, ocean uptake represents about 25% of the global carbon budget (Friedlingstein et al., 2022).

The most commonly used formulation for the gas transfer velocity only relies on the wind speed at 10 m, U_{10}, and the gas Schmidt number $Sc = \nu/D$ to scale exchange for different gases. Such formulations have initially been derived from historical carbon inventory, and yield empirical formulations of the form (Wanninkhof, 1992; Wanninkhof et al., 2009; Ho et al., 2011),

$$k_w = C_{gas} Sc^{-1/2} U_{10}^2 \tag{12.46}$$

The quadratic scaling with wind speed encompasses the sharp increase in gas exchange at high wind speed, due to the action of wind in enhancing turbulence in the water through direct forcing and through the amplification of the wave field, and eventually breaking of waves. The scaling with the Sc number can be justified by eddy renewal theory (as discussed in the next paragraph). $C_{gas} \approx 0.25$ (to obtain k_w in cm/hr) is a dimensional constant which is fitted to data. We discuss in the following section a physics-based formulation for the gas transfer velocity in various contexts.

Eddy renewal theory. The efficiency of gas exchange at an air-water interface and its link to water and air turbulence are discussed in detail in Levich (1962); Garbe et al. (2014); Farsoiya et al. (2023). As discussed earlier, the ocean-atmosphere interface is governed by the interaction of turbulent processes and waves, which will all contribute to the efficiency of the gas exchange. We note that the kinematic viscosity and thermal diffusivity of water differ by an order of magnitude, so that the thermal sublayer is within the viscous sublayer. Turbulent eddies close to the surface bring "fresh" bulk water to the surface from which gases can diffuse in or out of the water, therefore enhancing the gas exchange. This effect is referred to as eddy renewal of the surface. Following this idea, Higbie (1935) proposed that the transfer efficiency k_w can be written as

$$k_L \propto \sqrt{D/\theta} \tag{12.47}$$

where D is the gas diffusivity in the liquid (water) and θ is a contact time scale. The Higbie scaling theory will then apply once we have determined the appropriate time scale controlling the exchange of gas at the interface. Since the turbulent flow (assuming homogeneous and isotropic behavior) is characterized by multiple scales, finding the appropriate, or most effective, time scale associated with particular eddies is not obvious. Considering an anisotropic turbulent flow would further complicate the discussion. In an homogeneous and isotropic turbulent flow, the two extreme cases would be for the time factor to be driven by the large scale of the flow, i.e., the integral time scale $\theta \propto \tau_{L^\star} = L^\star/u_{\mathrm{rms}}$; or the smallest scale of the flow, i.e., the Kolmogorov time scale $\theta \propto \tau_\eta = (\nu_l/\epsilon)^{1/2}$.

At high Reynolds number, we expect the Kolmogorov scale to be the appropriate time scale so that the mass transfer coefficient scales as,

$$k_L/(\nu_l \epsilon)^{1/4} \propto \mathrm{Sc}^{-1/2} \tag{12.48}$$

In a non-breaking wave field, the near-surface water turbulence dissipation rate can be related (to first order) to the wind friction velocity, inputing energy and momentum into the water column, with

$$u_* \propto (\nu_l \epsilon)^{1/4} \tag{12.49}$$

so that we obtain the following scaling for an unbroken surface driven by a turbulent (shear) flow,

$$k_L \propto u_* \mathrm{Sc}^{-1/2} \tag{12.50}$$

Similar arguments can be used for the gas exchange at the bubble scale, with possible assumptions to describe the individual bubble gas exchange velocity, or in an equivalent way, the contact time scale θ, as recently summarized by Farsoiya et al. (2023). Assuming no water turbulence, the individual bubble will rise to the surface due to buoyancy, so that the contact time scale is given by the rise velocity in quiescent water w_b and the bubble size R_b, $\theta = R_b/w_b(R_b)$. This yields

$$k_L \propto \sqrt{D w_b / R_b} \tag{12.51}$$

the total amount of gas being exchanged is determined by k_L together with the residence time of the bubble underwater. The individual bubble gas transfer in a turbulent flow can then be obtained either by considering the modified rise velocity in turbulence (low turbulence) or by considering the eddy renewal high turbulence limit.

Bubble models for ocean-atmosphere interaction. Bubble-mediated gas transfer has been demonstrated to be important for low to intermediate gas solubility, with bubbles increasing the surface area and allowing for a more efficient gas exchange. For highly soluble gases, bubble effects remain small as equilibrium is reached quickly through the unbroken surface.

From knowledge of the bubble size distribution, the individual bubble rise velocity in the upper ocean, and the individual bubble exchange coefficient, integrated bubble-mediated exchange can be derived (Woolf and Thorpe, 1991; Keeling, 1993; Deike and Melville, 2018; Deike et al., 2025). The overall gas exchange will then be determined by the bubble lifetime in the water and the efficiency at transferring gas, which is a function of gas diffusivity (and solubility). The bubble residence time can be obtained from simple dimensional analysis by combining the depth of entrainment and the rise speed. In the case of underwater turbulence, bubble gas exchange will be impacted in two ways. First, the overall bubble rise speed will be decreased, leading to an increased residence time, while the gas exchange rate will then be controlled by the turbulence time scale of renewal. Under these hypotheses, turbulence would enhance bubble-mediated gas exchange in multiple ways: i) by increasing the bubble transfer velocity; ii) by increasing the residence time and allowing bubbles to get squeezed by hydrostatic pressure, hence increasing gas exchange; and iii) finally, bubbles can get trapped in the water column and get crushed by hydrostatic pressure and fully dissolve.

The implementation of bubble-mediated gas exchange models within large-scale Earth system modeling remains an active area of study, with empirical studies as a function of wind speed being mostly used. The path to include bubble-mediated gas exchange is to decompose the gas transfer velocity into breaking (bubble) and non-breaking terms,

$$k_w = k_{nb} + k_b \tag{12.52}$$

with k_{nb} given by Eq. (12.50), while k_b is a bubble mediated gas exchange gas transfer velocity. The bubble-mediated gas transfer coefficient can be related to the amount of air entrained under a breaking wave, which is described by Eq. (12.34). Empirical scaling of $\Lambda(c)$ yields a scaling $k_b \propto u_*^{5/3}\sqrt{gH_s}^{4/3} Sc^{-1/2}\alpha^{-0.35}$. The scalings of k_b in terms of the physico-chemical variables (solubility and diffusivity) depend on injection depth, bubble size distribution, and bubble dynamics used in the chemical model (Keeling, 1993; Deike and Melville, 2018; Deike, 2022).

We note that fully dissolving bubbles and bubbles getting squeezed by hydrostatic pressure also introduce an asymmetric flux going only into the water, which has recently been demonstrated to be a first order effect for the exchange of oxygen and the representation of oxygen in ocean models (Deike et al., 2025). Separately, surfactants, due to pollution or biological activity, have the ability to reduce gas exchange and the integration of their effect in large scale models remains to be completed.

12.5.3 Sea spray generation

At moderate to high wind speeds above the ocean surface (typically above 10 m/s), ocean spray droplets are generated by breaking waves. These droplets are transported in the atmospheric boundary layer, where they influence the transfer of mass and heat between the ocean and the atmosphere, contribute to the sensible heat flux, and may participate in the intensification of tropical cyclones (Veron, 2015).

Sea spray aerosols are composed of small liquid droplets formed through two main pathways following wave breaking. The spume drops, produced from the tearing of breaking wave crests by strong winds, consist mostly of drops larger than 10 microns (Veron, 2015). Drops produced by bursting bubbles are known as film drops and jet drops, with radii at formation ranging from sub-microns to hundreds of microns. These droplets may be transported to the upper atmosphere, where they evaporate, and may serve as cloud condensation nuclei Lewis and Schwartz (2004). The droplets can transport dissolved gases, salts, and biological materials, with their chemical composition affected by the production mechanisms (Wang et al., 2017; Cochran et al., 2017; Bertram et al., 2018). Knowledge of the size distribution of primary ocean spray aerosol particles and its dependence on meteorological and environmental variables is necessary for modeling cloud microphysical properties and the influence of aerosol on radiative processes (de Leeuw et al., 2011). Significant uncertainties remain regarding emissions as well as the transport of sea spray aerosols in the marine boundary layer.

Bubble bursting (reviewed by Lewis and Schwartz (2004); Veron (2015); Deike (2022)) generates sea spray droplets all the way from submicron to 500 μm; while wave breaking and wind shearing produce spume droplets mostly above 10 μm and up to 1 mm (Veron, 2015). Droplets from bubble bursting are separated into jet drops and film drops. Jet drops are produced when the bubble cavity collapses and forms a vertical upward jet that destabilizes into drops, leading to 1 to 10 drops. Film drops are generated by the rupture of the thin layer (film) of water that initially separates the bubble from the atmosphere. These beautiful fluid-mechanics phenomena have received extensive scrutiny over the past decades, in part thanks to the development of high-speed imaging and computational methods, which have enabled detailed physical descriptions and theories of the droplet formation and ejection processes. We refer the interested readers to reviews and books on the topic (Lewis and Schwartz, 2004; de Leeuw et al., 2011; Veron, 2015; Deike, 2022; Villermaux et al., 2022).

Larger droplets (typically above 10 microns) can be generated directly by wind shearing and breaking, resembling atomization processes (Villermaux, 2020) with a sequence of instabilities either driven by wind shear or wave dynamics. A definitive picture is still missing in parts due to the multi-scale nature of the interface deformation and breaking dynamics in the field. Most of our understanding on wind breaking waves generated drops come from laboratory experiments (and to some extent simulations), and typically lead to distributions showing a relatively shallow size dependency below a few hundreds microns $N(d) \propto d^{-1}$ to d^{-3} and a much steeper regime above a few hundred microns $N(d) \propto d^{-4}$ to d^{-5} (Veron et al., 2012; Erinin et al., 2019, 2022). The differences between the various experiments might be ascribed to the various heights above wave breaking, where the drops are being measured, or the various wind and wave conditions. Recent measurements in the droplets' velocity and acceleration have shown that the transport of inertial drops is highly transient, showing that the distribution being measured corresponds to a convolution of drops freshly emitted and drops being transported in the turbulent air flow above the wave (Erinin et al., 2022). Droplet size distributions near the surface have rarely been reported in the field, due to the difficulty in measuring droplets close to the surface and their relatively small residence time. Measurements several meters above the surface then require integrating the transient transport dynamics. Multiple studies have focused on the precise role of drop dynamics in heat exchange at high wind speed (Veron, 2015; Peng and Richter, 2017; Richter et al., 2019), but remain uncertain as it is highly dependent on the initial generation function in terms of size and velocity, as well as absolute magnitude.

While the sea spray fluxes are a key component in modeling air-sea interaction in the context of climate processes (in terms of sea salt aerosol emissions) or extreme weather forecast (role of large drops in enthalpy and drag during hurricane intensification), their precise dynamics and statistics when emitted at the sea surface, as well as their transport in the marine turbulent boundary layer remains largely an open research topic.

References

Ayet, A., Chapron, B., 2022. The dynamical coupling of wind-waves and atmospheric turbulence: a review of theoretical and phenomenological models. Boundary-Layer Meteorology 183, 1–33.

Belcher, S.E., Hunt, J.C.R., 1993. Turbulent shear flow over slowly moving waves. Journal of Fluid Mechanics 251, 109–148. https://doi.org/10.1017/s0022112093003350.

Belcher, S.E., 1999. Wave growth by non-separated sheltering. European Journal of Mechanics-B: Fluids 18 (3), 447–462. https://doi.org/10.1016/s0997-7546(99)80041-7.

Bertram, T.H., Cochran, R.E., Grassian, V.H., Stone, E.A., 2018. Sea spray aerosol chemical composition: elemental and molecular mimics for laboratory studies of heterogeneous and multiphase reactions. Chemical Society Reviews 47, 2374–2400.

Bosboom, J., Stive, M.J., 2021. Coastal Dynamics.

Cochran, R.E., Ryder, O.S., Grassian, V.H., Prather, K.A., 2017. Sea spray aerosol: the chemical link between the oceans, atmosphere, and climate. Accounts of Chemical Research 50, 599–604.

Deane, G.B., Stokes, M.D., 2002. Scale dependence of bubble creation mechanisms in breaking waves. Nature 418, 839–844.

Deike, L., 2022. Mass transfer at the ocean–atmosphere interface: the role of wave breaking, droplets, and bubbles. Annual Review of Fluid Mechanics 54, 191–224.

Deike, L., Melville, W., Popinet, S., 2016. Air entrainment and bubble statistics in breaking waves. Journal of Fluid Mechanics 801, 91–129.

Deike, L., Melville, W.K., 2018. Gas transfer by breaking waves. Geophysical Research Letters 45, 10482–10492.

Deike, L., Zhou, X., Rustogi, P., Stanley, R.H., Reichl, B.G., Bushinsky, S.M., Resplandy, L., 2025. A universal wind-wave-bubble formulation for air–sea gas exchange and its impact on oxygen fluxes. Proceedings of the National Academy of Sciences 122 (38), e2419319122.

Drazen, D.A., Melville, W.K., Lenain, L., 2008. Inertial scaling of dissipation in unsteady breaking waves. Journal of Fluid Mechanics 611, 307–332.

Duncan, J., 1981. An experimental investigation of breaking waves produced by a towed hydrofoil. Proceedings of the Royal Society of London. Series A, Mathematical and Physical Sciences 377, 331–348.

Edson, J., Jampana, V., Weller, R., Bigorre, S., Plueddemann, A., Fairall, C., Miller, S., Mahrt, L., Vickers, D., Hersbach, H., 2013. On the exchange of momentum over the open ocean. Journal of Physical Oceanography 43, 1589–1610.

Erinin, M., Néel, B., Ruth, D., Mazzatenta, M., Jaquette, R., Veron, F., Deike, L., 2022. Speed and acceleration of droplets generated by breaking wind-forced waves. Geophysical Research Letters 49, e2022GL098426.

Erinin, M.A., Wang, S.D., Liu, R., Towle, D., Liu, X., Duncan, J.H., 2019. Spray generation by a plunging breaker. Geophysical Research Letters 46, 8244–8251.

Falcon, E., Mordant, N., 2022. Experiments in surface gravity–capillary wave turbulence. Annual Review of Fluid Mechanics 54, 1–25.

Farsoiya, P.K., Magdelaine, Q., Antkowiak, A., Popinet, S., Deike, L., 2023. Direct numerical simulations of bubble-mediated gas transfer and dissolution in quiescent and turbulent flows. Journal of Fluid Mechanics 954, A29.

Fedorov, A.V., Melville, W.K., 1998. Non linear gravity-capillary waves with forcing dissipation. Journal of Fluid Mechanics 354, 1–42.

Friedlingstein, P., O'sullivan, M., Jones, M.W., Andrew, R.M., Gregor, L., Hauck, J., Le Quéré, C., Luijkx, I.T., Olsen, A., Peters, G.P., et al., 2022. Global carbon budget 2022. Earth System Science Data Discussions 2022, 1–159.

Garbe, C.S., Rutgersson, A., Boutin, J., de Leeuw, G., Delille, B., Fairall, C.W., Gruber, N., Hare, J., Ho, D.T., Johnson, M.T., Nightingale, P.D., Pettersson, H., Piskozub, J., Sahlée, E., Tsai, W.t., Ward, B., Woolf, D.K., Zappa, C.J., 2014. Transfer Across the Air-Sea Interface. Springer, Berlin Heidelberg, pp. 55–112.

Garrett, C., Li, M., Farmer, D., 2000. The connection between bubble size spectra and energy dissipation rates in the upper ocean. Journal of Physical Oceanography 30, 2163–2171.

Gemmrich, J.R., Farmer, D.M., 2004. Near-surface turbulence in the presence of breaking waves. Journal of Physical Oceanography 34, 1067–1086.

Hasselmann, K., 1962. On the nonlinear energy transfer in a gravity-wave spectrum. Journal of Fluid Mechanics 12, 481–500.

Higbie, R., 1935. The rate of absorption of a pure gas into a still liquid during short periods of exposure. Transactions of the AIChE 31, 365–389.

Hinze, J.O., 1955. Fundamentals of the hydrodynamic mechanism of splitting in dispersion processes. AIChE Journal 1, 289–295.

Ho, D.T., Wanninkhof, R., Schlosser, P., Ullman, D.S., Hebert, D., Sullivan, K.F., 2011. Toward a universal relationship between wind speed and gas exchange: gas transfer velocities measured with 3he/sf6 during the Southern Ocean gas exchange experiment. Journal of Geophysical Research: Oceans 116.

Janssen, P., 2004. The Interaction of Ocean Waves and Wind. Cambridge University Press.

Jeffreys, H., 1925. On the formation of water waves by wind. Proceedings of the Royal Society of London. Series A, Containing Papers of a Mathematical and Physical Character 107, 189–206.

Keeling, R.F., 1993. On the role of large bubbles in air-sea gas exchange and supersaturation in the ocean. Journal of Marine Research 51, 237–271.

Komen, G.J., Cavaleri, L., Donelan, M., Hasselmann, K., Hasselmann, S., Janssen, P., 1996. Dynamics and Modelling of Ocean Waves.

Kuznetsov, E.A., 2004. Turbulence spectra generated by singularities. Journal of Experimental and Theoretical Physics Letters 80 (2), 83–89.

Lamarre, E., Melville, W., 1991. Air entrainment and dissipation in breaking waves. Nature 351, 469–472.

de Leeuw, G., Andreas, E.L., Anguelova, M.D., Fairall, C.W., Lewis, E.R., O'Dowd, C., Schulz, M., Schwartz, S.E., 2011. Production flux of sea spray aerosol. Reviews of Geophysics 49.

Levich, V.G., 1962. Physicochemical Hydrodynamics.

Lewis, E.R., Schwartz, S.E., 2004. Sea salt aerosol production. Mechanisms, methods, measurements, and models. In: Geophysical Monograph 152. American Geophysical, Union, Washington, DC.

Li, T., Shen, L., 2022. The principal stage in wind-wave generation. Journal of Fluid Mechanics 934, A41.

Melville, W.K., 1996. The role of surface wave breaking in air-sea interaction. Annual Review of Fluid Mechanics 28, 279–321.

Miles, J.W., 1957. On the generation of surface waves by shear flows. Journal of Fluid Mechanics 3, 185–204.

Mostert, W., Popinet, S., Deike, L., 2022. High-resolution direct simulation of deep water breaking waves: transition to turbulence, bubbles and droplets production. Journal of Fluid Mechanics 942.

Nazarenko, S., 2011. Wave Turbulence, vol. 825. Springer.

Nazarenko, S., Lukaschuk, S., McLelland, S., Denissenko, P., 2010. Statistics of surface gravity wave turbulence in the space and time domains. Journal of Fluid Mechanics 642, 395–420.

Newell, A.C., Rumpf, B., 2011. Wave turbulence. Annual Review of Fluid Mechanics 43, 59–78.

Ni, R., 2024. Deformation and breakup of bubbles and drops in turbulence. Annual Review of Fluid Mechanics 56, 319–347.

Peng, T., Richter, D., 2017. Influence of evaporating droplets in the turbulent marine atmospheric boundary layer. Boundary-Layer Meteorology 165, 497–518.

Perlin, M., Choi, W., Tian, Z., 2013. Breaking waves in deep and intermediate waters. Annual Review of Fluid Mechanics 45, 115–145.

Perrard, S., Lozano-Durán, A., Rabaud, M., Benzaquen, M., Moisy, F., 2019. Turbulent windprint on a liquid surface. Journal of Fluid Mechanics 873, 1020–1054.

Phillips, O., 1977. The Dynamics of the Upper Ocean. Cambridge University Press.

Phillips, O., 1985. Spectral and statistical properties of the equilibrium range in wind-generated gravity waves. Journal of Fluid Mechanics 156, 505–531.

Phillips, O.M., 1957. On the generation of waves by turbulent wind. Journal of Fluid Mechanics 2, 417–445.

Phillips, O.M., 1958. The equilibrium range in the spectrum of wind-generated waves. Journal of Fluid Mechanics 4, 426–434.

Pizzo, N., Deike, L., Ayet, A., 2021. How does the wind generate waves? Physics Today 74, 38–43.

Plant, W.J., 1982. A relationship between wind stress and wave slope. Journal of Geophysical Research: Oceans 87, 1961–1967.

Rapp, R., Melville, W., 1990. Laboratory measurements of deep-water breaking waves. Philosophical Transactions of the Royal Society of London, Series A: Mathematical and Physical Sciences 331, 735–800.

Richter, D.H., Dempsey, A.E., Sullivan, P.P., 2019. Turbulent transport of spray droplets in the vicinity of moving surface waves. Journal of Physical Oceanography 49, 1789–1807.

Rivière, A., Mostert, W., Perrard, S., Deike, L., 2021. Sub-hinze scale bubble production in turbulent bubble breakup. Journal of Fluid Mechanics 917.

Rivière, A., Ruth, D.J., Mostert, W., Deike, L., Perrard, S., 2022. Capillary driven fragmentation of large gas bubbles in turbulence. Physical Review Fluids 7, 083602.

Romero, L., 2019. Distribution of surface wave breaking fronts. Geophysical Research Letters 46, 10463–10474.

Romero, L., Melville, W.K., 2010. Airborne observations of fetch-limited waves in the Gulf of Tehuantepec. Journal of Physical Oceanography 40, 441–465.

Romero, L., Melville, W.K., Kleiss, J.M., 2012. Spectral energy dissipation due to surface wave breaking. Journal of Physical Oceanography 42, 1421–1441.

Ruth, D.J., Aiyer, A.K., Rivière, A., Perrard, S., Deike, L., 2022. Experimental observations and modelling of sub-hinze bubble production by turbulent bubble breakup. Journal of Fluid Mechanics 951, A32.

Scapin, N., Wu, J., Farrar, J.T., Chapron, B., Popinet, S., Deike, L., 2026. Growth rate and energy dissipation in wind-forced breaking waves. Geophysical Research Letters 53 (1), e2025GL117901.

Snodgrass, F., Hasselmann, K., Miller, G., Munk, W.H., Powers, W., 1966. Propagation of ocean swell across the Pacific. Philosophical Transactions of the Royal Society of London, Series A: Mathematical and Physical Sciences 259, 431–497.

Sutherland, P., Melville, W.K., 2013. Field measurements and scaling of ocean surface wave-breaking statistics. Geophysical Research Letters 40, 3074–3079.

Sutherland, P., Melville, W.K., 2015. Field measurements of surface and near-surface turbulence in the presence of breaking waves. Journal of Physical Oceanography 45, 943–965.

Sverdrup, H.U., Munk, W.H., 1947. Wind, Sea and Swell: Theory of Relations for Forecasting, vol. 601. Hydrographic Office.

Takahashi, T., Feely, R.A., Weiss, R.F., Wanninkhof, R.H., Chipman, D.W., Sutherland, S.C., Takahashi, T.T., 1997. Global air-sea flux of co2: an estimate based on measurements of sea–air pco2 difference. Proceedings of the National Academy of Sciences of the United States of America 94, 8292–8299.

Terray, E.A., Donelan, M., Agrawal, Y., Drennan, W., Kahma, K., Williams, A.J., Hwang, P., Kitaigorodskii, S., 1996. Estimates of kinetic energy dissipation under breaking waves. Journal of Physical Oceanography 26, 792–807.

Thomson, J., Schwendeman, M.S., Zippel, S.F., Moghimi, S., Gemmrich, J., Rogers, W.E., 2016. Wave-breaking turbulence in the ocean surface layer. Journal of Physical Oceanography 46, 1857–1870.

Tian, Z., Perlin, M., Choi, W., 2010. Energy dissipation in two-dimensional unsteady plunging breakers and an eddy viscosity model. Journal of Fluid Mechanics 655, 217–257.

Toba, Y., 1972. Local balance in the air-sea boundary processes. Journal of Oceanography 28, 109–120.

Toba, Y., 1973. Local balance in the air-sea boundary processes: III. On the spectrum of wind waves. Journal of the Oceanographical Society of Japan 29, 209–220.

Toba, Y., 1978. Stochastic form of the growth of wind waves in a single-parameter representation with physical implications. Journal of Physical Oceanography 8, 494–507.

Veron, F., 2015. Ocean spray. Annual Review of Fluid Mechanics 47, 507–538.

Veron, F., Hopkins, C., Harrison, E., Mueller, J., 2012. Sea spray spume droplet production in high wind speeds. Geophysical Research Letters 39.

Villermaux, E., 2020. Fragmentation versus cohesion. Journal of Fluid Mechanics 898.

Villermaux, E., Wang, X., Deike, L., 2022. Bubbles spray aerosols: certitudes and mysteries. PNAS Nexus 1, 1–13.

Wang, X., Deane, G.B., Moore, K.A., Ryder, O.S., Stokes, M.D., Beall, C.M., Collins, D.B., Santander, M.V., Burrows, S.M., Sultana, C.M., 2017. The role of jet and film drops in controlling the mixing state of submicron sea spray aerosol particles. Proceedings of the National Academy of Sciences of the United States of America, 201702420.

Wanninkhof, R., 1992. Relationship between wind speed and gas exchange over the ocean. Journal of Geophysical Research: Oceans 97, 7373–7382.

Wanninkhof, R., Asher, W., Ho, D., Sweeney, C., McGillis, W., 2009. Advances in quantifying air-sea gas exchange and environmental forcing. Annual Review of Marine Science.

Woolf, D.K., Thorpe, S., 1991. Bubbles and the air-sea exchange of gases in near-saturation conditions. Journal of Marine Research 49, 435–466.

Wu, J., Popinet, S., Deike, L., 2022. Revisiting wind wave growth with fully coupled direct numerical simulations. Journal of Fluid Mechanics 951, A18.

Wu, J., Popinet, S., Deike, L., 2023. Breaking wave field statistics with a multi-layer model. Journal of Fluid Mechanics 968, A12.

Zakharov, V.E., Badulin, S.I., Hwang, P.A., Caulliez, G., 2015. Universality of sea wave growth and its physical roots. Journal of Fluid Mechanics 780, 503–535.

Zakharov, V.E., L'vov, V.S., Falkovich, G., 2012. Kolmogorov Spectra of Turbulence I: Wave Turbulence. Springer Science & Business Media.

Zakharov, V.E., L'vov, V.S., Falkovich, G., Zakharov, V.E., L'vov, V.S., Falkovich, G., 1992. Statistical description of weak wave turbulence. Kolmogorov Spectra of Turbulence I: Wave Turbulence, 63–82.

Turbulence and microphysics in atmospheric clouds

13

Raymond A. Shaw[a], *Kamal Kant Chandrakar*[b], *Prasanth Prabhakaran*[c,d,e], *Holger Siebert*[f], *and Fan Yang*[g]
[a]Department of Physics, Michigan Technological University, Houghton, MI, United States, [b]National Center for Atmospheric Research, Boulder, CO, United States, [c]Cooperative Institute for Research In Environmental Sciences (CIRES), University of Colorado, Boulder, CO, United States, [d]Chemical Sciences Laboratory, National Oceanic and Atmospheric Administration, Boulder, CO, United States, [f]Leibniz Institute for Tropospheric Research, Leipzig, Germany, [g]Environmental Science and Technologies Department, Brookhaven National Laboratory, Upton, NY, United States

13.1 Introduction

13.1.1 The cloud problem as a turbulence problem

Clouds have motivated artistic school children and poets for ages. Indeed, clouds bring an aesthetic quality to life, and we sympathize with the Cloud Appreciation Society manifesto: "Life would be dull if we had to look up at cloudless monotony day after day." It turns out that 'cloudless monotony' also results in a greatly simplified atmosphere and climate. In fact, one of the central challenges of weather forecasting and climate prediction is representing cloud processes that occur at scales much smaller than the grid resolution of the computational models (e.g., climate models may have a typical grid spacing of 100 km, whereas cloud processes occur on scales from the nanometer sizes of aerosol particles to the km to tens of km dimensions of cumulus clouds and clusters of clouds). From this perspective, the 'cloud problem' can be defined as follows: using coarse-resolution field variables such as temperature T and water vapor mixing ratio q_v as inputs, how can radiative, thermal-energy, and precipitation fluxes be calculated at each grid point? These fluxes and their spatial and temporal variability are strongly dependent on cloud properties and microphysical rates, and in turn directly influence the fluid motions of the atmosphere and thus are nonlinearly coupled to the formation of clouds themselves. The term 'microphysics' refers to the processes relevant to the formation and evolution of cloud particles, for example, the formation of cloud droplets on aerosol particles (activation), the growth of cloud droplets by water vapor condensation, the growth of drizzle or rain by the collision and coalescence of cloud droplets, the evaporation of droplets due to turbulent entrainment of dry air into a cloud, and even the breakup of raindrops to repopulate cloud and drizzle populations. The situation becomes much more complex when it is recognized that Earth's atmospheric conditions span temperatures and

[e] Now at the Centre for Atmospheric Sciences, Indian Institute of Technology Delhi.

Environmental Turbulence. https://doi.org/10.1016/B978-0-32-395873-8.00020-6
Copyright © 2026 Elsevier Inc. All rights are reserved, including those for text and data mining, AI training, and similar technologies.

pressures where water exists stably as ice, and therefore microphysical processes for the creation of the ice phase and for conversion between liquid and ice phases must be known. In most coarse-resolution models, these processes are represented by rate coefficients that attempt to capture the detailed microphysical and turbulence interactions. Specifically, accretion, condensation, deposition, evaporation, freezing, collection, melting, riming, and autoconversion rates need to be defined (Sui et al., 2020). In addition, fall speeds for all particle sizes, including nonspherical ice particles, must be specified in order to calculate precipitation fluxes.

The cloud problem, as envisioned above, is effectively a turbulence problem. One way to state the 'turbulence problem' is to predict fluxes (e.g., momentum, thermal-energy, and scalar constituents) given a knowledge of initial and boundary conditions for coarsely-resolved field variables (Wyngaard, 2010). As with turbulence, an Eulerian viewpoint of clouds results in a closure problem, for which 'Reynolds stresses' or covariances of dynamic and cloud microphysical variables must be known. Moreover, as with turbulence, there is a nonlinear coupling of scales that makes cloud parameterization challenging. In fact, it is widely recognized that "The principal sticking points in predicting climate are the subgrid-scale turbulent and convective motions in the oceans and atmosphere" (Schneider et al., 2021). This chapter provides an overview of cloud microphysical processes that are influenced by turbulence. It is necessarily broad-brushed, rather than exhaustive, meant to give the reader a general knowledge of fluid mechanics and a sense for cloud-turbulence interactions. At least three microphysical processes are directly influenced by turbulence: growth by condensation, including effects of entrainment, growth by collision-coalescence, and the removal of particles by sedimentation. To present a coherent picture with sufficient depth, we emphasize the condensation process. We cover, in turn, the dimensionless space representing cloud processes (Section 13.1.2), methods for measuring turbulence properties in clouds (Section 13.2), modeling approaches for turbulent clouds (Section 13.3), turbulence and the supersaturation field (Section 13.4), turbulence effects on activation and droplet growth (Section 13.5), and the role of turbulence in entrainment and its influence on cloud properties (Section 13.6). To limit the scope to what can be covered in a single chapter, our focus is on microphysical connections to turbulence in warm clouds; all of these topics can be extended to mixed-phase clouds that contain both supercooled liquid and ice (Korolev et al., 2017). Furthermore, we do not consider large-scale dynamical structure of cloud flows or the feedback to cloud circulation from turbulent entrainment, as important and fascinating as these topics are. We point the reader toward other sources for the larger-scale perspectives (Heintzenberg and Charlson, 2009; Siebesma et al., 2020).

13.1.2 Clouds and turbulence: the dimensionless space

We begin our discussion by considering the dimensionless variables relevant to cloud microphysical processes occurring in a turbulent flow. Perhaps surprisingly, it is not customary in the field of cloud physics to take this viewpoint, albeit with some exceptions (Pinsky et al., 2016; Fries et al., 2021; Thomas et al., 2022). For the purposes of a book on turbulence, however, we feel it is instructive to introduce some of the key variables and processes in this manner. The turbulent properties of the atmospheric convection that give rise to cloud formation can be characterized in the usual way, via the turbulence Reynolds number $Re = u'\ell/\nu$, where u' is the root-mean-square (RMS) velocity, ℓ is the large eddy length scale, and ν is the kinematic viscosity of air. The turbulence itself is typically generated through shear production or thermal convection. When in the presence of a stabilizing temperature inversion, such as at the top of a stratocumulus cloud, the shear production is often expressed through the Richardson number $Ri = (g/\theta)(\partial\theta/\partial z)(\partial U/\partial z)^{-2}$, where g is the gravitational acceleration, θ is the potential temperature, U is the horizontal wind speed, and z is the vertical coordinate. (It should be noted that in some communities Ri is used as a measure of the ratio of buoyancy to shear production, so it is thought of as a characteristic of buoyancy generation rather than shear; the context thus matters.) For thermal convection, the Rayleigh number $Ra = \Delta\theta g H^3(\theta\kappa\nu)^{-1}$ is relevant, where H is the depth of the convective layer and κ is the thermal diffusivity of air. Note that Ra can be written using forms of the potential temperature that include effects of water vapor density and phase changes (Niedermeier et al., 2018). Similar holds true for the resulting Nusselt number, the dimensionless heat flux for the turbulent convection (Thomas et al., 2022). In the Earth's atmosphere, turbulent flows are characterized by nearly fixed values of material coefficients, such that the Prandtl number $Pr = \nu/\kappa$ and Schmidt number $Sc = \nu/D$, where D is the water vapor diffusivity, are close to unity. Interestingly, the fact that the Lewis number $Le = \kappa/D = Sc/Pr$ departs slightly from unity is put to advantage in the design of instruments for studying aerosol activation and ice nucleation (Nenes et al., 2001; Stratmann et al., 2004). An additional dimensionless parameter that depends on material coefficients of central importance to turbulent clouds is the Stefan number $Ste = c_p\Delta T/L_v$, where c_p is the specific heat at constant pressure for air and L_v is the latent heat (enthalpy) of vaporization for water. This ratio of sensible and latent heat is responsible for the strong coupling between the equations for water vapor mass transfer and for thermal energy, thereby strongly influencing the buoyancy field and the turbulence itself.

The mechanics of a single droplet in a turbulent flow can be understood by considering the approximate equation of motion

$$\frac{dv_i}{dt} = \frac{u_i - v_i}{\tau_d} + g_i \tag{13.1}$$

where v_i is the ith component of the droplet velocity, u_i is the component of fluid velocity at the droplet position, and τ_d is the droplet viscous relaxation time (Tsai, 1957; Maxey and Riley, 1983; Shaw, 2003). For the large density ratio ρ_l/ρ_a for water droplets in air, it is expected that other terms in the equation of motion, such as added mass and Basset history force, can be neglected. Whether the history force plays a significant role in droplet collision rates is however still undetermined (Hill, 2005). Cloud droplets are within the $Re \ll 1$ creeping flow limit until the size at which collision growth becomes dominant. In that Stokes drag regime, the droplet response time is defined as $\tau_d = (2/9)(\rho_l/\rho_a)(r^2/\nu)$. The acceleration (or vorticity) spectrum peaks in the dissipation range so we non-dimensionalize Eq. (13.1) with the

Kolmogorov time and velocity scales, $\tau_\eta = (\nu/\varepsilon)^{1/2}$ and $u_\eta = (\nu\varepsilon)^{1/4}$, where ε is the turbulent kinetic energy dissipation rate (W/kg). The equation of motion, in terms of the Kolmogorov-normalized variables denoted by a double-tilde, then becomes

$$\mathrm{St}\frac{d\tilde{\tilde{v}}_i}{d\tilde{\tilde{t}}} = \tilde{\tilde{u}}_i - \tilde{\tilde{v}}_i - \mathrm{Sv}\hat{e}_i \tag{13.2}$$

where $\hat{e}_i$ is a unit vector, $\mathrm{St} = \tau_d/\tau_\eta$ is the Stokes number, and $\mathrm{Sv} = v_T/u_\eta$ is the settling parameter; here $v_T = \tau_d g$ is the Stokes terminal speed. In the limit $\mathrm{St} \ll 1$, which is valid for cloud droplets prior to collision growth,

$$v_i \approx u_i + \tau_d g_i - \tau_d a_i \tag{13.3}$$

or

$$\tilde{\tilde{v}}_i \approx \tilde{\tilde{u}}_i - \mathrm{Sv}\hat{e}_i - \mathrm{St}\tilde{\tilde{a}}_i \tag{13.4}$$

where a_i is the Lagrangian fluid acceleration. This suggests that cloud droplets follow the fluid flow, with a steady drift due to gravitational settling and a 'lag' resulting from the inability to perfectly respond to fluid accelerations.

To introduce dimensionless variables relevant to droplet growth, we use the equations for cloudy Rayleigh–Bénard convection, with the intention that the standard R-B equations are familiar and therefore this brief treatment will illuminate the cloud-specific terms. A more detailed treatment is provided by Thomas et al. (2022). The dimensionless equations for momentum and temperature have the expected dependence on Ra and Pr:

$$\frac{\partial \tilde{\mathbf{u}}}{\partial \tilde{t}} + \tilde{\mathbf{u}} \cdot \tilde{\nabla}\tilde{\mathbf{u}} = -\tilde{\nabla}\tilde{p} + \left(\tilde{T} + B_v\,\tilde{q_v} - B_l\right)\hat{e}_3 + \sqrt{\frac{Pr}{Ra}}\tilde{\nabla}^2\tilde{\mathbf{u}} \tag{13.5}$$

and

$$\frac{\partial \tilde{T}}{\partial \tilde{t}} + \tilde{\mathbf{u}} \cdot \tilde{\nabla}\tilde{T} = \frac{1}{\sqrt{Ra\,Pr}}\tilde{\nabla}^2\tilde{T} + \frac{\tau_t}{Ste}\,\dot{q}_l \tag{13.6}$$

where $\hat{e}_3$ denotes the vertical component of the unit vector (opposite the direction of the gravitational acceleration). The non-dimensional scaled variables (denoted by tildes) of length, time, velocity, temperature, and water vapor for moist Rayleigh–Bénard convection are $\tilde{L} = L/H$, $\tilde{t} = tw/H$, $\tilde{\mathbf{u}} = \mathbf{u}/w$, $\tilde{T} = (T - \overline{T})/\Delta T$, and $\tilde{q_v} = (q_v - \overline{q_v})/\Delta q_v$, where ΔT and Δq_v are the temperature and water vapor mixing ratio differences between the bottom and top boundaries, respectively. We have used the height of the convective layer (H) and the free-fall velocity ($w = \sqrt{g\beta\Delta T H}$) as the scales for length and velocity. The buoyancy term contains dimensionless parameters for water vapor $B_v = \epsilon\Delta q_v/(\beta\Delta T)$ and for liquid water $B_l = q_l/(\beta\Delta T)$, where β is the thermal expansion coefficient of air and $\epsilon = m_d/m_v - 1$, where m_d and m_v are the molecular weights of dry air and water vapor, respectively. The last term in the temperature equation represents the latent heating caused by phase changes, which directly influences the buoyancy term in the momentum equation according to the magnitude of the Stefan number Ste. We have written that term with a turbulence time scale $\tau_t = H/w$ and will next introduce the relevant time scale for the cloud condensation process, anticipating that this ratio of time scales will be of significance to the coupling between cloud growth and evaporation to the temperature field and, ultimately, to the buoyancy field.

The exponential 'phase relaxation' time for a cloud of droplets with number density n and mean radius $\bar{r}$ is $\tau_c = (4\pi\xi n\bar{r})^{-1}$, where ξ is a factor associated with thermal and vapor diffusion during the droplet growth (Rogers and Yau, 1989). A population of droplets will grow or evaporate depending on the water vapor supersaturation, which can be understood as the relative humidity difference from 100% (defined more precisely in Section 13.4). For an environment with supersaturation of s, the rate of change of liquid water mixing ratio is $\dot{q}_l = d/dt(4\pi r^3 n\phi/3) = 4\pi\xi nrs\phi$, where $\phi = \rho_l/\rho_a$, ρ_l is the density of water and ρ_a is the density of air, and we have assumed n is constant (Rogers and Yau, 1989). The ratio of the turbulence time scale to the phase relaxation time scale is a Damköhler number, $Da = \tau_t/\tau_c$ (Chandrakar et al., 2016). Therefore Eq. (13.6) can be rewritten as

$$\frac{\partial \tilde{T}}{\partial \tilde{t}} + \tilde{\mathbf{u}} \cdot \tilde{\nabla}\tilde{T} = \frac{1}{\sqrt{Ra\,Pr}}\tilde{\nabla}^2\tilde{T} + \frac{1}{Ste}\,\phi\,Da\,s \tag{13.7}$$

For $Da \gg 1$ cloud droplets respond efficiently to changes in the surrounding environment, whereas for $Da \ll 1$ the turbulent environment changes faster than the droplets can respond to it. We refer to these regimes as fast and slow microphysics, respectively. We note that the condensation term now consists of dimensionless quantities Ste, ϕ, Da, and s. Similarly, the equation for water vapor mixing ratio becomes

$$\frac{\partial \tilde{q}_v}{\partial \tilde{t}} + \tilde{\mathbf{u}} \cdot \tilde{\nabla}\tilde{q}_v = \frac{1}{\sqrt{Ra\,Sc\,Le}}\tilde{\nabla}^2\tilde{q}_v - \frac{1}{\Delta q_v}\,\phi\,Da\,s \tag{13.8}$$

Using $\tilde{q}_l = q_l/\Delta q_v$, the non-dimensional form of the equation for liquid water mixing ratio is,

$$\frac{\partial \tilde{q}_l}{\partial \tilde{t}} + \tilde{\mathbf{u}} \cdot \tilde{\nabla}\tilde{q}_l = Rou\tilde{\nabla}\,\tilde{q}_l\hat{e}_3 + \frac{1}{\Delta q_v}\phi\,Da\,s. \tag{13.9}$$

Here an additional dimensionless group appears, the ratio of the droplet terminal speed and the convection free-fall speed, which is sometimes referred to as the Rouse number $Rou = v_T/w$ (Richter and Chamecki, 2018). It can be related to the settling parameter Sv introduced previously.

In the following section, we consider measurements of cloud microphysical and turbulence variables that help define the range of the dimensionless space that must be considered for the cloud problem.

13.2 Cloud turbulence measurements, what we know about the parameter space

Most cloud types in the atmosphere are naturally embedded in turbulent flows, usually with strong wind shear and buoyancy as the main sources for turbulent kinetic energy. Enhanced turbulence often occurs especially at cloud edges and the larger turbulent structures are either visible to the naked eye or can be felt when flying through them. Fig. 13.1 shows a typical example of a section of a cloud field in which the filaments of turbulent flow are visualized by cloud droplets.

Figure 13.1 Filamentary cloud structures indicating strong turbulence and mixing. The photo was taken from a helicopter flying just above a cloud field over the Azores. Photo credit: H. Siebert.

The most direct insight into cloud turbulence has been gained by various types of aircraft observations using in-situ techniques (Wendisch and Brenguier, 2013). Recently, ground-based remote sensing methods such as LIDAR and RADAR have been increasingly used to describe the spatiotemporal evolution of turbulence in different cloud types (e.g., Shupe et al., 2012; Borque et al., 2016). In addition to the large-scale structure, such as regions of up- and downdrafts, turbulent processes on scales throughout the energy cascade down to cm-scales are now increasingly becoming the focus of current research (e.g., Gotoh et al., 2021). The aim of this section is to describe the parameter space and the spatio-temporal distribution of turbulence in different cloud types. In addition to purely dynamic aspects such as the spatial distribution of, for example, energy dissipation rates, the interaction of cloud droplets with a turbulent flow is also briefly discussed.

The mean turbulent kinetic energy dissipation rate ε, the measure of the extent to which kinetic energy is transferred from large to small vortices and ultimately dissipates into heat at the dissipation length scale $\eta = f(\varepsilon)$, is one of the key parameters describing the degree of turbulence. The dissipation rate can be approximated as

$$\varepsilon = u'^3/l \tag{13.10}$$

with u' being a typical fluctuation of the wind velocity field (RMS velocity) and l a large-eddy length scale. However, the definition of l is not straightforward and for atmospheric observations, often the integral length scale $\mathcal{L} = \int_0^\infty dr \rho_u(r)$ is used, with the auto-correlation function $\rho_u(r) = \langle u'(x+r) \cdot u'(x) \rangle_x / \sigma_u^2$ (Tennekes and Lumley, 1972). Integral length scales typically range from 10 m for more stratiform cloud types up to 1000 m for deep convective clouds and vertical wind velocities range from almost zero (non-turbulent cloud parts) up to $10\,\mathrm{m\,s^{-1}}$ in updrafts of deep convective clouds. Given those values for u' and l, Fig. 13.2 displays the expected parameter space for energy dissipation rates with ε ranging from about 10^{-4} to $10^{0}\,\mathrm{m^2\,s^{-3}}$ which is well in agreement with aircraft observations (e.g., MacPherson and Isaac, 1977; Caughey et al., 1982; Siebert et al., 2006). Given this range for ε we can finally estimate the Kolmogorov dissipation length scale $\eta = (\nu^3/\varepsilon)^{1/4}$, originally defined in Chapter 1, which can range from a few tenths of a millimeter to a few millimeters.

Besides ε, the Reynolds number Re is another main controlling parameter describing the general structure of a turbulent flow. For length scales where viscosity starts playing a dominant role, the Taylor microscale

$$\lambda = \sigma_u \sqrt{15\nu/\varepsilon} \tag{13.11}$$

is used as the appropriate length scale in combination with σ_u as the characteristic velocity, resulting in the Taylor–Reynolds number

$$Re_\lambda = \sigma_u^2 \sqrt{15/\nu\varepsilon} \tag{13.12}$$

In clouds, the high values of Re_λ ranging from 10^3 to 10^5 (e.g., Shaw, 2003; Siebert et al., 2006, 2010) are achieved primarily because of the large scales involved ($Re_\lambda \sim \varepsilon^{1/6} l^{2/3} \nu^{-1/2}$), whereas typically in the laboratory, the scale is confined, and the dissipation rate is high.

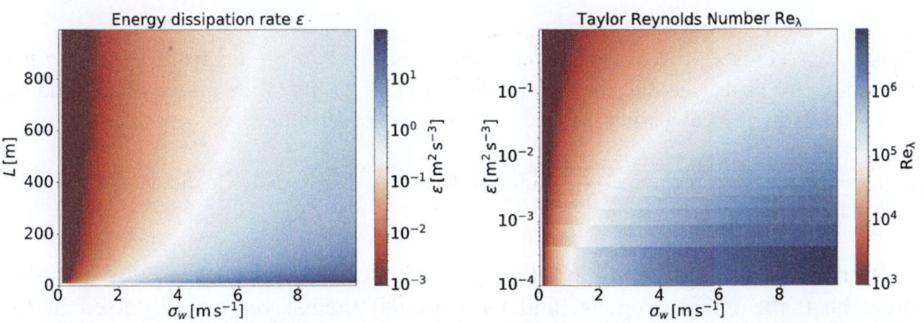

Figure 13.2 Energy dissipation rate $\varepsilon = \sigma_w^3/L$ (color-scale) over the (L, σ_w) parameter space (left), and the corresponding Taylor–Reynolds number Re_λ (color-scale) as a function of ε and σ_w (right).

The comparatively high Re_λ has direct consequences for the small-scale turbulence in a cloudy environment: with increasing Re_λ, a mean ε is no longer sufficient to describe the turbulent structure; instead a local dissipation rate ε_r becomes more important, where ε_r is defined as an independent parameter averaged over the length scale r. Laboratory flows suggest that ε_r is typically log-normally distributed in space and time, which is due to the fact that locally increased turbulence can occur, a phenomenon also called intermittency (Wyngaard, 2010; Kolmogorov, 1962). This concept of intermittency extends to scalar fields as well, which are relevant to processes like droplet growth (Warhaft, 2000). Fig. 13.3 shows probability functions (PDFs) of $\ln \varepsilon_\tau$ (here, dissipation is averaged over time intervals τ instead of r) measured in atmospheric clouds (Siebert et al., 2006). The PDFs can be well approximated by a lognormal fit, a fact also observed in well-defined laboratory flows, suggesting that findings from laboratory flows (without cloud droplets) can be applied to clouds to a reasonable extent (Siebert et al., 2010). This does not mean, however, that on small length scales, for example, the evaporation of cloud droplets and the latent heat thus released have no local effect on turbulence. Indications of such effects have been observed in laboratory experiments and in the field (Malinowski et al., 2008; Siebert et al., 2015a), and have been explored theoretically (Gotoh et al., 2021).

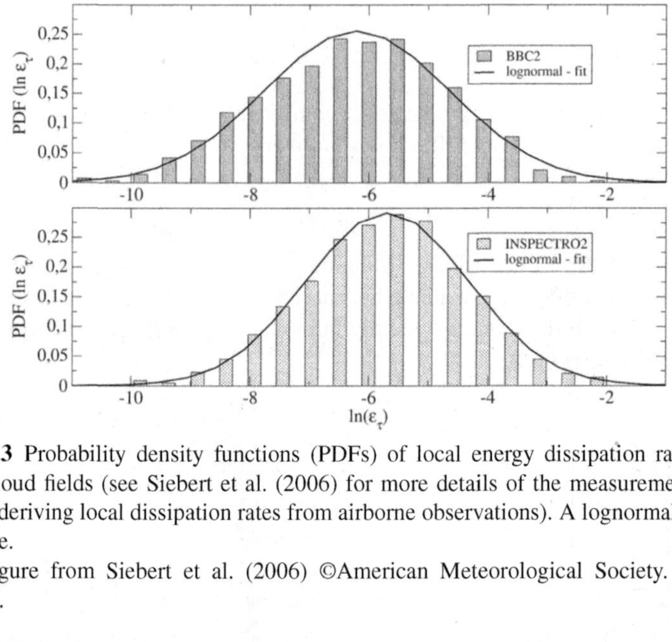

Figure 13.3 Probability density functions (PDFs) of local energy dissipation rates for two different cloud fields (see Siebert et al. (2006) for more details of the measurements and the method of deriving local dissipation rates from airborne observations). A lognormal fit is given as reference.
Source: Figure from Siebert et al. (2006) ©American Meteorological Society. Used with permission.

Besides the fact that cloud droplets can influence the turbulent flow by releasing latent heat, the cloud droplets (and ice crystals) themselves are dispersed in the turbulent flow, which they can only follow to a limited extent. The stronger the turbulence (high Re_λ), the stronger are the deviations of the droplet trajectories from the streamlines (as previously discussed, see Eq. (13.4)). To assess to what extent turbulence has an influence on the motion of the cloud droplets in the turbulent flow, it is necessary to estimate both the dimensionless parameters St and Sv and the Lagrangian acceleration a_i at the location of the droplet. Although previously defined, we repeat the Stokes and settling numbers here to emphasize the relevant variables measured in clouds, namely the droplet diameter and the turbulence energy dissipation rates. The Stokes number has a dependence

$$\mathrm{St} = \frac{\tau_d}{\tau_\eta} \propto d^2\sqrt{\varepsilon} \tag{13.13}$$

and the settling parameter has a dependence

$$\mathrm{Sv} = \frac{\tau_d \cdot g}{v_\eta} \propto \frac{d^2}{\varepsilon^{1/4}} \tag{13.14}$$

The variance of a_L can be calculated using ε, with the intermittency effects parameterized using Re_λ:

$$\sigma_{a_L}^2 = \varepsilon^{3/2}\nu^{-1/2}(2.5\mathrm{Re}_\lambda^{0.25} + 0.08\mathrm{Re}_\lambda^{0.11}) \tag{13.15}$$

which increases with Re_λ and ε (Hill, 2002).

In the absence of turbulence, cloud droplets are assumed to fall at their constant terminal velocity, and only droplets with significant differences in diameter collide. Since both dimensionless parameters are directly proportional to the square of the droplet diameter, it is logical that not only the degree of turbulence and thus essentially the cloud type is responsible for Sv and St, but also, given the same type, the cloud environment, which determines the mean droplet diameter. For a thermodynamically determined liquid water content, the mean droplet diameter depends on the droplet number concentration. A cloud in a clean environment will therefore always have a larger mean droplet diameter and a related increase in the Stokes and settling parameters. Fig. 13.4 shows a scatter plot of $\langle \mathrm{Sv}|\mathrm{St} \rangle$ points for several cloud samples, where each point represents an average over one second. The CARRIBA data represent trade wind clouds in comparatively clean environments with $\overline{d} \approx 15 - 30$ μm, whereas the Zugspitze data are more continental and therefore relatively more polluted resulting in $\overline{d} \approx 10 - 15$ μm.

To conclude, we consider the dimensionless number relevant to condensation growth in a turbulent medium, the Damköhler number, which appears in Eqs. (13.7)–(13.9). Desai et al. (2019b) used digital in-line holography to measure droplet number concentration and the droplet size distribution in marine boundary layer clouds. Using these measurements, distributions of phase relaxation times were calculated, and were observed to be quite broad for many clouds. Velocity measurements were used to estimate turbulence correlation times, which were then compared with phase relaxation times. It was found that clouds exist primarily in the $Da \gg 1$ limit, but that Da approaches and even is smaller than unity in some clouds, especially when drizzle is present. Signatures of stochastic condensation, meaning cloud droplet growth influenced by fluctuations in supersaturation, were observed and became more pronounced as Da approached unity. These signatures include increasing relative size dispersion and increasing droplet size with decreasing number density. These measurements are therefore consistent with the view that droplet growth due to stochastic condensation plays a role in broadening of cloud droplet size distributions and thereby

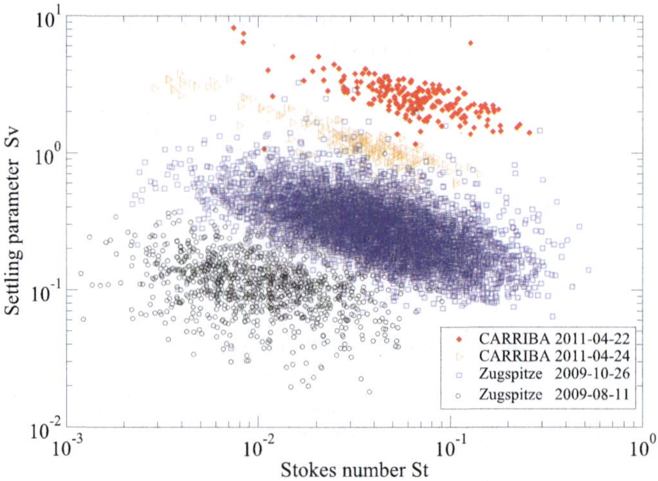

Figure 13.4 Dimensionless Stokes–settling parameter space. Each $\langle St, Sv \rangle$ point denotes a 1 s average of cloud data. CARRIBA data represent clean (red) and slightly polluted (yellow) trade-wind cumulus conditions (Siebert et al., 2015a), whereas 'Zugspitze' data (Siebert et al., 2015b) correspond to more polluted continental conditions with smaller droplets.

helping to jump the bottleneck between droplet growth due to diffusion and collision-coalescence. Further perspectives on the Damköhler number in the context of droplet size distribution response to entrainment are discussed in Section 13.6.

13.3 Modeling approaches for turbulent clouds

Numerical simulation has become a crucial approach to exploring cloud-turbulence interactions at various scales, from the core region of a turbulent cloud to the full depth of the planetary boundary layer. Successful simulation of turbulent clouds requires proper simulation of a turbulent flow, along with a suitable representation of hydrometeors and associated microphysical processes. There have been several comprehensive review papers and books on the simulation of turbulence (e.g., Fox and Lilly, 1972; Kajishima and Taira, 2017) and cloud (e.g., Khain et al., 2000; Khain and Pinsky, 2018; Morrison et al., 2020). In this section, we will give an overview of turbulent cloud simulation, mainly focusing on the introduction of underlying physical concepts, while details of numerical techniques and algorithms, although quite important, are not addressed here.

Air motion in turbulent clouds can be described by the Navier–Stokes (NS) equation with the Boussinesq approximation, in which the air density is treated as a constant except when it is coupled with gravity in the buoyancy term. The governing equations for cloudy Rayleigh–Bénard convection were already discussed in Eqs. (13.5)–(13.8), and here we show a general form:

$$\nabla \cdot \mathbf{u} = 0 \tag{13.16}$$

and

$$\frac{\partial \mathbf{u}}{\partial t} + (\mathbf{u} \cdot \nabla)\, \mathbf{u} = -\frac{1}{\rho_0} \nabla p + \nu \nabla^2 \mathbf{u} + \mathbf{B} \tag{13.17}$$

where $\mathbf{u}$ is the velocity, ρ_0 is the density, p is the pressure, and ν is the kinematic viscosity of air. The buoyancy term is

$$\mathbf{B} = -\mathbf{g} \left[\frac{T - T_0}{T_0} + \epsilon \left(q_v - q_{v0} \right) - q_l \right] \tag{13.18}$$

where T and q_v are air temperature and water vapor mixing ratio, and T_0 and q_{v0} are the reference temperature and water vapor mixing ratio. q_l is the liquid water mixing ratio, quantifying the condensed liquid mass per unit air mass. Governing equations for T and q_v in turbulent clouds are

$$\frac{\partial T}{\partial t} + (\mathbf{u} \cdot \nabla)\, T = \kappa \nabla^2 T + \frac{L_v}{c_p} \dot{q}_l \tag{13.19}$$

and

$$\frac{\partial q_v}{\partial t} + (\mathbf{u} \cdot \nabla)\, q_v = D \nabla^2 q_v - \dot{q}_l \tag{13.20}$$

where L_v is the enthalpy of vaporization, c_p is the specific heat of air at constant pressure, and κ and D are the diffusivities of temperature and water vapor, respectively. The condensation rate $\dot{q}_l = dq_l/dt$ has units of kg kg^{-1} s^{-1} and must be determined from the microphysics, as discussed below.

T and q_v are the two crucial scalars that drive cloud-turbulence interactions at the droplet level. T and q_v fields have a strong impact on the microphysical processes that will be discussed later, while they are also influenced by the microphysical processes. For example, condensation growth of cloud droplets is controlled by environmental supersaturation, which is a function of T and q_v. Meanwhile, condensation can affect T and q_v through latent heat release (last term in Eq. (13.19)) and consumption of water vapor (last term in Eq. (13.20)), respectively, which can further affect air motion through buoyancy (Eq. (13.17)).

Theoretically, if we know the boundary conditions and initial state of a turbulent flow precisely, we can predict the evolution of the turbulent flow by solving the governing equations (Eqs. (13.16)–(13.20)). However, directly solving the fully resolved equations is not practical for the simulation of atmospheric clouds due to the wide range of turbulence and microphysics length and time scales involved. For example, the Reynolds number varies 10^6–10^8 in atmospheric clouds (Shaw, 2003), implying that the length ratio between the largest and smallest turbulent eddies is $10^{4.5}$–10^6, which sets a challenge to fully resolve all eddies down to the Kolmogorov scale (a few mm) due to computational costs. Furthermore, even if the turbulence could be fully resolved, the droplet-scale interactions require orders of magnitude higher resolution, reaching the scale of $\sim$ 1 µm or even less to resolve the continuum breakdown relevant to droplet collision (How et al., 2021). Therefore, in practice, either small eddies or large eddies are resolved in turbulent models. Direct numerical simulation (DNS) and large eddy simulation (LES) are the two main types of turbulent models that have been used in the atmospheric science community to simulate turbulent clouds, as shown in Fig. 13.5. Specifically, DNS solves the Navier-Stokes equations directly to resolve the smallest turbulent eddies, while a forcing term is usually added in the momentum equation (Eq. (13.17)) to drive the flow, accounting for the energy input from unrepresented large scales (Moin and Mahesh, 1998). LES solves the spatially filtered governing equations to resolve larger turbulent eddies, while the impact of unresolved small eddies (i.e., smaller than the filter scale) on fluid motion is parameterized through subgrid-scale models (Mason, 1994). An idealized linear-eddy model, capable of representing turbulence over the full range of scales within the energy cascade, has been adapted to include cloud droplet growth and entrainment (Krueger, 1993; Krueger et al., 1997; Tölle and Krueger, 2014). It has now been implemented as a sub-grid parameterization within LES, thereby extending the microphysics-relevant fluctuations to the smallest scales (Hoffmann and Feingold, 2019).

One challenge of cloud simulation is to properly represent the population of cloud particles with various shapes and a wide range of sizes (Morrison et al., 2020). It is impossible to resolve all particles in atmospheric clouds due to the fact that there are numerous hydrometeors therein. For example, cloud droplet number concentration is on the order of about 100 cm^{-3}. Even in a small cumulus cloud with a size of 100 m $\times$ 100 m $\times$ 100 m (height), the total number of droplets is on the order of 10^{14}. If the property of each droplet (e.g., radius) is represented by only one real number (one Byte), at least 100 TB of memory is needed to record all droplets. Hydrometeors with various shapes, such as round cloud droplets, hexagonal ice plates, ice needles, and dendrite snowflakes, set another challenge for cloud simulations. Ice particles and the associated microphysical processes are still poorly represented in atmospheric models, and therefore, we mainly focus on the simulation of warm turbulent clouds here. Even in warm clouds, the size of liquid droplets varies over three orders of magnitude, from newly formed cloud droplets of a few μm to raindrops of a few mm.

Several approaches have been developed to represent cloud droplets in turbulent models, as illustrated in Fig. 13.5. The particle-resolved method is a way to directly calculate the motion and microphysical changes of individual cloud droplets. However, this method is computationally expensive, and it is only practical in a small simulation domain when using DNS, so that the total number of droplets is computationally manageable (e.g., Götzfried et al., 2017; Chen et al., 2018). The Lagrangian particle method (also named superdroplet method), which can be implemented in both DNS-based and LES-based turbulent models, is similar to the particle-resolved method, except that each Lagrangian particle (or superdroplet) represents a large number (called the multiplicity factor) of droplets with identical properties (e.g., size and mass) (e.g., Arabas and Shima, 2013; Chandrakar et al., 2021). By setting a proper multiplicity factor, this method can produce a good statistical representation of cloud properties with affordable computational costs. Bin and bulk microphysics methods are two Eulerian-based approaches, and they can only be implemented in LES, in which a sufficient amount of cloud droplets exists in the

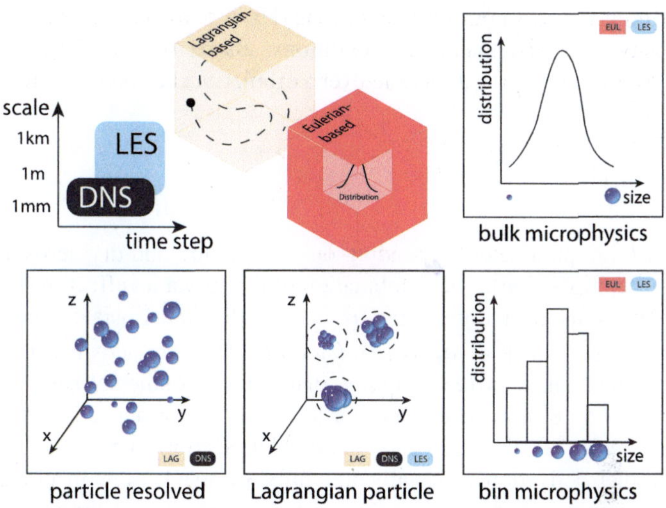

Figure 13.5 Schematic figure showing different levels of complexity in turbulent cloud simulation. The turbulent velocity and scalar fields are usually modeled through direct numerical simulation (DNS) and large-eddy simulation (LES), along with either Lagrangian-based (LAG) or Eulerian-based (EUL) approaches to simulate cloud microphysical properties and processes, including four different microphysical methods (from complex to simple): particle-resolved method, Lagrangian particle method, bin microphysics method, and bulk microphysics method. The first two can be used in DNS, while the last three can be applied in LES for cloud simulations.

model grid cell, allowing the use of small amounts of variables to adequately represent the statistics of droplet size distribution. The bin microphysics method employs a few tens of fixed Eulerian bins to represent the droplet size distribution. The droplet distribution (or number concentration or mass concentration) in each Eulerian bin is a prognostic variable, which can be affected by microphysical processes (e.g., condensation and collision). The bulk microphysics method usually assumes a specific shape of the droplet size distribution and calculates the evolution of moments of the full distribution, such as droplet number concentration (0th moment) and mean size (1st moment).

Another challenge in turbulent cloud simulation, as discussed by Morrison et al. (2020), is our incomplete understanding of aerosol-cloud-turbulence interactions (especially in cold clouds) at the process level, resulting in large uncertainties in the representation of those microphysical processes in cloud-resolving models. To limit the scope of this section, we mainly focus on the simulation of microphysical processes in warm turbulent clouds, including droplet activation from aerosol particles, the growth of cloud droplets by condensation, and the growth of cloud droplets by collision coalescence.

Droplets in warm clouds are formed from aerosol particles that commonly consist of water-soluble compounds, such as sodium chloride and ammonium sulfate. Those aerosol particles first absorb water and become aqueous particles, also called haze droplets, at high relative humidity conditions. Haze particles can then be activated as cloud droplets at supersaturated conditions. Based on the Köhler theory, the equilibrium water vapor saturation ratio over a droplet can be approximated as (Rogers and Yau, 1989),

$$S_{eq}(r) = 1 + \frac{a}{r} - \frac{b}{r^3} \tag{13.21}$$

where a and b are parameters depending on temperature and dry aerosol properties (e.g., size and composition). The combination of the curvature effect (a/r) and solute effect (b/r^3) results in a critical radius $r_c = \sqrt{3b/a}$ associated with a critical saturation ratio $S_c = 1 + \sqrt{4a^3/(27b)}$. Cloud droplets are activated if the particle radius r is larger than r_c, which requires that the environmental saturation ratio is larger than S_c. The Köhler theory and its implications for activation in a turbulent environment are further discussed and illustrated in Section 13.5.

Cloud droplet activation is usually not fully resolved, even in cloud-resolving models. For example, in bulk microphysics approaches, cloud droplet activation is usually parameterized based on Twomey-type parameterization (Twomey, 1959), in which the number of activated cloud droplets is a power-law function of environmental saturation ratio (e.g., $n = \alpha S^{\beta}$). In bin microphysics approaches, aerosol particles are represented in a few tens of bins. Particles in one specific aerosol bin are activated as cloud droplets if the environmental saturation ratio is larger than its corresponding S_c. Particle-resolved methods and Lagrangian particle methods can naturally resolve the activation process by directly calculating the growth of haze and cloud droplets, which will be discussed next.

The governing equation of the growth of an aqueous droplet (i.e., either haze or cloud) by condensation can be simplified as (Rogers and Yau, 1989)

$$\frac{dr}{dt} = \frac{1}{r}\xi(S - S_{eq}) \tag{13.22}$$

where r is droplet radius, ξ is the growth factor, S is the environmental water vapor saturation ratio, and S_{eq} is the equilibrium water vapor saturation ratio (Eq. (13.21)). Eulerian-based microphysics schemes usually parameterize droplet activation and only calculate the condensational growth of cloud droplets by ignoring the curvature and solute effects (i.e., $S_{eq} = 1$), while Lagrangian-based microphysics schemes can resolve the activation process by directly calculating the condensational growth of both cloud and haze particles via Eq. (13.22). Transport of cloud droplets in a turbulent field due to advection and gravitational sedimentation is also handled differently in Lagrangian-based and Eulerian-based microphysics approaches. Specifically, in Lagrangian-based approaches (i.e., particle-resolved method and Lagrangian particle method), the Lagrangian evolution of a particle is to solve Eq. (13.2). In Eulerian-based approaches (i.e., bin and bulk microphysics methods), advection schemes are used to transport all scalars related to hydrometeors, such as number concentration and mass concentration; a drawback here can be the numerical diffusion suffered from transport and sedimentation within the Eulerian space.

Cloud droplet growth by collision and coalescence is the key process responsible for rain precipitation in warm clouds. The impact of turbulence on droplet collision is still not fully understood, and the way to simulate droplet collision is not unified. In particle-resolved methods, trajectories of each droplet are calculated. Two droplets collide when their relative distance is smaller than the sum of their radii (Franklin et al., 2005). Such approaches simulate the collision of individual droplet pairs by assuming Stokes-flow interactions and do not resolve the full hydrodynamic interactions, and even so, they are computationally expensive. The way to calculate droplet collisions when using Lagrangian particle methods is to calculate the collision probability of all Lagrangian droplet pairs in each computational grid cell (Schwenkel et al., 2018). The collision probability between two Lagrangian droplets is proportional to the collision kernel, $K(m, m')$,

$$K(m, m') = \pi(r(m) + r(m'))^2 |v_T(m) - v_T(m')| E(m, m') \tag{13.23}$$

describing the collision rate between a droplet with mass m and a droplet with mass m'. Here, r and v_T are the droplet radius and terminal velocity, respectively, which are both functions of droplet mass. Collision efficiency, $E(m, m')$, is usually less than unity due to hydrodynamic interactions (Pinsky et al., 2001). $E(m, m')$ in a turbulent flow is typically larger than the efficiency for droplets approaching by differential sedimentation in a still environment (Grabowski and Wang, 2013). Particle-resolved DNS's have been applied to explore factors that can enhance $E(m, m')$ in a turbulent environment, including but not limited to, increasing droplet radial relative velocity, droplet clustering due to the effect of inertia (Ayala et al., 2008), and the effect of charge (Lu and Shaw, 2015). The enhancement of $K(m, m')$ can speed up drizzle onset (Onishi et al., 2015), which might explain the fast development of precipitation in warm clouds—still an open question in the cloud physics community.

Eulerian-based microphysics approaches do not calculate the collision between droplet pairs; instead, they try to capture the statistical evolution of the droplet population due to collision coalescence. Theoretically, if we know the cloud droplet mass distribution ($f(m)$) and the collection kernel between any two droplets ($K(m, m')$), the time evolution of $f(m)$ can be predicted by the stochastic collection equation (also called the Smoluchowski equation),

$$\begin{aligned}\frac{\partial f(m,t)}{\partial t} =& \frac{1}{2}\int_0^m K(m', m - m') f(m', t) f(m - m', t) dm' \\ &- \int_0^\infty K(m, m') f(m, t) f(m', t) dm'.\end{aligned} \tag{13.24}$$

The first term on the right represents the formation of droplets with mass m due to the collision between two smaller droplets with mass m' and $m - m'$, while the last term represents the loss of a droplet with mass m due to the collision between itself and other droplets. Eq. (13.24) is solved numerically when using the bin microphysics method, while the evolution of different moments of the distribution is parameterized when using the bulk microphysics method.

Caution should be taken in interpreting modeling results, and in particular, to what degree the impacts of turbulence on cloud properties are physical. One way to evaluate performance is through model intercomparison, in which different models are employed to simulate a turbulent cloud with the same initial and boundary conditions (Morrison et al., 2020). Due to complex interactions and feedback among microphysics, thermodynamics, and dynamics in a turbulent cloud, however, it is a challenge to determine how much of the model spread arises from the use of different bin microphysics approaches rather than other factors, such as different dynamic frameworks, advection schemes, and subgrid-scale parameterizations. Recent efforts for simulating turbulent clouds in a chamber with a well-controlled turbulent environment offer unique opportunities for the development and evaluation of cloud-resolving models (Chen et al., 2025; Thomas et al., 2019; Niedermeier et al., 2020; Yang et al., 2022). It has been shown that numerical convergence can be achieved between bin microphysics and Lagrangian microphysics for cloud chamber simulations if model setups as well as key equations to solve in the model are the same (Yang et al., 2023). Such synergistic simulation and laboratory approaches can enhance the interpretation of laboratory measurements and simulation results.

In short, model simulation of turbulent clouds is an important complement to in-situ measurements (Section 13.2) to explore cloud-turbulence interactions. Recent modeling approaches along with observational and theoretical approaches have advanced our understanding of supersaturation field in turbulent clouds (Section 13.4) and its impacts on droplet activation and growth (Section 13.5), as well as turbulence-induced entrainment and mixing and its impacts on cloud microphysical properties (Section 13.6).

13.4 Turbulence and the supersaturation field

Supersaturation or subsaturation of moist air surrounding cloud droplets is the primary driver for droplet formation, condensation growth, and evaporation. A joint scalar field, supersaturation is defined as the ratio of the water vapor pressure in excess of the equilibrium value over a flat surface, to its equilibrium vapor pressure, $s = (e_v - e_{v_s}(T))/e_{v_s}$. The nonlinear Clausius–Clapeyron equation defines the equilibrium vapor pressure: $e_{v_s} = e_{v_o} \exp(L_v/R_v(1/T_o - 1/T))$, where L_v is the latent heat of condensation, e_{v_o} is the reference vapor pressure at T_o, and R_v is the gas constant of water vapor. This nonlinear dependence of supersaturation on temperature and water vapor makes it an interesting quantity compared to independent passive scalars. For example, due to this nonlinearity, isobaric mixing of air at two different temperatures that is saturated (or nearly saturated) produces a mixture that is supersaturated. A mixing diagram shown in Fig. 13.6 illustrates this concept. During turbulent mixing (at times less than the turbulent mixing timescale), different mixing states can be realized in time and physical space between the two initial states of the unmixed air. Thus, supersaturation will also display spatial and temporal variability because of the nonlinearity of the equilibrium vapor pressure curve relative to linear mixing. It is also a foundational concept utilized for the cloud formation in the Pi Convection–Cloud Chamber (Chang et al., 2016). Similarly, the difference

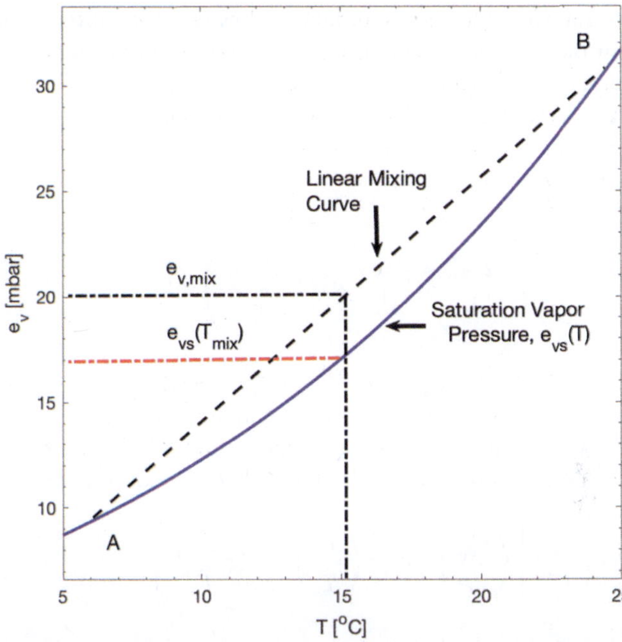

Figure 13.6 A phase diagram illustrating the formation of supersaturation through isobaric mixing at two different conditions. An idealized mixing process of two saturated air blobs at different temperatures, T_A and T_B (dashed line), and its comparison with the saturation (equilibrium) vapor pressure curve (blue line).

in diffusivity of water vapor and temperature drives a supersaturated condition in a particular region downstream when air passes through a water-saturated tube at a controlled temperature—a concept used for forming and measuring cloud droplets in instruments like the Cloud Condensation Nuclei Counter (CCNC, Nenes et al., 2001) and the Leipzig Aerosol Cloud Interaction Simulator (LACIS, Stratmann et al., 2004).

Turbulence drives fluctuations in independent scalars (water vapor and temperature) in moist convection and consequently causes variability in the supersaturation field. However, can we expect a response of the supersaturation field similar to passive independent scalars in a turbulent flow? The answer is not an unqualified yes. As mentioned earlier, nonlinear dependence on the two scalars causes deviation from a passive scalar response in a turbulent medium. The co-variability of water vapor and temperature fields is also critical since supersaturation is a combined scalar. Some past studies (e.g., Twomey, 1959; Lanotte et al., 2009; Sardina et al., 2015; Siewert et al., 2017; Gotoh et al., 2021) treat supersaturation as a passive scalar by solving the advection-diffusion equation similar to a passive independent scalar. Thus, a supersaturation fluctuation spectrum similar to a passive scalar is not surprising. However, such treatment can be misleading and creates challenges in understanding its interactions with cloud particles, especially in the presence of relative water vapor and temperature fluxes.

Supersaturation variability in a turbulent flow can be further understood by investigating its different contributions as a function of independent scalar statistics. A Taylor series expansion of saturation ratio ($S = (\overline{e_v} + e_v')/e_{v_s}(\overline{T} + T')$) up to the third order terms is (Kulmala et al., 1997; Siebert and Shaw, 2017):

$$S \approx S(\bar{e_v}, \bar{T})\Bigg(1 - a_1 T' + e_v'/\bar{e_v} + (a_1^2 - a_2/2)T'^2 - (a_1^3 + a_3/6 - a_1 a_2)T'^3 - a_1\frac{T'e_v'}{\bar{e_v}} + (a_1^2 - a_2/2)\frac{T'^2 e_v'}{\bar{e_v}}\Bigg) + \ldots \quad (13.25)$$

where, $a_1 = e_{v_s}^{-1}(\overline{T})\frac{\partial e_{v_s}(\overline{T})}{\partial T}$, $a_2 = e_{v_s}^{-1}(\overline{T})\frac{\partial^2 e_{v_s}(\overline{T})}{\partial T^2}$, and $a_3 = e_{v_s}^{-1}(\overline{T})\frac{\partial^3 e_{v_s}(\overline{T})}{\partial T^3}$. We can derive the second and higher-order supersaturation fluctuation statistics from this expansion. For example, the expression for supersaturation variance by neglecting the third and higher-order terms is (Kulmala et al., 1997; Chandrakar et al., 2020):

$$\sigma_s^2 \approx S(\bar{e_v}, \bar{T})^2\left[\left(\frac{\sigma_{e_v}}{\bar{e_v}}\right)^2 + \zeta^2\left(\frac{\sigma_T}{\bar{T}}\right)^2 - 2\zeta\frac{\overline{T'e_v'}}{\bar{T}\bar{e_v}}\right] \quad (13.26)$$

where $\zeta = L_v/(R_v\bar{T})$. Similarly, an expression for supersaturation skewness can be obtained as a function of the water vapor and temperature skewness and some other higher-order covariances. As mentioned earlier, the supersaturation fluctuation magnitude primarily depends on the normalized variance of water vapor pressure and temperature and their covariance. All three terms are typically of the same orders of magnitude and non-negligible in most scenarios. The higher-order contributions could become important near cloud boundaries where scalar gradients are sharp. An important point to note here is that the contribution of the covariance term to supersaturation variance is negative. Consequently, any flow condition that introduces relative fluxes of water vapor and temperature would lead to decorrelation of the two scalars and an increase in the supersaturation variability. For example, treating supersaturation as an independent passive scalar essentially ignores the possibility of the impact of decorrelation of water vapor and temperature fields.

13.4.1 Sources of supersaturation variability

In atmospheric clouds, there are multiple sources of supersaturation fluctuations: convective mixing and boundary fluxes, entrainment and mixing of dry air from outside to inside clouds (in cumulus clouds at side boundaries and in stratocumulus from free troposphere at the top) (e.g., Siebert and Shaw, 2017; Wyngaard et al., 1978), fluctuations in vertical velocity (e.g., Politovich and Cooper, 1988; Cooper, 1989; Abade et al., 2018), fluctuations in droplet/ice population (or equivalently variability in phase relaxation timescale) (e.g., Cooper, 1989; Desai et al., 2018; Beals et al., 2015), and co-variability of ice and droplet population in the mixed-phase cloud conditions (clouds with both ice and liquid phases) (e.g., Desai et al., 2019a; Korolev and Milbrandt, 2022). Several of these mechanisms can be understood through the equation for the evolution of supersaturation in an ascending parcel:

$$\frac{ds}{dt} = A_z v_z - \frac{s}{\tau_c} \tag{13.27}$$

where A_z is often assumed to be a constant that depends on the mean temperature and pressure, v_z is the vertical component of the velocity, and $\tau_c \propto (n\bar{r})^{-1}$ is the phase relaxation time, inversely proportional to the integral radius (Cooper, 1989). Above the cloud-base region where transient activation occurs, and far from the cloud edge where entrainment is important, n is often nearly constant and therefore a quasi-steady supersaturation $s_{qs} = A_z v_z \tau_c$ is approached with characteristic time τ_c (Kostinski, 2009). Turbulence can generate fluctuations in the supersaturation field through at least three mechanisms in Eq. (13.27). First, variation in the vertical velocity itself can lead to departures from the quasi-steady supersaturation, and therefore variability in the droplet size distribution; especially near cloud base, where $n\bar{r}$ is small, this mechanism may lead to significant broadening of the droplet size distribution (Korolev et al., 2013). Second, the supersaturation field itself can experience fluctuations due to entrainment and turbulent mixing with air having different temperatures and water vapor concentrations. Third, variation in τ_c due to n and $\bar{r}$, for example, due to changes in aerosol properties or to entrainment, will also change s_{qs}, as well as the rate at which the supersaturation field can respond. The response of s will depend on the characteristic time scale of the driving fluctuations compared to τ_c, which is effectively the Damköhler number $Da = \tau_t/\tau_c$ where τ_t is the turbulence correlation time. If $Da \gg 1$, the forcing is slow compared to the supersaturation response time, so the system evolves to different s_{qs} and droplet growth or evaporation may tend to buffer out the fluctuations; whereas if $Da \lesssim 1$, the fluctuations tend to persist. Recent work exploring these processes using stochastic differential equations, with the 'noise' term implemented through v_z or directly through s, has led to insights into the relevance for droplet size distribution broadening, and may serve as methods for parameterization in coarse-resolution models (McGraw and Liu, 2006; Paoli and Shariff, 2009; Field et al., 2014; Sardina et al., 2015; Chandrakar et al., 2016; Siewert et al., 2017; Chandrakar et al., 2018; Saito et al., 2019).

Some of the mechanisms discussed in the previous paragraph also introduce relative water vapor and temperature fluxes (e.g., the turbulent entrainment-mixing process) that decrease their covariance and increase the supersaturation variability, via the third term in Eq. (13.26). In the atmospheric boundary layer or clouds, entrainment of air from the free troposphere, condensation/evaporation, and surface inhomogeneities introduce relative fluxes of water vapor and heat. It can make the normalized covariance close to one near the surface but completely anticorrelated near the top of the boundary layer (Wyngaard et al., 1978; Wyngaard and LeMone, 1980; Williams et al., 1997; Katul et al., 2008). The mixing of entrained air in the boundary layer is a significant source of decrease in the correlation of the two scalars in a well-mixed layer. Via high-resolution measurements, Siebert and Shaw (2017) reported normalized covariance between water vapor and temperature significantly less than one in shallow cumulus clouds. Similarly, studies of idealized moist Rayleigh–Bénard convection (as in the Pi Convection Cloud Chamber) (e.g., Chandrakar et al., 2020; Thomas et al., 2021; Chandrakar et al., 2022) show how relative fluxes of water vapor and heat decrease the water vapor-temperature covariance and increase supersaturation fluctuations.

13.4.2 Turbulence statistics of supersaturation

Characteristics of turbulent fluctuations of supersaturation (e.g., power spectrum scalings, correlation functions, and probability distribution shape) are essential for understanding and modeling their interactions with droplet activation and growth processes. Observations of shallow cumulus clouds (e.g., Siebert and Shaw, 2017), radiation fog (e.g., Gerber, 1991), and laboratory moist convection (e.g., Anderson et al., 2021; Chandrakar et al., 2016) demonstrate that the measured probability density functions (PDFs) of supersaturation or relative humidity variability at different cloud and moist convection scenarios are closer to a Gaussian shape, and the skewness is apparently low. This PDF shape with weak skewness is only possible if relative water vapor and temperature fluxes are present in turbulent convection. Otherwise, in an adiabatic volume without any significant relative water vapor and temperature fluxes, the supersaturation PDFs are expected to be highly negatively skewed due to nonlinear dependence on temperature and water vapor field. Chandrakar et al. (2020) and Thomas et al. (2021) showed the impact of relative water vapor and temperature flux on supersaturation PDFs. The presence of water vapor or temperature fluxes affects the skewness of the water vapor and temperature PDFs (e.g., during the entrainment-mixing process, Mellado et al., 2017) and their higher-order covariances. Thus, the skewness of the supersaturation depends on the net contribution from primarily the third-order terms. An expression of the supersaturation skewness can be simply derived from Eq. (13.25), showing the contribution from all third-order terms. Condensation growth (and evaporation) of cloud droplets can also affect supersaturation PDFs since it involves latent heat exchange and a water vapor sink. The shape of the supersaturation PDF across various atmospheric scenarios remains an open topic and requires further investigation.

Another central statistic for characterizing turbulent supersaturation fluctuations is the variance or power spectrum. Similar to passive scalars, supersaturation is suggested to follow the classical $\kappa^{-5/3}$ Kolmogorov–Obukhov–Corrsin scaling in the inertial range by some past studies (e.g., Siewert et al., 2017; Gotoh et al., 2021). Gotoh et al. (2021) discusses two potential regimes with $\kappa^{-5/3}$ scaling in clouds separated by the transition wavenumber (κ) corresponding to a Damköhler number of 1, and a shallower spectrum following them in the viscous-convective range. This characterization of the supersaturation spectrum, considering supersaturation as an independent scalar, is stimulating. However, it requires further investigation since supersaturation is a dependent scalar. The presence of quasi independent scalar (water vapor and temperature) fluxes and their relative diffusivity could play a critical role at various scales. The supersaturation spectrum depends on the water vapor and temperature cospectrum. Wyngaard et al. (1978) showed that the one-dimensional cospectrum follows $\kappa^{-5/3}$ scaling in the lower part of the atmospheric boundary layer. However, in the upper half, large eddies and associated entrainment flux in the mixed layer severely impact the cospectrum.

Lagrangian scalar and supersaturation statistics along the droplet trajectories in clouds are also important for modeling the cloud-turbulence interactions. For example, a passive scalar field is expected to follow an exponential Lagrangian autocorrelation function in isotropic turbulence (proposed by G. I. Taylor, see Tennekes, 1979). It is a criterion for the Markovian process (a stochastic process that depends only on the current state), typically assumed in stochastic turbulence models. Can we expect such behavior in the supersaturation field? As discussed earlier, it would depend on the covariance function of water vapor and temperature. Apart from the deviation from the passive independent scalar statistics, droplet inertia and gravitational settling are expected to affect the supersaturation-droplet correlation and impact the supersaturation statistics somewhat, depending on droplet sizes. Lagrangian supersaturation statistics is still an open topic and requires detailed investigation to improve the representation of cloud processes in atmospheric models.

13.5 Turbulence effects on activation and droplet growth

The formation and growth of cloud droplets are key processes in the evolution of a cloud system. The classical models based on uniform supersaturation in an adiabatic parcel predict a narrow cloud droplet size distribution (DSD), whereas in situ measurements reveal a broad DSD. A lack of clarity in the underlying processes shaping the DSD is one of the oldest problems in cloud physics. Barring the effects of aerosol, broadening of the DSDs is due to turbulence, including saturation ratio variability and clustering of cloud droplets. Here, we focus on the role of supersaturation variability in shaping the DSD.

Similar to other fluid dynamical problems, to understand the evolution of the DSD (F) in a cloud system, one can either use the Eulerian or the Lagrangian framework. The Lagrangian framework is a more natural perspective for calculating cloud droplet formation and growth. However, it suffers from the complexity of tracking individual cloud droplets, which is nearly impossible during in-situ measurements. In the Eulerian framework, the DSD evolution equation (kinetic equation) is written as (Shaw, 2003)

$$\begin{aligned}\frac{\partial F}{\partial t} =& I - \nabla \cdot (F\mathbf{u}) + \frac{\partial}{\partial z}(v_T F) - \frac{\partial}{\partial r}(F\dot{r}) - \int_0^{\infty} k(r, r')F(r)F(r')dr' \\ &+ \frac{1}{2}\int_0^r \left(1 - \frac{r'^3}{r^3}\right)^{-2/3} k((r^3 - r'^3)^{1/3}, r')F((r^3 - r'^3)^{1/3})F(r')dr'\end{aligned} \tag{13.28}$$

Note that F in Eq. (13.28) is in r-space and is analogous to f in Eq. (13.24), which is in m-space. Additionally, Eq. (13.24) only represents collisional growth. The terms on the right side of Eq. (13.28) are the activation rate of aerosols I, transport due to the background flow, gravitational sedimentation of droplets, condensation growth (evaporation) due to supersaturation (subsaturation), and the last two terms represent

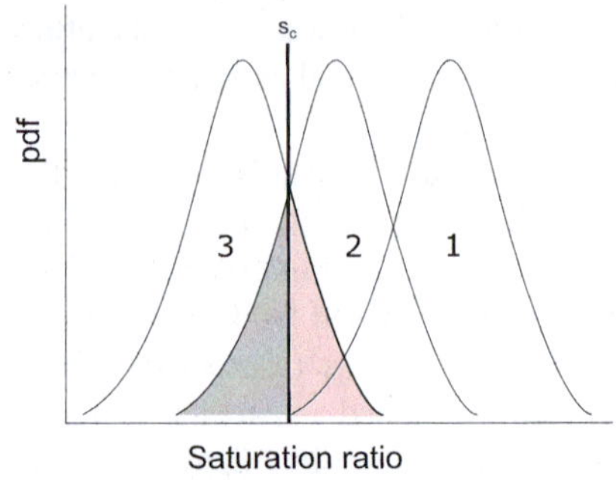

Figure 13.7 Three regimes of cloud droplet formation and growth: 1. Mean-dominant regime; 2. Fluctuation-influenced regime; 3. Fluctuation-dominant regime (reproduced from Prabhakaran et al., 2020b).

the stochastic collision–coalescence term. The collision terms are discussed earlier (Section 13.3).

The term $\partial(F\dot{r})/\partial r$ depicts advection in the radius space with a velocity $\dot{r} \sim 1/r$ (Eq. (13.22)). This decreasing growth rate with r results in the narrowing of the DSD over time in the presence of a uniform supersaturation field. However, in a turbulent environment, the supersaturation field is represented by a distribution with finite variance. Consequently, the droplets in a volume encounter different growth histories, which may result in a broad DSD. In the cloud physics literature, this process is referred to as stochastic condensation, originally hypothesized by Mazin (1966). A detailed review on this topic is presented in Mazin and Merkulovich (2008) and Jeffery et al. (2007). Early formulations based on adiabatic parcels concluded that broadening of the DSD through this mechanism is minimal. In these studies, the variability in supersaturation was linked to vertical velocity fluctuations. Later studies included the effects of entrainment/mixing, and supersaturation fluctuations were often treated as a random variable (e.g., Cooper 1989).

In a turbulent environment, based on the magnitude of the mean saturation ratio $\overline{S}$ and its variance σ_s, three regimes of droplet activation and growth are identified, namely: mean-dominant, fluctuation-influenced, and fluctuation-dominant regimes (Prabhakaran et al., 2020b; Shawon et al., 2021; Prabhakaran et al., 2022). This is best illustrated in Fig. 13.7 for monodisperse aerosol particles with critical saturation S_c. In the mean-dominant regime, $\overline{S} - S_C \gg \sigma_s$, i.e., all droplets activate and the activated droplets grow steadily over time. This regime is encountered in atmospheric clouds with strong updrafts and/or low droplet concentration. In the fluctuation-influenced regime, $|\overline{S} - S_C| \lesssim \sigma_s$ and $\overline{S} > S_C$, i.e., all activated droplets grow over time, whereas only the haze droplets that encounter a supersaturation above the critical value activate. In this regime, turbulence suppresses the activation rate of cloud droplets (gray shaded region in Fig. 13.7). In the fluctuation-dominant regime, $|\overline{S} - S_c| \lesssim \sigma_s$ and $\overline{S} < S_c$. In this regime, strong fluctuation in supersaturation is essential to activate cloud droplets and for their subsequent growth. Here, only a small fraction of the droplets in a volume encounter supersaturation values greater than the critical value (light-red

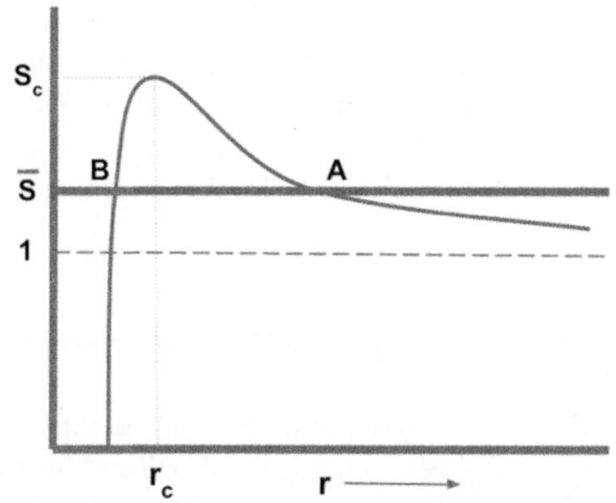

Figure 13.8 Schematic of a Köhler curve S_{eq} vs r. A denotes the unstable equilibrium point, and B denotes the equilibrium-stable point. r_c is the critical radius of a solute droplet, and S_c is the critical saturation ratio.

shaded region in Fig. 13.7). Additionally, depending on the value of $\overline{S}$ this regime has two sub-regimes: $1 < \overline{S} \leq S_c$ and $\overline{S} \leq 1$. This regime would be encountered in clouds with weak updrafts (e.g., fog), or polluted, or near cloud boundaries, where the saturation ratio is close to 1. In both regimes 2 and 3, there can be significant deactivation of cloud droplets and reactivation of haze droplets.

The discussion so far does not account for the kinetic limitations of droplet activation. Activation of a droplet, although a threshold process, requires a finite amount of time to overcome the threshold and attain equilibrium. The role of turbulent fluctuations in activation is of paramount importance in the fluctuation-dominant regime. For a value of $\overline{S}$ less than S_c, if $\overline{S} > 1$, there exist two equilibrium points: stable ($r < r_c$), denoted by B in Fig. 13.8, and unstable ($r > r_c$), denoted by A in Fig. 13.8. The fluctuations in the saturation ratio must be capable of transporting the unactivated droplets from the stable point to the unstable point. This would depend on the intensity of the supersaturation fluctuations and their correlation time. For instance, if the correlation time is very small compared to the activation time, the supersaturation fluctuations would not aid in the activation of the cloud droplet. A detailed discussion on this is provided in Arabas and Shima (2017) and Abade et al. (2018).

In a turbulent environment, using Reynolds decomposition, each variable can be decomposed into a mean and a fluctuation, e.g., $F = \overline{F} + F'$. Using this decomposition and Eq. (13.22) in Eq. (13.28), the evolution equation for $\overline{F}$ is written as

$$\frac{\partial \overline{F}}{\partial t} = -\nabla \cdot (\overline{F}\overline{\mathbf{u}}) - \nabla \cdot (\overline{F'\mathbf{u}'}) + \frac{\partial}{\partial z}(v_T \overline{F}) - \frac{\partial}{\partial r}\left(\xi \overline{F}\frac{(\overline{S} - Seq)}{r}\right) - \frac{\partial}{\partial r}\left(\xi \frac{\overline{F's'}}{r}\right) + I \tag{13.29}$$

where the first two terms represent the mean and turbulent transport, followed by gravitational sedimentation, mean-growth and turbulent-growth terms, respectively. Note that we do not consider the collision-growth terms here. We use a simplified form of this equation to understand the three regimes of activation. Here, we can ignore the transport terms as activation involves small droplets (haze and activated). Thus the equation reduces to

$$\frac{\partial \overline{F}}{\partial t} = I - \frac{\partial}{\partial r}\left(\xi \overline{F}\frac{(\overline{S} - S_{eq})}{r}\right) - \frac{\partial}{\partial r}\left(\xi \frac{\overline{F's'}}{r}\right) \tag{13.30}$$

In the mean-dominant regime, the contribution from the fluctuating-growth term is negligible. In the other two regimes, both mean-growth and fluctuating-growth terms are important. If we assume steady state (e.g., as in the Pi chamber, Prabhakaran et al. (2020b)), we obtain an expression for $\overline{F}$

$$\overline{F} = \frac{Ir - \xi \overline{F's'}}{\xi(\overline{S} - S_{eq})} \tag{13.31}$$

In the mean-dominant regime, $\overline{F's'} << \overline{F}(\overline{S} - S_{eq})$. In the fluctuation-influenced regime, $\overline{F's'} \sim \overline{F}(\overline{S} - S_{eq})$ and $\overline{S} > S_c$. This leads to $\overline{F's'} < 0$ for small droplets (including haze droplets). In the fluctuation-dominant regime, $\overline{F's'} \sim \overline{F}(\overline{S} - S_{eq})$ and $\overline{S} < S_c$. Since $\overline{F}$ is positive semi-definite, $\overline{F's'} > 0$ for small droplets (including haze droplets). In the fluctuation-influenced and fluctuation-dominant regimes, a finite magnitude of $\overline{F's'}$ in the haze droplet size range leads to the formation of a mode in this size range. This renders a bimodal distribution to the DSD (including the haze droplets). In other words, aerosol particles of the same size and composition can exist in both haze droplet and cloud droplet populations in a turbulent environment. Similar results were obtained in the numerical simulations of turbulent cloud parcels and in field measurements (Ditas et al., 2012; Abade et al., 2018).

The discussion about droplet activation can be extended to cloud droplet growth. However, the transport terms cannot be ignored while considering the growth of cloud droplets in a turbulent environment. The importance of transport was highlighted by Bartlett and Jonas (1972), where they showed that supersaturation variability introduced by vertical velocity fluctuations would not aid in broadening the DSD. This was formalized in the context of the Pi chamber in Prabhakaran et al. (2022). Traditionally, in the stochastic condensation framework, turbulent transport and sedimentation effects are ignored. In a Lagrangian framework, this leads to equal time for all cloud droplets in a given volume. However, this violates the dispersive properties of the underlying turbulent flow. Like every other turbulent flow, transport is an important part of cloud microphysics and should be appropriately represented to determine the shape of the DSD. Accounting for turbulent transport tends to diminish the broadening effect of supersaturation fluctuations in the large tail of the DSD. Thus to determine the mean DSD, all the terms in Eq. (13.29) are required.

To predict the shape of the mean DSD using Eq. (13.29), a closure is required for the covariance terms involving F', s', and $\mathbf{u}'$. The closure problem in cloud microphysics is more complex than that of shear flows, partly due to the lack of high-resolution measurements and numerical simulations. Additional complexities arise once the effect of collision-coalescence is included. The most popular closure scheme for the $\overline{F's'}$ is based on the gradient transport model, which is ubiquitous in turbulent flows despite limitations on its success (Narasimha, 1990). The gradient transport model is closely associated with the Fokker–Planck theory, which is used for predicting the evolution of distribution functions in non-equilibrium systems. The key assumption is that the RMS supersaturation fluctuation s' is an externally imposed Gaussian random variable. Thus, using the gradient transport model in the r^2 space, the mean size distribution $\overline{g}(r^2, t) = \overline{F}(r)/2r$ evolution equation is written as a drift-diffusion equation

$$\frac{\partial \overline{g}}{\partial t} = I - 2\xi(\overline{S} - S_{eq})\frac{\partial \overline{g}}{\partial r} + D_T \frac{\partial^2 \overline{g}}{\partial r^2} \tag{13.32}$$

where $D_T \propto \sigma_s^2 \tau$, is the turbulent diffusivity in the r^2 space and τ is the s' correlation time scale. Interested readers can refer to Jeffery et al. (2007) for a detailed derivation and discussion. The second term on the right side represents drift (advection), and the third term represents diffusion in the r^2 space. A recent LES study of the Pi chamber has shown the limitations of this drift-diffusion model in representing the DSD in a cloud system. One limitation is the assumption that s' is a Gaussian random variable, whereas the s' obtained from a simple mixing process is non-Gaussian (Thomas et al., 2021). This suggests that the turbulent fluctuations and the covariance terms have a structure, and thus modeling them as random fluctuations akin to molecular fluctuations in kinetic theory may be misleading. With the advent of faster computers, algorithms, and improvements in the in-situ measurement resolution, we should be able to estimate the turbulent covariance terms in Eq. (13.29) accurately. This would aid in formulating a more realistic closure for these covariance terms.

13.6 Turbulence and entrainment

The formation and persistence of stratocumulus and cumulus clouds are strongly influenced by the entrainment of subsaturated air (and detrainment of cloudy air) caused by turbulent mixing between clouds and their environments. Capturing or representing the entrainment process occurring at spatial scales far below the resolution of weather and climate models therefore remains an enduring challenge of cloud physics (De Rooy et al., 2013; Cohen et al., 2020).

A rigorous determination of the mixing of air from the undisturbed environment and the undiluted core of the cloud can be made by using variables that are conserved even in the presence of a phase change. Assuming removal of condensate by precipitation can be neglected, the total water content ($q_t = q_v + q_l$) is conserved. This allows the final result of a mixture of cloudy and environmental air, with cloud mixing fraction χ_c, to be determined as

$$q_{t,m} = (1 - \chi_c)q_{t,e} + \chi_c q_{t,i} \tag{13.33}$$

where the subscripts i and m denote the initial and mixed clouds, and e denotes the environment; χ_c ranges from 0 to 1, corresponding to a mixture consisting of only environmental air to one with only the initial cloud. Another conserved variable (for reversible phase transitions and no external diabatic processes) is liquid water potential temperature, approximated as $\theta_l = \theta[1 - q_l L_v/(c_p T)]$, where θ is the potential temperature (Rogers and Yau, 1989). For a mixture of air from the initial cloud $\theta_{l,i}$ and the environment $\theta_{l,e}$, the result is simply

$$\theta_{l,m} = (1 - \chi_c)\theta_{l,e} + \chi_c\theta_{l,i} \tag{13.34}$$

Following the derivation of Yang et al. (2016), we consider a mixture having final properties $q_{l,m}$ and T_m. The resulting change $\Delta q_l = q_{l,i} - q_{l,m}$ depends on the change $\Delta T = T_i - T_m$, and therefore the conservation equations for q_t and θ_l must be solved simultaneously (Gerber et al., 2008). Alternatively, an approximate analytical solution can be obtained by using a linearized Clausius–Clapeyron equation

$$q_{v,i} - q_{v,m} \approx \frac{S_i e_s(T_i) L_v \epsilon}{p_i R_v T_i^2}(T_i - T_m) = C_2 (T_i - T_m) \tag{13.35}$$

where C_2 is a constant depending on initial thermodynamic conditions. The conservation equations for total water content and liquid water potential temperature, coupled using Eq. (13.35), result in

$$q_{l,m} \approx \chi_c q_{l,i} - (1 - \chi_c)\frac{q_{v,i} - q_{v,e} - C_2(T_i - T_e)}{1 + L_v C_2/c_p}. \tag{13.36}$$

This equation allows the final liquid water mixing ratio of the mixture to be determined using only the initial properties of the cloudy and environmental air. For example, the function $q_{l,m}(\chi_c)$ allows the critical cloud fraction at which $q_{l,m} = 0$ to be determined.

The microphysical implications of a reduction in liquid water mixing ratio due to entrainment can be appreciated by recognizing that $q_l = n(\rho_l/\rho_a)(4/3)\pi r^3$. Thus a reduction of $q_{l,m}$ can be achieved in a continuum of different ways, with extremes being reducing n while holding r constant, or by reducing r while maintaining the (globally defined) n constant. In the latter case, the cloud n only varies according to the dilution, i.e., without complete evaporation of a subset of droplets. Thermodynamics tells us what $q_{l,m}$ exists for a given set of cloud and environment conditions, but the microphysical response to turbulent mixing dictates how that change is achieved between n and r. The droplet number density and the shape of the size distribution are crucial for understanding precipitation formation rates and cloud optical properties, so the turbulent mixing problem is central to cloud microphysics. We take cloud optical properties as an illustrative example. In the geometric optics limit, the optical thickness of a cloud scales approximately as the cross-sectional area of the droplet population, $\tau \sim nr^2$. Cloud liquid water mixing ratio, on the other hand, scales as $q_l \sim nr^3$. The mixing limit with constant r leads to $d\ln\tau = d\ln q_l$. A hypothetical limit with constant n leads to $d\ln\tau = (2/3)d\ln q_l$ (Jeffery, 2007). Therefore the 'susceptibility' of cloud optical thickness to a change in liquid water mixing ratio differs significantly depending on the microphysical response to mixing.

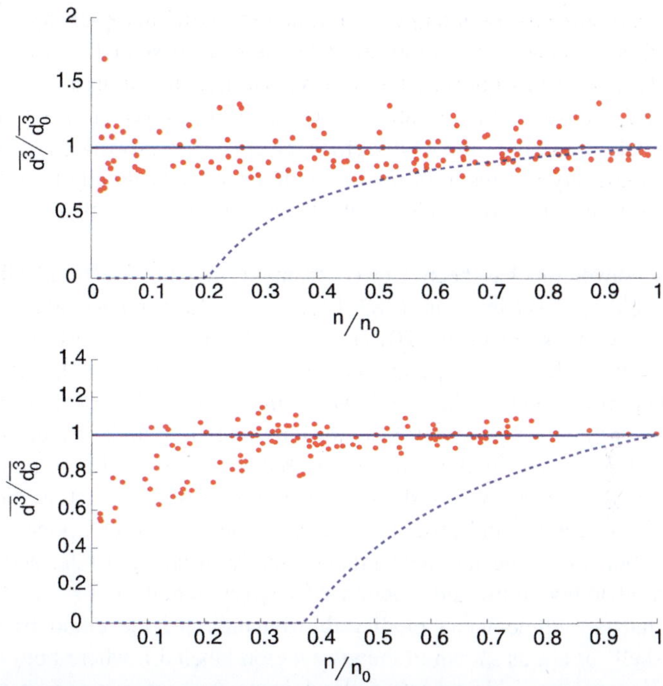

Figure 13.9 Examples of mixing diagrams showing measurements of mean-cubed diameter versus cloud droplet number density for two cumulus clouds. Each red dot is derived from droplets in a single hologram recorded by the holographic detector for clouds (HOLODEC), deployed on two different research aircraft. The solid blue lines represent the expected behavior for the extreme inhomogeneous mixing limit, and the dashed blue curves represent homogeneous mixing based on the measured temperature and relative humidity outside the cloud.
Source: Figure from Beals et al. (2015). Reprinted with permission from AAAS.

The mixing extremes defined above can be interpreted as resulting from two limits of the Damköhler number Da. When $Da \ll 1$, the mixing is rapid compared to the phase relaxation time, and therefore all droplets are exposed to a uniform, well-mixed humidity field, and the adjustment to lower q_l is achieved by all droplets being reduced in radius, with none (or very few) completely evaporating. The $Da \gg 1$ limit, in contrast, suggests that mixing is slow compared to the phase relaxation time or droplet evaporation time (Lehmann et al., 2009), so complete droplet evaporation occurs at the boundaries between filaments of clear and cloudy air. This conceptual picture was developed based on laboratory observations (Latham and Reed, 1977) and extensions of turbulence theory (Baker et al., 1984), and eventually led to in situ cloud observations in which the n and d^3 contributions to reductions in q_l were depicted in 'mixing diagrams' (Jensen et al., 1985). Examples of mixing diagrams for two cumulus-cloud transects made with an airborne digital holographic instrument are shown in Fig. 13.9. The measurements are unique because each data point corresponds to the mean-cube diameter and the droplet number density as estimated from single holograms, and therefore do not require long spatial averaging that can artificially bias toward inhomogeneous mixing (Burnet and Brenguier, 2007). These measurements strongly support the extreme inhomogeneous limit for the mixing process in small cumulus clouds in dry environments. Recent work suggests that this strong inhomogeneity is a result of mixing with air that has already been humidified during previous mixing events (Fries et al., 2021), and a mechanism for this is discussed next.

Cumulus convection has been viewed from the perspective of idealized rising plume or bubble models (Simpson, 1983; Blyth, 1993; Narasimha et al., 2011; Pinsky et al., 2022). Schmeissner et al. (2015) defined the active stage of cumulus cloud growth as having both positive updraft velocity and buoyancy: $w > 0$ and $B > 0$. The maximum height reached by the cloud and its microphysical properties are strongly influenced by entrainment of dry air from the environment into the cloud. An example of the microphysical, thermodynamic, and turbulent structure of a cumulus cloud is shown in Fig. 13.10. The panels show, from top to bottom, the cloud liquid water content, the temperature and absolute humidity, the vertical component of the air velocity, the buoyancy, the turbulent kinetic energy dissipation rate, and the liquid water potential temperature and fraction of environmental air ($\chi \equiv 1 - \chi_c$). It is evident that entrainment can modify the dynamics of the cloud by forming a 'subsiding shell' of air, as shown in the gray region labeled I, where both $w < 0$ and $B < 0$. This is a remarkable aspect of cloud–turbulence interactions, that a mixture of clear and cloudy air can be negatively buoyant relative to both starting fluids, as a result of evaporative cooling (Burnet and Brenguier, 2007). Noting that ε is plotted on a logarithmic scale, it is also evident that the turbulence is locally enhanced in this subsiding shell region, which has interesting possible implications for nonlinear enhancement of entrainment. The gray region labeled II has $w < 0$ and $B \approx 0$, suggesting that the subsiding shell also influences the environment. The two conserved variables q_t, which is determined from measurements of q_v and q_l, and θ_l, which is determined from measurements of T, p, and q_l, give a consistent view of the degree of mixing, as shown in the bottom panel of Fig. 13.10. As anticipated, in the active growth stage, the cloud interior appears to be relatively uniform and (presumably) undiluted, whereas the subsiding shell (gray region I) varies sporadically from the cloud value to the environment value. The outer gray region (region II) appears to consist of environmental air only, and may be a result of the compensating downward acceleration due to effective buoyancy (Tarshish et al., 2018).

13.7 Concluding remarks

We conclude with a brief discussion of what we view as some of the chief outstanding challenges related to turbulence-microphysics interactions.

First, as suggested in Section 13.4, direct measurement of supersaturation fluctuations in cloudy conditions remains a challenge. A few decades ago, it was noted by Bradshaw (1994) that pressure was one of the "Great Unmeasurables" of

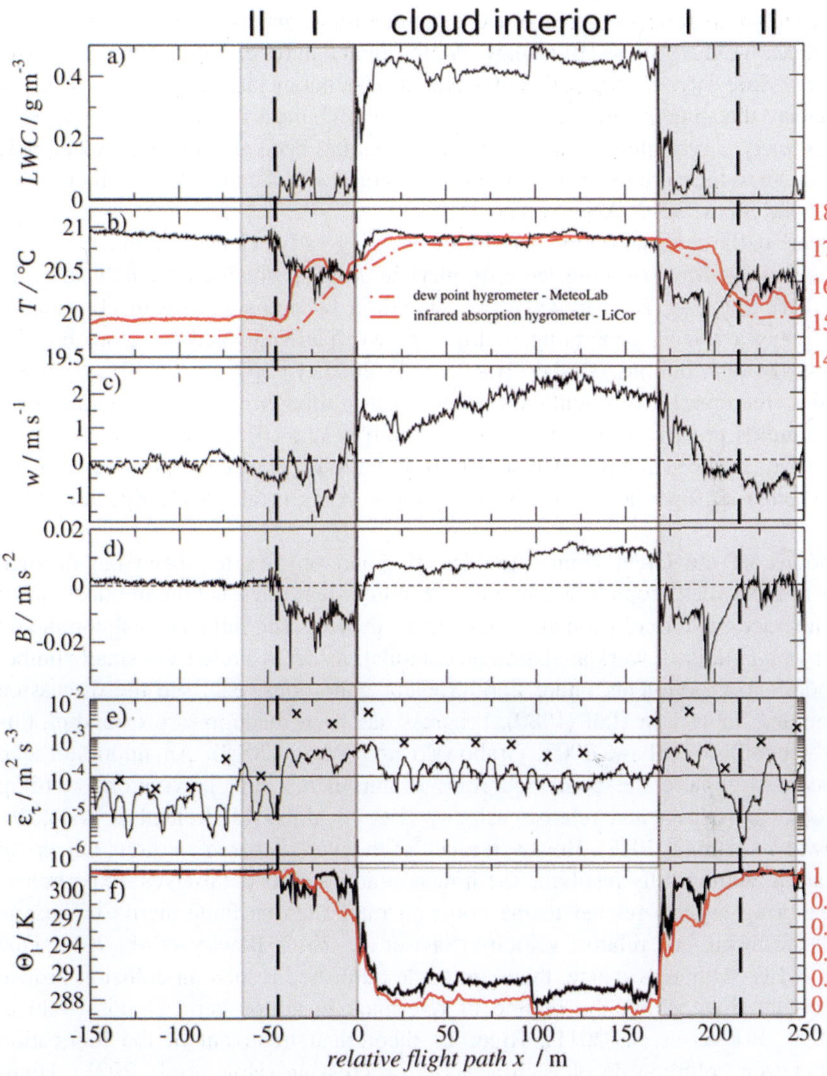

Figure 13.10 Example of the structure of a single cumulus cloud and its environment. (a) Liquid water content, (b) temperature and absolute humidity (the red line denotes measurements with a LiCor infrared-absorption hygrometer and the dot-dashed red line with a relatively slow dewpoint hygrometer), (c) vertical component of air velocity, (d) buoyancy, (e) turbulence kinetic energy dissipation rate [the black line denotes a direct method and the crosses mark absolute values calculated based on the second-order structure function], and (f) liquid water potential temperature and mixing fraction. The white region between the two gray regions represents the cloud interior, and the two gray regions denote the subsiding shell at the edges of the cloud. The vertical black dashed line divides the subsiding shell into the inner shell (denoted I) and the outer shell (denoted II).
Source: Figure from Katzwinkel et al. (2014) ©American Meteorological Society. Used with permission.

turbulence. It is central to turbulence dynamics, e.g., pressure-strain terms are of importance for shear stress (Bradshaw, 1994), but at that time, and to some extent still today, pressure was only accessible via computation, not by measurement. By analogy, we can say that supersaturation is one of the Great Unmeasurables of cloud physics. It is routinely accessible through computation, but has been and remains exceedingly challenging to measure reliably, especially in natural cloud conditions (Gerber, 1991; Siebert and Shaw, 2017; Anderson et al., 2021).

Access to the supersaturation distribution would greatly enhance our understanding of mechanisms for producing large droplets in clouds, which are thought to be of importance for triggering growth of precipitation by collision growth. Turbulence likely plays a role in generating multiple growth pathways and therefore broader droplet size distributions (Lasher-Trapp et al., 2005; Cooper et al., 2013). This is related to the concept of circulation mixing that is thought to be of importance for stratocumulus clouds (Wang et al., 2009; Yeom et al., 2021). Indeed, entrainment and subsequent growth have been shown, from various perspectives, to be a possible mechanism for 'super-adiabatic' droplet growth (Yang et al., 2016; Korolev et al., 2013).

Another of the Great Unmeasurables of cloud physics has been the detailed interaction between droplets as they interact in turbulent flow, and the implications of these interactions for collision and coalescence growth. The 'efficiencies' commonly used in computational work are based on compilations from a relatively small number of laboratory experiments under non-turbulent conditions, e.g., see the discussion surrounding Table 1 in Hall (1980). Progress has been made toward extending this to turbulent flows (Shaw, 2003; Grabowski and Wang, 2013). An important step has been to explore the combined roles of turbulence and gravitational settling on droplet clustering and relative velocity (Bec et al., 2014; Ireland et al., 2016; Dhariwal and Bragg, 2018). However, much of the work is based on direct numerical simulation without fully resolving the flow around droplets themselves. Experiments are exploring aspects related to the collision rate, such as finite-inertia effects on spatial clustering and relative velocity (Saw et al., 2012; Bewley et al., 2013; Dou et al., 2018). Although sparse, there are some published results on collision growth in turbulent flow within the regime of relevance to atmospheric clouds (Vohl et al., 1999; Bordás et al., 2011). Rigorous theoretical treatment of the lubrication flow between colliding droplets in a gas is progressing (How et al., 2021). High-resolution experimental exploration of the collision efficiency is also stimulating deeper theoretical understanding of the droplet collision process (Magnusson et al., 2022).

An aspect of droplet collision rates that may be tied to turbulence is its stochastic nature. Kostinski and Shaw (2005) suggested that fluctuations may allow 'lucky drops' to dominate the growth process for precipitation due to the strong $\sim r^6$ radius dependence of the collision rate. Related concepts have been explored from several perspectives (Bec et al., 2016; Wilkinson, 2016; Li et al., 2022), but it remains unresolved to what extent this perspective differs from the standard Smoluchowski approach discussed in Section 13.3, and to what extent turbulence may play a role in causing the colliding lucky droplets. Looking to the future, there is some discussion of a large laboratory chamber for exploring collision growth in turbulent flow (Ren et al., 2026; Shaw et al., 2020; Wang et al., 2024).

Finally, much work remains to advance our understanding of turbulence in mixed-phase and ice clouds. For example, turbulence likely plays a role in sustaining the coexistence of supercooled liquid water and ice in mixed-phase clouds (Field et al., 2014; Desai et al., 2019a). There is growing evidence that turbulence also affects the collision growth of ice hydrometeors by ice-ice and ice-droplet collisions in clouds (Jost et al., 2019; Sheikh et al., 2022). More exotic connections also may be of relevance, for example, turbulent mixing in the wake of riming or melting hydrometeors may play a role in droplet activation and ice formation (Prabhakaran et al., 2020a).

Acknowledgments

RAS thanks the Climate Modeling Alliance (CliMA) at Caltech for hosting him during the time this chapter was written. RAS acknowledges support from NSF grant AGS-2133229. KKC was supported by the National Center for Atmospheric Research, which is a major facility sponsored by the National Science Foundation under Cooperative Agreement No. 1852977. PP acknowledges support from the U.S. Department of Commerce, Earth's Radiation Budget grant, NOAA CPO Climate and CI no. 03-01-07-001. FY was supported by the Office of Biological and Environmental Research in the Department of Energy, Office of Science, through Contract No. DE-SC0012704 to Brookhaven National Laboratory. Comments from a careful reading by Nadir Jeevanjee (GFDL) are appreciated and have improved the text.

References

Abade, G.C., Grabowski, W.W., Pawlowska, H., 2018. Broadening of cloud droplet spectra through eddy hopping: turbulent entraining parcel simulations. Journal of the Atmospheric Sciences 75, 3365–3379.

Anderson, J.C., Thomas, S., Prabhakaran, P., Shaw, R.A., Cantrell, W., 2021. Effects of the large-scale circulation on temperature and water vapor distributions in the π chamber. Atmospheric Measurement Techniques 14, 5473–5485.

Arabas, S., Shima, S., 2017. On the CCN (de)activation nonlinearities. Nonlinear Processes in Geophysics 24, 535–542. https://doi.org/10.5194/npg-24-535-2017.

Arabas, S., Shima, S., 2013. Large-eddy simulations of trade wind cumuli using particle-based microphysics with Monte Carlo coalescence. Journal of the Atmospheric Sciences 70, 2768–2777.

Ayala, O., Rosa, B., Wang, L.P., Grabowski, W.W., 2008. Effects of turbulence on the geometric collision rate of sedimenting droplets. Part 1. Results from direct numerical simulation. New Journal of Physics 10, 075015.

Baker, M., Breidenthal, R., Choularton, T., Latham, J., 1984. The effects of turbulent mixing in clouds. Journal of the Atmospheric Sciences 41, 299–304.

Bartlett, J., Jonas, P., 1972. On the dispersion of the sizes of droplets growing by condensation in turbulent clouds. Quarterly Journal of the Royal Meteorological Society 98, 150–164.

Beals, M.A., Fugal, J.P., Shaw, R.A., Lu, J., Spuler, S.M., Stith, J.L., 2015. Holographic measurements of inhomogeneous cloud mixing at the centimeter scale. Science 350, 87–90.

Bec, J., Homann, H., Ray, S.S., 2014. Gravity-driven enhancement of heavy particle clustering in turbulent flow. Physical Review Letters 112, 184501.

Bec, J., Ray, S.S., Saw, E.W., Homann, H., 2016. Abrupt growth of large aggregates by correlated coalescences in turbulent flow. Physical Review. E 93, 031102.

Bewley, G.P., Saw, E.W., Bodenschatz, E., 2013. Observation of the sling effect. New Journal of Physics 15, 083051.

Blyth, A.M., 1993. Entrainment in cumulus clouds. Journal of Applied Meteorology and Climatology 32, 626–641.

Bordás, R., Hagemeier, T., Wunderlich, B., Thévenin, D., 2011. Droplet collisions and interaction with the turbulent flow within a two-phase wind tunnel. Physics of Fluids 23, 085105.

Borque, P., Luke, E., Kollias, P., 2016. On the unified estimation of turbulence eddy dissipation rate using Doppler cloud radars and lidars. Journal of Geophysical Research. Atmospheres 121, 5972–5989. https://doi.org/10.1002/2015JD024543.

Bradshaw, P., 1994. Turbulence: the chief outstanding difficulty of our subject. Experiments in Fluids 16, 203–216.

Burnet, F., Brenguier, J.L., 2007. Observational study of the entrainment-mixing process in warm convective clouds. Journal of the Atmospheric Sciences 64, 1995–2011.

Caughey, S.J., Crease, B.A., Roach, W.T., 1982. A field study of nocturnal stratocumulus. II: turbulence structure and entrainment. Quarterly Journal of the Royal Meteorological Society 108, 125–144.

Chandrakar, K.K., Cantrell, W., Chang, K., Ciochetto, D., Niedermeier, D., Ovchinnikov, M., Shaw, R.A., Yang, F., 2016. Aerosol indirect effect from turbulence-induced broadening of cloud-droplet size distributions. Proceedings of the National Academy of Sciences of the United States of America 113, 14243–14248.

Chandrakar, K.K., Cantrell, W., Krueger, S., Shaw, R.A., Wunsch, S., 2020. Supersaturation fluctuations in moist turbulent Rayleigh–Bénard convection: a two-scalar transport problem. Journal of Fluid Mechanics 884.

Chandrakar, K.K., Cantrell, W., Shaw, R.A., 2018. Influence of turbulent fluctuations on cloud droplet size dispersion and aerosol indirect effects. Journal of the Atmospheric Sciences 75, 3191–3209.

Chandrakar, K.K., Grabowski, W.W., Morrison, H., Bryan, G.H., 2021. Impact of entrainment mixing and turbulent fluctuations on droplet size distributions in a cumulus cloud: an investigation using Lagrangian microphysics with a subgrid-scale model. Journal of the Atmospheric Sciences 78, 2983–3005.

Chandrakar, K.K., Morrison, H., Grabowski, W.W., Bryan, G.H., Shaw, R.A., 2022. Supersaturation variability from scalar mixing: evaluation of a new subgrid-scale model using direct numerical simulations of turbulent Rayleigh–Bénard convection. Journal of the Atmospheric Sciences 79, 1191–1210.

Chang, K., Bench, J., Brege, M., Cantrell, W., Chandrakar, K., Ciochetto, D., Mazzoleni, C., Mazzoleni, L., Niedermeier, D., Shaw, R., 2016. A laboratory facility to study gas–aerosol–cloud interactions in a turbulent environment: the π chamber. Bulletin of the American Meteorological Society 97, 2343–2358.

Chen, S., Yau, M.K., Bartello, P., Xue, L., 2018. Bridging the condensation–collision size gap: a direct numerical simulation of continuous droplet growth in turbulent clouds. Atmospheric Chemistry and Physics 18, 7251–7262.

Chen, Sisi, Krueger, Steven K., Dziekan, Piotr, Enokido, Kotaro, MacMillan, Theodore, Richter, David, Schmalfuß, Silvio, et al., 2025. A model intercomparison study of aerosol-cloud-turbulence interactions in a cloud chamber: 1. Model results. Journal of Advances in Modeling Earth Systems 17 (7), e2024MS004562.

Cohen, Y., Lopez-Gomez, I., Jaruga, A., He, J., Kaul, C.M., Schneider, T., 2020. Unified entrainment and detrainment closures for extended eddy-diffusivity mass-flux schemes. Journal of Advances in Modeling Earth Systems 12, e2020MS002162.

Cooper, W.A., 1989. Effects of variable droplet growth histories on droplet size distributions. Part I: theory. Journal of the Atmospheric Sciences 46, 1301–1311.

Cooper, W.A., Lasher-Trapp, S.G., Blyth, A.M., 2013. The influence of entrainment and mixing on the initial formation of rain in a warm cumulus cloud. Journal of the Atmospheric Sciences 70, 1727–1743.

De Rooy, W.C., Bechtold, P., Fröhlich, K., Hohenegger, C., Jonker, H., Mironov, D., Pier Siebesma, A., Teixeira, J., Yano, J.I., 2013. Entrainment and detrainment in cumulus convection: an overview. Quarterly Journal of the Royal Meteorological Society 139, 1–19.

Desai, N., Chandrakar, K., Kinney, G., Cantrell, W., Shaw, R., 2019a. Aerosol-mediated glaciation of mixed-phase clouds: steady-state laboratory measurements. Geophysical Research Letters 46, 9154–9162.

Desai, N., Chandrakar, K.K., Chang, K., Cantrell, W., Shaw, R., 2018. Influence of microphysical variability on stochastic condensation in a turbulent laboratory cloud. Journal of the Atmospheric Sciences 75, 189–201.

Desai, N., Glienke, S., Fugal, J., Shaw, R., 2019b. Search for microphysical signatures of stochastic condensation in marine boundary layer clouds using airborne digital holography. Journal of Geophysical Research. Atmospheres 124, 2739–2752.

Dhariwal, R., Bragg, A.D., 2018. Small-scale dynamics of settling, bidisperse particles in turbulence. Journal of Fluid Mechanics 839, 594–620.

Ditas, F., Shaw, R.A., Siebert, H., Simmel, M., Wehner, B., Wiedensohler, A., 2012. Aerosols-cloud microphysics-thermodynamics-turbulence: evaluating supersaturation in a marine stratocumulus cloud. Atmospheric Chemistry and Physics 12, 2459–2468.

Dou, Z., Bragg, A.D., Hammond, A.L., Liang, Z., Collins, L.R., Meng, H., 2018. Effects of Reynolds number and Stokes number on particle-pair relative velocity in isotropic turbulence: a systematic experimental study. Journal of Fluid Mechanics 839, 271–292.

Field, P., Hill, A., Furtado, K., Korolev, A., 2014. Mixed-phase clouds in a turbulent environment. Part 2: analytic treatment. Quarterly Journal of the Royal Meteorological Society 140, 870–880.

Fox, D.G., Lilly, D.K., 1972. Numerical simulation of turbulent flows. Reviews of Geophysics 10, 51–72. https://doi.org/10.1029/RG010i001p00051.

Franklin, C.N., Vaillancourt, P.A., Yau, M., Bartello, P., 2005. Collision rates of cloud droplets in turbulent flow. Journal of the Atmospheric Sciences 62, 2451–2466.

Fries, J., Sardina, G., Svensson, G., Mehlig, B., 2021. Key parameters for droplet evaporation and mixing at the cloud edge. Quarterly Journal of the Royal Meteorological Society 147, 2160–2172.

Gerber, H., 1991. Supersaturation and droplet spectral evolution in fog. Journal of the Atmospheric Sciences 48, 2569–2588.

Gerber, H.E., Frick, G.M., Jensen, J.B., Hudson, J.G., 2008. Entrainment, mixing, and microphysics in trade-wind cumulus. Journal of the Meteorological Society of Japan 86, 87–106.

Gotoh, T., Saito, I., Watanabe, T., 2021. Spectra of supersaturation and liquid water content in cloud turbulence. Physical Review Fluids 6, 110512.

Götzfried, P., Kumar, B., Shaw, R.A., Schumacher, J., 2017. Droplet dynamics and fine-scale structure in a shearless turbulent mixing layer with phase changes. Journal of Fluid Mechanics 814, 452–483.

Grabowski, W.W., Wang, L.P., 2013. Growth of cloud droplets in a turbulent environment. Annual Review of Fluid Mechanics 45, 293–324.

Hall, W.D., 1980. A detailed microphysical model within a two-dimensional dynamic framework: model description and preliminary results. Journal of the Atmospheric Sciences 37, 2486–2507.

Heintzenberg, J., Charlson, R.J., 2009. Clouds in the perturbed climate system. MIT press.

Hill, R.J., 2002. Scaling of acceleration in locally isotropic turbulence. Journal of Fluid Mechanics 452, 361–370.

Hill, R.J., 2005. Geometric collision rates and trajectories of cloud droplets falling into a Burgers vortex. Physics of Fluids 17, 037103.

Hoffmann, F., Feingold, G., 2019. Entrainment and mixing in stratocumulus: effects of a new explicit subgrid-scale scheme for large-eddy simulations with particle-based microphysics. Journal of the Atmospheric Sciences 76, 1955–1973.

How, M.L.S., Koch, D.L., Collins, L.R., 2021. Non-continuum tangential lubrication gas flow between two spheres. Journal of Fluid Mechanics 920. https://doi.org/10.1017/jfm.2021.308.

Ireland, P.J., Bragg, A.D., Collins, L.R., 2016. The effect of Reynolds number on inertial particle dynamics in isotropic turbulence. Part 2. Simulations with gravitational effects. Journal of Fluid Mechanics 796, 659–711.

Jeffery, C.A., 2007. Inhomogeneous cloud evaporation, invariance, and Damköhler number. Journal of Geophysical Research 112. https://doi.org/10.1029/2007JD008789.

Jeffery, C.A., Reisner, J.M., Andrejczuk, M., 2007. Another look at stochastic condensation for subgrid cloud modeling: adiabatic evolution and effects. Journal of the Atmospheric Sciences 64, 3949–3969.

Jensen, J., Austin, P., Baker, M., Blyth, A., 1985. Turbulent mixing, spectral evolution and dynamics in a warm cumulus cloud. Journal of the Atmospheric Sciences 42, 173–192.

Jost, A., Szakáll, M., Diehl, K., Mitra, S.K., Hundertmark, A., Klug, B.S., Borrmann, S., 2019. The effect of turbulence on the accretional growth of graupel. Journal of the Atmospheric Sciences 76, 3047–3061.

Kajishima, T., Taira, K., 2017. Numerical simulation of turbulent flows. In: Computational Fluid Dynamics. Springer, pp. 207–235. https://doi.org/10.1115/1.4005282.

Katul, G.G., Sempreviva, A.M., Cava, D., 2008. The temperature–humidity covariance in the marine surface layer: a one-dimensional analytical model. Boundary-Layer Meteorology 126, 263–278.

Katzwinkel, J., Siebert, H., Heus, T., Shaw, R.A., 2014. Measurements of turbulent mixing and subsiding shells in trade wind cumuli. Journal of the Atmospheric Sciences 71, 2810–2822.

Khain, A., Ovtchinnikov, M., Pinsky, M., Pokrovsky, A., Krugliak, H., 2000. Notes on the state-of-the-art numerical modeling of cloud microphysics. Atmospheric Research 55, 159–224. https://doi.org/10.1016/S0169-8095(00)00064-8.

Khain, A.P., Pinsky, M., 2018. Physical Processes in Clouds and Cloud Modeling. Cambridge University Press. https://doi.org/10.1017/9781139049481.

Kolmogorov, A.N., 1962. A refinement of previous hypotheses concerning the local structure of turbulence in a viscous incompressible fluid at high Reynolds number. Journal of Fluid Mechanics 13, 82–85.

Korolev, A., McFarquhar, G., Field, P.R., Franklin, C., Lawson, P., Wang, Z., Williams, E., Abel, S.J., Axisa, D., Borrmann, S., et al., 2017. Mixed-phase clouds: progress and challenges. Meteorological Monographs 58, 5.

Korolev, A., Milbrandt, J., 2022. How are mixed-phase clouds mixed? Geophysical Research Letters 49, e2022GL099578.

Korolev, A., Pinsky, M., Khain, A., 2013. A new mechanism of droplet size distribution broadening during diffusional growth. Journal of the Atmospheric Sciences 70, 2051–2071.

Kostinski, A., 2009. Simple approximations for condensational growth. Environmental Research Letters 4, 015005.

Kostinski, A.B., Shaw, R.A., 2005. Fluctuations and luck in droplet growth by coalescence. Bulletin of the American Meteorological Society 86, 235–244.

Krueger, S.K., 1993. Linear eddy modeling of entrainment and mixing in stratus clouds. Journal of the Atmospheric Sciences 50, 3078–3090.

Krueger, S.K., Su, C.W., McMurtry, P.A., 1997. Modeling entrainment and finescale mixing in cumulus clouds. Journal of the Atmospheric Sciences 54, 2697–2712.

Kulmala, M., Rannik, Ü., Zapadinsky, E.L., Clement, C.F., 1997. The effect of saturation fluctuations on droplet growth. Journal of Aerosol Science 28, 1395–1409.

Lanotte, A.S., Seminara, A., Toschi, F., 2009. Cloud droplet growth by condensation in homogeneous isotropic turbulence. Journal of the Atmospheric Sciences 66, 1685–1697.

Lasher-Trapp, S.G., Cooper, W.A., Blyth, A.M., 2005. Broadening of droplet size distributions from entrainment and mixing in a cumulus cloud. Quarterly Journal of the Royal Meteorological Society 131, 195–220.

Latham, J., Reed, R., 1977. Laboratory studies of the effects of mixing on the evolution of cloud droplet spectra. Quarterly Journal of the Royal Meteorological Society 103, 297–306.

Lehmann, K., Siebert, H., Shaw, R.A., 2009. Homogeneous and inhomogeneous mixing in cumulus clouds: dependence on local turbulence structure. Journal of the Atmospheric Sciences 66, 3641–3659.

Li, X.Y., Mehlig, B., Svensson, G., Brandenburg, A., Haugen, N.E., 2022. Collision fluctuations of lucky droplets with superdroplets. Journal of the Atmospheric Sciences 79, 1821–1835.

Lu, J., Shaw, R.A., 2015. Charged particle dynamics in turbulence: theory and direct numerical simulations. Physics of Fluids 27, 065111. https://doi.org/10.1063/1.4922645.

MacPherson, J.I., Isaac, G.A., 1977. Turbulent characteristics of some Canadian cumulus clouds. Journal of Applied Meteorology 16, 81–90.

Magnusson, G., Dubey, A., Kearney, R., Bewley, G., Mehlig, B., 2022. Collisions of micron-sized charged water droplets in still air. Physical Review Fluids 7, 043601.

Malinowski, S.P., Andrejczuk, M., Grabowski, W.W., Korczyk, P., Kowalewski, T.A., Smolarkiewicz, P.K., 2008. Laboratory and modeling studies of cloud-clear air interfacial mixing: anisotropy of small-scale turbulence due to evaporative cooling. New Journal of Physics 10, 075020.

Mason, P.J., 1994. Large-eddy simulation: a critical review of the technique. Quarterly Journal of the Royal Meteorological Society 120, 1–26. https://doi.org/10.1002/qj.49712051503.

Maxey, M.R., Riley, J.J., 1983. Equation of motion for a small rigid sphere in a nonuniform flow. Physics of Fluids 26, 883–889.

Mazin, I.P., 1966. On temperature and humidity stratification in clouds. Trudy TsAO 71, 3–15.

Mazin, I.P., Merkulovich, V.M., 2008. Stochastic condensation and its possible role in liquid cloud microstructure formation. In: Ivanov, A.A., Mazin, I.P., Strunin, M.A., Bakhanov, V.P., Korolev, A.V., Leshkevich, T.V., Postnov, A.A. (Eds.), Some Problems of Cloud Physics: Collected Papers. National Geophysical Committee, Russian Academy of Science, pp. 217–267.

McGraw, R., Liu, Y., 2006. Brownian drift-diffusion model for evolution of droplet size distributions in turbulent clouds. Geophysical Research Letters 33, L03802.

Mellado, J.P., Puche, M., van Heerwaarden, C.C., 2017. Moisture statistics in free convective boundary layers growing into linearly stratified atmospheres. Quarterly Journal of the Royal Meteorological Society 143, 2403–2419.

Moin, P., Mahesh, K., 1998. Direct numerical simulation: a tool in turbulence research. Annual Review of Fluid Mechanics 30, 539–578. https://doi.org/10.1146/annurev.fluid.30.1.539.

Morrison, H., van Lier-Walqui, M., Fridlind, A.M., Grabowski, W.W., Harrington, J.Y., Hoose, C., Korolev, A., Kumjian, M.R., Milbrandt, J.A., Pawlowska, H., et al., 2020. Confronting the challenge of modeling cloud and precipitation microphysics. Journal of Advances in Modeling Earth Systems 12, e2019MS001689. https://doi.org/10.1029/2019MS001689.

Narasimha, R., 1990. The utility and drawbacks of traditional approaches. In: Whither Turbulence? Turbulence at the Crossroads. Springer, pp. 13–48.

Narasimha, R., Diwan, S.S., Duvvuri, S., Sreenivas, K., Bhat, G., 2011. Laboratory simulations show diabatic heating drives cumulus-cloud evolution and entrainment. Proceedings of the National Academy of Sciences 108, 16164–16169.

Nenes, A., Chuang, P.Y., Flagan, R.C., Seinfeld, J.H., 2001. A theoretical analysis of cloud condensation nucleus (ccn) instruments. Journal of Geophysical Research 106, 3449–3474.

Niedermeier, D., Chang, K., Cantrell, W., Chandrakar, K.K., Ciochetto, D., Shaw, R.A., 2018. Observation of a link between energy dissipation rate and oscillation frequency of the large-scale circulation in dry and moist Rayleigh-Bénard turbulence. Physical Review Fluids 3, 083501.

Niedermeier, D., Voigtländer, J., Schmalfuß, S., Busch, D., Schumacher, J., Shaw, R.A., Stratmann, F., 2020. Characterization and first results from lacis-t: a moist-air wind tunnel to study aerosol–cloud–turbulence interactions. Atmospheric Measurement Techniques 13, 2015–2033. https://doi.org/10.5194/amt-13-2015-2020.

Onishi, R., Matsuda, K., Takahashi, K., 2015. Lagrangian tracking simulation of droplet growth in turbulence–turbulence enhancement of autoconversion rate. Journal of the Atmospheric Sciences 72, 2591–2607.

Paoli, R., Shariff, K., 2009. Turbulent condensation of droplets: direct simulation and a stochastic model. Journal of the Atmospheric Sciences 66, 723–740.

Pinsky, M., Eytan, E., Koren, I., Khain, A., 2022. Convective and turbulent motions in nonprecipitating cu. Part II: les simulated cloud represented by a starting plume. Journal of the Atmospheric Sciences 79, 793–813.

Pinsky, M., Khain, A., Korolev, A., 2016. Theoretical analysis of mixing in liquid clouds–part 3: inhomogeneous mixing. Atmospheric Chemistry and Physics 16, 9273–9297.

Pinsky, M., Khain, A., Shapiro, M., 2001. Collision efficiency of drops in a wide range of Reynolds numbers: effects of pressure on spectrum evolution. Journal of the Atmospheric Sciences 58, 742–764. https://doi.org/10.1175/1520-0469(2001)058<0742:CEODIA>2.0.CO,2.

Politovich, M.K., Cooper, W.A., 1988. Variability of the supersaturation in cumulus clouds. Journal of the Atmospheric Sciences 45, 1651–1664.

Prabhakaran, P., Kinney, G., Cantrell, W., Shaw, R.A., Bodenschatz, E., 2020a. High supersaturation in the wake of falling hydrometeors: implications for cloud invigoration and ice nucleation. Geophysical Research Letters 47, e2020GL088055.

Prabhakaran, P., Shawon, A.S.M., Kinney, G., Thomas, S., Cantrell, W., Shaw, R.A., 2020b. The role of turbulent fluctuations in aerosol activation and cloud formation. Proceedings of the National Academy of Sciences 117, 16831–16838.

Prabhakaran, P., Thomas, S., Cantrell, W., Shaw, R.A., Yang, F., 2022. Sources of stochasticity in the growth of cloud droplets: supersaturation fluctuations versus turbulent transport. Journal of the Atmospheric Sciences 79, 3145–3162.

Ren, Y., Chandrakar, K.K., Yang, F., Shaw, R.A., 2026. Evaluating the collision-coalescence process in idealized cloud convection using large-eddy simulations with Lagrangian microphysics. Geophysical Research Letters 53, e2025GL118873.

Richter, D., Chamecki, M., 2018. Inertial effects on the vertical transport of suspended particles in a turbulent boundary layer. Boundary-Layer Meteorology 167, 235–256.

Rogers, R., Yau, M., 1989. A Short Course in Cloud Physics. Pergamon.

Saito, I., Gotoh, T., Watanabe, T., 2019. Broadening of cloud droplet size distributions by condensation in turbulence. Journal of the Meteorological Society of Japan, Ser. II 97, 867–891.

Sardina, G., Picano, F., Brandt, L., Caballero, R., 2015. Continuous growth of droplet size variance due to condensation in turbulent clouds. Physical Review Letters 115, 184501.

Saw, E.W., Shaw, R.A., Salazar, J.P., Collins, L.R., 2012. Spatial clustering of polydisperse inertial particles in turbulence: ii. comparing simulation with experiment. New Journal of Physics 14, 105031.

Schmeissner, T., Shaw, R., Ditas, J., Stratmann, F., Wendisch, M., Siebert, H., 2015. Turbulent mixing in shallow trade wind cumuli: dependence on cloud life cycle. Journal of the Atmospheric Sciences 72, 1447–1465.

Schneider, T., Jeevanjee, N., Socolow, R., 2021. Accelerating progress in climate science. Physics Today 74, 44–51.

Schwenkel, J., Hoffmann, F., Raasch, S., 2018. Improving collisional growth in Lagrangian cloud models: development and verification of a new splitting algorithm. Geoscientific Model Development 11, 3929–3944. https://doi.org/10.5194/gmd-11-3929-2018.

Shaw, R.A., 2003. Particle-turbulence interactions in atmospheric clouds. Annual Review of Fluid Mechanics 35, 183–227. https://doi.org/10.1146/annurev.fluid.35.101101.161125.

Shaw, R.A., Cantrell, W., Chen, S., Chuang, P., Donahue, N., Feingold, G., Kollias, P., Korolev, A., Kreidenweis, S., Krueger, S., et al., 2020. Cloud–aerosol–turbulence interactions: science priorities and concepts for a large-scale laboratory facility. Bulletin of the American Meteorological Society 101, E1026–E1035.

Shawon, A.S.M., Prabhakaran, P., Kinney, G., Shaw, R.A., Cantrell, W., 2021. Dependence of aerosol-droplet partitioning on turbulence in a laboratory cloud. Journal of Geophysical Research. Atmospheres 126, e2020JD033799.

Sheikh, M., Gustavsson, K., Lévêque, E., Mehlig, B., Pumir, A., Naso, A., 2022. Colliding ice crystals in turbulent clouds. Journal of the Atmospheric Sciences 79. https://doi.org/10.1175/JAS-D-21-0305.1.

Shupe, M.D., Brooks, I.M., Canut, G., 2012. Evaluation of turbulent dissipation rate retrievals from Doppler cloud radar. Atmospheric Measurement Techniques 5, 1375–1385. https://doi.org/10.5194/amt-5-1375-2012. https://amt.copernicus.org/articles/5/1375/2012/.

Siebert, H., Lehmann, K., Wendisch, M., 2006. Observations of small scale turbulence and energy dissipation rates in the cloudy boundary layer. Journal of the Atmospheric Sciences 63, 1451–1466.

Siebert, H., Shaw, R.A., 2017. Supersaturation fluctuations during the early stage of cumulus formation. Journal of the Atmospheric Sciences 74, 975–988.

Siebert, H., Shaw, R.A., Ditas, J., Schmeissner, T., Malinowski, S.P., Bodenschatz, E., Xu, H., 2015a. High-resolution measurement of cloud microphysics and turbulence at a mountaintop station. Atmospheric Measurement Techniques 8, 3219–3228.

Siebert, H., Shaw, R.A., Ditas, J., Schmeissner, T., Malinowski, S.P., Bodenschatz, E., Xu, H., 2015b. Schneefernerhaus as a mountain research station for clouds and turbulence – part 2: cloud microphysics and fine-scale turbulence. Atmospheric Measurement Techniques Discussions 8, 569–597. https://doi.org/10.5194/amtd-8-569-2015.

Siebert, H., Shaw, R.A., Warhaft, Z., 2010. Statistics of small-scale velocity fluctuations and internal intermittency in marine stratocumulus clouds. Journal of the Atmospheric Sciences 67, 262–273.

Siebesma, A.P., Bony, S., Jakob, C., Stevens, B., 2020. Clouds and Climate: Climate Science's Greatest Challenge. Cambridge University Press.

Siewert, C., Bec, J., Krstulovic, G., 2017. Statistical steady state in turbulent droplet condensation. Journal of Fluid Mechanics 810, 254–280.

Simpson, J., 1983. Cumulus clouds: interactions between laboratory experiments and observations as foundations for models. In: Mesoscale Meteorology—Theories, Observations and Models. Springer, pp. 399–412.

Stratmann, F., Kiselev, A., Wurzler, S., Wendisch, M., Heintzenberg, J., Charlson, R., Diehl, K., Wex, H., Schmidt, S., 2004. Laboratory studies and numerical simulations of cloud droplet formation under realistic supersaturation conditions. Journal of Atmospheric and Oceanic Technology 21, 876–887.

Sui, C.H., Satoh, M., Suzuki, K., 2020. Precipitation efficiency and its role in cloud-radiative feedbacks to climate variability. Journal of the Meteorological Society of Japan. Ser. II 98, 261–282.

Tarshish, N., Jeevanjee, N., Lecoanet, D., 2018. Buoyant motion of a turbulent thermal. Journal of the Atmospheric Sciences 75, 3233–3244.

Tennekes, H., 1979. The exponential Lagrangian correlation function and turbulent diffusion in the inertial subrange. Atmospheric Environment 13, 1565–1567.

Tennekes, H., Lumley, J.L., 1972. A First Course in Turbulence. The MIT Press, Cambridge, MA, USA. 300 p.

Thomas, S., Ovchinnikov, M., Yang, F., van der Voort, D., Cantrell, W., Krueger, S.K., Shaw, R.A., 2019. Scaling of an atmospheric model to simulate turbulence and cloud microphysics in the pi chamber. Journal of Advances in Modeling Earth Systems 11, 1981–1994.

Thomas, S., Prabhakaran, P., Cantrell, W., Shaw, R.A., 2021. Is the water vapor supersaturation distribution Gaussian? Journal of the Atmospheric Sciences 78, 2385–2395.

Thomas, S., Prabhakaran, P., Yang, F., Cantrell, W.H., Shaw, R.A., 2022. Dimensionless parameters for cloudy Rayleigh-Bénard convection: supersaturation, Damköhler, and Nusselt numbers. Physical Review Fluids 7, 010503.

Tölle, M.H., Krueger, S.K., 2014. Effects of entrainment and mixing on droplet size distributions in warm cumulus clouds. Journal of Advances in Modeling Earth Systems 6, 281–299.

Tsai, S.T., 1957. The resistance on a small sphere moving in non-uniform flow. Acta Physica Sinica 13, 388–398.

Twomey, S., 1959. The nuclei of natural cloud formation part II: the supersaturation in natural clouds and the variation of cloud droplet concentration. Pure and Applied Geophysics 43, 243–249.

Vohl, O., Mitra, S., Wurzler, S., Pruppacher, H., 1999. A wind tunnel study of the effects of turbulence on the growth of cloud drops by collision and coalescence. Journal of the Atmospheric Sciences 56, 4088–4099.

Wang, J., Daum, P.H., Yum, S.S., Liu, Y., Senum, G.I., Lu, M.L., Seinfeld, J.H., Jonsson, H., 2009. Observations of marine stratocumulus microphysics and implications for processes controlling droplet spectra: results from the marine stratus/stratocumulus experiment. Journal of Geophysical Research 114. https://doi.org/10.1029/2008JD011035.

Wang, A., Ovchinnikov, M., Yang, F., Schmalfuss, S., Shaw, R.A., 2024. Designing a convection-cloud chamber for collision-coalescence using large-eddy simulation with bin microphysics. Journal of Advances in Modeling Earth Systems 16, e2023MS003734.

Warhaft, Z., 2000. Passive scalars in turbulent flows. Annual Review of Fluid Mechanics 32, 203–240.

Wendisch, M., Brenguier, J.L., 2013. Airborne Measurements for Environmental Research: Methods and Instruments. John Wiley & Sons.

Wilkinson, M., 2016. Large deviation analysis of rapid onset of rain showers. Physical Review Letters 116, 018501.

Williams, A.G., Hacker, J.M., Kraus, H., 1997. Transport processes in the tropical warm pool boundary layer. Part II: vertical structure and variability. Journal of the Atmospheric Sciences 54, 2060–2082.

Wyngaard, J., LeMone, M., 1980. Behavior of the refractive index structure parameter in the entraining convective boundary layer. Journal of the Atmospheric Sciences 37, 1573–1585.

Wyngaard, J., Pennell, W., Lenschow, D., LeMone, M.A., 1978. The temperature-humidity covariance budget in the convective boundary layer. Journal of the Atmospheric Sciences 35, 47–58.

Wyngaard, J.C., 2010. Turbulence in the Atmosphere. Cambridge University Press.

Yang, F., Hoffmann, F., Shaw, R.A., Ovchinnikov, M., Vogelmann, A.M., 2023. An intercomparison of large-eddy simulations of a convection cloud chamber using haze-capable bin and Lagrangian cloud microphysics schemes. Journal of Advances in Modeling Earth Systems 15, e2022MS003270.

Yang, F., Ovchinnikov, M., Thomas, S., Khain, A., McGraw, R., Shaw, R.A., Vogelmann, A.M., 2022. Large-eddy simulations of a convection cloud chamber sensitivity to bin microphysics and advection. Journal of Advances in Modeling Earth Systems 14, e2021MS002895.

Yang, F., Shaw, R., Xue, H., 2016. Conditions for super-adiabatic droplet growth after entrainment mixing. Atmospheric Chemistry and Physics 16, 9421–9433.

Yeom, J.M., Yum, S.S., Shaw, R.A., La, I., Wang, J., Lu, C., Liu, Y., Mei, F., Schmid, B., Matthews, A., 2021. Vertical variations of cloud microphysical relationships in marine stratocumulus clouds observed during the ace-ena campaign. Journal of Geophysical Research 126, e2021JD034700.

Index

9780323958738